Edwin Hubble Centennial Symposium

EVOLUTION OF THE UNIVERSE OF GALAXIES

A SERIES OF BOOKS ON RECENT DEVELOPMENTS IN ASTRONOMY AND ASTROPHYSICS

Printed by BookCrafters, Inc.

First published 1990

Library of Congress Catalog Card Number: 90-81491
ISBN 0-937707-28-7

D. Harold McNamara, Managing Editor of Conference Series
293 ESC Brigham Young University
Provo, UT 84602

ASTRONOMICAL SOCIETY OF THE PACIFIC
CONFERENCE SERIES

Volume 10

EVOLUTION OF THE UNIVERSE OF GALAXIES

Edwin Hubble Centennial Symposium

Edited by
Richard G. Kron

Table of Contents

Preface v

Conference Registrants vi

Part 1 — Edwin Hubble

Self-Made Cosmologist: The Education of Edwin Hubble
Donald E. Osterbrock, Ronald S. Brashear, and Joel A. Gwinn 1

Shapley, Hubble, and Cosmology *Owen Gingerich* 19

Edwin Hubble's Cosmology *Norriss S. Hetherington* 22

Part 2 — The Nearby Universe

Thin Discs and Dense Bulges: What Do They Tell Us about Galaxies?
Jeremiah P. Ostriker 25

Morphology and Dynamics of Galaxies in the Hickson Compact Groups
Vera C. Rubin, W. Kent Ford, Jr., and Deidre Hunter 30

Scaling Laws for Dark Matter in Late-Type Galaxies
John Kormendy 33

Halo Model Mass Limits for Spiral Galaxies
Lance K. Erickson, Stephen T. Gottesman, and James H. Hunter, Jr. 45

The Dark Matter Content of Disk Galaxies *Michael J. Pierce* 48

The Dynamics of the Local Group and the Mass of the Galaxy
Dennis Zaritsky 51

Surface Brightness and Bandpass Bias in the Hubble
Morphological Sequence *Gregory D. Bothun* 54

Blue Compact Dwarf Galaxies in Virgo and Nearby Groups
G. Lyle Hoffman, E.E. Salpeter, and George Helou 67

S0 Galaxies and the Hubble Sequence *Sidney van den Bergh* 70

Elliptical Galaxies *Garth D. Illingworth and Marijn Franx* 82

CCD Surface Photometry of Corona Borealis Cluster Galaxies
Donald H. Gudehus 103

The Mass-Metallicity-Luminosity Density Relation for Elliptical Galaxies
R. de Carvalho and S. Djorgovski 106

Dark Matter in Elliptical Galaxies *Stephen M. Kent* 109

Stellar Populations in Elliptical Galaxies *William A. Baum* 119

Stellar Content of Nearby Galaxies: II. The Local Group Dwarf Elliptical Galaxy M32 *Wendy L. Freedman* 131

Dwarf Cepheids in the Carina Dwarf Galaxy
James Nemec and Mario Mateo 134

The Globular Cluster System of the Edge-On Sb Galaxy NGC 5170
Philippe Fischer, James E. Hesser, Hugh C. Harris, and Gregory D. Bothun 138

The Evolution of Disk Galaxies at $z = 0$ *Robert C. Kennicutt, Jr.* 141

Nearby Luminous Blue Galaxies *John S. Gallagher III* 157

A Search for Wolf-Rayet Stars at the Galactic Center
M.W. Werner, J.R. Stauffer, and E.E. Becklin 167

Profuse Central Star Formation in 'S0' Galaxies
L.L. Dressel, R.W. O'Connell, C.M. Telesco, and R. Decher 170

A Non-Steady Cooling Flow Model for NGC 1275 (Perseus)
Avery Meiksin 173

High Resolution CO Images of Seyfert Galaxies
Margaret Meixner, Melvyn Wright, Rich Puchalsky, and Leo Blitz 176

CCD Photometry of the Ring Galaxy Arp 146
L.D. Spight, A.B. Schultz, P.T. Colegrove, M.A. DiSanti, and U. Fink 179

Color Maps of Arp 146 *A.B. Schultz, L.D. Spight, P.T. Colegrove, M.A. DiSanti, and U. Fink* 182

Part 3 — The Distant Universe

Extracting Cosmological and Evolutionary Parameters from Distant Galaxy Observations *Gustavo Bruzual A.* 185

Spectral Evolution of Cluster Galaxies *Alan Dressler and James E. Gunn* 200

The Brightest Galaxies in Clusters *Suketu P. Bhavsar* 209

Galaxy Evolution and Formation *Lennox L. Cowie and Simon J. Lilly* 212

The Cosmic Submillimeter Background: A Signature of the Initial Burst of Galaxy Formation? *S. Djorgovski and W.N. Weir* 228

Dark Matter, Galaxies, and Background Radiation *Eric R. Wollman* 231

Voids in the Lyman α Forest *Jill Bechtold* 234

The HI Column Density Spectrum: A Simulation Provides Evidence for Tytler's "Single Population" *Curtis V. Manning* 245

Spectroscopy of Large Numbers of Faint Galaxies *Richard S. Ellis* 248

UBC/Laval 2.7m Liquid Mirror Telescope (LMT): Progress Report *Brad K. Gibson and Paul Hickson* 265

The Evolution of Field Galaxies: Is $\Omega = 1$? *David C. Koo* 268

K Band Photometry of a Complete Sample of Field Galaxies with Spectroscopic Redshifts *Matthew Bershady, Mark Hereld, Richard Kron, and David Koo* 286

The Study of Galaxy Evolution Using a Large Area 2μ Survey *Richard Elston, George Rieke, and Marcia Rieke* 289

Imaging Faint Galaxies *J.A. Tyson* 292

Deep CCD Imaging of Field Galaxies in U, B_J, and R: Constraints on Galaxy Evolution *P. Guhathakurta, J.A. Tyson, and S.R. Majewski* 304

Lyman α Emission From Young Galaxy Disks *Ross D. Cohen and Harding E. Smith* 307

A Search for Primeval Galaxies at $z = 2$
C.J. Pritchet and F.D.A. Hartwick 319

The Evolution of Galaxies and Galaxy Clusters Associated with Quasars
H.K.C. Yee 322

The Evolution of Quasars in Galaxy Cluster Environments
E. Ellingson 334

Testing the Binary Black Hole Model of OJ 287
George J. Corso, Ronald W. Harris, Richard Fox, and Joseph Schultz 337

Redshifts and the SMF Model for Quasars and AGNs
Howard D. Greyber 340

High Redshift Radio Galaxies: Evidence for Early Galaxy Formation
Simon J. Lilly 344

Evolutionary Model Uncertainties in the K-Band Hubble Diagram
W.N. Weir, S. Djorgovski, and Gustavo Bruzual A. 356

Relations Between Radio and Optical Properties in Distant Radio Galaxies
Wil J.M. van Breugel and Patrick J. McCarthy 359

Infrared Images of Distant 3C Radio Galaxies
P. Eisenhardt, A. Chokshi, M. Dickinson, S. Djorgovski, P. McCarthy, and H. Spinrad 371

High Redshift Radio Galaxies and the Alignment Effect
K.C. Chambers and G.K. Miley 373

The Evolution of Weak Radio Galaxies at Radio and Optical Wavelengths
Rogier Windhorst, Doug Mathis, and Lyman Neuschaefer 389

Preface

The *Evolution of the Universe of Galaxies Symposium*, held at the University of California, Berkeley, on June 21, 22, and 23, 1989, celebrated both the centennial of the founding of the Astronomical Society of the Pacific, and the 100th anniversary of Edwin Hubble's birth. With this heritage in mind, we look forward to another century of discovery using powerful new tools such as the Hubble Space Telescope.

The Symposium consisted of invited papers according to a program designed by the Scientific Organizing Committee, J.E. Gunn, R.G. Kron (chair), M. Schmidt, and H. Spinrad. The invited papers were organized according to three sub-themes: "The Form and Content of Nearby Galaxies: Clues to Their Past"; "Galaxy Formation: Theory and Observational Constraints"; and "Distant Galaxies: Direct Evidence for Galaxy Evolution." Contributed poster papers were displayed during the course of the meeting, and many of these related directly to the organized program.

The contents of these Proceedings are a combination of the invited papers, the contributed poster papers in the field of extragalactic astronomy, and additional contributions relating to the career of Edwin Hubble. The historical papers were not part of the *Evolution of the Universe of Galaxies Symposium*, but in the spirit of the ASP Centennial Meeting, with its *Astronomy Over the A.S.P.'s Century: A History Symposium*, these contributions have been included along with the scientific papers.

Part 1 of these Proceedings consists of one major paper on the early life of Edwin Hubble, and two shorter contributions that were presented at the *History Symposium*. Part 2 contains papers generally concerned with galaxies at relatively small redshifts, and Part 3 extends the discussion to galaxies and QSO absorbers at larger redshifts.

I would like to thank the Local Organizing Committee and the staff of the Astronomical Society of the Pacific for their extensive efforts on behalf of the Symposium. I also thank the Observatories of the Carnegie Institution of Washington for permission to reproduce the photograph of Hubble used as the frontispiece. Finally, I am grateful to Suzanne Swanson and Richard Dreiser for their assistance in assembling these Proceedings.

Richard G. Kron

Conference Registrants

Harold Ables
Peter Adams
Alane Alchorn
J.R. Allington-Smith
Robert Antonucci
Gordon Augason
Fred Aves
William Baum
Jill Bechtold
Matthew Bershady
Suketu Bhavsar
Michael Bolte
Kirk Borne
Greg Bothun
Jean Brodie
William Brown
Gustavo Bruzual
Margaret Burbidge
Miriam Cacolin
Duane Carbon
Adeline Caulet
Kenneth Chambers
Stephane Charlot
Arati Chokshi
Ross Cohen
Todd Colegrove
Jack Connors
Peter Conti
George Corso
Len Cowie
Stephane Courteau
Anne Cowley
Clarence Custer
Arthur Davidsen
Peter Dawson
Lara Descartes
Claudia Mendes de Oliveira
Michael De Robertis
Mark Dickinson
S.G. Djorgovski
Linda Dressel
Alan Dressler
Gordon Drukier
Ed Duckworth
Frank Edmondson
Eiichi Egami
Peter Eisenhardt
Erica Ellingson
Richard Ellis
Nancy Ellman
Richard Elston
Lance Erickson
Robert Ewing
Michael Fall
Wendy Freedman
Howard French
Carlos Frenk
Jay Gallagher
Greg Giannoni
Brad Gibson
Kirk Gilmore
Peter Goldreich
Bob Goodrich
Howard Greyber
Donald Gudehus
Puragra Guhathakurta
Jim Gunn
David Hartwick
Timothy Heckman
Karl Henize
Mark Hereld
Lyle Hoffman
Robert Hornbeck
Irwin Horowitz
Jack Houng
Esther Hu
Charles Huff
Garth Illingworth
Mary Lou Jewett
Doug Johnstone
Vijay Kipahi
Robert Kennicutt
Stephen Kent
Ivan King
John Kolena
David Koo
John Kormendy
Kevin Krisciunas
Gerald Kron
Katherine Kron
Richard Kron
Gordon Langlois
John Lattanzio
Simon Lilly
Tod Lauer
Steven Lord
Barry Madore
Steve Majewski
Curtis Manning
Mark Mattox
Patrick McCarthy
Avery Meiksin
T.K. Menon
John Miche
Marcos Montes
Leonidas Moustakas
Jeff Munn
James Nemec
Lyman Neuschaefer
Richard Nolthenius
Michael Nowak
Bob O'Connell
Donald Osterbrock
J.P. Ostriker
Yichuan Pei
James Peters
Sig Peterson
Andrew Phillips
John Phillips
Jeff Pier
Pere Planesas
Marc Postman
Joel Primack
Chris Pritchet
Gibson Reaves
Michael Rigler
Lloyd Robinson
Brigitte Rocca-Volmerange
Joshua Roth
Vera Rubin
Toshiyuki Sasaki
James Schombert
Alfred Schultz
Joe Silk
Albert Smith
Hyron Spinrad
Thomas Statler
Tom Steiman-Cameron
Brian Swimme
Wei-Hsin Sun
C. Tadhunter
Chris Thompson
Tony Tyson
Barbara Uchida
William Vacca
Francisco Valdes
Wil van Breugel
Sidney van den Bergh
N.V. Vidal
Steve Warren
Wayne Warren
Nicholas Weir
Jon Weisheit
Michael Werner
Simon White
Albert Whitford
Eric Wilcots
Tom Williams
Andrew Wilson
Rogier Windhorst
Eric Wollman
James Wright
Rosemary Wyse
Mark Yates
Howard Yee
Dennis Zaritsky
Lin Zuo

SELF-MADE COSMOLOGIST: THE EDUCATION OF EDWIN HUBBLE

DONALD E. OSTERBROCK
University of California Observatories/Lick Observatory, University of California, Santa Cruz, CA 95064

RONALD S. BRASHEAR
Department of Manuscripts, Henry E. Huntington Library, 1151 Oxford Road, San Marino, CA 91108

JOEL A. GWINN
Department of Physics, University of Louisville, Louisville, KY 40292

ABSTRACT The youth, education, early life, and first research accomplishments of Edwin Hubble, the outstanding American observational cosmologist, are described. This biography is based very largely on written and published sources contemporary with the events described.

INTRODUCTION

Edwin Hubble changed our view of the universe in which we live more than any astronomer since Galileo. Many cosmologists believe that he first observed the high radial velocities of the galaxies, or that he discovered the expansion of the universe. In fact he did not; these discoveries were made by V. M. Slipher and Carl Wirtz, respectively.[1] But Hubble's drive, scientific ability, and communication skills enabled him to seize the problem of the whole universe, make it peculiarly his own, contribute more to it than anyone before or since, and become the recognized world expert of the field. He was a great scientist. He was also a great writer and a great speaker. No one who read one of his books, or heard one of his lectures, describing in simple yet vibrant terms his research and his views on the universe, could ignore him. Few tried to do so. He had an exciting, compelling personality, far different from most astronomers, much more like those of the movie stars and writers who became his friends and companions in the later years of his life.

Hubble's early life, like those of the stars, or of the hero kings of ancient wars, has become a myth. He has been the subject of fiction, and of supposedly factual biographies which bear little relation to his own real life. This is not strange for he himself, in life, constantly played the hero's role and his wife, who idolized him, after his death glorified him even further. Hubble's official biography was written by his protege, admirer, and friend, Nicholas U. Mayall, while the great man's widow was still living.[2] There was no reason for Mayall to probe deeply into Hubble's early life; his main task was to tell of his mentor's scientific career, and for the rest he accepted at face value the stories that Hubble had told him and the embellishments that his widow had provided. Why should he go beyond them? She in her turn devoted years of her life to writing a Parson Weems biography of a George Washington of American science that is demonstrably false in many of its details, and that omits whole areas of his life (and even more of her own) but which has become the inspiration and source for books and articles that purport to tell the story of Edwin Hubble.[3]

In fact Hubble's life was an interesting one. His youth and education were peculiarly American. Many contemporary sources of information do still exist. We have based our story as much as we can on them, taking from Grace Hubble's account only the parts that ring true in their context, or that can otherwise be verified.

FAMILY AND YOUTH

Edwin Powell Hubble (he did not use his middle name ever, nor his middle initial after he could avoid it) was born November 20, 1889 in Marshfield, Missouri. His father, John P. Hubble, was an insurance agent who had been born near Springfield in the same state; his mother, Virginia Lee James Hubble, had been born in Nevada.[4] Both of them could trace their ancestry back through many generations of Southern yeoman families. In Grace Hubble's words:

> His ancestors came from England, Ireland and Wales, with no strains of foreign blood. The first of them to come to America, in the middle of the 17th Century, was an officer of the Royalist Army. In the Revolutionary War and Civil War they were soldiers; in times of peace, pioneers, living on the land. Tall, well-made, strong, their bodily inheritance had come down to him, even to the clear, smooth skin that tanned in the sun, and the brown hair with a glint of reddish gold. They had handed down their traditions as well, integrity, loyalty as citizens, loyalty to their families ... and a sturdy reliance on their own efforts.[5]

The Hubbles had a large family, with seven children who survived. Edwin was the third; his brother Henry and his sister Lucy Lee were older than he, his brother William was the next after him, each separated by two or three years, while his sisters Helen, Emma Jane and Elizabeth ranged down to fifteen years younger than he.[6] He started school in Marshfield in 1895. Three years later his father transferred to the Chicago agency of his fire insurance firm and moved his family to nearby Evanston. Two years later they moved to Wheaton, a village west of Chicago, also within commuting distance by train of John Hubble's office in the Loop. Edwin completed grade school in Wheaton.[7]

His seventh grade teacher was Harriet Grote, later, after she married, the mother of Grote Reber, the great pioneer of radio astronomy, who also grew up in Wheaton and erected his first radio telescope there.[8] She recognized Hubble as a "bright boy," and later followed his career with pride and held him up to her son as a model.[9] Hubble's school records are still available for eighth grade, and for his four years of high school, which was in the same "Old Red Castle" in Wheaton as the grade school.[10] They show that in eighth grade his averages in different classes were typically 85 to 90, but that in high school he blossomed. There his grades were nearly all between 90 and 100, and particularly in English, mathematics, sciences (he took chemistry, biology and physics) and languages (he took four years of Latin and two of German) were mostly all between 95 and 100. He was not the best student in his class, but he was consistently in the top quarter. He could relax however, for his scores in "application" and "deportment," though quite adequate, were typically lower than were those of the other good students. Throughout high school Hubble was always one of the youngest students in his class, usually about two years younger than the average, and he was only sixteen years old when he graduated at the end of the winter term in 1906.[11]

Yet Hubble was in no sense a grind. He led a healthy outdoor life, with frequent summer vacations at his grandfather's dairy farm near Marshfield. He was big and strong, not only for his age but in comparison with the other boys in his classes, and he became a star athlete for Wheaton High School, especially on its track team. He was a high jumper, broad jumper, shot putter and discus thrower, and ran on the relay team.[12] One of the few surviving clippings from his high-school days shows that as a senior in the spring of 1906 he won the high jump at the Northwestern Interscholastic meet with a leap of 6 feet 1/4 inch.[13]

UNIVERSITY OF CHICAGO

On graduation Hubble received an academic scholarship to the University of Chicago, which had first gone into operation only fourteen years before. He entered as a sixteen-year-old freshman in September 1906, and again was a good student, particularly in his first two years. In addition to mathematics, chemistry, physics and astronomy, he took French, Greek and more Latin. At the end of his sophomore year he received the two-year Associate in Science degree, as was then standard at the University of Chicago, and won a scholarship as the best student in physics getting the degree that year. All of the advanced astronomy courses Hubble took, except for one quarter of observing with a transit and small telescopes, were in analytical mechanics and celestial mechanics, the standard fare in the campus department dominated by Kurt Laves, Forest R. Moulton, and William D. MacMillan. Hubble's overall work slipped a little, and his average for his last two years was only B −, well over passing however, and he received his Bachelor of Science degree at the end of March 1910, with enough credits to skip the spring quarter of his senior year.[14]

Again, in spite of the fact that Hubble was two years younger than most of his classmates, he was a very good athlete who won letters in track and basketball. He was tall for the intercollegiate basketball teams of his time (6 feet 3 inches) and very well coordinated, excelling on defense. He could score too, sometimes making two or four points, sometimes as many as eight or even twelve, as Chicago won by scores like 27 to 2 over Minnesota, 17 to 15 over Illinois, and 28 to 4 over Northwestern. The Maroons won the mythical national intercollegiate championship in Hubble's junior year, and the Big Ten basketball championship in his senior year as well. In track, coached by the famous Amos Alonzo Stagg until spring football practice started, Hubble was a good but not outstanding shot putter and high jumper. He usually placed in the Big Ten dual meets in which Chicago participated, though he seldom won.[15]

Hubble, like most University of Chicago student athletes of his time, belonged to a fraternity. His was Kappa Sigma. On one occasion in his sophomore year, he and some of his fraternity brothers got into trouble for throwing raw eggs from their windows at passing black-clad divinity students. In his junior year Hubble was one of the appointed university marshals, the student leaders who served as guides at the graduation of the class ahead of them. In his senior year he was elected vice president of his class. In that same year he took the examination for the Rhodes Scholarship. The British-born South African diamond king Cecil Rhodes, in establishing these scholarships, wanted to strengthen the connections between the United States and the United Kingdom, by bringing American students to Oxford University. One scholarship was awarded in each state each year, and Rhodes had specified in his will that each successful candidate should be a manly chap who was a combined good student (but not a "mere bookworm"), athlete, and leader. The Rhodes Scholars were to be between 19 and 25 years old when chosen, and in practice the successful candidates had usually completed their undergraduate degrees in America before they went to Oxford. Hubble was supremely qualified in every way. Six Illinois candidates took the examination in December 1909; Hubble and one other, a student from Greenville College, passed. The Illinois Rhodes Scholarship Committee, consisting of the presidents of the University of Chicago, the University of Illinois, Northwestern University, and Jacksonville College, was to make the final choice between them.[16] Robert A. Millikan, for whom as a senior Hubble had worked as a laboratory assistant in the elementary physics course, wrote a testimonial letter for him. In it Millikan described Hubble as a "man of magnificent physique, admirable scholarship, and worthy and lovable character," and concluded "I have seldom known a man who seemed to be better qualified to meet the conditions imposed by the founder of the Rhodes scholarship than is Mr. Hubble."[17] Not surprisingly, Hubble got the scholarship.

OXFORD UNIVERSITY

When Hubble "went up" to Oxford in October 1910, he was one of an outstanding group of American students. Some of the other Rhodes Scholars who began with him that year and later became well known public figures were Elmer Davis, later a famous reporter and newscaster, Christopher Morley, a prolific author, John Crowe Ransom, a poet and writer, and Robert Hale, a member and speaker of the Maine House of Representatives, and later a United States congressman.[18] At Oxford Hubble entered Queens College, and "read" (studied) jurisprudence. Both his father and grandfather wanted him to become a lawyer, and he had probably decided to try that career route. At the end of his first year he described his study methods in a letter to his father:

> As you know, I have been studying Law for six months. Real Property and Contracts are now completed and a large slice taken out of Roman Law. The method of study is very different from home. I do not attend lectures for they are mere dictations which we are supposed to take down verbatim. The lectures will not have to be typewritten, so I simply read other men's notes. This permits me to work undisturbed 7:30 or 8:00 to 1:00 at a stretch. That is about all on Law as there are many other things which are worth while. Most of the long afternoon is given to sports. Then tea, then discussions until Hall, which in summer is at 7:30. After Hall there is generally some Club meeting or other gathering to attend or else I read History, Economics &c in my room ... Of c[our]se I havent done as much studying as I meant to do, but have kept up with what my tutor has planned for me and he says that my work has been very satisfactory.[19]

By these methods Hubble completed the Jurisprudence course in the standard two years, and received "second class honors" in 1912. He stayed a third year, beginning the work needed for a bachelor's degree in law, but dropped it and switched to Spanish instead. In Hubble's Rhodes Trust record, the Warden (head scholastic official) wrote "Considerable ability. Manly. Did quite well here. I didn't care v[ery] much for his manner — but he was better than his manner. Will get a."[20] A fellow Rhodes Scholar, Jakob A. O. Larsen of Luther College, Iowa, thought that Hubble in one year had become "very British," aping the language and manners of the Oxford upper crust, and perhaps this is what the Warden meant by "his manner."[21]

As he had written his father, Hubble participated actively in sports at Oxford. At Queen's College he was a star in the high jump and broad jump, and better than most in "putting the weight" (shot put) and running events from a quarter mile to a mile. He was on the Oxford University Athletic Club team in the shot put and hammer throw, and also swam on the water polo team. During his vacations he traveled widely on the continent with college friends, mostly by train and bicycle. His trips included visits to France, Belgium, Switzerland, Germany, Austria-Hungary, and Spain, as well as several long tours of England. At the end of three years, the maximum allowed for a Rhodes Scholar, Hubble returned to the United States with the announced intention of practicing law.[22]

LOUISVILLE

During Hubble's senior year at the University of Chicago, his family had moved to Kentucky, where his father took over the National Fire Insurance Company agency in Louisville. At first they lived in nearby Shelbyville and he commuted to his office, but in 1911 they moved in to Louisville.[23] John Hubble, however, suffered from a kidney disease, from which he died at the age of 52 on January 19, 1913, during Edwin's last year at Oxford.[24] The family moved from South Brook Street to Everett Avenue, both located in a pleasant, tree-lined, middle-class residential area. Edwin's older brother Henry began work for the Kentucky Actuarial Bureau, and later became an insurance agent himself.[25]

Soon after Hubble returned home, he passed the Kentucky bar examination.[26] At the time this was not a written examination; the process consisted of the candidate, traditionally bringing a bottle of Kentucky bourbon or a box of cigars as a gift, visiting a circuit court judge of a neighboring county, who would quiz him and then advise him in the duties of an attorney, often in his home over dinner.[27] However, there is absolutely no record that Hubble ever actually practiced law in Kentucky.[28] This is not surprising; the Oldham County judge in La Grange, a 22-mile interurban trolley ride from Louisville would have been glad to do a favor for the grieving widow of an insurance agent he probably knew, and pass the handsome, attentive son who obviously was an expert in the general principles of jurisprudence after three years of Oxford, but he had no knowledge whatever of the details of Kentucky laws or practice, and he had to make a living.

What Hubble did do was get a job teaching Spanish and physics in the high school at New Albany, Indiana, a suburb just across the Ohio River from Louisville. Probably he commuted daily by street car, though this is not certain. In addition to teaching, he coached the high school basketball team, taking them through an undefeated season and all the way to the state tournament in Bloomington, where they won their first two games before they were eliminated. Hubble was a very popular teacher and coach, and the students of the Class of 1914 "lovingly" dedicated their year book to him.[29]

YERKES OBSERVATORY

Hubble, however, had decided to go back to astronomy; he knew by then that the school-teaching life was not for him. As the end of the school year approached he wrote Moulton, his astronomy professor at the University of Chicago, to ask about coming back as a graduate student. Moulton immediately recommended him very strongly to Edwin B. Frost, the director of Yerkes Observatory, where the graduate work was actually given:

> Personally, [Hubble] is a man of the finest type. Physically he is a splendid specimen. In his work here, altogether, and especially in science, he showed exceptional ability. I feel sure you would find him just the sort of man you would wish to have.[30]

Hubble sounded good to Frost, especially because he had experience with scientific instruments in the physics lab and as a surveyor in the North Woods during his summer vacations. Yerkes needed graduate students who were well qualified to work as assistants on the observing program with the 40–inch refractor. Though the fellowships for the next year were already taken, Frost could offer Hubble a "service scholarship" covering tuition ($120 a year), plus $30 a month for his living expenses. The observatory was located in the little village of Williams Bay, on Lake Geneva in southern Wisconsin, about seventy miles northwest of Chicago. Hubble jumped at the offer, borrowed W. W. Campbell's *Stellar Motions* and Agnes Clerke's *History of Astronomy in the Nineteenth Century* from Frost, and wrote that he would come to Yerkes in the fall. However, the American Astronomical Society was going to meet at nearby Evanston, Illinois at the end of August, and its members would visit Yerkes for one day. Frost suggested that Hubble come north in time for the meeting, and said he would recommend him for membership if he wished. The dues were $2 a year, and Hubble quickly joined up.[31]

Therefore he was present at the meeting on the Northwestern University campus at which V. M. Slipher, the quiet, modest Lowell Observatory astronomer, presented his latest spectrographic observations of "nebulae." He had been the first to obtain well-exposed, well-calibrated spectrograms of spiral "nebulae" which not only showed their absorption-line spectra, but when measured revealed their large Doppler shifts and thus their large radial velocities. These are much larger than the typical velocities of stars, and when Slipher had presented his first results at Atlanta the previous summer, they had created a sensation. Some astronomers, including the renowned Henry Norris Russell, had expressed skepticism. Now, in the paper Hubble heard, Slipher had additional observational results, all of which confirmed and extended his earlier measurements.[32] Hubble must have felt the excitement in the air.

In contrast, at Yerkes Observatory Hubble found a near-moribund institution. Founded by George Ellery Hale and built around the largest refracting telescope in the world, it had gone into operation with great fanfare in 1897. But in 1904 Hale had left permanently for Southern California, to build Mount Wilson Observatory, taking with him George W. Ritchey, Walter S. Adams, Ferdinand Ellerman and Francis G. Pease. Frost, who inherited the Yerkes directorship, had originally been a competent spectroscopist, but without any creative ideas of his own he continued carrying on the same routine programs unchanged for years. He lost the sight of one eye while Hubble was at Yerkes, and stopped observing; he was to become completely blind in 1921 but nevertheless remained director until he retired in 1932. The senior faculty member, E. E. Barnard, had started his astronomical career as an amateur, and had made spectacular discoveries as a visual observer in his youth, and an outstanding series of wide-field photographs of the Milky Way later, but he had no theoretical training or ideas and did not advise or teach students. The other faculty members were astronomical lightweights, John A. Parkhurst, and Storrs B. Barrett, who had been hired originally as Hale's assistants and left behind when he took the "first team" to Pasadena, and Oliver J. Lee, a longtime Yerkes computer and assistant, who had just become an instructor.[33]

No formal courses were offered at Williams Bay; in each quarter of his two years there Hubble would register, with the few other students, for "Research at Yerkes Observatory."[34] He observed regularly on the radial-velocity program with the 40–inch telescope, and except for some other minor duties the rest of his time was his own, for study and research under the general supervision of Frost. The 24–inch reflector, built by Ritchey, had been standing mostly idle since its maker had left for Mount Wilson. Somehow, perhaps inspired by Slipher's paper and encouraged by

Frost, Hubble began a program of nebular photography with it. Soon he made his first discovery; comparing his direct plates of NGC 2261 with an earlier plate taken by Frank Jordan and with other plates from other observatories, he could see that definite changes had occurred within the nebula in a few years. It was an important result, not only because it demonstrated that changes can occur in nebulae, but also because it showed that this nebula must be small and relatively near. Two other variable nebulae were known, and Hubble noted that all three of these objects are associated with "dark occulting matter" (interstellar dust), an important step toward understanding them.[35] Frost included these results in the paper he presented at the meeting of the National Academy of Sciences in Washington in April 1917.[36]

Hubble's program grew into his Ph.D. thesis, "Photographic investigations of faint nebulae." It contains many foreshadowings of his later research on galaxies and the universe. He described and classified the numerous small, faint "nebulae," stated correctly that most of them are not spirals (as was then widely believed), but what we now call ellipticals. He confirmed and emphasized that their distribution in the sky avoids the Milky Way, and that many of them occur in clusters. "Suppose them to be extra-sidereal and perhaps we see clusters of galaxies; suppose them within our system, their nature becomes a mystery," he wrote. If the typical spiral "nebula" has the same size as our Milky Way system, he said, estimated by Arthur S. Eddington as 2000 pc in radius, they must be at distances measured in millions of light years.[37] Hubble's thesis is not very good technically, contains few references to earlier work, and is decidedly confused in its theoretical ideas, but it shows clearly the hand of a great scientist groping toward the solution of great problems. Hubble was never an outstanding technical observationalist, like Walter Baade and Milton L. Humason, but he always had the drive, energy, and enough skill to use available instruments so as to get the most out of them. He was aware of the 24–inch reflector's limitations, and of his own. But he recognized the right questions to ask, and he had the self-confidence to see what was on his plates, and describe it, where others who had perhaps seen it before had ignored it, or worse, tried to ignore it, because it did not fit the current pictures of the universe that they had in their minds.

Hubble was clearly a very good student and research worker, but in the context of the Yerkes Observatory of that time he was unique. A large fraction of its students were teachers at colleges who came only in the summers, typically only one student completed the Ph.D. every two or three years beginning in 1904, and except for Hubble none of them was ever heard of again before Otto Struve, who started in 1920 and finished in 1923.[38] For his second year Hubble received a fellowship which paid $320, and for his third year Frost recommended him for the top fellowship, worth $520, but there was not enough money and he got $320 again. Following the normal pattern, Hubble returned to the campus for the fall and winter quarters of that last year, to take the required graduate courses in mathematics, physics and astronomy, the latter all celestial mechanics and "theories of applied differential equations" (numerical methods of integrating orbit equations), and to begin writing his thesis. He also became the graduate resident head of Snell Hall, one of the men's dormitories, a perfect post for a natural leader of his abilities, which gave him free room and board.[39]

In October 1916 Hubble met Hale, the director of Mount Wilson Observatory, who often visited Chicago, where his brother lived and where he himself had been born and grown up. He had evidently heard through the astronomical grapevine of the bright young student doing an observational thesis on nebulae with the 24–inch reflector. Hale was looking for future staff members, for the 100–inch reflector was nearing completion on Mount Wilson. Henry Gale, Hale's friend and collaborator in the Chicago physics department, recommended Hubble highly, and when Hale met him he offered him a job, conditional on completion of his Ph.D.[40]

Frost had planned that Hubble follow the normal program and return to Yerkes for his last quarter, to complete his thesis and take the final oral examination.[41] However by January 1917 the United States was close to entering World War I against Germany, a preparedness campaign gripped the nation, and patriotic young Americans, particularly those with deep roots in Great Britain who had been Rhodes Scholars, were itching to get into the fight. Hubble, probably partly to keep closer to the center of the action and partly because he liked life on the campus better than in the isolated Williams Bay, decided to remain in Chicago for his last quarter. Moulton had offered him the chance to teach an elementary astronomy course and he accepted it. At the end of March Hubble went out to Yerkes and took his last plate of the variable nebula. He still hoped to finish his thesis in June and take the job in Mount Wilson.[42] But on April 4 President Woodrow Wilson went to Congress and asked for a declaration of war; on April 6 he got it and America was in the fight. Four days later Hubble sent Frost a report on his work of the previous quarter, together with a second paper on NGC 2261 and asked him to recommend him for a commission in the army! Under the preparedness laws then in force, any college graduate like Hubble could apply for a commission, and needed only letters of recommendation from five worthy citizens to prove his qualifications. Frost of course obliged, but urged Hubble to complete his thesis and take his examination before going off to war.[43] Hubble also wrote Hale, telling him of the step he was taking, requesting a letter of recommendation, and asking him if he knew any reason he should not join the army. Hale, already deeply engaged in organizing the U. S. wartime scientific effort at the National Research Council in Washington replied that "under the circumstances it would be natural for you to apply for a commission" and added that he "hope[d] to renew" his offer of a position at Mount Wilson as soon as Hubble had his degree and the war ended. He enclosed a letter recommending Hubble "as one most worthy of receiving a commission ... who will, I am sure, be a credit to the service."[44]

By May 1 Hubble had learned that the officers' training camp to which he was to be sent would open in two weeks. He had completed the draft of his thesis and was brushing it up, but it still seemed "scrimpy" to him. Frost advised him to fatten it up by including his paper on NGC 2261 in it. In normal times he would probably have insisted that Hubble rewrite and expand the thesis, but under the circumstances that was impossible. On Saturday May 12 Hubble took his final oral examination at Yerkes, before Frost, Barnard, Parkhurst, MacMillan, and Lee, and passed with flying colors. The committee awarded him his degree *magna cum laude*. On Tuesday May 15 he reported for duty at nearby Fort Sheridan.[45]

ARMY

The "fort" was actually a military reservation on Lake Michigan, north of Evanston and Chicago, used as a training area in World War I (and later again in World War II). Hubble and the other college graduates who started officer training the same week he did made up Company 10, Illinois Training Camp. They took a quick one-month course in everything from signalling to bayonet fighting with plenty of close-order drill, and military courtesy and discipline as well, no doubt. Hubble thrived on it and reported to Frost, "this military game seems to be a nitch [niche] into which I fit. I was the fourth man to be made student captain — we have a new one each week . . . " After one month most of the candidates were sent on to the artillery, in which they would become first or second lieutenants, but Hubble volunteered for the infantry, in which he could expect to be commissioned as a captain. At his commanding officer's request, he instructed his fellow trainees in night marching by the stars, using books borrowed from the Yerkes library. After two months more training in the 11th Provisional Regiment at Fort Sheridan, Hubble was commissioned a captain on 15 August 1917.[46]

He was ordered almost immediately to Camp Grant, near Rockford, Illinois, about 80 miles west of Chicago. It had been a corn field only two months before and there Hubble and the other newly commissioned officers from the 11th Provisional Regiment were to bring into existence the 86th Division. Its commanding general, the senior staff officers, the regimental commanders and the sergeants came from the regular army, while the rest of the initial cadre were new reserve officers. Two weeks after the officers had arrived, the first "selects" (draftees) began pouring in. Most of the men came from northern Illinois, Wisconsin and Minnesota, and the unit was called the Black Hawk Division for the Sauk Indian chief who had battled the Americans in the region a century before.[47] Hubble became commanding officer of the 2nd Battalion, 343rd Infantry Regiment as the 86th Division was being organized, or very soon thereafter.[48] He gloried in it, and wrote to Frank Aydelotte, a fellow Rhodes Scholar who later became president of Swarthmore College:

> They have given me a Battalion to work my will upon, and, we hope, to lead to the front. Stirring times — I can't picture myself missing the gathering, as it were, of the clans.[49]

New men kept arriving as the training continued. In October a large draft was sent to Texas to fill vacancies in the 33rd Division, the Illinois National Guard division, being readied for shipment to France. Hubble's 86th Division became in reality a training division, with new recruits continually streaming in, and half-trained veterans of a few months departing for other divisions headed for France. Throughout the bitterly cold winter of 1917–18 Hubble trained his 2nd Battalion. With his large stature, athlete's body, perfect physical condition, outdoor experience, enthusiasm, intelligence and leadership ability, he was undoubtedly an excellent commander, and the letter of commendation he received for his service "during the very trying days while we were training at Camp Grant" was surely well deserved.[50] He was promoted to major in January 1918.[51] Frost, who had decided Hubble's thesis was not suitable for publication in the *Astrophysical Journal,* but should go into the *Publications of the Yerkes Observatory,* sent him the proofs, but he evidently did not receive them, or ignored them until after the war had ended.[52] His paper was finally published in 1920.

By June 1918 nearly 90,000 men had passed through the Black Hawk Division to other units. Finally it received its orders to move to Hoboken, the port of embarkation, in August. However, the officers and men soon learned that their division was to be broken up, and that they were to become replacements in the divisions already committed in France. Most of the batallion, company and platoon commanders, including Hubble, were ordered from their units to the Advance School Detachment consisting of 140 officers and an approximately equal number of noncoms. They sailed out of New York harbor on September 9 on the *Walmar Castle,* across the Atlantic Ocean, around the north of Ireland, and up the Firth of Clyde to Glasgow, where Hubble debarked on September 19. He and the rest of the detachment went by troop train to Romsey, near Southampton, where they were reunited with the rest of the division, which had landed at Liverpool. In Romsey they were hit by a raging influenza epidemic, but most of them survived and were ferried across the English Channel in paddlewheel steamers to Le Havre by night. From there Hubble and his fellow officers split off to the advanced combat training schools that General John J. Pershing's American Expeditionary Force had set up in France. First Hubble was sent to the Third Corps officers' school in Clamecy, far behind the lines, 200 km southeast of Paris and 100 km west of Dijon. In the first week of October he went on to the advanced school at Langres, about 35 km southeast of Pershing's headquarters at Chaumont, and closer to the Alsace-Lorraine front.[53] It is probable that Hubble was taken from Langres, with other officer trainees, to observe the fighting in the Meuse-Argonne offensive, for soon after he returned to the United States after the war, he wrote Frost "I barely got under fire," and this was the only occasion on which he could have done so.[54]

From the training schools Hubble and the other officers were sent back to rejoin the 86th Division at its camps in the vicinity of Bordeaux. Most of the enlisted men had gone to other divisions as replacements, but a skeleton cadre had been retained in each company, battalion and regiment of the Black Hawk Division. When the officers returned, they believed that their division would be filled with fresh replacements, retrained, and sent to the front, but the Armistice ended the war before this could begin. The 86th Division never got into combat, or even into the reserve lines. On November 12 it was broken up, and most of its officers and men became the nucleus of the unit which operated the embarkation center at Le Mans, devoted to preparing their comrades for the journey back to the States and demobilization. Other officers were sent to other commands and headquarters throughout the A. E. F.[55]

Frost, who had been following Hubble's and other former Yerkes students' wartime experiences with pride and concern, wrote him immediately after the Armistice. He gave him the news from the observatory, enclosed a recent paper by Slipher from Lowell Observatory on Hubble's variable nebula, and also sent him a letter of introduction that would help him to visit observatories in Europe, now that the war was over.[56] Hubble had many friends among the officers of the 86th Division, of whom several, particularly the West Pointers, had important staff jobs in commands throughout France and Germany. With his handsome good looks, stern military bearing, general legal background, and ability to speak French and German, he was a good prospect for them. By Christmas 1918 Hubble was at U. S. Army of Occupation headquarters in Trier, Germany. In January he was sent back to the Combat Officers Depot in Gondrecourt, France, and then to Paris to serve briefly with the American Peace Commission. After a month there, he went back to Gondrecourt, no doubt for further processing, and then on to England.[57] He had managed to find out about an Army program for officers and men to study in British universities while awaiting shipment back to the United States, and to have himself assigned to it.

CAMBRIDGE

The Army had set up, in record time, a vast system of schools, ranging from post and divisional schools to the A. E. F. University at Beaune, on the Cote d'Or in France. In addition, arrangements were made for qualified students to attend French and British universities. Candidates could volunteer for these programs, and if they were lucky or had the right connections, be accepted. In all about 6,000 American officers and enlisted men attended universities in France, and 2,000 more, universities in England, in the spring and early summer of 1919.[58]

Hubble was one of the two hundred army students who arrived at Cambridge at the end of March.[59] Probably about ten or twenty percent of them were officers, the rest enlisted men. Hubble stated in a letter that he was "in charge of American Army Students in a dozen British Universities," and while this could not have been true from the start (a picture of the group at Sheffield University shows a colonel nearer the middle of the front row, clearly outranking Hubble, who is at one end of the same row) it may have been correct by the time he wrote it in May. As a major, Hubble probably was the ranking officer in Cambridge. He wrote that he was "availing myself of the opportunity of studying the school of statistical analysis as expounded by [Professor Arthur S.] Eddington." By this Hubble probably meant that he was studying notes taken by a student in the Michelmas (fall) term, when Eddington had lectured on "Combination of Observations," or readings from his syllabus for that course. Eddington himself was gone for almost the entire Easter term, when Hubble was in Cambridge, on the famous solar eclipse expedition to Principe, West Africa to measure the gravitational deflection of light.

Hubble made these statements in a letter to Hale, whom he wrote to remind of his pre-war job offer, and to ask if it still held. He told Hale that he often enjoyed the hospitality of H. F. Newall, the wealthy English astronomer who entertained regally in his great house Madingley Rise, close to the Cambridge Observatory.[60] Hubble need not have worried, for Hale, and Adams who was by now his trusted lieutentant and the acting director during his frequent absences, had kept the promising prospective staff member in mind all during the war.[61] Hale assured Hubble that he could have a staff position as soon as he could get out of the army, at a salary of $1500 and the promise of "advancement as rapidly as your work and the funds at our disposal will warrant. Please come as soon as possible," he added, "as we expect to get the 100–inch telescope into commission very soon, and there should be abundant opportunity for work by the time you arrive."[62]

Newall had Hubble proposed for membership in the Royal Astronomical Society at its regular monthly meeting in London on 9 May, 1919, "from personal knowledge." Probably Hubble attended the meeting himself and met many of the English astronomers present; certainly he was on travel status in England for nearly a week including that date. Slipher had been proposed by the Council as an "Associate," the equivalent of an honorary membership reserved for distinguished foreign scientists, at the previous meeting of the Society.[63]

Two months later Hubble certainly attended the special meeting of the Society held to meet and honor the American astronomers passing through England on their way to the conference in Brussels, which led to the founding of the International Astronomical Union. W. W. Campbell, director of Lick Observatory, headed the American delegation, which also included Adams, Frank Schlesinger, S. A. Mitchell, Charles St. John, Benjamin Boss, Joel Stebbins and F. H. Seares. Many of them gave papers at the Royal Astronomical Society meeting. Afterward all of them, and Hubble, joined their hosts at the Royal Astronomical Society Club dinner. The Club, with its restricted, self-perpetuating membership, included all the leaders of British astronomy; many of them were present for this informal meeting with the leaders of American astronomy. The tall, handsome American major sat between astrophysicist Arthur Schuster and Frank Dyson, the Astronomer Royal, close to Campbell, who was at the end of the head table.[64] Hubble must have discussed his future at Mount Wilson with Adams, St. John and Seares, all of whom were on its staff, but no record has survived.

A few weeks later Hubble was on his way to the port of embarkation at Brest. He sailed for America at the end of July and landed in New York on 10 August. While Hubble was still a student at Yerkes Observatory, his mother had moved with her family to Madison, Wisconsin, only about 70 miles from Williams Bay (and not much further from Fort Sheridan or Camp Grant).[65] However, he did not go there, but straight to California, with only a one-day stop in Chicago. He was discharged at the Presidio in San Francisco on 20 August 1919, and went immediately to

Pasadena to begin work. He summed up his military career to Frost in the words:

> I am still a major. I barely got under fire and altogather I am disappointed in the matter of the War. However, that Chapter is finished and another is opening.[66]

MOUNT WILSON

Hubble had come to Mount Wilson at a good time. The 60–inch reflector had been in use since 1908 and the 100–inch, first assembled and pointed at the sky from Mount Wilson in 1917 but held up by the war, finally went into general use by the staff on 11 September 1919, just a few days after Hubble's arrival.[67] From his Yerkes experience he knew how to use reflectors effectively, and he knew about nebulae and problems that needed solving. He went right to work. Milton L. Humason, later an outstanding staff member and Hubble's collaborator in his studies of the distant universe, broke in the newly appointed astronomer on 25 October 1919, the first time he used the Mount Wilson telescopes. Years later, Humason's memories of Hubble's first observing at Mount Wilson were still clear:

> I received a vivid impression of the man that night that has remained with me over the years. He was photographing at the Newtonian focus of the 60–inch, standing while he did his guiding. His tall, vigorous figure, pipe in mouth, was clearly outlined against the sky. A brisk wind whipped his military trench coat around his body and occasionally blew sparks from his pipe into the darkness of the dome. "Seeing" that night was rated as extremely poor on our Mount Wilson scale, but when Hubble came back from developing his plate he was jubilant. "If this is a sample of poor seeing conditions," he said, "I shall always be able to get usable photographs with the Mount Wilson instruments." The confidence and enthusiasm which he showed on that night were typical of the way he approached all his problems. He was sure of himself — of what he wanted to do, and of how to do it.[68]

The proofs of Hubble's thesis finally caught up with him and he corrected them and returned them; soon afterward his paper appeared in the *Publications of the Yerkes Observatory*.[69] For a year or two he corresponded politely with Frost, giving him news of Mount Wilson and expediting visits for astronomers who came west from the Wisconsin observatory.[70] Soon he had completed his first Mount Wilson research program, a comprehensive study of diffuse galactic nebulae (within our Galaxy). He published the results, separating reflection nebulae and emission nebulae physically, and showing how their nature and observed properties depended completely upon the type of stars within them, which are the sources of their luminosity. Already in these papers he indicated he was also studying deeply the nature of the "non-galactic nebulae." This name he stated, "does not mean that the latter class must be considered as 'outside' our galaxy, but that its members tend to avoid the galactic plane and concentrate in high galactic latitudes."[71] Clearly he had not yet fully accepted the conclusions of Edward A. Fath, Roscoe F. Sanford and Heber D. Curtis, who had provided observational evidence that these objects are "island universes" or galaxies outside our own, as Hubble himself was soon to confirm brilliantly.[72]

MARRIAGE

On 26 February 1924 Hubble married Grace Burke Leib. She had been born in

Iowa, the daughter of John P. Burke and his wife Luella.[73] Her father, as a young man, had been the cashier and then manager of a small bank in Walnut, Iowa. When Grace was two years old the family moved to San Jose, California, where John Burke became the manager of a street railway company, then an attorney, and then manager of the Bank of San Jose. In 1906 when Grace was 17, they moved on to Los Angeles, where John Burke was a very successful banker, a vice president and director of the First National Bank of Los Angeles.[74]

After her family moved south, Grace Burke finished high school at Marlborough, a private, prestigious school for girls in west Los Angeles. Then she went to Stanford University, from which she graduated in the class of 1912. She was a good student who was elected to the scholastic honor society, Phi Beta Kappa; she was also a member of Alpha Phi sorority and of the Senior Prom Committee. Just before Christmas after her graduation she married her Stanford classmate, Earl R. Leib, the son of Judge Samuel F. Leib, a prominent San Jose attorney. Earl Leib was a geologist who worked for the Southern Pacific Company, often in the field, assaying coal deposits and mines, and the couple lived with her parents in Los Angeles.[75]

Hubble first met Grace Leib at Mount Wilson. Her husband's older sister, Elna, was the wife of Lick Observatory astronomer William H. Wright. They had gone on many camping trips and climbing expeditions in the Sierras together. In June 1920 Hale invited Wright to bring his ultraviolet spectrograph to Mount Wilson and observe with it on the 100–inch telescope, and when the Lick astronomer drove south his wife came with him. They stopped in Los Angeles and urged Grace to come with them to the mountain, and stay with her sister-in-law at the Kapteyn Cottage, the little wooden summer house where wives and friends could live, while the astronomers observed at night and slept by day in the male-only "Monastery." As they drove up to the observatory, Wright described for his wife and sister-in-law the new Mount Wilson astronomer who had such great promise, Hubble. One day the two women walked over to the laboratory building, where the astronomers worked in the afternoons, to borrow some books from the little library it contained. It was there that Grace first saw Hubble. As she remembered it years later, after his death,

> When I first saw him on Mt. Wilson that summer afternoon in 1920, he was standing at the laboratory window, looking at a plate of Orion. This should not have seemed unusual, an astronomer examining a plate against the light. But if the astronomer looked an Olympian, tall, strong and beautiful, with the shoulders of the Hermes of Praxiteles, and the benign serenity, it became unusual. There was a sense of power, channeled and directed in an adventure that had nothing to do with personal ambition and its anxieties and lack of peace. There was hard concentrated effort and yet detachment. The power was controlled.

Probably it is a very good description of what she thought she saw, although Hubble had certainly not photographed Orion the previous June night, for it was behind the sun in the daytime sky. During the rest of Wright's observing run, they met frequently at the Kapteyn Cottage, and when they left the mountain, Hubble rode down to Pasadena with them.[76]

One year after that electrifying meeting, Grace Leib's husband Earl died. Alone, he had gone down into the shaft of an isolated coal mine in Amador County without a mask, and had been overcome by an unexpected accumulation of gas. He fell to his death. The State Industrial Accident Commission, which investigated the accident, ruled that he had died as a result of asphyxiation.[77]

After Leib's death, Grace and Hubble met again, according to her account, just after the total solar eclipse of 10 September 1923. Lick Observatory sent a party to Baja California, near Ensenada, headed by Wright. His wife accompanied him, and again he asked Grace to come along with them. She was not enthusiastic until Hubble urged her to do so (evidently they were seeing each other or writing before this); then she decided to go. He accompanied the Mount Wilson party, headed by Adams, which was located near San Diego.

Both groups, and all the other American astronomers in the vicinity, most of whom were on Catalina Island, were clouded out at the time of the eclipse, but afterward Grace and Hubble met in San Diego. "On February 26, 1924 we were married," is her next statement; more than this we cannot know. It was a Catholic ceremony, the faith in which she had been born and raised. They honeymooned at her family's private cottage near Pebble Beach in Carmel, then went abroad for a tour of Europe before he returned to his work at the observatory.[78]

GALAXIES

By this time Hubble was making very rapid progress on understanding the nature of the spiral "nebulae." John C. Duncan, a frequent visitor to Mount Wilson Observatory from Wellesley College, had used the 60–inch and 100–inch telescopes to discover three variable stars in M 33, one of the largest spiral "nebulae."[79] Hubble quickly took over this program and found many more variable stars in M 33, and in M 31. From a long series of observations he was able to derive light curves of these variables and show that many of them are cepheid variables. These stars, of a recognizable type known in our Galaxy, in addition to the novae that George W. Ritchey, H. D. Curtis, and Harlow Shapley had previously discovered in M 31 and other spirals, convinced all but a handful of the most skeptical astronomers that these objects are actually galaxies of stars. Hubble's work soon became well known, and he shared in a $1,000 prize for the outstanding paper read at the Washington meeting of the American Association for the Advancement of Science at the end of 1924. His paper listed 22 cepheids in M 33 and 12 in M 31, and from them he derived the distances of these two spiral galaxies.[80] A series of papers on NGC 6822, M 33 and M 31 quickly followed, along with Hubble's papers on the classification of galaxies and their absolute magnitudes. Before long Humason was working with him, measuring the radial velocities of galaxies for Hubble's discussions of the velocity-distance relationship.[81] Hubble was soon established as *the* outstanding observational cosmologist, and he dominated the field until his death in 1953.

CONCLUSION

Hubble had begun with a high-quality education at a very good academic high school, Wheaton, and at an excellent university, Chicago. He was a good student and an intelligent person. But he had very little training in research at Yerkes Observatory or anywhere else, except what he taught himself by doing it. He was exceptionally self-reliant, and had tremendous drive. Partly those were his inheritance from his insurance-agent father, and they were amplified by his strength and athletic ability. His handsome features and physical stature made him instantly attractive, and got him off on the right foot wherever he went and whatever he started. His experiences at Oxford and in the army, where he suddenly thrust himself into strange, different situations, and succeeded, provided the finishing touches for an exceptional research worker.

ACKNOWLEDGEMENTS

We are very grateful to many helpful friends, colleagues and correspondents who provided information without which this study of Edwin Hubble's early life would have been incomplete: to Diane Batson (Wheaton Central High School), Judith L. Bausch (Yerkes Observatory), Barbara Beury (New Albany Senior High School), Lawrence Blakeé (Pasadena), Robert C. Bless (University of Wisconsin), D. W. Dewhirst (Cambridge University), R. A. Fletcher (Rhodes Trust), John Hale (New Albany), Mark Harris (Louisville Free Public Library), Rufus C. Harrod (Shelbyville), John Lankford (University of Missouri), Roxanne Nilan (Stanford University), Andrew and Carl Osterbrock (Cincinnati), Leon E. Panetta (U. S. Congress), Grote Reber (Tasmania), John H. Rhodehamel (Huntington Library), John J. Sloanaker (Carlisle Barracks), Joseph L. Spradley (Wheaton College), Maxine H. Sullivan and Jillian Warmund (University of Chicago) and Wendy A. Whitfield (West Point).

REFERENCES

1. See for instance W. G. Hoyt, *Bio. Mem. N.A.S.*, **52,** 411, 1980; D. E. Osterbrock, *Modern Cosmology in Retrospect*, ed B. Bertotti and S. Bergia (Cambridge, in press); W. C. Seitter and H. W. Dürbeck, *ibid.*
2. N. U. Mayall, *Bio. Mem. N.A.S.*, **41,** 175, 1970. A shorter obituary article, written soon after Hubble's death by a colleague and close friend, is H. P. Robertson, *Pub. A.S.P.*, **56,** 120, 1954. Like Mayall's biography, it is based completely on information provided by Edwin and Grace Hubble.
3. [G. B. Hubble, *Edwin Powell Hubble, a Biographical Memoir*], manuscript, HHL, afterward referred to as "Memoir" in this paper.
4. Thirteenth Census of the United States, Shelbyville, Shelby Co., Kentucky, Precinct No. 3.
5. [G.B. Hubble] to N.[U. Mayall], [$\sim$ 1 Jan 1957], SLO.
6. See reference 4; [N. U. Mayall], "Genealogy of Edwin Powell Hubble" and "Genealogy and Schooling," undated manuscript notes, SLO, afterward referred to as "Notes" in this paper. Mayall's notes were partly based on Grace Hubble's statements to him, but partly also on family papers she let him see in San Marino after Hubble's death.
7. Notes.
8. J. L. Spradley, *Sky and Telescope*, **76,** 28, 1988.
9. G. Reber to D. E. Osterbrock, 24 Oct. 1988.
10. R. E. Morgan, "Partial history of Longfellow School," [Wheaton, Ill.], manuscript, 1979.
11. Wheaton School District records, 1901–1905; Edwin P. Hubble report card, Wheaton High School, 1905–06, HHL.
12. Notes.
13. Undated clippings, Hubble scrapbook, HHL.
14. Edwin P. Hubble, Undergraduate course book, University of Chicago, HHL; Edwin Powell Hubble, University of Chicago transcript, afterward referred to as "Transcript."
15. *Chicago Maroon*, many daily sports articles January 1909 through March 1910. Two first-place ribbons for the running high jump in dual meets with Wisconsin and Purdue, both in May 1908, are in the HHL. Hubble's height is from an undated [1908–10] clipping from a Springfield, Mo. newspaper, HHL.
16. Helen Hubble Lane to L. Blakeé, 6 July 1985; *Chicago Maroon*, 16 Dec. 1909, 26 Jan. 1910.
17. R. A. Millikan to E. James, 8 Jan. 1910, HHL.

18. Edwin Powell Hubble, Matriculation card, Oxford University, 18 Oct. 1910, HHL; *Register of Rhodes Scholars 1903–1950* (London, 1950).
19. E. Hubble to [J. P. Hubble], 17 June 1911, HHL.
20. Edwin Powell Hubble, Rhodes Trust record, Rhodes House, Oxford.
21. J. Lankford to D. E. Osterbrock, 28 Nov. 1988. Larsen, who became a professor of Roman and Greek history at Ohio State University and the University of Chicago, never met Hubble again after they left Oxford. He told this reminiscence to Prof. Lankford in 1979 or 1980, unaware that it was a criticism which many of Hubble's contemporary American astronomers had later made also.
22. *Queen's College Athletic Sports, Monday, February 13, 1911,* [Oxford, 1911], *Rhodes Scholar Class 1910–1913,* [Oxford, 1913], HHL (both are printed pamphlets); Medals for high jump (first), hammer throw (second), weight throw (second), Oxford University Athletic Club, 1912, HHL.
23. Caron's Louisville City Directories, Louisville, 1910, 1911, 1913; also reference 4.
24. *Louisville Courier Journal,* 20, 21 Jan. 1913; *Louisville Herald,* 20, 21 Jan. 1913.
25. Caron's Louisville City Directories, Louisville, 1913, 1914, 1915.
26. Reference 20; E. H. Davis, *American Oxonian,* **1,** 119, 1914.
27. Acts of the General Assembly of the Commonwealth of Kentucky, 1918, Chapter 131; *Louisville Courier Journal,* 20 Sep. 1919; R. F. Bossmeyer to J. A. Gwinn, 15 May 1989.
28. J. A. Gwinn, *Filson Club History Quarterly,* **56,** 415, 1982. The Kentucky Bar Association has records going back only to 1933, but the Kentucky Supreme Court in Frankfort has records of the names of all attorneys who practiced in Kentucky back to just after the Civil War. Hubble is not among them. B. K. Wilson to J. A. Gwinn, 30 May 1979; G. L. Sego to J. A. Gwinn, 12 Aug. 1981; J. C. Scott to J. A. Gwinn, 25 May 1989.
29. *The Senior Blotter,* New Albany High School (yearbook), 1914; E. H. Davis, *American Oxonian,* **1,** 119, 1914.
30. F. R. Moulton to E. B. Frost, 27 May 1914, YOA.
31. Hubble to Frost, 19 May, 6 June, 17 June, 20 July 1914; Frost to Hubble, 2 June, 15 June, 8 July, 24 July 1914; Frost to Moulton, 1 June 1914, YOA.
32. W. G. Hoyt, *Bio. Mem. N.A.S.,* **52,** 411, 1980; *Pop. Astron.* **22,** 551, 631, 1914.
33. E. B. Frost, *An astronomer's life* (Boston and New York, 1933).
34. Transcript.
35. Hubble, *Proc. N.A.S.,* **2,** 230, 1916; Hubble, *Ap. J.,* **44,** 190, 1916.
36. Frost to Hubble, 7 Apr. 1917, YOA.
37. Hubble, *Pub. Yerkes Obs.,* **4,** Part 2, 1, 1920.
38. Department of Astronomy and Astrophysics, University of Chicago, "Recipients of the Degree of Doctor of Philosophy 1904–1979," [1979], manuscript.
39. Frost to Moulton, 14 Feb., 9 Mar. 1915; 6 Mar., 17 Mar. 1916; Moulton to Frost, 1 Mar., 9 Mar. 1916; Frost to R. D. Salisbury, 4 Apr. 1916; Salisbury to Frost, 7 Apr. 1916; Frost to Hubble, 4 Oct. 1916, YOA; E. H. Davis, *American Oxonion,* **4,** 99, 1917; Transcript.
40. G. E. Hale to W. S. Adams, 1 Nov., 9 Nov. 1916; [W. S. Adams] to H. G. Gale, 20 Dec. 1916; Gale to Adams, 31 Dec. 1916, George Ellery Hale Papers, Microfilm Edition, Carnegie Institution of Washington and California Institute of Technology, 1918, afterward referred to as "Hale Microfilm."
41. Frost to Hubble, 17 Jan., 15 Feb., 1 Mar. 1917, YOA.
42. Gale to Adams, 4 Apr. 1917; [Adams] to Hale, 13 Apr. 1917; Hale Microfilm.
43. Hubble to Frost, 10 Apr. 1917; Frost to Hubble, 12 Apr. 1917; Frost to

Commanding General, Central Department, U. S. Army, 12 Apr. 1917, YOA; Hubble, *Ap. J.*, **45**, 351, 1917.
44. Hubble to Hale, 10 Apr. 1917; Hale to Hubble, 19 Apr. 1917; Hale to O. W. Bell, 20 Apr. 1917; Hale to Adams, 20 Apr. 1917, Hale Microfilm.
45. Hubble to Frost, 1 May 1917; I. L. Carter to Frost, 4 May 1917; Frost to W. D. MacMillan, 4 May 1917; Frost to Hubble, 7 May 1917; W. P. Payne, [~ Apr 27, 1917], undated formal notice of Hubble Ph.D. examination; Report of Hubble Ph.D. final examination, 12 May 1917, YOA; Transcript.
46. Hubble to Frost, 12 June 1917; Frost to Hubble, 15 June 1917; S. B. Barrett to Moulton, 1 Oct. 1917, YOA; Commanding Officer Training Camp, Fort Sheridan, Ill. to listed officers (including Hubble), 7 Aug. 1917; HQ Military Training Camp, Fort Sheridan, Ill., General Orders No. 30, 15 Aug. 1917, HHL.
47. J. G. Little, *The official history of the eighty-sixth division* (Chicago, 1921). This book is the source for all the statements about the 86th Division in the remainder of the paper; Hubble's part is fitted into it by the letters, orders and other documents cited. He is mentioned by name only in the overseas shipment roster, p. 105.
48. *American Oxonion*, **4**, 158, 1917.
49. Hubble to [F.] Aydelotte, 17 Sep. 1917, SLO. Aydelotte soon afterward became the long-term American secretary of the Rhodes Scholars.
50. Hubble, Army physical examination reports, 21 May 1918, 9 July 1918, HHL; C. R. Howland to Hubble, 4 Nov. 1918, HHL.
51. War Department Washington, D. C., Special Orders No. 72, 1918 Mar 27, HHL.
52. Frost to Hubble, 4 June 1917, 19 Feb. 1918, 20 Aug. 1919; Hubble to Frost, 3 Nov. 1918, YOA.
53. The various dates and places where Hubble served are from his Quartermaster voucher record, National Personnel Records Center, St. Louis, Mo., and conform to the divisional history of reference 47. See also O. Paisley to C. R. Howland, 5 Nov. 1918, 6 Nov. 1918, HHL.
54. Hubble to Frost, 14 Aug. 1919, YOA.
55. Reference 47.
56. Frost to Hubble, 15 Nov. 1918 [2 letters], YOA.
57. Hubble quartermaster voucher record; A. E. F., H.Q. District of Paris, Special Orders No. 31, 31 Jan. 1919; American Commission to Negotiate Peace, War Damages, Paris, Special Orders No. 47, 13. Feb. 1919, HHL.
58. R. I. Rees, "Educational and Vocational Training," July 7, 1919, in *Reports of Commander-in-Chief, A.E.F., Staff Sections and Services*, Historical Division, Department of the Army (Washington), 1948.
59. Address of the retiring vice-chancellor, *Cambridge University Reporter*, 2 Oct. 1919.
60. Hubble to Hale, 12 May 1919, Hale Microfilm.
61. Adams to Hale, 14 Jan 1918; Hale to Adams, 23 Dec. 1918, Hale Microfilm.
62. Hale to Hubble, 9 June 1919, Hale Microfilm.
63. *Observatory*, **42**, 225, 1919; *M.N.R.A.S.*, **79**, 408, 467, 1919; Hubble quartermaster voucher record.
64. *Observatory*, **42**, 297, 1919; R. A. S. Club dinner lists, (744) July 11, 1919 (privately printed).
65. Frost to F. J. Gurney, 16 Aug. 1917, YOA; Madison City Directories 1916, 1917, 1919, 1921.
66. Hubble to Frost, 14 Aug. 1919, YOA; Hubble honorable discharge certificate, 20 Aug. 1919, HHL; Hubble quartermaster voucher record.
67. 100-inch telescope observing log book 1, 1916–1922, Carnegie Observatories, Pasadena.

68. M. L. Humason, *M.N.R.A.S.*, **114,** 201. 1954. This eyewitness account is undoubtedly basically true, and perhaps even the quotation, in which Hubble speaks in correct British English in the middle of the night to a poorly educated assistant, was remembered correctly. The other nonscientific parts of this obituary article that describe events at which Humason was not present are based on stories which Hubble and Grace Hubble told him, and which she approved for publication; they must be read with a high degree of skepticism. The date is from 60–inch log book 4, 1918–1919, Carnegie Observatories, Pasadena.
69. Reference 37.
70. Hubble to Frost, 3 Nov. 1919, 4 Apr., 28 Nov. 1920; Frost to Hubble, 21 May, 17 Nov. 1920, YOA.
71. Hubble, *Ap. J.*, **56,** 162, 1922; Hubble, *Ap. J.*, **56,** 400, 1922.
72. See, *e.g.*, Osterbrock, reference 1.
73. *Los Angeles Times,* 26 Feb. 1924; Certificate of death, Grace B. Hubble, 22 Mar. 1980, State of California.
74. *Who's Who in the Southwest* (Los Angeles, 1913).
75. Marlborough School records, Los Angeles; *Stanford Quad* (Palo Alto 1912); *Stanford University Alumni Directory and Ten Year Book IV 1891–1931* (Palo Alto, 1932); *Fifty Years of Phi Beta Kappa at Stanford,* ed. E. Mirrielees, (Palo Alto, 1941); *Pasadena Star News,* "Obituary notices," 25 Mar. 1980.
76. Memoir, "The missing years" and "The astronomer 1913–1917," HHL.
77. *San Jose Mercury Herald,* 17 June, 19 June 1921.
78. Memoir, "The missing years," HHL; W. H. Wright, *Pub. A.S.P.*, **35,** 275, 1923; Hubble-Leib Marriage Certificate, 26 Feb. 1924, Los Angeles.
79. J. C. Duncan, *Pub. A.S.P.* **34,** 291, 1922.
80. G. E. Hale, *Carnegie Year Book,* **23,** 81, 1924; *Pop. Astron,* **23,** 158, 1925; Hubble, *Pop. Astron.*, **33,** 252, 1925.
81. See *e.g.* Osterbrock, reference 1.

HHL = Edwin Hubble Papers, Henry Huntington Library, San Marino, California.

SLO = Mary Lea Shane Archives of the Lick Observatory, University Library, University of California, Santa Cruz.

YOA = Director's Papers, Yerkes Observatory Archives, Williams Bay, Wisconsin.

SHAPLEY, HUBBLE, AND COSMOLOGY

OWEN GINGERICH
Harvard-Smithsonian Center for Astrophysics, Cambridge, MA 02138

ABSTRACT Working for several years around 1918 at Mount Wilson Observatory, Harlow Shapley almost single-handedly delineated the vast scope of the Milky Way, but in 1921 he left California for Harvard. Edwin Hubble, who provided the primary observational foundation for the concept of the expanding universe (which is now considered the major astronomical idea of our century), spent his entire career in southern California. From opposite sides of the continent the two men sparred over the homogeneity of the observable universe.

Harlow Shapley's seven years at Mount Wilson (1914-21) were the most productive of his research career. He wrote over 100 notes and papers (30 in the *Publications of the A.S.P.*), and he radically altered the concept of the Milky Way. Early in 1914 Shapley wrote to the Dutch astronomer J.C.Kapteyn that "the work goes on monotonously, but the results are a continual pleasure. Give me time enough and I shall get something out of the problem yet." Barely a year later he was able to report to Arthur Eddington that "with startling suddenness [the cluster studies] seem to have elucidated the whole sidereal structure. . . . To be brief, the globular clusters outline the sidereal system." Walter Baade later remarked, "I have always admired the way in which Shapley finished this whole problem in a very short time, ending up with a picture of the Galaxy that just about smashed up all the old school's ideas about galactic dimensions."

These researches at Pasadena, combined with Adriaan van Maanen's measurements of the apparent rotation of spiral nebulae, led to a very different view of the universe than the one favored farther north at Lick Observatory. There, Heber D. Curtis was convinced that the spirals were island universes, but he was skeptical of Shapley's use of cepheid variables to delineate the scale of the Milky Way. George Ellery Hale, director of the Mount Wilson Observatory and an energetic member of the National Academy of Sciences, arranged for the two astronomers to present their divergent views in Washington in April of 1920. The famous "great debate" took place near the end of both of their West Coast careers.

In 1919 Hale had added another Missourian to his staff. Edwin Hubble had been an undergraduate at the University of Chicago, and after a sojourn in Oxford studying law on a Rhodes scholarship, he eventually returned to Chicago for a doctorate in astronomy. When Hubble arrived in Pasadena, Shapley, and perhaps others, found his carefully cultivated Oxford accent a little off-putting. The tall, athletic, urbane Hubble was a study in contrasts with Shapley, who remained at the core a wide-eyed farm boy, and I can well imagine that Shapley may have harbored more than a twinge of jealousy for his new junior colleague.

Hale brought Hubble to Mount Wilson specifically to work on nebulae, just as Shapley's job was to study clusters. Shapley was later sometimes criticized, by Baade among others, for not carrying the hunt for cepheids to the spirals, but in addition to the psychological barrier, it was a question of turf, and clearly from the time of his arrival, Hubble had the nebulae as his mandate. In any event, as far as the spiral nebulae were concerned, the party line at Mount Wilson was to reject the Lick astronomers' notion that they were island universes, and this was so even after Shapley had gone east. Thus, in July of 1922 Hubble wrote to Shapley at Harvard saying that, "My own preference concerning the [star-like] objects clustering [around] M87 is to call them stars until they may be definitely shown not to be stars." With our 20-20 hindsight we know these objects are particularly luminous globular clusters around the giant spherical galaxy M87, and to mistake them for single stars could only lead to a conclusion that the Virgo nebulae were too close to be considered as island universes.

At best the situation was quite muddled until the beginning of 1924, when Edwin Hubble wrote to Shapley that, using the 100-inch at Mt Wilson, he had found a faint cepheid variable in M31. Application of Shapley's calibration of the period-luminosity law for cepheids then placed the spiral at nearly a million light-years. Shapley promptly responded that "Your letter telling of the . . . variable stars in the direction of the Andromeda nebula is the most entertaining piece of literature I have seen for a long time," and he promptly capitulated to the island universe viewpoint.

This still left the problem that our Milky Way was vastly larger than the Andromeda spiral. One approach was to consider that the Milky Way and its associated system of globular clusters was actually a supergalaxy, a collection of many systems perhaps something like the great congregation in Virgo. Shapley laid the foundation for such a discussion with paper published in 1926 on the Virgo cluster. Writing jointly with Adelaide Ames, he described the cluster as a cloud of galaxies placed at ten million light-years. By 1930 Shapley suggested as a working hypothesis the notion that our "unexpectedly extensive and populous" Milky Way system was "an amalgamation of star clusters and star clouds." His earlier insight that the Milky Way was far vaster than previously believed had set him at odds with Kapteyn, whose statistical measurements led to a much smaller sun-centered system. Shapley now proposed that our local system (which would roughly correspond to the dimensions found by Kapteyn) was similar to one of the Magellanic Clouds, and that the Scutum star cloud, the Cygnus star cloud, and perhaps half a dozen others were typical galaxies "in the sense that the average spiral nebula is a galaxy."

Meanwhile, other developments were afoot, namely, the discovery by Hubble and his associates that the fainter the nebula, the greater its spectral red shift, a discovery that led to the concept of the expanding universe.

Shapley, lacking the large light-gathering power and spectrographic facilities of the Mount Wilson Observatory, was locked out of the game, but with the legacy of survey instruments inherited from Pickering, he could mount large scale surveys, which began to show the irregularities of galaxy distributions. The joint paper with Adelaide Ames on the Virgo cluster was a typical fruit of this approach. There they tossed out the challenge that "Much remains unknown concerning the distribution over the sky of the nebulae of the spiral family, notwithstanding the rather confident assertions and generalizations that are frequently made." In a semi-popular series of lectures at Rice University, Shapley threw down the gauntlet: "Taking the whole sky, we find, in agreement with Hubble, that the increase of numbers with decreasing brightness is approximately of the order of magnitude appropriate to uniform density [that is, about four times more galaxies for each fainter magnitude]. But," he continued, "for various large sections of the sky the uniformity criterion fails conspicuously."

Hubble responded privately to Shapley, saying, "I do not place so much significance in the non-uniformity of distribution as you, for the data tend to smooth out when fainter limits are used." Shapley countered, "I sometimes feel you have done harm by emphasizing too much the uniformity, thus leading theorists into too much optimism over their highly simplified set-ups. On the other hand, you may think I emphasize the irregularity too much. But there is a deep meaning in non-uniformity in the star distribution. . . . These irregularities in the distribution of nebulae are highly important structural phenomena."

In retrospect we can sympathize with Hubble's desire for the existence of large-scale homogeneities, which made possible an initial exploration of cosmological issues. The concept of the expanding universe is undoubtedly the major astronomical idea of the twentieth century. Both theoreticians and observers helped fashion this majestic cosmic view. Even Shapley had played a part with his initial calibration of the period-luminosity law of the cepheids. Yet one name above all is linked with the expansion of the universe: Edwin Hubble's. He is rightfully considered the prime architect of this powerful cosmological view. However, I don't think it detracts from his enormous credit to notice that the irregular distribution of galaxies, which Shapley was beginning to delineate, today adds an important dimension to our understanding of the large-scale structure of the universe.

REFERENCES

The controversy between Shapley and Hubble on the distribution of galaxies is considered in greater detail, together with the archival references, in my article "Through Rugged Ways to the Galaxies" in *Journ. Hist. Astron.*, **21**, no. 1, (1990).

EDWIN HUBBLE'S COSMOLOGY

NORRISS S. HETHERINGTON
Office for the History of Science and Technology, 470 Stephens Hall, University of California, Berkeley CA 94720

ABSTRACT An analysis and appreciation of Edwin Hubble's science touches on many things. Characterizing Hubble's extraordinary achievements is the combination of elements that he alone connected, investigated, developed, and demonstrated persuasively. He cooperated with others, was persistent, and had an excellent sense of strategy in pursuing problems and in presenting solutions.

HUBBLE'S FIRST STEPS

Hubble was born in 1889, studied mathematics and astronomy at the University of Chicago, went to Oxford in 1912 as a Rhodes Scholar to study law, may have practiced briefly upon his return to the United States, and then took his PhD at the Yerkes Observatory in 1917.

Their shapes and high radial velocities suggested that spiral nebulae were island universes. Arguing against an extragalactic nature was the zone of avoidance, bright novae in spirals, until consistently much fainter novae were detected beginning in 1917, and Adriaan van Maanen's purported measurement of internal motions in spirals. Were they distant and comparable in size to our galaxy, van Maanen's rotation period would translate into a speed greater than the speed of light.

Hubble's doctoral dissertation was based on photographs of clusters of nebulae. To the 76 known nebulae in clusters he added 512 more. His classification scheme would be announced in the 1920s, by the 1930s his tuning-fork diagram would be famous, and the Hubble atlas of galaxies would appear posthumously in the early 1960s.

Regarding the nature of spiral nebulae, Hubble had no conclusive evidence, but he thought that "the great spirals, with their enormous radial velocities and insensible proper motions apparently lie outside our system." Characteristic of Hubble's science is an early focus upon crucial issues and an unrelenting pursuit of them, for decades if necessary.

DISTANCES TO SPIRAL NEBULAE

Following service in World War I, Hubble went to the Mount Wilson Observatory. Studying photographs of nebulae, he found Cepheid variable stars. Then he used the period-luminosity relation calibrated by Harlow Shapley to determine distances to the nebulae, distances that placed them far beyond the boundary of our galaxy.

Shapley could have done the same, had he not accepted George Ritchey's belief that the images beginning to be resolved on photographs of spiral nebulae were not stellar, as also might astronomers at the Lick Observatory. But it was Hubble, inspired by the island universe theory, who focused his attention on a consequential problem, who examined potential observational consequences, searched for variable stars in spiral nebulae, found them, and used them to measure distances.

In retrospect, the work appears inevitable. But considerable boldness, even lack of normal scientific prudence, was required. Hindsight justifies Hubble's work, but also reveals that what he thought were stars around a spiral were globular clusters, and Cepheid variables in spiral nebulae are not comparable with Cepheids in our galaxy. Yet each assumption was necessary. Justification lies not in any continuing agreement with advancing factual knowledge but in their necessity for extrapolations from the know to the possibly knowable.

Another aspect of Hubble's genius is found in his sense of strategy in presenting results. When his distances contradicted van Maanen's measurements, Hubble seemingly faced unsatisfactory alternatives of ignoring van Maanen's work or challenging it in an ungentlemanly, unscientific, and unacceptable manner. He ingeniously resolved the dilemma, making new measures and turning a former witness for van Maanen against him. Drawing on his legal training, Hubble skilfully employed trial tactics to attain a favorable verdict from the court of science.

THE VELOCITY-DISTANCE RELATION

Hubble's distance determination for spiral nebulae was the missing piece in a long-continuing puzzle of great significance. He used it to open a new phase of astronomical investigation, the relation between distance and radial velocity which he soon established, and to overthrow the centuries-old opinion that the universe was static.

An earlier, injudicious attempt to establish a velocity-distance relation had created a climate of suspicion, which Hubble took care to counter. Ignoring theories, he emphasized the empirical nature of his work. Also, he dragged into the discussion critics of the earlier effort, unnecessarily so except that their presence implied support for Hubble's work.

A RELATIVISTIC, EXPANDING, HOMOGENEOUS UNIVERSE

Hubble at first left interpretation of the velocity-distance relation to others, but then obtained help from Richard Tolman. Mrs. Hubble remembered: "I think it was in the 1930s when about every two weeks some of the men from Mount Wilson and Cal Tech came to the house in the evening. They brought a blackboard and put it up on the living room wall. In the dining-room were sandwiches, beer, whiskey and sodawater; they strolled in and helped themselves. Sitting around the fire, smoking pipes, they talked over various approaches to problems, questioned, compared and contrasted their points of view - someone would write equations on the blackboard and talk for a bit, and a discussion would follow."

Science deals with judgments secured by observation and experiment, and Hubble proposed to choose among cosmological theories. Theoretical relations were calculated for different models and complications in treatment of data resolved. Problems remained, however, and conclusions were tentative.

Cosmology, for centuries consisting of speculation based on a minimum of observational evidence and a maximum of philosophical predilection, had become an observational science. But philosophical values were not banned. Hubble, enamored of relativity theory, refused to accept falsification of that model of the universe when preliminary data argued against it.

CONCLUSION

World War II interrupted Hubble's work on cosmology and a heart attack ended his life in 1953, soon after the 200-inch telescope was completed on Palomar Mountain and too soon for conclusive answers from the research program planned by Hubble.

He chose crucial problems and pursued them intelligently, diligently, and unrelentingly. Hubble made assumptions necessary for extrapolating from the known to the possibly knowable, taking bold leaps of faith. He practiced cooperative research. He skilfully orchestrated scientific evidence, to overcome prejudice and to attain a fair hearing. He made cosmology an observational science but was not a scientific automaton, and when evidence contradicted a theory cherished for its philosophical value, Hubble persisted in its testing.

Some aspects of Hubble's science might be grouped under the labels "insight" or "intuition". These terms, however, have an unsatisfactory mystical connotation; they suggest acquisition of knowledge and understanding without any evident intervention of rational thought and inference. Analysis of Hubble's cosmology in this centennial year clears away some of the mystery of his success and shows it to be one of the great accomplishments of the human intellect.

THIN DISCS AND DENSE BULGES: WHAT DO THEY TELL US ABOUT GALAXIES?

JEREMIAH P. OSTRIKER
Princeton University, Peyton Hall, Princeton, NJ 08544

ABSTRACT Two basic facts about galaxies, the high density of the bulges and the thinness of the discs, are analyzed dynamically with the following conclusions: a) The mass added in stellar lumps after the formation of the major part of the disc is less than 3% of the Galactic mass, and b) spiral bulges are not made from mergers of bulgeless spirals, *i.e.*, not made out of disc material.

I) INTRODUCTION

Hubble recognized that, in visual appearance, typical spiral galaxies are comprised of two primary components: a bulge or spheroid, and a flat disc. The former is quite centrally concentrated, with scale length for the galactic spheroid estimated at 0.1 kpc (Caldwell and Ostriker 1987) and a shape which may be truly spheroidal, or triaxial, or even somewhat "boxy" but, in any case, is crudely speaking "round" in that the smallest axis is never considered to be smaller than the largest by as much as a factor of two. The disc (and its embedded Pop I component) is extended with scale length typically ~ 4 kpc and is quite thin with half light thickness ~ .3 kpc at a diameter of 15 kpc giving an axis ratio of ≥ 50.

The ratio of light from the two components in a Sb system of L_* luminosity is typically 0.6 (Schechter and Dressler 1987) which, if we assume the mass-to-light ratio differs by a factor of 2, makes the mass ratio in the two components 1.2, *i.e.*, they are comparable in mass.

A most important yet infrequently addressed question is how did the typical galaxy form with two such dynamically distinct components, and how is that distinction maintained over the lifetime of the system in the face of numerous perturbative forces?

Typical galaxies live in groups similar to the local group: the Galaxy, Andromeda, the Magellanic Clouds, M32, and a few other miscellaneous objects. Close approaches by companions can exert strong dynamical perturbations. Further, the typical velocity dispersion in the groups of 100-200 km/s is comparable to the internal velocity dispersion within galaxies, so that such close passages are ideal for dynamical interactions.

II) MASSIVE HALOS, DYNAMICAL FRICTION AND MERGERS

The flat rotation curves of galaxies and other lines of evidence indicate that the mass density falls as r^{-2} far more slowly than the light emissivity for normal spirals. If this dark halo material extends as far as the orbits of satellites surrounding a normal spiral, a condition satisfied for both M32 orbiting Andromeda and the Clouds orbiting our Galaxy, then straightforward Newtonian theory predicts that dynamical friction should cause the satellites to spiral into the parent system in a time $\approx T_{orbit}(M_{gal}/M_{sat})$ with the omitted dimensionless constant somewhat less than unity. This conclusion, reached by Tremaine, Ostriker and Spitzer (1975), implies that the two local giant spirals will cannibalize their companions in the foreseeable future with attendant increase in mass by several percent and presumably some accompanying dynamical indigestion.

Tremaine (1980), after systematically examining the situation for the average giant spiral, concluded that the mass accreted in a Hubble time was approximately

$$\Delta M_{acc} = 4 \times 10^8 \, M_\odot \left[H_o t \left(\ln\Lambda\right)/\sigma_{100}\right]^{0.6} (M/L)^{1.6} h^{-2.6}$$

$$\doteq 4 \times 10^9 \, M_\odot \quad ,$$

i.e., of the same order as the mass of the local satellite systems. That is, our situation is typical.

What will be the consequences of such satellite capture? The Magellanic Clouds are embedded in a stream of gas which is presumably a tidal plume and evidence for ongoing stripping. As they spiral in, we can expect most of their metal-poor gas ($\sim 10^8 \, M_\odot$) to be added to the gaseous disc of our Galaxy increasing its mass by ~10%. The stellar component will cause a gravitational stirring of the disc. A low density, relatively fluffy satellite such as the SMC will be tidally disrupted at a moderately large radius (the Roche limit for it being $R_{Roche} = R_{cloud} (5/3 \, v^2_{rot}/v^2_{cl,rms})^{1/2}$ kpc), and its stars added to the low metallicity stars of the spheroid at radii comparable to the Roche limit. But, a dense satellite like M32 will remain essentially intact and probably spiral all the way down to the core of its parent M31, where its relatively metal-rich stars will be added to the nucleus of M31 (which has a comparable metallicity).

Let us now examine these physical processes in a somewhat more quantitative fashion. Suppose that the gas is added to a very thin layer at the center of the disc on a timescale slower than the vertical period of a typical disc star. Then the vertical change in gravity for the typical star is

$$\frac{\langle \delta g \rangle}{\langle g \rangle} = 2 \frac{\Delta \Sigma_{gas}}{\Sigma_{tot}} \quad ,$$

with the factor of 2 coming from the fact that the gas is added at the center of the disc rather than (on average) at the half-mass point. The virial theorem for a self-gravitating planar system of scale height H_* can be written

$$\frac{\langle \delta v^2 \rangle}{\langle v^2 \rangle} = \frac{\langle \delta g \rangle}{\langle g \rangle} + \frac{\delta H_*}{H_*} ,$$

where an exponential stellar density distribution has been assumed and $<v^2>$ is the vertical velocity dispersion at the plane. This can be combined with the adiabatic invariant

$$\frac{\langle \delta v^2 \rangle}{\langle v^2 \rangle} = -\frac{2\delta H_*}{H_*} ,$$

to give

$$\frac{\langle \delta v^2 \rangle}{\langle v^2 \rangle} = \frac{2}{3}\frac{\langle \delta g \rangle}{\langle g \rangle} = \frac{4}{3}\frac{\Delta \Sigma_{gas}}{\Sigma_{tot}} ,$$

and

$$\frac{\Delta H_*}{H_*} = -\frac{2}{3}\frac{\Delta \Sigma_{gas}}{\Sigma_{tot}} .$$

Next let us imagine that the extra gas is turned into stars which share energy with the existing stars. This further reduces the thickness of the stellar disc to give

$$\frac{\langle \Delta H_* \rangle}{\langle H_* \rangle} = -\frac{10}{9}\frac{\Delta M_{gas}}{M_{disc}} ,$$

where we have now averaged H_* over the gaseous disc. More accurate modeling leads to a smaller numerical coefficient in the last equation. Thus, adding gas to a disc does, as expected, reduce its thickness in proportion to the added mass. The above calculation treats a limiting case where radiative losses are considered to be maximal, no gas is ejected, etc. Thus, adding a mass in gas equal to 10% of the disc mass will cause the star layer to contract by less than 12%.

The stellar component of the satellite orbiting through the disc heats it by three distinct processes. The first two involve a transfer of energy from the satellite orbit to random motion of the disc stars as investigated by Quinn and Goodman (1986) and Hernquist and Quinn (1987). Crudely, a loss of satellite energy of order $M_{sat}v^2_{orb}$ will have an energy change in the disc of order $M_{disc}\,\Delta v^2_{rand}$ so that

$$\Delta v^2_{ran} \approx v^2_{orb}\,(M_{sat}/M_{disc}))$$

Thus, infall of a satellite orbiting with a velocity of 250 km/s and mass 10% of the disc mass produces, for the disc, an energy equivalent to a random velocity of 80 km/s, which is larger than the value observed in the solar neighborhood. Alternatively, if we take a disc with an initial vertical velocity dispersion of 30 km/s and one third of this energy were deposited in z motions, it would increase the z velocity dispersion to 40 km/s and increase the scale height by about 80%. This calculation underestimates the heating effect, since it neglects the velocity dispersion increase at the expanse of the circular velocity produced by scattering of disc stars from the gravitational potential fluctuations caused by the infalling satellite.

This example shows two things: that gaseous infall is less effective at making discs thin than is stellar infall effective at thickening them, and that infall in the past of stellar systems in excess of a few percent of the mass of the Galactic disc is excluded on grounds that it would have left us with a thicker and hotter disc than we now have.

A detailed calculation by Toth and Ostriker (1990) indicates that if all of the current local disc thickness were due to satellite infall, then the maximum mass permitted to have fallen in since the disc was turned is 0.05 m_d/η where η is the ratio of the total heating rate to the satellite energy loss rate. Taking η = 1.8 from Carlberg and Hartwick (1989), we then estimate that 3% infall is the maximum amount tolerable. As noted, correction for gas dynamic effects will be small (perhaps increasing the limit to 3.3%). A similar but more stringent limit can be obtained by noting that the Toomre Q parameter limits the velocity dispersion components parallel to stellar discs if spiral structure is to be maintained ($Q<2$). Since the typical galaxy shows some significant spiral structure, it is evident that not more than a few percent of the disc mass can have been accreted in stellar lumps as dense as the spiral galaxies (mean density, not disc density).

Let us turn now to the origin of galactic bulges or spheroids Two salient facts emerge. First, they are far denser than the mean density of spirals in both real space and phase space density. Second their masses are comparable to that in the disc for a typical Hubble type Sb spiral. From the fact that they are denser than the mean density of the spiral component, it follows that, had they been accreted, they would not have suffered tidal disruption during infall. This argues that, if bulges have accreted, then the arguments of the last section apply and the mass accreted could not have exceeded more than a few percent of the disc masses. Thus bulges of Sc galaxies may have been made by accretion, but bulges of Sa and Sb systems are too massive to have formed by this process. Second, for the low mass very dense bulges found in Sb and Sc systems, the phase space density is orders of magnitude larger than that in spiral galaxies (Carlberg 1989). Since phase space density can only be reduced by mixing and violent relaxation, this implies that the bulges of these systems, if formed by the accretion of stellar systems, were produced by accreting other stellar spheroids far more dense than normal spirals.

III) CAVEATS AND CONCLUSIONS

We have argued that elementary dynamical arguments, when combined by the best known properties of the Hubble specification of galaxies, *i.e.* most galaxies have spiral patterns, thin discs and dense bulges, allow surprisingly strong conclusions to be reached about the origins of typical systems. Specifically, (1) the central spheroids were not made from merging ordinary spiral discs, (2) nor were they made from accreting dense bulges after the discs were formed. This does not, of course, prevent bulges having been formed from accumulation of other stellar bulges *prior* to disc formation, nor does it prevent their having been made by purely gas dynamical processes at any epoch. In so far as the bulge mass to total galaxy mass is greater than the gas mass to total mass of normal spirals, it would be necessary to use as building blocks systems more rich in gas than normal spirals if this expedient is to be utilized. The last conclusion (3) is that the total mass added to our Galaxy and other normal spirals in stellar components (as dense as 10^{-1} stars/pc^3) has not exceeded a few percent of the disc mass ($< 2 \times 10^9\ M_\odot$), since most of the disc was formed. This of course does not at all limit the infall of gas as proposed by Gunn (1983) and others, provided that infall is sufficiently gradual so as to avoid both stirring the disc and producing too high an X-ray accretion luminosity.

REFERENCES

Caldwell and Ostriker, J. P. 1987, *Ap.J.*, **251**, 67.
Carlberg, R. G. 1989, preprint.
Carlberg, R. G. and Hartwick, F. V. A. 1989, *Ap.J.*, **345**, 196.
Gunn, J. E. 1983, "The Formation of Galaxies," in *Internal Kinematics and Dynamics of Galaxies*, IAU Symposium, ed. E. Athanassoula (Dordrecht: Reidel), p. 379.
Hernquist, L. and Quinn, P. J. 1987, *Ap.J.*, **312**, 1.
Quinn, P. J. and Goodman, J. 1986, *Ap. J.*, **309**, 472.
Schechter, A. and Dressler, A. 1987, *A.J.*, **94**, 563.
Toth, G. and Ostriker, J. P. 1990, in preparation.
Tremaine, S., Ostriker, J. P. and Spitzer, L. 1975, *Ap.J.*, **196**, 407.
Tremaine, S. 1980, in *The Structure and Evolution of Normal Galaxies*, ed. S. M. Fall and D. Lynden-Bell (Cambridge: Cambridge University Press), p. 67.

MORPHOLOGY AND DYNAMICS OF GALAXIES IN THE HICKSON COMPACT GROUPS

VERA C. RUBIN, W. KENT FORD, JR.
Dept. of Terrestrial Magnetism, Carnegie Institution of Washington, Washington, D.C. 20015

DEIDRE HUNTER
Lowell Observatory, Flagstaff, Arizona 86001

The formation and evolution of galaxies in clusters and compact groups must be significantly more complex than for galaxies isolated in the field. We are presently studying the morphology and dynamics of spiral and elliptical galaxies in the Hickson (1982) compact groups in order to understand their past and future evolutionary history. We have obtained red and Hα images for 20 compact groups with the Kitt Peak 36-in telescope. In addition, we have obtained long slit spectra for 45 galaxies in 16 groups with the Palomar 200-inch telescope and double spectrograph + CCD.

In the spectroscopic sample, 33 of the 45 galaxies are classified SO/a or later type. These 33 spirals have rotational properties as follows: 9: too abnormal to classify, 13: peculiar or irregular, 11: normal. Two common types of peculiar rotation curves are seen; sinusoidal forms, with velocities going to zero at large radial distances, and asymmetrical forms, where velocities from one side of the major axis are flat, and those from the other side are rising or falling.

Within the groups we have observed, the distribution of normal rotation curves is not random. The dominant group member generally has a peculiar rotation curve (6 peculiar of 7 observed), in contrast to lower luminosity galaxies (5 peculiar of 17) which tend to have more normal rotation curves. Rotation curves in which velocities for the two sides of the major axes are significantly different are observed in 3 dominant galaxies.

Twelve galaxies for which we have long slit spectra are of types E and SO. Most (10 of 12) exhibit remarkable nuclear emission of great individuality. This high frequency of nuclear emission exceeds that previously reported (Phillips et al. 1986). In some E and S0's, nuclear emission arises in multiple nuclear knots. In others, emission extends beyond the nucleus; irregular and regular rotation patterns occur. In Hickson 23C (S0, NGC 1216) a gas disk rotates counter to the stellar population (Rubin et al. 1989).

Compact groups must be short lived and evolve rapidly. The peculiar dynamics may be the signature of a merger event. The nuclear gas is likely to be a recent acquisition. From the deep R-band imaging of the Hickson groups, we observe that some groups exhibit a common envelope which contains all of the member galaxies. The total group luminosity is consistent with the suggestion of Barnes (1989) that rapid dynamical evolution within a compact group will produce a dynamically ordinary elliptical galaxy as the central merger remnant within only a few orbital periods.

Hickson 31 is an especially remarkable group; Hickson identified galaxies A-D (Fig. 1). Hα images reveal at least three other strong emission-line objects (E,F,G; Fig. 1 right) not previously identified as extragalactic. The extended HI cloud found from VLA observations (Williams et al. 1989) shows peak HI intensities at the positions of galaxies H31C, F, and G. Peak-to-peak measured velocities within galaxies A,B,D,E, and F range only from 3970 to 4200 km/sec; we have not yet obtained a spectrum of G. The separation on the sky C-G is only 50 kpc (H=50). H31C (Mk1089) has known Wolf-Rayet spectral features, implying the presence of 19,500 WR stars (Kunth and Schild 1986). Our spectra of H31A also show WR features near λ4600; [FeIII] λ4987 and HeI λ5016 are also prominent. All of these characteristics suggest that H31 is a collection of young objects which are unlikely to retain their individuality for very long in the high density environment. The tidal tails and starburst nuclei of A and C indicate that two gas rich systems have already collided with consequent enhanced star formation; the resulting object strongly resembles IIZw40 (Baldwin et al. 1982; see also Schweizer 1989).

ACKNOWLEDGEMENTS

We thank the Directors of Palomar Observatory and Kitt Peak National Observatory for telescope time, Barbara Williams and Jacquline van Gorkom for the 21-cm observations of H31, and François Schweizer for valuable conversations.

REFERENCES

Baldwin, Spinrad, and Terlevich, 1982, *MNRAS*, **198**, 535.
Barnes, 1989, *Nature* **338** 123.
Hickson, 1982, *Ap.J.* **255**, 382.
Kunth and Schild, 1986, *Astron. and Astrophys.* **169**, 71.
Phillips, Jenkins, Dopita, Sadler, and Binette, 1986, *A.J.* **91**, 1062.
Rubin, Ford, and Hunter, 1989, STScI Workshop, May 1989.
Schweizer, 1989, in *Dynamics and Interactions of Galaxies*, ed. R. Wielen (Springer-Verlag, Heidelberg), in press.
Williams, McMahon, and van Gorkom, 1989, STScI Workshop, May 1989.

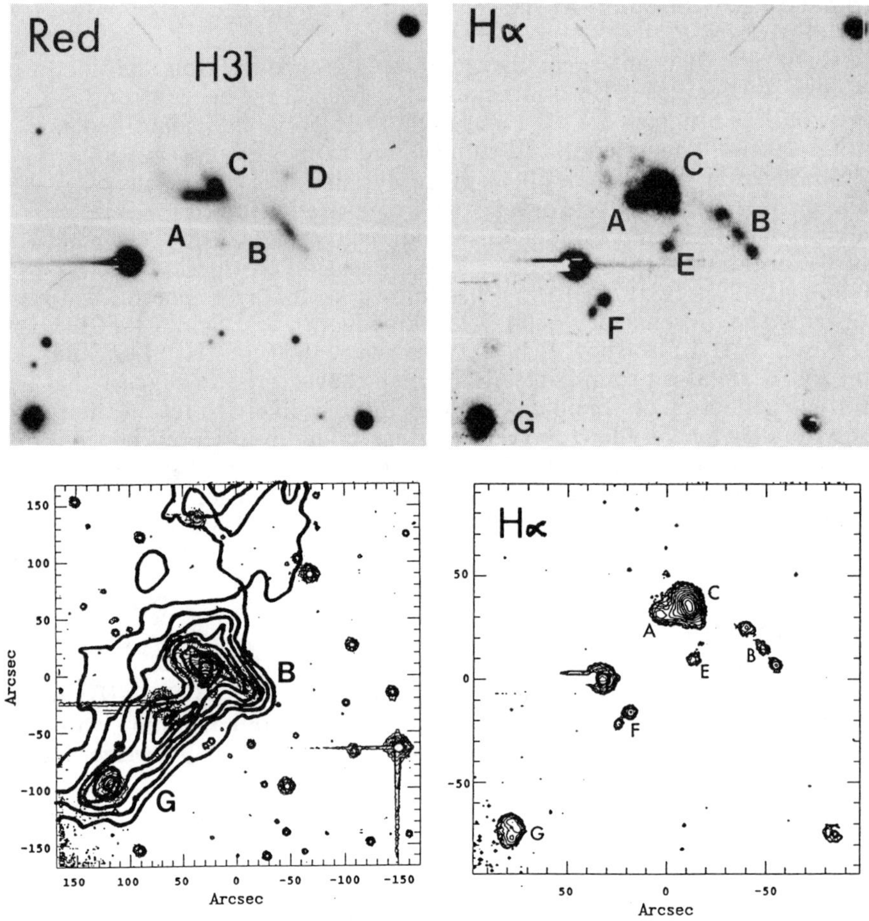

Fig. 1. Hickson31, from red and Hα frames taken with the KPNO 36-in telescope; N is up, E to left. At H31, 50" = 20 kpc. Of the four galaxies identified by Hickson (top left), only A,B,C are strong in Hα. The stong newly identified Hα emission line objects we have named E,F, and G (top right). It is not surprising but very satisfying that these three objects are located at peaks of the VLA 21-cm HI contours (left lower; Williams et al. 1989), peaks with no previously known optical counterparts. Emission line velocities of E and F are similar to those of A,B,C. Note change of origin and change of scale between red and Hα contour plots.

SCALING LAWS FOR DARK MATTER IN LATE-TYPE GALAXIES

JOHN KORMENDY
Dominion Astrophysical Observatory, Herzberg Institute of Astrophysics, 5071 W. Saanich Road, Victoria, British Columbia V8X 4M6, Canada

ABSTRACT Published mass models fitted to galaxy rotation curves are used to study the systematic properties of dark matter (DM) halos. DM in Sc – Im galaxies satisfies well defined scaling laws: halos in less luminous galaxies have smaller core radii, higher central densities, and higher central velocity dispersions. These correlations provide new constraints on the nature of DM and on galaxy formation and evolution. For example: (1) Galaxies are increasingly dominated by dark matter at lower luminosities. In the smallest dwarfs, the visible matter is almost negligible. There may exist a population of dwarf halos that are completely dark. (2) The surprisingly high DM densities found in the dwarf spheroidal galaxies Draco and UMi by Aaronson and Olszewski are nearly normal for galaxies of such low luminosity. These appear to be real galaxies and not just tidal fragments. (3) The correlation between central density and velocity dispersion (the "cooling diagram") is consistent with baryonic DM. If DM is not baryonic, then the cooling diagram provides a measure of the nature and amplitude of primordial fluctuations. (4) In the cooling diagram, the fact that elliptical galaxies are much denser than DM halos shows that low-luminosity ellipticals formed by dissipative collapse whether or not mergers were involved.

I. INTRODUCTION

Progress on a variety of extragalactic problems is slowed by the fact that we know so little about the main mass constituent in the Universe. At least 90 % of the mass in the Universe is believed to be dark matter, but until recently we knew only general things about its distribution (see Kormendy and Knapp 1987 and Trimble 1987 for reviews). However, there were already hints that galaxy halos have important systematic properties. Bahcall and Casertano (1985) found that halo densities at the 26.6 B mag arcsec^{-2} isophote and the ratio 1.08 ± 0.09 of dark to visible matter inside this isophote are almost the same in eight late-type spirals. In this paper, I use published mass models of 48 galaxies for a more detailed investigation of the systematic properties of dark halos.

My main aim is to look for relationships between scale parameters of halos, i.e., central densities ρ_0, core radii r_c at which projected densities have fallen by a factor of two, and central velocity dispersions σ. These are measured by decomposing rotation curves into bulge, disk, and halo components. Much of this paper is based on mass models by Athanassoula, Bosma, and Papaioannou (1987), because these use spiral structure instability criteria to constrain the range of disk mass-to-light ratios. Using these data, Kormendy (1988, 1989*c*) discovered a set of halo scaling laws and gave a preliminary discussion of their implications. This paper summarizes that work.

II. MEASUREMENTS OF DM HALO PARAMETERS

Many authors emphasize that mass modeling is an uncertain procedure. Lake and Feinswog (1989) point out that if measurement errors are interpreted strictly, few observations of rotation curves are comprehensive enough to determine halo parameters. I therefore begin with a review of the measurement procedure, to explain why I use a particular set of decompositions and to summarize the remaining assumptions.

Consider the rotation curve of an isothermal sphere in the ideal case where we can measure a massless disk embedded in it. Then at $r \ll r_c$,

$$V \simeq \left(\frac{4\pi G\rho_0}{3}\right)^{1/2} r, \tag{1}$$

and at $r \gg r_c$,

$$V \simeq \sqrt{2}\sigma = \sqrt{2}\left(\frac{4\pi G\rho_0 r_c^2}{9}\right)^{1/2}, \tag{2}$$

where G is the gravitational constant. If we observe only the $V \propto r$ part of the rotation curve, we can measure ρ_0 but not r_c or σ. In contrast, if the measurements reach far enough into the $V =$ constant part of the rotation curve, then all three parameters can be measured. Low-luminosity Sc – Im galaxies come closest to the above ideal: visible matter turns out to contribute only a small fraction of the total mass in the smallest dwarfs.

More generally, visible matter dominates the rotation curve near the center. Then a multicomponent mass model is required. The rotation curve $V_{\rm vis}$ of the visible matter is calculated from the observed brightness distribution assuming that the mass-to-light ratio M/L of each component is constant with radius. Values of M/L are adjusted to fit as much of the inner rotation curve as desired. Extreme cases bracketing the correct answer are the "maximum disk" solution in which the disk explains as much of the rotation curve as possible, and the "minimum disk" solution with the smallest possible M/L. A model density distribution is then fitted to the halo rotation curve $V_{\rm DM}(r) = (V^2 - V_{\rm vis}^2)^{1/2}$ to derive the halo maximum velocity $V_{\rm max} = 1.12\sqrt{2}\sigma$ (for an isothermal), r_c, and ρ_0.

Uncertainties in this procedure are discussed by van Albada *et al.* (1985), Freeman (1987), Skillman *et al.* (1987), Kormendy (1987*a*, 1988, 1989*c*), and Lake and Feinswog (1989). The main problem is the unknown mass-to-light ratio of the visible matter. The ratio of visible to dark mass can be varied greatly while preserving a good fit to the data; as the visible mass is reduced, the central density of the halo must be increased, and its core radius must be decreased. The extreme models (e.g., van Albada *et al.* 1985, Fig. 4 and 8) are usually the maximum disk case and a solution with $(M/L)_D = 0$ in which the halo explains all of the rotation curve. In massive galaxies, these solutions are very different. If we knew nothing more, we could say little about halo properties. Fortunately, we have other constraints. The maximum disk model is unrealistic. If the disk accounts for all of the rising part of the rotation curve, then the halo has a hollow core. So $(M/L)_D$ is first reduced until the halo density nowhere decreases toward the center. I redefine this to be the maximum disk solution. We also have a stronger constraint. We know that $(M/L)_D$ cannot be too small because disks contain bars and spiral density waves. If $(M/L)_D$ were zero, such self-gravitating structure would not be possible. The problem has been to turn this qualitative remark into a practical constraint on $(M/L)_D$.

The necessary breakthrough has been provided by Athanassoula, Bosma, and Papaioannou (1987, hereafter ABP). They apply Toomre's (1981) swing amplifier instability criterion and demand that the disk have the proper mass-to-light ratio to give the observed spiral structure. First, a minimum-halo "error bar" is given by the maximum-disk decomposition. Then the adopted best model is derived by reducing $(M/L)_D$ until one-armed instabilities disappear (otherwise these would dominate). Finally, $(M/L)_D$ is reduced further until two-armed modes disappear, too; this provides a maximum-halo error bar.

The above procedure is still uncertain. The swing amplifier criterion is preliminary. Also, Athanassoula and collaborators assume that all galaxies have two arms; the observed variety is difficult to quantify and to use. And it is difficult to judge the errors, since the fits are not illustrated. Finally, except for the best-determined cases, the procedure depends somewhat on the assumption that rotation curves that flatten out to $V \simeq$ constant stay flat outside the radius range measured (Lake and Feinswog 1989). Clearly the technique deserves further development. However, the basic idea that self-gravitating structure rules out small $(M/L)_D$ is certainly correct. For bright galaxies, the ABP analysis is the best available effort to circumvent modeling uncertainties. It also implies disk mass-to-light ratios that are similar to those measured directly (Bahcall 1987; van der Kruit and Freeman 1984, 1986).

III. DWARF GALAXIES ARE MOSTLY DARK MATTER

The first systematic property of halos that I want to discuss is the increasing domination of DM in smaller galaxies (see also Tinsley 1981; Kent 1987; Persic and Salucci 1988). Bahcall and Casertano's (1985) result that visible and dark matter contribute equally inside the optical radius appears to

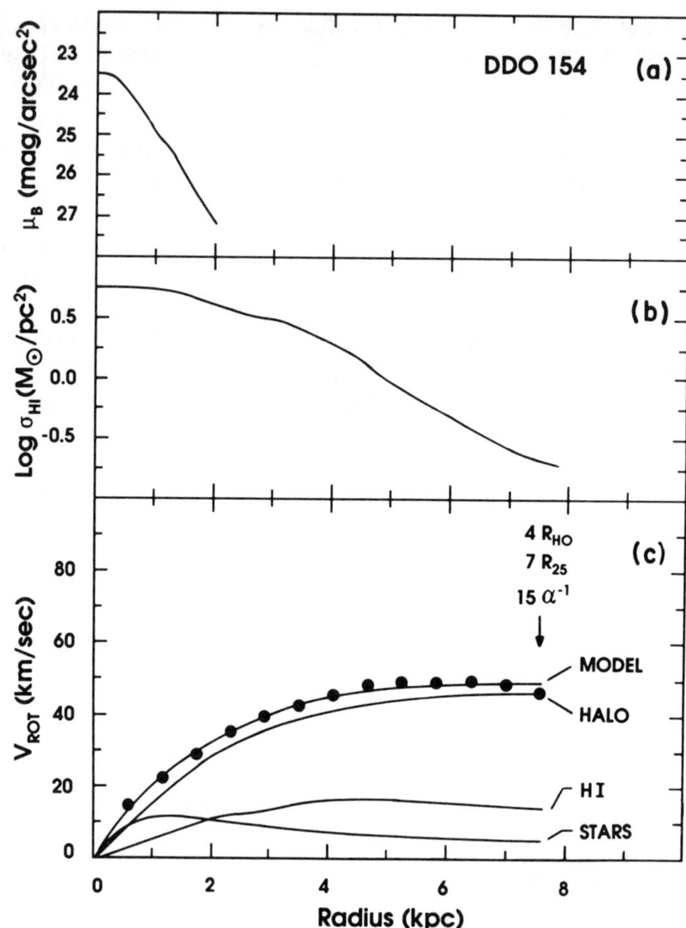

Fig. 1. Brightness profile (*top*), H I surface density (*middle*), and rotation curve and mass model for DDO 154 (Carignan and Freeman 1988).

be valid only for bright galaxies. Already at $M_B \simeq -17$, DM is becoming more important (Skillman *et al.* 1987; Lake, Schommer, and van Gorkom 1989). The faintest galaxies with reliable mass models are DDO 154 ($M_B = -15.1$; Carignan and Freeman 1988) and DDO 127 ($M_B = -14.4$; Bosma, Kormendy, and Souviron 1989). In these galaxies (e.g., Fig. 1) and in DDO 170 ($M_B = -16.4$; Lake, Schommer and van Gorkom 1989), it is no longer possible to find a maximum disk solution that explains the rising part of the rotation curve. The reason is that the scale length of the visible matter is too short: if $(M/L)_D$ is chosen large enough to explain the amplitude of the rotation curve, then $V(r)$ rises too rapidly. Therefore the rotation curve is produced mostly by DM even near the center. In DDO 154, DM accounts for 92 % of the matter inside the radius of the H I measurements (Carignan and Freeman 1988).

This trend continues in two of the dwarf spheroidal (dSph) companions of our Galaxy. (I compare dSph and late-type galaxies because structural similarities and other arguments suggest a generic relationship between them; see Kormendy 1987*b* for a review.) Aaronson (1983) and Aaronson and Olszewski (1987) have shown that the central stellar velocity dispersions in Draco and UMi are ~ 10 km s^{-1}, much larger than predicted if the galaxies are in equilibrium and if they consist of stars with globular cluster mass functions. They deduce that DM is present in such large quantities that the central mass-to-light ratios are $\sim 10^2$. This result has important implications, so it has been tested carefully (see the above papers; Tremaine 1987). The results have only strengthened the case for DM. Godwin and Lynden-Bell (1987) are unconvinced until independent velocity measurements of the same stars are shown to agree; the beginnings of such a comparison for the Aaronson and Olszewski data are encouraging (Kormendy 1987*a*).

The halo parameters of Draco and UMi require some discussion. They are derived in Kormendy (1987*a*) by assuming King (1966) models for the light and mass distributions. Central volume densities are estimated; this minimizes the dependence on the assumption of a King model. Then the central mass density is $\rho_0 = 166\, \sigma^2/r_c^2\ \mathrm{M}_\odot\ \mathrm{pc}^{-3}$, where σ is in km s^{-1} and r_c is in pc. But what is the core radius of the mass distribution? Most authors use the core radius of the visible stars. However, DM is likely to be more extended than the stars, as it is in spirals. Underestimating r_c in the above equation results in an overestimate of ρ_0. There may be a way to deal with this difficulty, below, but I will conclude at the end that the halo properties of Draco and UMi are still puzzling.

First assume that dark and visible matter have the same core radius. This leads to the first puzzle. The central DM densities in Draco and UMi are ~ 0.6 and $\sim 0.9\ \mathrm{M}_\odot\ \mathrm{pc}^{-3}$, the largest measured in any galaxy (Fig. 2). As emphasized in Kormendy (1987*a*), it is the absolute DM densities, and not the large mass-to-light ratios, that are surprising (large M/L ratios can result from the ejection of baryons by supernova-driven winds; Dekel and Silk 1986).

Could the answer be that I have underestimated the DM core radii? The most plausible assumption is that the halos have large core radii. If they nevertheless dominate the potential, it is easy to show that I overestimated ρ_0 by only a factor of 2 (Pryor and Kormendy 1989). But the density profile of isothermal test particles in a constant-density halo is Gaussian. The observed brightness distributions are exponential (Faber and Lin 1983). Within the errors of the star counts, Gaussian profiles cannot be excluded, but a small core radius gives a better fit. In general, the visible matter density should cut off at about the radius where DM starts to control the potential (Peebles 1984). The relatively weak cutoff in Draco is most consistent with a DM core radius that is not too different from that of the stars. Based on our experience with more luminous galaxies, this is surprising. The halos of Draco and UMi are discussed in more detail in Pryor and Kormendy (1989); here I adopt the above densities. Modeling uncertainties in ρ_0 are about a factor of 2 if the stellar velocities are isotropic. Other uncertainties are discussed in Kormendy (1987*a*).

Then the central DM densities in Draco and UMi are one or two orders of magnitude larger than those of the visible matter. That is, Draco and UMi are almost completely dark. These galaxies support the trend of increasing DM domination in smaller dwarfs. Dekel and Silk (1986) predict such a trend, because smaller galaxies have shallower potential wells and so are expected to lose more of their baryons through supernova-driven galactic winds.

Four dwarf galaxies are reported to have small DM densities that disagree with this trend. Lake and Skillman (1989) find that IC 1613 ($M_B = -14.8$) has a central halo density of ~ 0.0009 M$_\odot$ pc^{-3}, one of the smallest measured. However, since the axial ratio is 0.8, the inclination is uncertain, especially for a galaxy classified as IABm. While the Lake and Skillman results are important, it is not yet clear that IC 1613 is an exception to the above correlation. The other extreme dwarfs that are not DM dominated are the dSph galaxies Fornax, Sculptor, and Carina (see Kormendy 1987*a* for a review). It is interesting to speculate that these may not be real galaxies, but tidal fragments pulled out of the Magellanic Clouds during the formation of the Magellanic Streams (Lynden-Bell 1976, 1983).

IV. HALO SCALING LAWS

In this section, I combine the results of §§ II and III to look for halo scaling laws. For bright galaxies, I use the "no $m = 1$" solutions of ABP. Additional dwarf spiral and irregular galaxies are from Skillman *et al.* (1987); Carignan, Sancisi, and van Albada (1988); Carignan and Freeman (1988); Lake, Schommer, and van Gorkom (1989); and Bosma, Kormendy, and Souviron (1989). These dwarfs are sufficiently dominated by DM that spiral structure constraints are not needed and that halo parameters are reasonably well determined even without reaching far into the V = constant part of the rotation curve (unless $V(r)$ rises again outside the radius where it appears to have leveled off, Lake and Feinswog 1989). The correlations between r_c, ρ_0, σ, and M_B are shown in Figures 2 and 3.

Figure 2 suggests that halo parameters satisfy a set of scaling laws like those for visible matter, but with many discrepant galaxies of anomalously small r_c, high ρ_0, and high σ. These are precisely the discrepancies that we would expect if halos were pulled inward by the luminous matter. Most of the galaxies involved are early in type. Earlier-type galaxies have higher central densities of luminous matter, e. g. in the form of bulges. This suggests that they depart from intrinsic DM parameter correlations because of gravitational compression of their halos. ABP find other correlations that are also consistent with theoretical predictions (Blumenthal *et al.* 1986; Barnes 1987; Ryden and Gunn 1987; Ryden 1988) that such compression must occur.

Figure 3 is a first attempt to derive intrinsic halo scaling laws without this compression. All galaxies earlier in type than Sc are omitted. Sc galaxies are flagged; they are more likely than later-type galaxies to have dense disks. Draco and UMi are retained. In these, the detection of DM seems secure, and the baryonic densities are low enough so that the halos are not compressed.

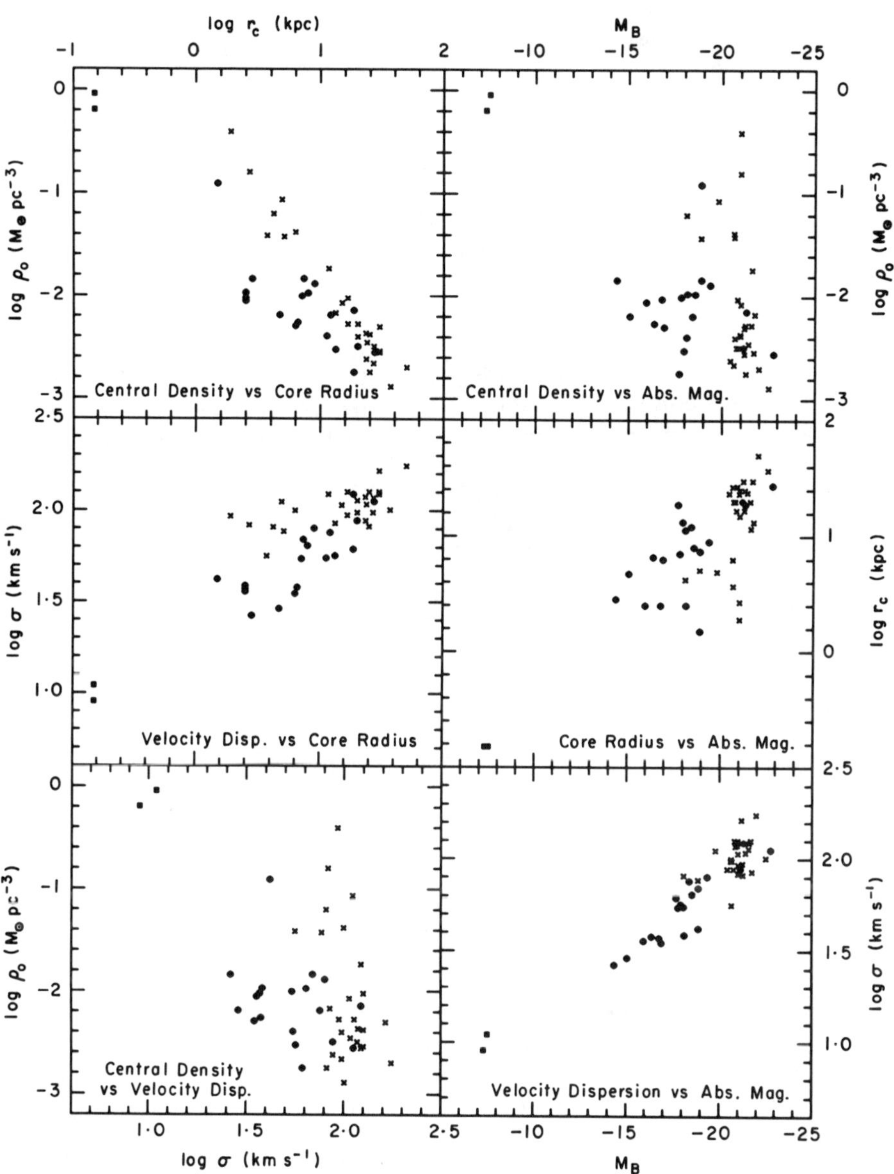

Fig. 2. – Correlations of DM halo parameters. This figure shows the whole sample divided up by Hubble type: dSph galaxies (*filled squares*), S0 – c galaxies (*crosses*), and Scd – Im galaxies (*filled circles*). For Draco and UMi, the optical σ and r_c are adopted. In Figures 2 – 4, $H_0 = 50$ km s^{-1} Mpc^{-1}.

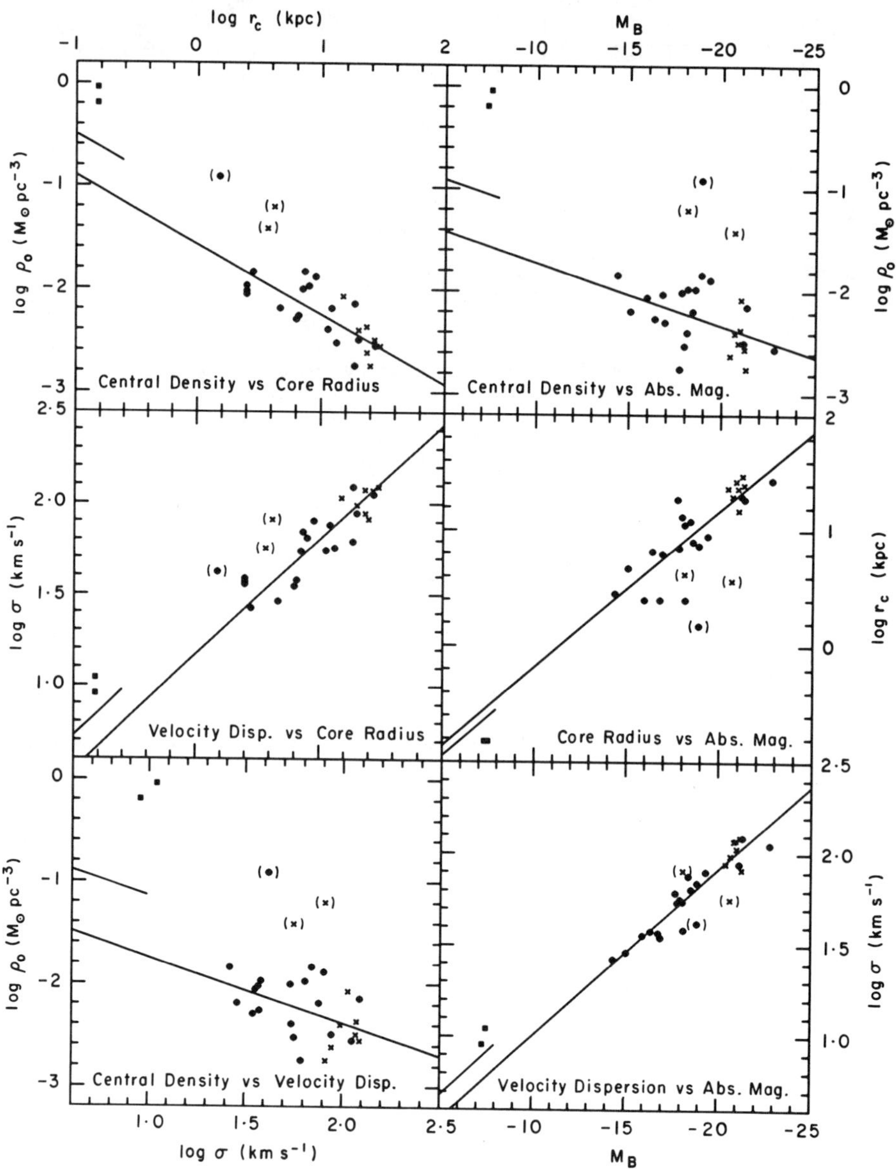

Fig. 3. – Halo parameter correlations for Draco and UMi (*filled squares*), Sc galaxies (*crosses*), and Scd – Im galaxies (*filled circles*). The straight lines are least-squares fits (equations 3 – 9), omitting Draco, UMi, and the points in parentheses. Line segments show the zeropoints for $H_0 = 100$ km s^{-1} Mpc^{-1}.

Figure 3 shows that DM halos in lower-luminosity galaxies are smaller, denser, and lower in velocity dispersion. The least-squares fits are:

$$\rho_0 = 0.0071\ M_\odot\ \mathrm{pc}^{-3} \left(\frac{L_B}{10^9\ L_{B\odot}}\right)^{-0.15}; \qquad (0.023) \qquad (3)$$

$$r_c = 5.9\ \mathrm{kpc} \left(\frac{L_B}{10^9\ L_{B\odot}}\right)^{0.34}; \qquad (4.7) \qquad (4)$$

$$\sigma = 44\ \mathrm{km\ s}^{-1} \left(\frac{L_B}{10^9\ L_{B\odot}}\right)^{0.23}; \qquad (60) \qquad (5)$$

i.e., $L_B \propto \sigma^{4.4}$; (6)

$$\rho_0 = 0.027\ M_\odot\ \mathrm{pc}^{-3} \left(\frac{r_c}{1\ \mathrm{kpc}}\right)^{-0.67}; \qquad (0.067) \qquad (7)$$

$$\rho_0 = 0.0041\ M_\odot\ \mathrm{pc}^{-3} \left(\frac{\sigma}{100\ \mathrm{km\ s}^{-1}}\right)^{-0.64}; \qquad (0.017) \qquad (8)$$

$$\sigma = 15\ \mathrm{km\ s}^{-1} \left(\frac{r_c}{1\ \mathrm{kpc}}\right)^{0.63}, \qquad (23) \qquad (9)$$

($H_0 = 50$ km s^{-1} Mpc^{-1}). Approximate zeropoints for $H_0 = 100$ km s^{-1} Mpc^{-1} are given in parentheses, although rotation-curve decompositions do not scale exactly with H_0 (H I and dynamical masses scale differently with distance).

These relations are still very uncertain. The similarity to scaling laws for visible components is suspicious. Could rotation-curve decomposition falsely give results that are tied to the optical parameters? A careful investigation of such effects will be important. However, the dwarf spirals are relatively immune to these problems, because DM dominates so close to the center. And in some bright galaxies, rotation curves are observed to very large radii, giving good leverage on the DM parameters. Therefore I believe that the sense of the correlations is correct. However, the slopes could be in error, especially if the bright galaxies still have undergone some halo compression.

Halo scaling laws provide new constraints on DM and on galaxy formation. Their application will in general require more detailed theoretical predictions than are now available. However, some simple conclusions follow immediately.

1. – The high central DM densities in Draco and UMi are at one extreme of a general correlation; they are not as anomalous as they appeared in § III. If $H_0 = 100$ km s^{-1} Mpc^{-1}, they are higher than the ridge line of the correlation for Sc – Im galaxies by about an order of magnitude. I suspect that ρ_0 has been overestimated in these galaxies, e. g. if the stellar orbits are nearly radial.

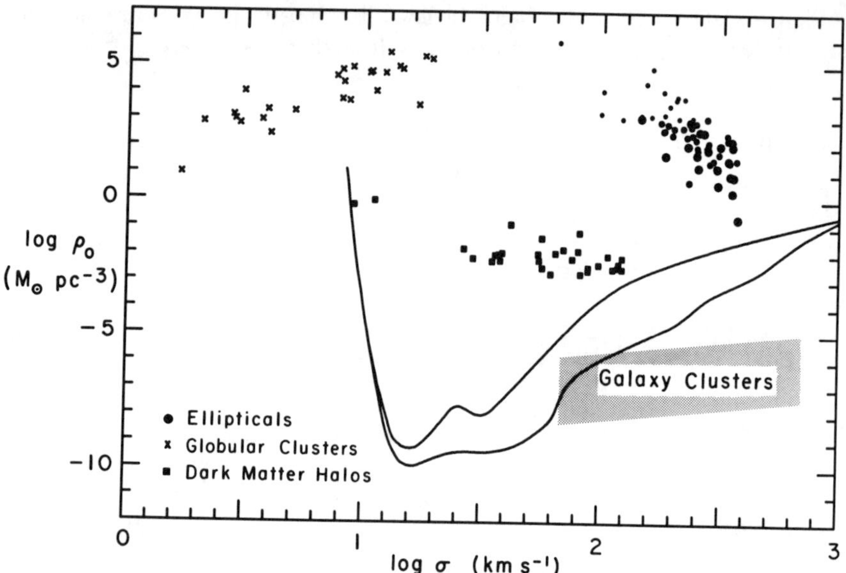

Fig. 4. – Cooling diagram for cores of halos and of various kinds of stellar systems. Protogalactic gas cooling curves are shown for solar metal abundance (*lower*) and for no metals (Blumenthal *et al.* 1984). Below the curves, the cooling time is longer than the collapse time; baryonic objects that have not dissipated lie here. Above the curves, the cooling time is shorter than the collapse time; larger distances above the curves imply more dissipation during formation. If halos are non-baryonic, they did not cool and can be anywhere in the diagram.

2. – Draco and UMi appear to be genuine galaxies, with substantial DM halos. This argues against the hypothesis that they are tidal debris of brighter galaxies: tides would further reduce the progenitors' already low DM densities. None of this is inconsistent with the observation that they are in the Magellanic Stream (Lynden-Bell 1976, 1983); they may once have been satellites of the Magellanic Clouds.

3. – The fact that, as luminosity decreases, dwarf galaxies become both more numerous and more nearly dominated by DM raises the possibility that there is a large population of objects that are completely dark. If so, these "empty halos" are likely to be small and dense, and to have small total masses.

4. – The density–velocity dispersion correlation is the classic "cooling diagram" for protogalaxies (Rees and Ostriker 1977; Silk 1977, 1987; Faber 1982; Blumenthal *et al.* 1984). It is shown in more detail in Figure 4. The positions of DM halos imply that they could have formed dissipatively if they are baryonic. If they are not baryonic, then their positions provide information about the nature and magnitude of primordial fluctuations. Comparing Figure 4 with Figure 3 in Blumenthal *et al.* (1984) suggests that the fluctuations that formed the cores of DM halos were in the $\sim 3\,\sigma$ range. However, such a comparison is not straightforward, because virial parameters are predicted, but core parameters are

observed. We know almost nothing about halos outside the radius of rotation-curve measurements. Further progress requires a theory of galaxy formation that is detailed enough to predict DM core properties.

5. – Finally, the cooling diagram shows that small elliptical galaxies formed by dissipative collapse whether or not this was triggered by mergers (Kormendy 1989*a*, *b*). Suppose that ellipticals did form by mergers of galaxy disks (Toomre 1977). If disks had formed with no dissipation, they would now appear below the distribution of their halos in Figure 4 by an order of magnitude in ρ_0 (there is $\sim 10\,\%$ as much visible matter as dark matter). Dissipation during disk formation moved these points upward and toward larger σ; disks are ~ 10 times as dense as their halos (APB, Fig. 11). Cores of low-luminosity ellipticals are far above the distribution of disks because of further dissipation. I conclude that even if small ellipticals formed by mergers of disks, most of the required dissipation took place after disk formation and during the merger.

In conclusion, I emphasize again that the results of this section are very preliminary. Parameter measurement errors are still large (see Kormendy 1989*c* for estimates). The present discussion gives only a hint of what we can learn. I want to close by emphasizing the importance of further work to measure halo parameters accurately and to exploit them to study DM physics.

REFERENCES

Aaronson, M. 1983, *Ap. J. (Letters)*, **266**, L11.

Aaronson, M., and Olszewski, E. 1987, in *IAU Symposium 117, Dark Matter in the Universe*, ed. J. Kormendy and G. R. Knapp (Dordrecht: Reidel), p. 153.

Athanassoula, E., Bosma, A., and Papaioannou, S. 1987, *Astr. Ap.*, **179**, 23 (ABP).

Bahcall, J. N. 1987, in *IAU Symposium 117, Dark Matter in the Universe*, ed. J. Kormendy and G. R. Knapp (Dordrecht: Reidel), p. 17.

Bahcall, J. N., and Casertano, S. 1985, *Ap. J. (Letters)*, **293**, L7.

Barnes, J. E. 1987, in *Nearly Normal Galaxies: From the Planck Time to the Present*, ed. S. M. Faber (New York: Springer Verlag), p. 154.

Blumenthal, G. R., Faber, S. M., Flores, R., and Primack, J. R. 1986, *Ap. J.*, **301**, 27.

Blumenthal, G. R., Faber, S. M., Primack, J. R., and Rees, M. J. 1984, *Nature*, **311**, 517.

Bosma, A., Kormendy, J., and Souviron, J. 1989, in preparation.

Carignan, C., and Freeman, K. C. 1988, *Ap. J. (Letters)*, **332**, L33.

Carignan, C., Sancisi, R., and van Albada, T. S. 1988, *A. J.*, **95**, 37.

Dekel, A., and Silk, J. 1986, *Ap. J.*, **303**, 39.

Faber, S. M. 1982, in *Astrophysical Cosmology*, ed. H. A. Brück, G. V. Coyne and M. S. Longair (Vatican City: Pontifical Academy of Sciences), p. 191.

Faber, S. M., and Lin, D. N. C. 1983, *Ap. J. (Letters)*, **266**, L17.

Freeman, K. C. 1987, in *IAU Symposium 117, Dark Matter in the Universe*, ed. J. Kormendy and G. R. Knapp (Dordrecht: Reidel), p. 119.
Godwin, P. J., and Lynden-Bell, D. 1987, *M. N. R. A. S.*, **229**, 7P.
Kent, S. M. 1987, *A. J.*, **93**, 816.
King, I. R. 1966, *A. J.*, **71**, 64.
Kormendy, J. 1987*a*, in *IAU Symposium 117, Dark Matter in the Universe*, ed. J. Kormendy and G. R. Knapp (Dordrecht: Reidel), p. 139.
Kormendy, J. 1987*b*, in *Nearly Normal Galaxies: From the Planck Time to the Present*, ed. S. M. Faber (New York: Springer-Verlag), p. 163.
Kormendy, J. 1988, in *Origin, Structure and Evolution of Galaxies*, ed. Fang Li Zhi (Singapore: World Scientific), p. 252.
Kormendy, J. 1989*a*, *Ap. J. (Letters)*, **342**, L63.
Kormendy, J. 1989*b*, in *Dynamics and Interactions of Galaxies*, ed. R. Wielen (New York: Springer-Verlag), in press.
Kormendy, J. 1989*c*, *Ap. J.*, in preparation.
Kormendy, J., and Knapp, G. R. (ed.) 1987, *IAU Symposium 117, Dark Matter in the Universe* (Dordrecht: Reidel).
Lake, G., and Feinswog, L. 1989, *A. J.*, **98**, 166.
Lake, G., and Skillman, E. D. 1989, *A. J.*, in press.
Lake, G., Schommer, R. A., and van Gorkom, J. H. 1989, *A. J.*, in press.
Lynden-Bell, D. 1976, *M. N. R. A. S.*, **174**, 695.
Lynden-Bell, D. 1983, in *IAU Symposium 100, Internal Kinematics and Dynamics of Galaxies*, ed. E. Athanassoula (Dordrecht: Reidel), p. 89.
Peebles, P. J. E. 1984, *Ap. J.*, **277**, 470.
Persic, M., and Salucci, P. 1988, *M. N. R. A. S.*, **234**, 131.
Pryor, C., and Kormendy, J. 1989, in preparation.
Rees, M. J., and Ostriker, J. P. 1977, *M. N. R. A. S.*, **179**, 541.
Ryden, B. S. 1988, *Ap. J.*, **329**, 589.
Ryden, B. S., and Gunn, J. E. 1987, *Ap. J.*, **318**, 15.
Silk, J. 1977, *Ap. J.*, **211**, 638.
Silk, J. 1987, in *IAU Symposium 117, Dark Matter in the Universe*, ed. J. Kormendy and G. R. Knapp (Dordrecht: Reidel), p. 335.
Skillman, E. D., Bothun, G. D., Murray, M. A., and Warmels, R. H. 1987, *Astr. Ap.*, **185**, 61.
Tinsley, B. M. 1981, *M. N. R. A. S.*, **194**, 63.
Toomre, A. 1977, in *The Evolution of Galaxies and Stellar Populations*, ed. B. M. Tinsley and R. B. Larson (New Haven: Yale University Obs.), p. 401.
Toomre, A. 1981, in *The Structure and Evolution of Normal Galaxies*, ed. S. M. Fall and D. Lynden-Bell (Cambridge: Cambridge University Press), p. 111.
Tremaine, S. 1987, in *Nearly Normal Galaxies: From the Planck Time to the Present*, ed. S. M. Faber (New York: Springer-Verlag), p. 76.
Trimble, V. 1987, *Ann. Rev. Astr. Ap.*, **25**, 425.
van Albada, T. S., Bahcall, J. N., Begeman, K., and Sancisi, R. 1985, *Ap. J.*, **295**, 305.
van der Kruit, P. C., and Freeman, K. C. 1984, *Ap. J.*, **278**, 81.
van der Kruit, P. C., and Freeman, K. C. 1986, *Ap. J.*, **303**, 556.

HALO MODEL MASS LIMITS FOR SPIRAL GALAXIES

LANCE K. ERICKSON
Embry-Riddle University, Daytona Beach, Florida 32114

STEPHEN T. GOTTESMAN
Department of Astronomy, University of Florida, Gainesville, Florida 32611

JAMES H. HUNTER JR.
Department of Astronomy, University of Florida, Gainesville, Florida 32611

ABSTRACT Compact galaxy groups with a gravitationally dominant spiral galaxy and orbiting dwarf satellites were measured with HI observations for both dynamical and disk mass. The ratios of the central galaxy masses were compared to simulated values produced with N-body integrations to determine limits on any unobserved halo mass. The incremental simulations included isothermal halos with mass and radial limits to 10 times the central galaxy mass and 150 kpc respectively. A comparison of the two distributions of dynamical to disk mass ratios establishes limits for the halo models similar to those suggested in the binary galaxy study of van Moorsel (van Moorsel 1982, 1987), but with distinctly different observed values.

INTRODUCTION

Evidence for dark matter surrounding isolated spiral galaxies is seen in the flat rotation curves of many spirals which extend to the edge of the observed disk, although this can be duplicated with a relatively small halo mass (van der Kruit 1979). More distant gravitational probes of the spiral galaxies are needed for a more complete measurement of the mass beyond the disk. However, the orbital projection for satellites not associated with the disk becomes a problem since it cannot be measured and results in a dynamical mass estimate which is a projected function. A solution to this problem was developed for the study of binary systems by Gustav van Moorsel (van Moorsel 1982, 1987).

Disk and exterior mass measurements of the galaxy pairs were compared by van Moorsel using the dynamical to disk mass ratio, or chi. Random orientations of the pairs are assumed, providing random projections for the dynamical masses which measure a greater volume than the disk rotation curve. The differences in the model and observed mass ratios reflect the effects of halo mass, making it useful for measuring model halo limits.

Our choice of compact groups for a similar study included single dominant spirals and dwarf satellites to provide a more isolated system, increasing the probability of bound membership. This is critically important since unbound objects can produce the same effects as significant halo mass.

OBSERVATIONS AND SIMULATIONS

HI observations of selected spirals were made with the NRAO VLA radiotelescope, providing the central galaxy rotation curves and dwarf satellite positions with radial velocities (Erickson 1987, Erickson, Gottesman and Hunter 1987). These observations provided the two different measurements of the central galaxy masses and the corresponding chi mass ratios for each primary-satellite pair. Supplemental group measurements were also included from other HI observations (Erickson, Gottesman and Hunter 1987).

N-body simulations of dwarf objects orbiting the central galaxy were made using the time difference method of Aarseth (Aarseth 1971). Logarythmic haloes with frictional dissipation were included in the $2x10^{10}$ y integrations (H_0 = 100km $sec^{-1}Mpc^{-1}$). Corresponding model mass ratios (chi model) were produced from these group simulations. Because orbital eccentricity affects the distribution of chi values, separate simulations were made for three eccentricity ranges. These three were circular, with the eccentricity between 0 and 0.3, moderately eccentric, with the eccentricity between 0.3 and 0.6, and eccentric, with the eccentricity between 0.6 and 0.9.

DISCUSSION

Tests on the model and the observed chi mass ratios identified a range of acceptable models; those with moderately eccentric orbits and halos less than approximately 60 kpc in radius with three to four times the central disk mass. Circular orbits were allowed but only for massive, extended halos (radius > 90 kpc and mass > 10 x disk mass). This is not supported by our data since the HI and visible observations show a rapidly decreasing number of satellites at increasing separation.

The success of these similarity tests was limited to the observed chi mass ratios which represented bound systems. In addition, for moderately eccentric orbits, no external halo mass is required to produce a reasonable model with the bound restriction. This limitation duplicates the removal of peculiar and very massive systems included in the data.

The halo limits found by the Kolmogorov-Smirnov tests are of the same order as those discussed by Sandage and Fouts (Sandage and Fouts 1987), Little and Tremaine (Little and Tremaine 1987) and van Moorsel (van Moorsel 1982, 1987). Although the additional halo mass required to fit the van Moorsel binary data was similar to our group data, in van Moorsel's case a minimum of 5 times the disk masses, the binary data show a substantially different distribution of chi mass ratio values than those found in this study. The significant difference in distributions could be indicative of interlopers or perhaps physical differences in galaxy formation or evolution.

REFERENCES

Aarseth, S. J. 1971, *Astrophysics and Space Science*, **14**, 118.
Erickson, L. K. 1987, Ph.D. Dissertation, University of Florida.
Erickson, L. K., Gottesman, S. T., and Hunter, J. H., Jr. 1987, *Nature*, **325**, 779.
Little, B. and Tremaine, S. 1987, *Astrophysical Journal*, **320**, 493.
Sandage, A. and Fouts, G. 1987, *Astronomical Journal*, **92**, 74.
van der Kruit 1979, *Astronomy and Astrophysics*, **99**, 289.
van Moorsel, G. 1982, Ph.D. Dissertation, University of Groningen, Netherlands.
Van Moorsel, G. 1987, *Astronomy and Astrophysics*, **176**, 13.

THE DARK MATTER CONTENT OF DISK GALAXIES

Michael J. Pierce
Doninion Astrophysical Observatory
5071 W. Saanich Rd.
Victoria, BC V8X4M6 CANADA

ABSTRACT The global structural properties and M/L_I of disk galaxies are found to abruptly change at log $L_I \sim 9.8\ L_\odot$. The most plausible interpretation is that lower luminosity systems have a larger dark matter content than do higher luminosity systems. These observations are compared to the model of dwarf galaxy formation of Dekel and Silk.

BACKGROUND AND DATA

The role played by nonluminous matter in the formation of galaxies is only now being fully appreciated. Unfortunately, the morphology, luminosities, sizes, and kinematics of galaxies are the only visible relics of the formation process. However, the mass of disk systems can be inferred using high quality measurements of the above parameters with some simple assumptions and enables an estimate of the dark mater content of disk systems.

The data consist of B, R, and I CCD images of all late-type galaxies within the Virgo and Ursa Major Clusters with $B_T \lesssim 14.0$ and with good S/N HI data available from from a variety of sources, primarily Fisher and Tully (1981) and Helou *et al.* (1981, 1984). Virgo Cluster ellipticals with ${}_T \lesssim 12.0$ are also included for comparison. The combined sample consists of 117 galaxies. The data are wide field images ensuring good sky subtraction and this enables full growth curves to be obtained for each galaxy. Total magnitudes and an effective radius containing 80% of the light were obtained from the growth curves. Details of the photometric procedures as well as the corrections for extinction and projection can be found in Pierce (1988). The distance to the two clusters is essentially the same ~ 15 Mpc (Pierce and Tully 1988; Pierce 1989) so the data from both clusters will be combined. There is background contamination in the Virgo data (~ 10%) and the data for these systems have been corrected accordingly (see Pierce and Tully 1988).

RESULTS AND DISCUSSION

Figure 1 illustrates the change in the structural scaling of disk systems at log $L_I \sim 9.8\ L_\odot$. There is a corresponding change in slope of the luminosity–

line-width relations (Tully and Fisher 1977) at a similar luminosity. Assuming that the mass internal to the optical radius is given by $m_i \sim V^2R/G$, then figure 1c shows the dependence of M/L_I on L_I. The use of the I band greatly reduces the sensitivity of the results to the star formation history of a particular galaxy. Systems with log L $\gtrsim 9.8\ L_\odot$ have $M/L \sim 3$ while less luminous systems have values of M/L a factor of two larger. This result is similar to that first obtained by Tinsley (1981) but the abrupt onset is much more dramatic and the interpretation less model dependent. This critical luminosity is also near the transition between organized and more chaotic spiral structure. Since the disk of a galaxy must be self gravitating in order to sustain spiral density waves (Athanossoula, Bosma, and Papaioannou 1987) these data would suggest that the luminous matter in the lower luminosity systems is not self gravitating. Thus the critical luminosity would be the demarcation between systems which are luminous or dark matter dominated within their optical radius.

An explanation for the change in structural scaling and M/L_I for disk systems at the critical luminosity can be found in the model for the formation of dwarf galaxies proposed by Dekel and Silk (1986). They suggested that supernova winds are capable of effectively sweeping gas from protogalaxies below a critical mass, resulting in an increase in M/L for lower mass (luminosity) systems. This critical mass corresponds to the luminosity at which the scaling properties and M/L_I of disk systems abruptly change. These results are in remarkable agreement with the model. One important aspect which the model does not address is the fact that there is a very small dispersion in the mass-luminosity (Tully-Fisher) relation for dwarfs. The model suggests that up to 90% of the gas mass of a dwarf galaxy is removed via supernova winds. In such a circumstance it is difficult to understand how mass and luminosity remain coupled. Perhaps some subtle mechanism acts a control to the star formation rate during galaxy formation in order to maintain this coupling.

ACKNOWLEDGEMENTS

Much of this work was done as part of a thesis at the University of Hawaii under the direction of Brent Tully.

REFERENCES

Athanassoula, E., Bosma, A,and Papaioannou, S. 1987, Astr. Ap., 179, 23.
Dekel, A., and Silk, J. 1986, Ap. J., 303, 39.
Fisher, J. R., and Tully, R. B. 1981, Ap. J. Suppl., 47, 139.
Helou, G., Giovanardi, C., Salpeter, E., E., and Krumm, N. 1981, Ap. J. Suppl., 46, 267.
Helou, G., Hoffman, G. L., and Salpeter, E. E. 1984, Ap. J. Suppl., 55, 433.
Pierce, M. J., Ph D. Thesis, University of Hawaii.
Pierce, M. J., and Tully, R. B. 1988, Ap. J., 330, 579.
Pierce, M. J. 1989, Ap. J. *Letters*, in press.
Tinsley, B. M. 1981, M.N.R.A.S., 194, 63.
Tully, R. B., and Fisher, J. R. 1977, Astr. Ap., 54, 661.

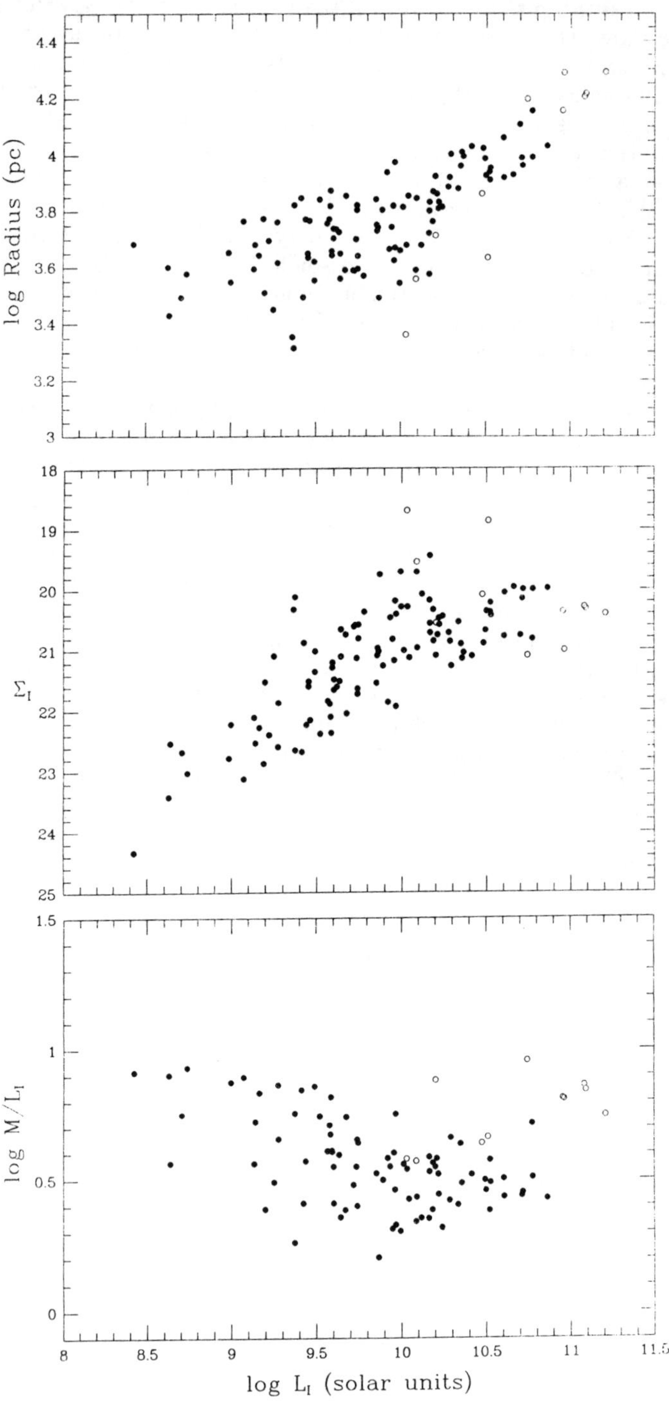

4.4
4.2
4
3.8
3.6
3.4
3.2
3
log Radius (pc)
18
19
20
21
22
23
24
25
Σ_I
1.5
1
0.5
0
log M/L_I
8
8.5
9
9.5
10
10.5
11
11.5
log L_I (solar units)

THE DYNAMICS OF THE LOCAL GROUP AND THE MASS OF THE GALAXY

DENNIS ZARITSKY
Steward Observatory, University of Arizona, Tucson, AZ, 85721

ABSTRACT Timing arguments using Leo I, M31, and more distant members of the Local Group are all in agreement and imply that the Galaxy has a mass $\geq 13 \times 10^{11}\,M_\odot$ for an adopted age of the universe of 14 Gyr and an M31/Milky Way mass ratio of 1.8. The "best-fit" models of the remote Local Group members imply either that there is a Local Group dark matter halo of roughly $20 \times 10^{11}\,M_\odot$ out to radii of 1 Mpc, or that the mass of the Milky Way is $\geq 17.5 \times 10^{11}\,M_\odot$. If the Milky Way has a mass of $17.5 \times 10^{11}\,M_\odot$, the M31-Milky Way timing arguments can only be satisfied if the Milky Way has a tangential velocity component with respect to M31 of approximately 55 km s^{-1}.

INTRODUCTION

Timing arguments are a powerful yet simple technique with which to determine limits for the mass of the Local Group (LG). In the analysis, one presumes that M 31 and the Milky Way galaxy (MW) were "ejected" from the center-of-mass by the universal expansion and were thereafter influenced solely by their mutual gravitational attraction. The LG, when modelled as consisting of only the MW and M 31, can be described by simple two-body orbit equations (*e.g.* Kahn and Woltjer 1959) and the results from such an analysis predict that the LG has at least a mass of $3 \times 10^{12}\,M_\odot$ for standard estimates of the age of the universe (Kahn and Woltjer 1959; Einasto and Lynden-Bell 1982). Given a value of the mass ratio between M 31 and the MW, this directly produces an estimate of the mass of the MW.

To constrain the LG mass distribution out to radii of 1 Mpc and to test the results presented by Zaritsky *et al.* (1989; hereafter Z89), *i.e.* $M_{MW} \gtrsim 13 \times 10^{11}\,M_\odot$, I examine the dynamics of the outer LG systems. Timing arguments have been applied exclusively to two-body systems (*e.g.* Kahn and Woltjer 1959 and Z89); however, the basic premise of the timing argument should be applicable to appropriate few-body systems, like the remote members of the LG. Restricted three-body simulations of the orbits of LG dwarf galaxies have been done (Mishra 1985), but the mass of the LG has been determined from the two-body analysis of the M31-MW system. In the analysis presented here, a mass distribution for the LG is assumed and the orbits of the galaxies are integrated backward in time. For plausible models the galaxies will reach the LG barycenter at a time that is approximately the assumed age of the universe.

THE MODELS

For a particular simulation, the model LG consists of M31, the MW, a dark matter component (DMC), and the satellite. Assuming a constant M/L and adopting luminosities for M31 and the MW from van den Bergh (1980), the mass ratio is determined to be $\sim$ 1.8. The density of the DMC and the mass of the MW are free parameters, and the age of the universe is taken to be 14 $\times$ 10^9 years. Because most of the orbits of the LG dwarf members are severely influenced by M31 or the MW, only three of the most remote and least perturbed satellites, Leo A, Aquarius, and Sagittarius, were used in the simulations.

Timing arguments can now be applied to systems at all relevant scales in the LG. The orbit of Leo I can be used to constrain the mass of the MW, the orbits of M31 and the MW can be used to constrain the combined mass of the MW, M31 and the DMC between the two galaxies, and the orbits of Leo A, Aquarius, and Sagittarius can be used to constrain the combined mass of the MW, M31, and the DMC out to radii of $\sim$ 1 Mpc. Tabulated below are the results from two types of models. The first is the "light" galaxy model, where the mass of the MW is taken to be 4 $\times$ $10^{11}\,M_\odot$. The second is the "heavy" galaxy model, where the mass of the MW is taken to be $>10^{12}\,M_\odot$.

TABLE I Timing Argument Models

No.	Satellite	M_{MW} $10^{11} M_\odot$	M^a_{DMC} $10^{11} M_\odot$	t^b 10^9 yrs	t^c 10^9 yrs
1	Leo A(= DDO 69)	4.0	0	...	>> 19
	Aquarius(= DDO210)	4.0	0	...	>> 19
	Sagittarius	4.0	0	...	>> 19
2	Leo A	4.0	37	14	14
	Aquarius	4.0	12	14	14
	Sagittarius	4.0	35	14	14
3	Leo A	13.7	0	14	> 19
	Aquarius	13.7	0	14	15
	Sagittarius	13.7	0	14	16
4	Leo A	12.4	20	14	14
	Aquarius	13.7	6	14	14
	Sagittarius	12.7	16	14	14
5	Leo A	17.5	0	14	14
	Aquarius	17.5	0	14	14
	Sagittarius	17.5	0	14	14

[a] Mass of DMC interior to satellite.
[b] Time required for M31 and MW to reach LG barycenter.
[c] Time required for satellite to reach LG barycenter.

CONCLUSIONS

I have used outer members of the LG to place limits on the distribution of mass in the LG. The results of these models lead to the following conclusions:

1) Based on the agreement between models using *six* different galaxies (MW, M31, Leo I, Aquarius, Leo A, and Sagittarius), the hypothesis that M31 has a random velocity, *i.e.* not predominantly caused by the gravitational interaction with the MW, is incorrect.

2) The assertion that the LG consists primarily of two galaxies, *i.e.* M31 and the MW (Model 1), which have masses of *only* a few $\times$ $10^{11}\,M_\odot$ is *inconsistent* with the observations of Leo I, M31, Leo A, Aquarius, and Sagittarius.

3) The "light" galaxy models are only plausible with the inclusion of an additional dark matter component with a density dependence contrived to account for the dynamics of M31 and the outer systems (Model 2). However, this model is *ad hoc* and does not explain the dynamics of Leo I (for Leo I analysis see Z89).

4) The "heavy" galaxy models are much more self-consistent than the "light" galaxy models. The results from simulations using LG satellites suggest that the LG is more massive than implied by the M31-MW timing argument using radial orbits (Model 3). The extra mass required may be located in M31 and the MW, if the MW has a transverse velocity of $\sim 55\ \mathrm{km\,s^{-1}}$(Model 5), or it may be located in a diffuse DMC that has little consequence on the dynamics of M31, the MW, and Leo I (Model 4). However, the uncertainties involved, primarily in the distances to the outer systems, weaken the extra mass argument. A *conservative* model of the LG has the $M_{MW} \sim 13 \times 10^{11} M_\odot$ and $M_{31} \sim 25 \times 10^{11} M_\odot$. These results confirm previous determinations of the LG mass from M31-MW timing arguments and confirm the results presented by Z89 for the mass of the MW. *Both* the MW and M31 contain massive halos that extend out to radii of *hundreds* of kpc and contain at least 1 to 2 $\times$ $10^{12} M_\odot$.

ACKNOWLEDGMENTS

The author gratefully acknowledges support from an NSF Graduate Fellowship.

REFERENCES

Einasto, J., and Lynden-Bell, D. 1982, *M.N.R.A.S.*, **199**, 67.

Kahn, F. D., and Woltjer, L. 1959, *Ap. J.*, **130**, 705.

Mishra, R. 1985, *M.N.R.A.S.*, **212**, 163.

van den Bergh, S. 1980, in NATO ASI *The Structure and Evolution of Normal Galaxies*, ed. S.M. Fall and D. Lynden-Bell (Cambridge University Press: Cambridge), p. 201.

Zaritsky, D., Olszewski, E.W., Schommer, R.A., Peterson, R.C., and Aaronson, M. 1989, *Ap. J.* in press.

SURFACE BRIGHTNESS AND BANDPASS BIAS IN THE HUBBLE MORPHOLOGICAL SEQUENCE

GREGORY D. BOTHUN
Department of Astronomy
University of Michigan
Ann Arbor MI 48109

ABSTRACT In this contribution I discuss some bandpass and surface brightness biases that are associated with our perception of the known population of galaxies. I further discuss some properties of low surface brightness galaxies and what they tell us about the efficiency of disk formation and the importance of threshold gas densities in the star formation histories of disk galaxies.

INTRODUCTION

The field of extragalactic astronomy was born immediately after it was established that the Universe indeed was much larger than the Milky Way. If we use history as a guide, most scientific disciplines take many generations to develop and the first steps in any newly opened field of inquiry is the arduous task of cataloguing and classifying objects for future study. In most all disciplines, classification is performed on the objects which are the most accessible. Thus while the modern day botanist can jet his/her way to most terrestrial locales in order to discover, catalog, and classify a new species of plant or animal life, the rich bottanical variety of the ocean floor remains virtually uncatalogued and it would be foolhardy of any biologist to formulate a comprehensive evolutionary theory based on such limited cataloguing of species. Any attempt to construct such a theory must adopt the optimistic view that a *representative* sample of species has been catalogued.

The parallel with galaxy classification and subsequently developed theories of galaxy evolution should be clear. To date, astronomers have only catalogued the most conspicuous galaxies; those which exhibit the highest contrast with respect to the night sky background. It is these galaxies which define the Hubble sequence. Yet, can we be sure that these galaxies are representative of the general population? If we are secure that the censoring of the night sky brightness is not very severe, then perhaps in the 75 years since they were first discovered, we have fully sampled the true spectrum of galaxy types that exist in the universe. However, in these proceedings I hope to raise the level of insecurity and offer the probability that our knowledge of the true population of galaxy types is still somewhat limited. A dramatic confirmation of this limitation is provided by the serendipitous discovery of at least two, nearby, extremely large, low surface brightness (LSB) disk galaxies that are among the most luminous disk galaxies yet observed but both have remain undiscovered and uncatalogued until just recently.

In discussing the process of galaxy discovery and classification it seems appropriate to begin with the following quote from Hubble (1922):

"Subdivision of non-galactic nebulae is a much more difficult problem. At present and for many years to come, their classification must rest solely upon the simple inspection of photographic images, and will be confused, by the use of telescopes of widely differing scales and resolving powers. Whatever selection of types is made, longer exposures and higher resolving powers will surely cause a reclassification of many individual nebulae, although the true spiral once recognized as such will maintain its class with all more powerful instruments ..."

In this quote Hubble is prophetic in that galaxy classification is confusing (especially to non-practioneers!), has outlined the essence of the classification process as simple inspection of a photographic image, and has discussed its fundamental flaw in that classification is highly dependent upon observing equipment and seeing. Thus, while it is possible to produce an internally consistent classification system, there is no real external verification of the process as it involves no physics and few, if any, quantitative measurements. This makes the process difficult to duplicate generations later which simply illustrates that there is no substitute for quantitative measurement.

FORM AND CONTENT

The basic morphological elements of the Hubble-Sequence of spirals have been set forth by Sandage (1961). The following is my biased summary of these elements:

S0: no spiral structure but has varying degrees of dust (makes one wonder how edge-on S0's can be classified as such!)
Sa: smooth arms (period)
Sb: somewhere between an Sa and Sc
Sc: multiple arms; prominent H II regions; no bulge
Sd: Like an Sc but lacking rotational symmetry

In principle barred versions of the above sequence also exist though SBc's and SBd's are somewhat rare.

Given the basic classification system as originally defined by Hubble, a logical question to ask is "What can I physically learn about a spiral galaxy from its appearance on a photographic plate?". I offer four possibilities depending on your point of view:

a) everything - the galaxy classifier
b) nothing since it is all dark matter anyway - the physicist
c) only its relative star formation rate - the extragalactic observer
d) its bulge-to-disk ratio - the extragalactic theorist

Alternative a) represents the answer in the case of an internally consistent classification system where all one needs to really know about the galaxy is its type. As I have been reminded on several occasions by other astronomers, perhaps it is sufficient to just know that these type galaxies *exist*. As an example, if one morphologically recognizes that a four legged mammal is, in fact, a cow (renegade or otherwise), then one can walk away satisfied at making this discovery and understanding the object. The evolutionary history of mammals that leads to the formation of cows is then relegated to another discipline. Alternative b) is the physical reality thrust upon us by the theoretical appeal of inflationary cosmologies and serves as a reminder that if stellar populations aren't even fundamental properties of galaxies then surely morphological classification has little use. Alternative c) strikes me as the most reasonable answer because it is clear that arm texture is the basic criterion for the classification of spiral galaxies into the Hubble sequence (see Sandage 1961). In this case, classification becomes directly linked to the relative star formation rate (SFR) via the illumination of the spiral pattern by young stars. Spiral galaxy morphology is therefore highly transient as only 1-2 % of the disk mass actually participates in current star formation (see Kennicutt 1983, von Hippel and Bothun 1990). An illustration of this transient nature is provided by the following slides (which unfortunately I can not reproduce in these proceedings and hence much of my impact on this subject is lost) where I show what the Digital Hubble Atlas might look like. The major point of this now virtual slide show is to explicitly demonstrate that the stellar population that best defines the morphology is merely painted on an extensive underlying disk that contains the real secrets regarding the star formation history of that particular galaxy. Comparison of the B and I images reinforces the notion that it is only the relative star formation rate which is measured by morphology. Finally, alternative d) is frequently confused (e.g., Meisels and Ostriker 1984) as the main element of the Hubble Sequence. Part of this confusion is inherent in the ambiguous use of the classification system whereby face-on spirals are classified using different criterion than edge-on spirals. A conscientious attempt to remove this ambiguity was made by van den Bergh (1976), but that classification system (RDDO) was never embraced by observers and hence the ambiguity is perpetuated. Indeed, Kennicutt (1981) shows that bulge-to-disk (B/D) ratio is only very loosely correlated with arm pitch angle and/or degree of resolution. Tighter correlations (e.g., Lake and Carlberg 1988) can of course occur if one uses B/D as the principal element of the classification system. In that case, however, one is no longer following Hubble's (1935) original intention.

Once a classification system is developed its validity and usefulness can be tested by searching for correlations between measurable parameters and classification type. There have been numerous investigations of this type (e.g., Roberts 1975, de Vaucouleurs 1977, Bothun 1982, Lake and Carlberg 1988) and the results can be succinctly summarized as: *a mean trend with large scatter.* So while an average Sa has a red disk, low H I content, low relative SFR and an average Sc has a blue disk, high H I content and high relative SFR, exceptions in both cases can be found. This merely illustrates that spiral disks, in reality, span a continuum of properties and the observed mean trends with type are reflections of varying degrees of bulge light contamination. This is especially true for integrated U-B and B-V colors. A typical Sb spiral (e.g., M31 I guess) has B/D $\approx$ 0.2 (see Boroson 1981, Kent 1985). However, most photoelectric aperture photometry of galaxies is obtained through apertures which are at most 1-2 scale lengths (α) in radius. The canonical spiral disk has a central blue surface brightness of 21.65 mag arcsec^{-2} (Freeman 1970). In this case the disk surface brightness at 1α is 22.74 mag arcsec^{-2}, which is at or below the blue night sky brightness and it is difficult to obtain good signal-to-noise data without the use of a CCD to obtain a simultaneous measurement of the sky. For an exponential disk, 26 % of the total disk luminosity is contained within 1α and thus even for an Sb galaxy, the effective B/D ratio through the typical photoelectric aperture is quite large (0.5 - 1.0). Thus, without an extensive CCD survey, I don't think it is possible to know the UBV colors of the disk components of spiral galaxies unless they have very small B/D ratios. A case in point comes from the work of Bothun and Gregg (1990) who show that after the bulge light is properly removed, the photometric properties of S0 disks and the disks of Sc I galaxies are remarkably similar even though they form the respective ends of the morphological sequence.

In the CCD era it may be possible to develop a more quantitative system of galaxy classification which is based on measured luminosity profiles and rotation curves for a large number of spiral galaxies. This is similar to the approach taken by Rubin, Burstein and colleagues who have constructed various "mass type" classes of galaxies and is an excellent step toward developing a *physical* classification system. I encourage all workers in the field to expand upon it where possible. The construction of a catalog of surface brightness profiles and rotation curves would also be a productive step to take.

DIFFERENT WAVELENGTH WINDOWS

Morphological classification is highly bandpass dependent since most of the information is provided by young stars. Since it is now possible to image galaxies from 1,000 Å to 100 μm one might expect that significant morphological differences would arise when inspecting images of the same galaxy taking at different wavelengths. In particular, some bandpasses would tend to exaggerate morphological differences. An excellent example of this is provided by the debate over galaxy type as an important second parameter in the Tully-Fisher (TF) relation (see Rubin *et al.* 1985, Aaronson and Mould 1983, Pierce and Tully 1988). In the blue TF, is a clear type dependence. This is probably a reflection of differences in relative SFR (that help to define the morphological class) which leads to differences in the blue mass-to-light ratio. In contrast, the infrared TF relation using H-band aperture magnitudes through a standard aperture that encloses 1 blue scale length of light (confusing isn't it?) shows no type dependence which indicates that the inner regions of spiral galaxies are very similar at these longer wavelengths.

However, although morphological features are bandpass dependent the actual structure of the galaxy is not. A case in point is provided by M33. While the blue image of the galaxy is heavily resolved into regions of star formation, the 60 μm image obtained with *IRAS* is quite smooth. Surface photometry performed on that image by myself (see also Rice *et al.* 1989) reveals a strongly exponential profile with a fitted α of 6.3 $'$. This structure is very similar to that observed in the optical. This result should not be surprising since 60 μm emission is the result of dust heated by the old underlying disk population (see Walterbos and Schwering 1987, Bothun *et al.* 1989a), and therefore its luminosity profile should be like that of the underlying disk. Furthermore, since the structure of galaxies is most well defined by the luminous component that contains most of the mass (e.g., low mass stars), we would also expect that the disk structure of galaxies that comprise the Hubble sequence is unrelated to their type. That is to say, all spiral galaxies illustrated in the Hubble Atlas should occupy similar ranges of disk α and central surface brightness (B(0)) and this appears to be the case. Thus different morphological type does not necessarily imply different disk structure

The launch of the Hubble Space Telescope will finally open up a wavelength window which is right at the heart of morphology. Imaging galaxies at 1500 angstroms will best portray that component of the galaxy which has the most contrast and yet the shortest lifetime. A comparison of that image with one taken at 2.2 μm with the next generation of large format IR imagers will be most illuminating and will echo the wisdom in Hubble's words of 1922.

SURFACE BRIGHTNESS BIAS

If you see a galaxy you can catalog and classify it. If you don't see it you can't. This is the bias in optical catalogs that was first emphasized by Disney (1976) but was known at the time of Messier. Galaxies are, in general, quite diffuse and low contrast objects. A simple thought experiment, based on the Copernican principle, immediately suggests the bias exists. Suppose that we lived on a planet that was located in the inner regions of an elliptical galaxy. The high stellar density would produce a night sky background that would not be conducive for the discovery of galaxies. Similarly, suppose our planet had two small moons, at least one of which was visible at all times. Again, the sky background would be quite high. Although these are absurd situations, they do point out that it is merely a statistical accident that Magellan, Messier, and Hubble lived on a planet with 50 % dark time per orbital cycle and was located in the outer part of a typical spiral galaxy which allowed them to discover faint diffuse objects at high galactic latitude. For all observers, the brightness of the night sky acts as a visibility filter which, when convolved with the true population of galaxies, produces the population that appears in catalogs. So, how severe is this bias?

The surface brightness bias actually assumes two forms: a bias against discovering either compact galaxies or giant, very LSB galaxies. This is best illustrated by considering the following three hypothetical galaxies listed below. All three have pure exponential light distributions and the same total luminosity.

Galaxy A $\alpha = 0.5$ kpc $B(0) = 16.0$ mag arcsec^{-2}
Galaxy B $\alpha = 5.0$ kpc $B(0) = 21.0$ mag arcsec^{-2}
Galaxy C $\alpha = 50.0$ kpc $B(0) = 26.0$ mag arcsec^{-2}

Let's assume that the space density of these three galaxies are equal and let's suppose that we conduct a survey to catalog galaxies which have diameters measured at the B= 25.0 mag arcsec^{-2} level of greater than 1 $'$. Galaxy B, a typical spiral, could be discovered out to a distance of 125 Mpc. Galaxy A, on the other hand, is quite compact (ratio of 1/2 light diameter to $D_{25} = 0.33$) and would fall below the catalog limit beyond D = 60 Mpc. Galaxy C, of course, would **never** be discovered in such a survey. Thus, from this intrinsic distribution of galaxy types, our catalog would contain 90 % of type B galaxies and 10 % of type A galaxies. This is a severe bias which would lead us to erroneously conclude that there is predominately 1 type of disk galaxy. Although this is chosen for illustrative purposes to be an extreme example, the point remains that galaxies of type A or C can be easily missed.

Personally, I do not think that we have missed many of the compact type galaxies. Such high surface brightness systems are often quite luminous at other wavelengths (Einstein, IRAS - see Margon *et al.*

1988) and are generally emission line galaxies which are easily detected in objective prism surveys (e.g., Salzer 1987, Bothun *et al.* 1989b). However, it is not at all clear that galaxies of type C don't exist. Over the last 3 years I and mostly my collaborators (C. Impey, D. Malin, J. Schombert, N. Caldwell, S. Schneider) have been conducting various surveys for LSB galaxies in hopes of finding type C objects. The serendipitous discovery of the type C galaxy Malin-1 (e.g., Bothun *et al.* 1987) seems to provide confirmation that large LSB galaxies do indeed exist and are awaiting discovery. I will summarize the properties of this and another giant, LSB disk galaxy in the next section. For now I just ask two rhetorical questions concerning Malin-1: what is its Hubble type and where would it fall on the TF relation?

LOW SURFACE BRIGHTNESS GALAXIES

a) General Properties

The following is a brief summary of what I think we have established based on our limited surveys:

- The space density of LSB (defined to have B(0) fainter than 23.0 mag arcsec^{-2}) galaxies which have α's $\geq$ 1 kpc does not appear to be larger than galaxies of normal surface brightness (see Schombert and Bothun 1988). Based on our surveys I estimate that $\approx$ 25 % of disk galaxies would fall into the LSB category. This is a significant percentage in terms of theories of disk galaxy evolution, but is not large enough to claim that most of the galaxy population has been missed.
- The space density of LSB dwarf galaxies (i.e., α $\leq$ 1 kpc) appears to continually increase with decreasing B(0) (see also Bingelli *et al.* 1985, Ferguson 1989, Irwin *et al.* 1989). Malin's technique of photographic amplification finds galaxies down to a limit of B(0) = 27.0 mag arcsec^{-2}. So far these searches have been conducted in the Virgo, Fornax and Centaurus clusters. Without redshifts, it is only guilt by association that these LSB galaxies are indeed dwarfs. Some may be additional huge, background LSB galaxies like Malin-1. In general, these very LSB galaxies either have exponential or King profiles (see Impey *et al.* al 1988). If they are indeed members of these clusters and if they have normal (M/L), the implied central mass densities are $\leq$ 10 $M_\odot$ pc^{-2}. As such, their formation would seem to be a real challenge to theory and their survivability in a cluster environment over a Hubble time seems incongruous. Perhaps this argues that such systems must contain large amounts of dark matter just to survive.
- The very LSB galaxies discovered by Impey *et al.* (1988) violate the surface brightness vs. absolute magnitude relation discovered by Caldwell (1983) in the sense that they are too bright relative to their surface brightness. This occurs because their α's are fairly large (e.g., 0.5-1 kpc). This means that completeness corrections to the galaxy luminosity function (LF) based solely on limiting apparent

magnitude are incorrect. Galaxies are always selected by surface brightness. The discovery of very LSB but large galaxies suggests that the incompleteness can occur even at relatively bright apparent magnitudes. Formally, a correction based on SB can yield a faint end slope that is divergent in luminosity (see Impey *et al.* 1988).

• For LSB disk galaxies (i.e., $\alpha \geq 1$ kpc) there is no relation between B(0) and disk color. Thus, these LSB galaxies are not faded remnants of their former selfs. To be sure, some LSB disks are quite red and may be the signature of episodic star formation in spiral galaxies (see Schommer and Bothun 1983, Bothun *et al.* 1984a), but many are quite blue as well. Since the blue surface brightnesses of galaxies depend both on stellar population and surface mass density, the similarity of disk colors between low and high SB disk galaxies suggests that similar stellar populations have formed in dissimilar surface mass density environments. This again seems paradoxical as the character of star formation is thought to be tied to local density.

• There is no suggestion from our surveys that LSB galaxies tend to be also low mass. This is best demonstrated in Figure 1 which shows a plot of diameter vs surface brightness for various H I selected samples. The data are for late type galaxies and come from observations by S. Schneider and collaborators as well as myself. The dashed line represents the boundary between normal and low SB in these units of average SB. Clearly, rather large (and sometimes extreme) examples of LSB galaxies can be found. The rotational velocity of some of these objects approaches 350 km s^{-1} and their α's are 5-10 kpc. These galaxies are then ≈ 5 times more massive than the Milky Way.

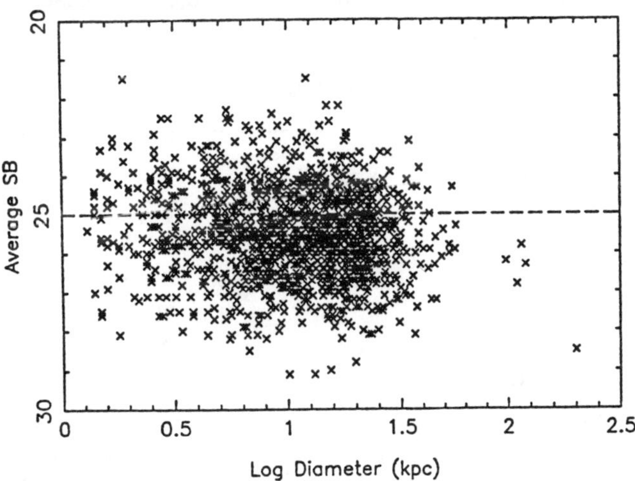

• Narrow band imaging at Hα shows that LSB disk galaxies often contain a large number of H II regions, the brightest of which have luminosities that rival the large H II regions in M101. However, unlike M101, these galaxies have little or no spiral structure. They also appear to be quite deficient in molecular gas (see Schombert *et al.* 1989). Thus at least some LSB disk galaxies are actively star forming in the absence of pronounced spiral structure and molecular gas. Presumably, this is similar to the first phases of disk galaxy evolution.

• Although the data is limited (e.g., van der Hulst *et al.* 1987, Skillman *et al.* 1987, van Gorkom *et al.* 1989), aperture synthesis observations at 21-cm have shown that LSB galaxies always have low peak column densities of H I. The peak averages $\approx 10^{20.5}$ atoms cm^{-2} which is below the canonical threshold value for the conversion of atomic hydrogen to molecular hydrogen (see Gallagher and Hunter 1984, Skillman 1987, Kennicutt 1989).

b) Very Luminous but Uncatalogued LSB Galaxies

Bothun *et al.* (1987,1989) have reported the discovery of two very large (diameters $\geq$ 100 kpc) but LSB galaxies. Nonetheless, their total luminosities place them at the extreme end of the disk galaxy LF and yet these galaxies have remained uncatalogued. We thus should feel somewhat insecure in the knowledge that some of the most luminous disk galaxies can go undetected! The properties of these two galaxies are summarized in Table 1. Since this is the centennial of Hubble's birth, all distance dependent quantities assume $H_o = 100$ km s^{-1} Mpc^{-1}.

Table 1. Some Properties of Huge LSB Disks

Parameter	Malin 1	Malin 2
Position	α=12 34 27.3	α = 10 37 08.5
	δ = 14 36 15	δ = 21 06 24
Velocity	24750 km s^{-1}	13830 km s^{-1}
H I Mass	1.1 x 10^{11} $M_\odot$	3.1 x 10^{10} $M_\odot$
α	55 kpc	16 kpc
$B(0)_{disk}$	26.5	23.5
Bulge Mag.	M_B = -18.6	M_B = -19.5
Disk Mag.	M_B = -20.8	M_B = -21.1
$(B\text{-}V)_{disk}$	0.80 $\pm$ 0.1	0.55 $\pm$ 0.05
Dynamical Mass	2.2 x 10^{12} $M_\odot$	0.9 x 10^{12} $M_\odot$

These two galaxies show a great deal of similarity in additional respects. Both are found in relatively low density regions and have large H I extents. In fact, van Gorkom *et al.* (1989) measure an H I size of $\approx$ 0.5 h^{-1} Mpc for Malin-1! Both galaxies have high dynamical masses and (M/L) ratios. This means that the fractional H I contents are, in fact, normal and implies that perhaps the IMF is skewed toward producing a high (M/L) stellar population which makes them, in essence, "dark" galaxies (see also Carignan and Freeman 1988). Both galaxies have normal bulge components so galaxy formation did begin at these sites but disk formation has been quite inefficient. Both galaxies have some kpc sized H II regions with luminosity of $\approx 10^{41}$ ergs sec^{-1} which, in each case, are located at radii of 50 - 100 kpc from the nucleus! Both galaxies exhibit weak, but very broad Hα in their nuclear spectrum along with rather strong [S II]/Hα and could be classified as low luminosity Seyfert I nuclei. Both galaxies have low peak column densities of H I but very large cross-section and thus represent exactly the kinds of galaxies which are capable of producing damped Ly_α absorption toward distant quasars. These galaxies then may be relics of disks which formed at $z \approx 2$ and have remained mostly quiescent to the present day (see Impey and Bothun 1989). If they were not in regions of low galaxy density, it would be unlikely that their extended, large, low density H I reservoir would have been retained and only the remnant bulge would be apparent.

DISK FORMATION TIMESCALES AND THRESHOLD EFFECTS

I would like to conclude by mentioning 3 important results which have been derived for disk galaxies and which point to the fact that disk formation occurs over a long timescale.

- Measurements of colors, surface brightnesses and gas contents for hundreds of disk galaxies show that they occupy a wide range of evolutionary states at the present epoch. One can now find examples of mostly *unevolved* disks (like Malin-1) to examples of disk galaxies which have exhausted more than 99 % of their gas supply (e.g., Eder 1989). Thus, disk galaxy evolution is still occurring at the present epoch.
- There appears to be an age-mass relation in the sense that lower mass galaxies have younger mean ages (see Bothun *et al.* 1984b). This does not seem to depend on the B/D ratio. Thus, low mass disk galaxies appear to evolve more slowly than their high mass counterparts. Conceivably, this is due to differences in mass distributions documented by Rubin and collaborators.
- The disks of S0 galaxies are younger than their bulges by 3-5 Gyr. In some cases, AGB light is seen in the disk indicating a period of active star formation $\approx$ 2-3 Gyr ago (see Bothun and Gregg 1990). The fact that S0 disks are gas poor at $z = 0$ probably reflects an initial underabundance of disk gas (see also Sandage, Freeman and Stokes 1970) due to the formation of a large bulge. Thus S0 disks are predicted to be actively star forming at $z \approx 0.5$ and provide the most natural explanation for the Butcher-Oemler effect (see also Couch and Sharples 1987).

These observations are all consistent with a simple model in which disk evolution is mostly independent of Hubble type with an evolutionary clock that is fundamentally controlled by the mean surface density of H I (see also Kennicutt 1989). Galaxy dynamics, ISM metallicity, the dark matter distribution, the presence of an AGN, interactions among galaxies, are all secondary influences whose amplitude likely depends on the ambient gas density of the disk. Thus when the surface density of H I is high, the disk evolves rapidly and efficiently. When the surface density of H I falls below some threshold value (probably around 10^{21} atoms cm^{-2}) the formation of molecular clouds is inhibited and disk evolution slows. It is quite possible that this behavior is episodic and thus it is dangerous to derive gas depletion timescales based on current SFRs. If this episodic behavior actually occurs, one might expect similar fluctuations in UV/blue surface brightness which would lead to important selection effects when discovering faint galaxies at high redshift (see also Gallagher 1989).

If the H I remains below the threshold value the disk can become quiescent. Observations by Wolfe *et al.* (1986) show a large number of damped Lyα systems at $z \approx 2$, which they suggest are proto-disk galaxies. Interestingly, the inferred column density is similar to the suspected threshold value. Narrow-band imaging to detect Lyα in emission by Cohen (1989) has revealed no cases of active star formation. In view of the threshold model, this result is not surprising. Galaxies with surface gas densities above the threshold will rapidly produce metals and dust thereby making the disk optically thick to Lyα photons. Galaxies below the threshold will not make many stars, metals, or dust and if they are located in relatively low density regions of the universe it is possible that this low density gas disk has remained intact over a Hubble time, resulting in galaxies with the morphological characteristics of Malin-1 or Malin-2.

In summary, the discovery of LSB galaxies has helped to recover the full range of disk formation efficiencies that operate in the universe. Their existence also vividly illustrates that our present day catalogs of galaxies is censored by the brightness of the night sky. The severity of this bias is difficult to estimate. Certainly it is not overwhelming but it is significant. In general, these LSB galaxies can not be placed into the Hubble sequence due to the lack of a discernible spiral pattern. Nonetheless, their stellar populations appear to be the same as standard Hubble sequence spirals. This further weakens the already crumbling link between galaxy form and galaxy content. This coupled with the sensitivity of morphology to bandpass advises caution when generalizing results obtained from morphologically selected samples. Finally, LSB galaxies serve to emphasize the importance of threshold effects in the evolution of disk galaxies. It may very well be that the disk component of all galaxies evolves in the same manner, if the H I surface density is above the threshold value. In that case, the astration clock is set by the percentage of gas used up initially in the formation of non-disk components.

ACKNOWLEDGEMENTS

I would like to thank the Scientific Organizing Committee for the inviting me to speak and for arranging some glorious California weather. I would also like to thank my many collaborators for assistance in data acquisition and the generation of ideas. I hope they are not too embarrassed by my remarks.

REFERENCES

Aaronson, M. and Mould, J. 1983, *Ap. J.*, **265**, 1.

Bingelli, B., Sandage, A., and Tammann, G. 1986, *A.J.*, **90**, 1681.

Boroson, T. 1981, *Ap. J. Suppl.*, **46**, 177.

Bothun, G. 1982, *Pub. A.S.P.*, **94**, 774.

Bothun, G., Schommer, R., Aaronson, M., Mould, J., and Huchra, J. 1984a, *Ap. J.*, **278**, 475.

Bothun, G., Romanishin, W., Strom, S., and Strom, K. 1984b, *A.J.*, **89**, 1300.

Bothun, G., Impey, C., Malin, D., and Mould, J. 1987, *A.J.*, **94**, 23.

Bothun, G., Londsale, C., and Rice, W. 1989a, *Ap. J.*, **341**, 129.

Bothun, G., Halpern, J., Impey, C., Lonsdale, C. and Schmitz, M. 1989b, *Ap. J. Suppl.*, **70**, 271.

Bothun, G., Impey, C., Schombert, J. and Schneider, S. 1989 preprint.

Bothun, G., and Gregg, M. 1990 *Ap. J.* in press.

Caldwell, N. 1983, *A.J.*, **88**, 804.

Carignan, C. and Freeman, K. 1988, *Ap.J.*, **332**, 33.

Cohen, R. 1989 this conference.

Couch, W., and Sharples, R. 1987, *M.N.R.A.S.*, **229**, 423.

de Vaucouleurs, G. 1977 in *The Evolution of Galaxies and Stellar Populations* ed. B. Tinsley and R. Larson (Yale University Press).

Disney, M. 1976, *Nature*, **263**, 573.

Eder, J. 1989, Ph. D. Thesis, Yale University.

Ferguson, H. 1989, *A.J.*, **98**, 367.

Freeman, K. 1970, *Ap. J.*, **160**, 811.

Gallagher, J. and Hunter, D. 1984, *ARAA*, **22**, 37.

Gallagher, J. 1989 - this conference.

Hubble, E. 1922, *Ap.J.*, **56**, 162.

Hubble, E. 1935 *The Realm of the Nebulae.*

Impey, C., Bothun, G., and Malin, D. 1988, *Ap. J.*, **330**, 634.
Impey, C. and Bothun, G. 1989, *Ap. J.*, **341**, 89.
Irwin, M., Davies, J., Disney, M., and 1989 preprint.
Kennicutt, R., 1981, *A.J.*, **86**, 1487.
Kennicutt, R., 1983, *Ap.J.*, **272**, 54.
Kennicutt, R., 1989 preprint.
Kent, S. 1985, *Ap. J. Suppl.*, **59**, 115.
Lake, G. and Carlberg, R. 1988, *A.J.*, **96**, 1581.
Margon, B., Anderson, S., Mateo, M., Fich, M. and Massey, P. 1988, *Ap. J.*, **334**, 597.
Meisels, A. and Ostriker, J. 1984, *A.J.*, **89**, 1451.
Pierce, M., and Tully, B. 1988, *Ap. J.*, **330**, 579.
Rice, W., Boulanger, F., Viallefond, F., Soifer, T. and Freedman, W. 1989 preprint.
Roberts, M. 1975 in *Stars and Stellar Systems* ed. A. Sandage, M. Sandage and J. Kristian, Chap. 9.
Rubin, V., Burstein, D. Ford, K. and Thonnard, N. 1985, *Ap. J.*, **289**, 81.
Salzer, J. 1987 Ph. D. Thesis, Univ. of Michigan.
Sandage, A. 1961 *The Hubble Atlas of Galaxies* Carnegie Institute of Washington, Pub. No. 618.
Sandage, A., Freeman, K., and Stokes, N. 1970, *Ap. J.*, **160**, 831.
Schombert, J. and Bothun, G. 1988, *A.J.*, **95**, 1392.
Schombert, J., Mundy, L., Bothun, G. and Impey, C. 1989 preprint.
Schommer, R. and Bothun, G. 1983, *A.J.*, **88**, 577.
Skillman, E. 1987 in *Star Forming Dwarf Galaxies and Related Objects* ed. D. Kunth, T. Thuan and J. Van (Editions Frontiers, Paris).
Skillman, E., Bothun, G., Murray, M., and Warmels, R. 1987, *Astr. Ap.*, **185**, 61.
van den Bergh, S. 1976, *Ap.J.*, **206**, 883.
van der Hulst, T., Skillman, E., Kennicutt, R., and Bothun, G. 1987, *Astr. Ap.*, **177**, 63.
von Hippel, T. and Bothun, G. 1989 preprint.
Walterbos, R. and Schwering, P. 1987, *Astr. Ap.*, **180**, 27.
Wolfe, A., Turnshek, D., Smith, H., and Cohen, R. 1986, *Ap. J. Suppl.*, **61**, 249.

BLUE COMPACT DWARF GALAXIES IN VIRGO AND NEARBY GROUPS

G. LYLE HOFFMAN
Lafayette College, Dept. of Physics, Easton, PA 18042

E. E. SALPETER
Center for Radiophysics and Space Research, Space Sciences Building, Cornell University, Ithaca, NY 14853

GEORGE HELOU
Infrared Processing and Analysis Center, Mail Code 100-22, Caltech, Pasadena, CA 91125

Abstract We report the results of a neutral hydrogen (HI) survey of blue compact dwarf (BCD) galaxies in the Virgo Cluster area and some nearby groups of galaxies. The BCDs are comparable to other dwarf irregular galaxies in rotation speed, blue and HI luminosities, but are significantly smaller in diameter. The dearth of protogalactic HI clouds with no accompanying stellar luminosity leads us to conclude that the BCDs, for the most part, must have experienced star formation prior to the current burst. The distribution of the BCDs among the subclusters within Virgo and among groups further out in the supercluster is not uniform; typically 10% of all dwarf irregulars are BCDs, but in two "special" clouds the fraction of BCDs is more than twice as high.

Blue compact dwarf galaxies are high surface brightness dwarf irregular galaxies, resembling one or a few isolated giant HII regions in a galaxy which otherwise has exceedingly low surface brightness. In some cases *no* surrounding luminosity is visible even on CCD frames. Metalicities are generally low — $\frac{1}{3}$ to $\frac{1}{30}$ of solar abundances — and there have been speculations that BCDs result from recent (short-lived) bursts of star formation in a gas-rich dwarf galaxy that has previously had very little star formation.

COMPARISON TO OTHER MORPHOLOGICAL TYPES

In a recent paper (Hoffman *et al.* 1989, *Ap. J.*, **339**, 812), we studied the BCDs in the Virgo Cluster Catalog (VCC) of Binggeli, Sandage and Tammann (1985, *A. J.*, **90**, 1681), analyzing our own HI and co-added IRAS data for the BCDs,

along with optical data from other sources, in comparison to comparable data on the sequence of low surface brightness "normal" dwarf irregulars (Im) in the cluster. We also obtained Palomar optical redshifts (Helou and Schombert, 1989, in preparation) for those BCDs that were undetected in HI. Finally, we used the many HI reference beams and signal-free portions of on-source spectra, acquired in the process of our Arecibo mapping and detection surveys of all spiral and late-type dwarf galaxies in the cluster area, to set stringent upper limits on the number of HI clouds comparable in mass to the smallest dwarf irregular galaxies, but which have so far failed to produce enough stars to be visible as Im galaxies on the du Pont plates used to compile the VCC. We concluded that:
(1) All BCDs at redshifts < 3000 km s^{-1} have significant (easily detectable) quantities of HI; all undetected BCDs in the VCC turned out to have redshifts $> 10{,}000$ km s^{-1}.
(2) In general, the BCDs are comparable to other dwarf irregular galaxies in dynamical mass (as measured by rotation speed), blue and HI luminosities, and those quantities are reasonably well correlated.
(3) However, the optical diameters of BCDs are smaller by a factor of 2 or 3 than those of Im galaxies of comparable luminosity or HI profile width.
(4) BCDs are more readily detected by IRAS than are Im galaxies of comparable luminosity or profile width, and the interstellar medium of BCDs is deficient in dust by only a modest factor of 2 or 3 compared to normal spiral galaxies.
(5) There are very few, if any, protogalactic HI clouds in the well-surveyed Virgo area; we detected none with properties that would be appropriate to a "quiescent" proto-BCD still awaiting its first burst of star formation.
(6) Since previous star formation is needed to make interstellar dust, and since there is no significant population of quiescent gas clouds to serve as progenitors for the (presumably short-lived) BCD phenomenon, the current burst of star formation cannot be the first experienced by most, if not all, BCDs.

DISTRIBUTION AMONG SUBCLUSTERS WITHIN VIRGO

A dendogramatic analysis (modified from Tully, 1987, *Ap. J.*, **321**, 280) of subclustering in the Virgo area finds four major clouds within the cluster complex: the main core or "M87 Cloud;" the Southern Extension, including a southern portion of what is traditionally considered to be the cluster core; the M and W groups which lie at ~55% greater distance; and a "Low Velocity Cloud" (LVC) of spiral, irregular and dwarf elliptical galaxies superimposed on the core of the cluster but at substantially lower (even negative) heliocentric velocity. The assignment of individual galaxies to these clouds is largely confirmed by the method of Dressler and Shechtman (1988, *A. J.*, **95**, 985). The Southern Extension would continue on beyond the celestial equator if an all-sky catalog were available, but the other three clouds are essentially complete as confirmed by a dendogram generated from the CfA redshift catalog (extending over a wider area, but restricted to brighter galaxies).

A Tully-Fisher (blue magnitude — HI velocity width) analysis on the spiral galaxies (Sb through Sm) in these clouds confirms previous claims that the W & M Cloud lies at 55% greater distance than the major part of the cluster, but we cannot resolve any difference in distance between the other three clouds.

The BCDs are *not* uniformly distributed among the four clouds. The M87 Cloud has the lowest population of BCDs, comprising only 10% of all its dwarf irregulars; the other three clouds have more than 25% BCDs. However the distribution of BCDs is even more spotty that these percentages suggest:
(1) The M87 Cloud BCDs are mostly at low velocities, only just high enough to have been grouped into this cloud rather than the LVC; and the NE third of the M87 Cloud is completely devoid of BCDs.
(2) The Southern Extension BCDs lie mainly toward the western edge and at higher than average velocities, in the vicinity of the M & W Cloud.

Tully-Fisher plots of the BCDs in these four clouds do not clearly show the separation in distance of the M & W Cloud from the other three. No doubt this is partly due to our inability to assign inclinations to the BCDs on the basis of optical information alone; in fact, we have assigned a single average inclination to all BCDs. In the B_T — $\log(\Delta V_c)$ plane there is some indication of the BCDs in the M & W Cloud being somewhat fainter than the rest at constant profile width, but an alternative explanation would be that a handful of Southern Extension BCDs are anomolously bright. Note also that all of the LVC BCDs (only 4 have measured profile widths; the other two are confused with Galactic HI) have smaller profile widths than all but one of the M & W Cloud BCDs. This may be simply due to small number statistics, or to a selection effect arising from the greater distance of the M & W Cloud.

BCDs morphologically selected on the same scheme have been found by Binggeli, Sandage and coworkers in the Leo, Ursa Major and Canes Venaticii complexes of groups. We have surveyed these in HI as well (Hoffman *et al.*, 1989, *Ap. J. Suppl. Series*, in press), obtaining redshifts to determine which are groups members and which are background objects. These group complexes all have BCDs comprising only 10% by number of their dwarf irregulars.

We speculate that 10% is the typical fraction, and that the BCD phenomenon is somehow favored in the LVC and M & W Cloud (or high velocity portion of the Southern Extension), perhaps by tidal or hydrodynamic processes as these clouds fall into the cluster core.

SO GALAXIES AND THE HUBBLE SEQUENCE

SIDNEY VAN DEN BERGH
Dominion Astrophysical Observatory, National Research Council of Canada, 5071 West Saanich Road, Victoria, B.C., V8X 4M6, Canada

ABSTRACT It is shown that the SO class comprises a number of physically quite distinct types of objects. Many luminous SO's are probably truly intermediate between galaxies of types E and Sa, whereas most less-luminous SO's might be rotationally supported disk-like lenticulars. A third class of SO galaxies, comprising such objects as NGC5128 and NGC5195, probably owe their present appearance to recent close tidal encounters or mergers. The results obtained in this paper suggest that SO galaxies, even though they exhibit superficial similarities, probably arrived at their present morphological state via a variety of quite different evolution tracks.

I. GALAXY CLASSIFICATION

More than 60 years ago Lundmark (1926) and Hubble (1926) recognized two basic types of galaxies *viz*. ellipticals and spirals. Ten years later Hubble (1936) postulated SO galaxies as a hypothetical transitional type spanning the chasm between ellipticals and spirals. Hubble's final ideas on this fascinating subject are described in considerable detail by Sandage (1961).

For reasons that are still not entirely clear spirals come in two flavors; normal spirals and barred spirals. On the basis of their stellar population content Magellanic irregular galaxies fit naturally at the blue end of the morphological sequence Sa-Sb-Sc-Ir. The dichotomy between normal and barred forms also applies to irregulars. The LMC is a fine example of a barred irregular. A basic difference between spiral and irregular galaxies is that all spirals, that are close enough to be observed at high resolution, appear to have nuclei. On the other hand no irregular galaxy is known to have a nucleus.* It is noted in passing that this implies that quasar/AGN-type phenomena cannot have played a role in the evolution of Magellanic irregulars.

An important caveat (cf. van den Bergh 1977) is that the Hubble classification system is defined in terms of giant and supergiant class examples. Except, perhaps, for Magellanic irregulars Hubble type is therefore a meaningless concept for galaxies fainter than $M_V \approx -16$. Spheroidal galaxies are an example of an entire class that falls completely outside the original Hubble classification scheme. The first examples of this class (which, on the basis of their high frequency of occurrence in the Local Group, are now believed to be the most common type of galaxy in the Universe) were discovered by Shapley (1938). Quite recently it has been recognized (Wirth and Gallagher 1984, Kormendy 1985) that objects such as NGC205 are bright examples of the spheroidal class, not dwarf ellipticals like M32. It is not yet clear how (or if) the bright end of the spheroidal sequence merges with that of spiral and S0 galaxies. A highly tentative schematic for galaxy classification types is shown in Fig. 1.

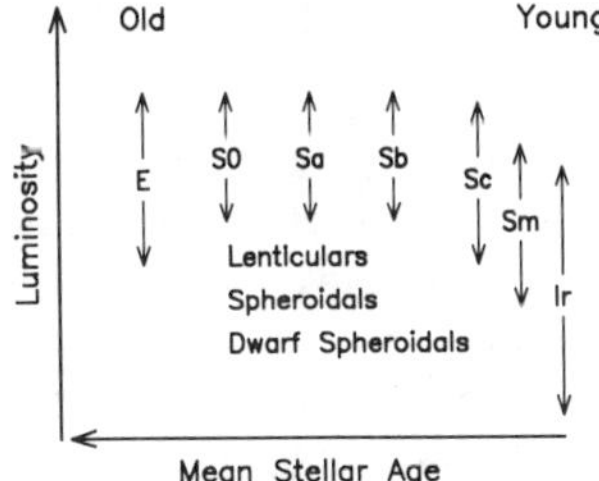

Fig. 1 Tentative schematic showing relative positions of various galaxy classification types.

An alternate scheme, in which low-luminosity galaxies are simply classified as "active dwarfs", "less active dwarfs" and "inactive dwarfs", was proposed by van den Bergh (1977).

It is now generally accepted that typical ellipticals and spheroidals have had quite different star forming histories. Almost all stars in dense elliptical galaxies appear to have been formed during an early short-lived burst of star formation. This contrasts with the situation in spheroidal galaxies, in many of which a low level of star formation still persists to the present day. NGC205 and the Phoenix dwarf (Ortolani and Gratton 1988, Caldwell and Schommer 1988) are examples of bright and faint spheroidals, respectively,

---

* I do not here make the distinction (cf. Kormendy and Djorgovski 1989) between dynamically distinct nuclei like that in M31, and cusps in galaxy cores, such as the one that is seen in M32.

in which some star formation is still taking place at the present time. In the dwarf spheroidal companions to the Galaxy and M31 the existence of C stars attests to the presence of intermediate-age populations.

Elliptical galaxies are a quite homogeneous class of objects. In the mean, round ellipticals are, however, slightly redder than more flattened ones (van den Bergh 1979). Possibly this is a metallicity effect, but more probably it is due to the presence of weak star-forming disks in some objects of types E4-E7 (Carter 1987, Ebneter, Djorgovski and Davis 1988). The most luminous ellipticals are often embedded in huge envelopes. The nature of the outer envelopes of such cD galaxies (Matthews, Morgan and Schmidt 1964) is still not understood. They might either have grown by accretion of tidally stripped material (Malumuth and Richstone 1984), or they could be due to early star formation in a cooling flow (Fabian and Nulsen 1977). Alternatively these envelopes might reflect the structure of the cluster potential well (van den Bergh 1977, Schombert 1988) or they could be of primordial origin (Merritt 1984).

The Hubble (1936) classification scheme is defined in terms of giant and supergiant type examples. Catalogs of galaxies selected on the basis of their apparent magnitude are dominated by objects of high luminosity. As a result our thinking about galaxies, and galaxy evolution, has been dominated by mental images of intrinsically luminous galaxies. Nevertheless much can be learned from objects of lower luminosity that are, per unit volume, much more common than relatively rare supergiant galaxies. A good sample of such relatively faint objects are the Shapley-Ames galaxies with $-18.0 \leqslant M^{0,i}_{B_T} < -19.0$, which will subsequently be referred to as galaxies with $M_B = -18$. Such objects are bright enough that they can be seen well beyond the Local Group, yet they are more than 3 mag fainter than the average Shapley-Ames galaxy. In Table 1 the frequency distribution of galaxies with $M_B = -18$ is compared with that of all galaxies in _A Revised Shapley-Ames Catalog of Bright Galaxies_ (Sandage and Tammann 1981). The table shows the following:

1. Galaxies of types Sa-Sbc are much less frequent (5%) among dwarfs than they are for all Shapley-Ames galaxies (25%). This suggests that many less active low luminosity disk galaxies may have been classified as S0 or as "amorphous" galaxies.

2. Among dwarfs objects of types Scd-Im account for 29% of all galaxies, compared to only 5% for all Shapley-Ames galaxies. This is, no doubt, due to the well-known fact that Sd and Sm galaxies galaxies are, on average, much fainter than spirals of types Sa-Sc.

TABLE I Frequency Distribution of Galaxy Classification Types

Type*	All SA galaxies n	%	$-18.0 \leq M_B < -19.0$ n	%
E+E/SO	173	14	5	8
SO+SO/a	190	15	11	18
Sa+Sab	165	13	1	2
Sb+Sbc	283	22	2	3
Sc	370	29	17	27
Scd+Sd	34	3	12	19
Sm+Im	22	2	6	10
Amorphous	13	1	4	6
Other	26	2	4	6
Total	1276		62	

* Include both normal and barred subtypes

II. DISKS AND BULGES

Disks and bulges are the two main structural components of galaxies (de Vaucouleurs 1959). Disk-to-bulge ratio (or integrated color) is therefore a natural classification parameter. The second obvious classification parameter is galaxy luminosity. The Hubble classification sequence E → Sa → Sb → Sc → Ir is a sequence of increasing disk-to-bulge ratio. It is also a sequence of decreasing average stellar age. An old red population predominates in ellipticals, whereas young blue stars dominate the integrated light of irregulars. Among spirals luminous objects are generally "pretty" with long well-developed spiral arms, whereas spirals of lower luminosity have less well-developed spiral structure and a more chaotic appearance. This phenomenon was used by van den Bergh (1960abc) to set up a system of luminosity classification for spiral galaxies. Such classifications make it possible to estimate the luminosity of a spiral to within a factor of ~2.

As was already noted by Holmberg (1958) luminous spirals have a higher surface brightness than do intrinsically faint disk systems. The opposite effect is seen in ellipticals in which the lowest luminosity systems have the highest surface brightness. Since the collapse timescale of a galaxy is $\sim\sqrt{G\rho}$ it follows that dense dwarf ellipticals, such as M32, may have been among the first objects in the Universe to form stars. The observation that the least luminous spheroidal galaxies have the <u>lowest</u> surface brightness distinguishes such objects from ellipticals and is consistent with the notion that spheroidals are, in fact, disk-like rather than bulge-like systems (Kormendy 1985).

It was first noted in van den Bergh (1960b) that spiral galaxies with fuzzy arms, that appear to exhibit a rather anemic rate of star formation, are particularly frequent in the Virgo cluster. At first it was thought that this peculiar morphology might be due to the higher frequency of interactions among spirals in rich clusters. More recently (van den Bergh 1976) it has become clear that the rather anemic rate of star formation in many Virgo spirals is more like to be due to a deficiency of hydrogen gas. Radio observations, the most detailed of which are by Huchtmeier and Richter (1989), strongly confirm this suggestion. These authors observe an enormous hydrogen deficiency for spirals in the inner 3° core of the Virgo cluster, and a lesser deficit of HI extending to 6° from the cluster core. Radio observations (van Gorkom and Kotanyi 1984, van Gorkom, Balkowski, Kotanyi 1984) show that the hydrogen has mainly been stripped from the _outer_ parts of spirals in the core of the Virgo cluster. Chamaraux, Balkowski and Fontanelli (1986) have suggested that the spirals in the center of the Virgo cluster are just now evolving into SO's. The depletion pattern observed in Virgo spirals is consistent with that which would be expected from ram-pressure stripping (Gunn and Gott 1972). The fact that HI is most deficient in the inner regions of many nearby field SO galaxies (van Driel 1987) strongly indicates that neutral hydrogen has been removed from such objects by a different process from that which stripped gas from big-bulge spirals in rich clusters. These results suggest that there might be two physically distinct kinds of SO galaxies.

III. TWO POPULATIONS OF SO GALAXIES?

A Revised Catalog of Shapley-Ames Galaxies (Sandage and Tammann 1981) constitutes the largest homogeneous sample of accurate galaxy classifications that is currently available. In addition to morphological types this catalogue also lists absolute magnitudes ($H = 50$ km s^{-1} Mpc^{-1} assumed) of galaxies that have been corrected for internal absorption. These data are ideally suited for a comparison of the relative frequencies with which galaxies of various absolute magnitude M_B occur in classification types E, SO, Sa and Sb. _If_ SO galaxies are, as suggested by Hubble (1936), intermediate between ellipticals and spirals, then one would expect SO galaxies to have a luminosity function (and hence an absolute magnitude distribution in the _Shapley-Ames Catalog_) that is intermediate between those of elliptical and spiral galaxies. The data which are plotted in Fig. 2, show

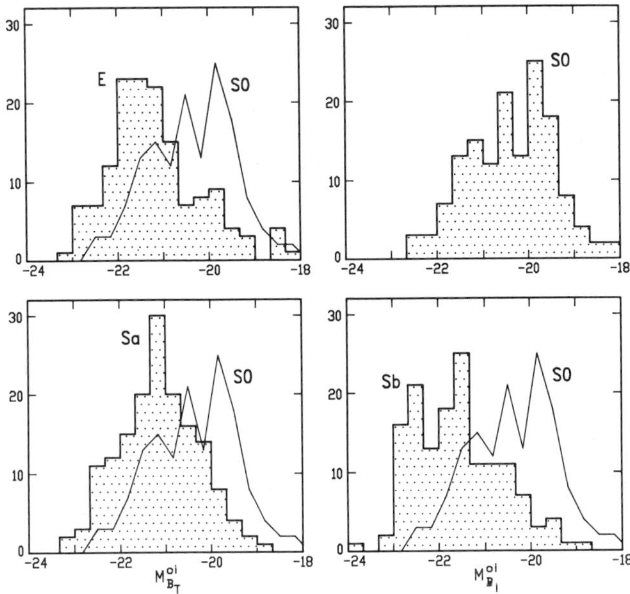

Fig. 2 Histograms of the frequency distribution of absorption-free magnitudes M_B among Shapley-Ames galaxies. The figure shows that the magnitude distribution (and hence the luminosity function) of S0 galaxies differs significantly from that of both elliptical and spiral galaxies.

that this is not the case. The histograms exhibited in Fig. 2 show that E, Sa (Sa + SBa + Sab + SBab), and Sb (Sb + SBb) galaxies have a luminosity distribution that differs significantly from that of S0 (S0 + SB0) galaxies. A Kolmogoroff-Smirnoff test shows that there is $\ll 0.1\%$ probability that E and S0 galaxies were drawn from the same parent population. The same conclusion holds from a comparison between S0 and Sa galaxies. Note, however, that the magnitude distributions of E and Sa galaxies do not differ significantly. Morphological differences between E and S0 galaxies are sometimes quite subtle. It is therefore quite likely that a significant number of E and S0 galaxies have been misclassified. The real difference between the magnitude distributions (and hence between the luminosity functions) of elliptical and S0 galaxies may therefore be even greater than that shown in Fig. 2. This suspicion is strengthened by the observation that 14 out of 25 (56%) of all Shapley-Ames "ellipticals" fainter than $M_B = -20.0$ are flattened objects of types E4-E7, whereas only 36 out of 124 (29%) of the ellipticals brighter than $M_B = -20.0$ are of types E4-E7. This suggests the possibility that the Shapley-Ames sample of E galaxies fainter than $M_B = -20.0$ may be heavily contamined by disk systems. The lower panel of Fig. 3 shows a plot of the distribution of flattening values $\varepsilon = 10(a - b)/a$ for faint SO galaxies with $-20.0 < M_B < -17.0$. The figure shows a surprising absence of SO galaxies with small ε values. The most plausible

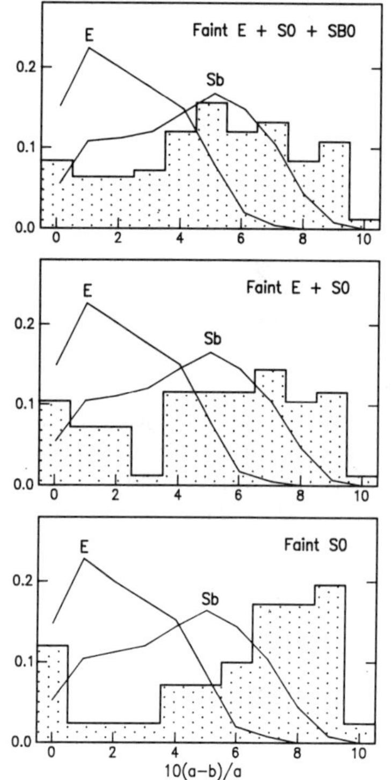

Fig. 3 Frequency distribution of ellipticities of lenticular galaxies with $-20.0 < M_B < -17.0$ compared to that of E and Sb galaxies given by Sandage, Freeman and Stokes (1970). The deficiency of lenticulars with small ε values in the lower panel of the figure can be accounted for by assuming that most nearly pole-on lenticulars have been misclassified as ellipticals by Sandage and Tammann (1981).

explanation for this is that faint SO galaxies, that are viewed almost pole-on, are missing from the sample because they have been misclassified as ellipticals in A Revised Shapley-Ames Catalog of Bright Galaxies.

Modern photometry has only been published for 3 Shapley-Ames ellipticals with $-20.0 < M_B < -17.0$ that have morphological types in the range E0-E4. Capaccioli, Piotto and Rampazzo (1988) conclude that NGC1426 is, in fact, an SO whereas NGC1439 appears to be a true elliptical. Prugniel, Nieto and Simien (1987) find that NGC4478 does not follow a

$r^{0.25}$ law. It may, however, have been tidally truncated by its massive neighbor M87 (NGC4486). CCD photometry for additional Shapley-Ames E galaxies with $-20.0 < M_B < -17.0$, having types E0-E4, has recently been obtained on Mauna Kea and will be discussed in Pierce and van den Bergh (1990).

The upper panels of Fig. 3 show that the observed distribution of ε values for faint S0 + E galaxies is quite similar to that for galaxies of type Sb (Sandage, Freeman and Stokes 1970). This indicates that faint S0's, which will subsequently be referred to as lenticulars, are highly flattened objects. This is consistant with spectroscopic observations by Davies _et al._ (1983) which show that faint "ellipticals" are rotationally supported. Fig. 4 shows that flattened S0 galaxies in the Shapley-Ames Catalog, with axial ratios $a/b \geqslant 7$ are, in the mean, fainter than more rounded

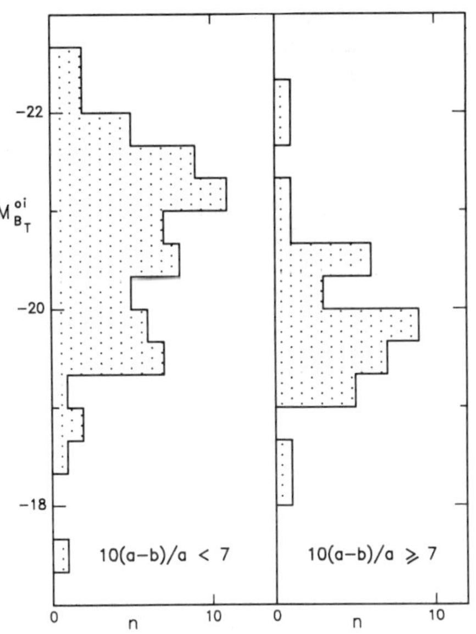

Fig. 4 Comparison of the luminosity distributions of flattened ($a/b \geqslant 7$) and more rounded ($a/b < 7$) S0 galaxies. The figure shows that flat S0's are, in the mean, less luminous than are more rounded ones.

S0's with $a/b < 7$. A Kolmogoroff-Smirnoff test shows that there is only a <0.1% probability that the flattened and more rounded S0 galaxies were drawn from the same parent population. A small part of this difference may be due to the fact that edge-on S0 galaxies will be most affected by internal dust absorption. The major portion of this effect must, however, be due to the fact that flattened S0's are, in the mean, intrinsically much less luminous than rounder ones.

The observed difference between the magnitude distribution of S0 galaxies on the one hand, and of elliptical and spiral galaxies on the other, can be explained in two different ways. SO galaxies might be a distinct class of objects that are not intermediate between ellipticals and spirals. Alternatively (and more probably) the SO classification type encompasses at least two physically distinct types of objects: (1) Bright galaxies that are truly intermediate between ellipticals and spirals, and (2) faint lenticulars. The latter interpretation is supported by the data plotted in Fig. 5 which shows a plot of M_B versus log σ, in which σ is the central velocity dispersion in km s^{-1} (Whitmore, McElroy and Tonry 1985), for Shapley-Ames galaxies (Sandage and Tammann 1981) of types E, SO + SBO and Sa + SBa. The figure shows that SO galaxies brighter than M_B = -20.0 follow the same Faber-Jackson (1976) relation as do ellipticals, whereas lenticulars fainter than M_B = -20.0 appear to behave more like galaxies of type Sa. The observations plotted in Fig. 5 therefore favor the hypothesis that the Shapley-Ames SO galaxies are a composite population, with the objects fainter than M_B = -20.0 being most closly related to galaxies of type Sa. This interpretation is consistent with the data plotted in Fig. 4, which show that highly flattened SO galaxies are, in the mean, less luminous than rounder ones. In Fig. 5 there is no systematic difference between the distributions of SO and SBO galaxies.

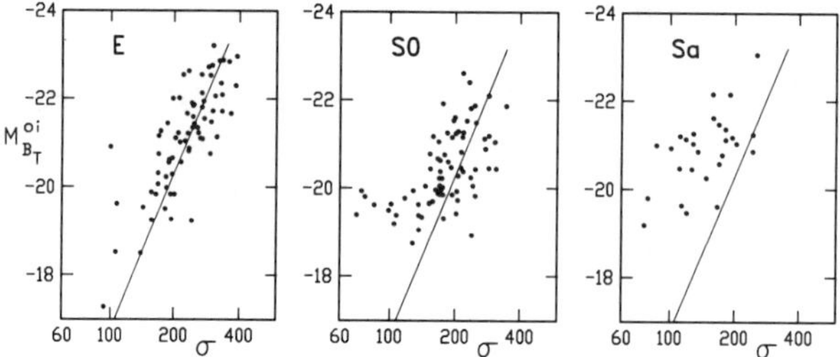

Fig. 5 Faber-Jackson relations for Shapley-Ames galaxies of types E, SO and Sa. SO galaxies fainter than M_B = -20 appear to behave like spirals of type Sa, whereas bright SO's seem to follow the same L-σ relationship as ellipticals.

To guide the eye the Faber-Jackson relation

$$M^{oi}_{B_T} = 7.08 - 11.91 \log \sigma \quad (1)$$

has been plotted in Fig. 5. This line passes through the positions of the faintest (M32) and the mean of the two most luminous (NGC2832, NGC4889) ellipticals in the Shapley-Ames Catalog.

IV. WHAT ARE SO GALAXIES?

Two major problems associated with studies of SO galaxies are: (1) Weak stellar disks exist in most rapidly rotating ellipticals (Carter 1987), which blurs the distinction between E and SO galaxies (Ebneter, Djorgovski and Davis 1988) and (2) the superficial morphological similarities between SO galaxies do not necessarily imply that all of these objects arrived at their present state along similar evolutionary paths. SO galaxies in rich clusters might, for example, have had their gas removed by ram pressure stripping (Gunn and Gott 1972), evaporation of gas (Cowie and Songaila 1977), or the cluster environment might have cut short the infall of disk material (Larson, Tinsley and Caldwell 1980). On the other hand the gas in field SO galaxies could have been depleted by a burst of very rapid star formation or by ejection of gas during violent AGN/quasar-like activity in their nuclei. Presently available data are not yet sufficient to establish whether lenticular galaxies might be generically related to spheroidal galaxies and gas-poor anemic spirals (van den Bergh 1976).

Sandage (1983) has emphasized the well-known fact that most E and SO galaxies have a significantly higher effective surface brightness than do spirals of types Sa, Sab and Sb. This observation implies that SO galaxies have a higher space density of stars than do spirals. Proto-SO galaxies will therefore have cooled and collapsed (Gunn 1983) faster, and used up more of their gas, than typical spirals. Such a scenario accounts for the fact that nearby lenticulars have depleted the gas in their inner regions, but are often embedded in a ring of neutral hydrogen that is largely located outside the main body of the optical galaxy. For luminous SO galaxies in rich clusters ram-pressure stripping still remains by far the most attractive alternative. The pattern of gas depletion in the inner regions of SO galaxies outside great cluster (van Driel 1987) agrees with that expected from fast gas depletion by rapid star formation (or gas ejection by nuclear activity), but <u>not</u> with that expected from ram pressure stripping, which would preferentially remove the less tightly bound gas in the outer regions of spirals.

Arguments that have been used against the stripping hypothesis are; (1) the spheroidal components of SO galaxies are substantially more luminous than the bulges of spirals (Dressler 1980), and (2) the central velocity dispersion in SO's is intermediate between that of E and Sa galaxies (Dressler 1983). The first argument may be countered by assuming that the spirals that originally formed in the dense cluster environment had larger bulges than those that are now seen in field spirals. Perhaps M104 (the Sombrero) is one of the rare survivors of this class of big bulge cluster spirals; most of which were stripped of their interstellar gas long ago. The second argument loses much of its force if SO galaxies are, as suggested by Fig. 5, a mixture of physically distinct types of objects.

Sandage and Tammann (1981) have classified the early-type component of NGC5128 (Cen A) as a galaxy of type SO. Since this object appears to have undergone a recent merger it seems improbable that it attained its present morphology by evolving along a track similar to those of either (1) bright cluster SO's, or (2) faint flattened lenticulars. Galaxies such as NGC3718 and the M51 companion NGC5195 may also owe their present SO/SBO-like morphology to recent interactions. This suggests that there may be more than two physically distinct types of SO galaxies. This conclusion supports Morgan's speculation (Matthews, Morgan and Schmidt 1963) that the SO class contains physically quite distinct types of objects that only have a superficial morphological similarity.

REFERENCES

Caldwell, N., and Schommer, R. 1988, in The Extragalactic Distance Scale, ed. S. van den Bergh and C.J. Pritchet (Provo: Young University Press), p.77.

Capaccioli, M., Piotto, G., Rampazzo, R. 1988, Astr.Ap., **96**, 487.

Carter, D. 1987, Ap.J., **312**, 514.

Chamaraux, P., Balkowski, C., and Fontanelli, P. 1986, Astr.Ap., **165**, 15.

Cowie, L.L., and Songaila, A. 1977, Nature, **266**, 501.

Davies, R.L., Efstathiou, G., Fall, S.M., Illingworth, G., and Schechter, P.L. 1983, Ap.J., **266**, 41.

de Vaucouleurs, G. 1959, Handbuch der Physik, **53**, 311.

Dressler, A. 1980, Ap.J., **236**, 351.

Dressler, A. 1983, in IAU Symposium 100, Internal Kinematics and Dynamics of Galaxies, ed. E. Athanassoula (Dordrecht: Reidel), p.271.

Dressler, A., and Sandage, A. 1983, Ap.J., **265**, 664.

Ebneter, K., Djorgovski, S., and Davis, M. 1988, A.J., **95**, 422.

Faber, S.M., and Jackson, R.E. 1976, Ap.J., **204**, 668.

Fabian, A.C., and Nulsen, P.E.J. 1977, M.N.R.A.S., **180**, 479.

Gunn, J.E. 1983, in IAU Symposium 100, Internal Kinematics and Dynamics of Galaxies, ed. E. Athanassoula (Dordrecht: Reidel), p. 379.

Gunn, J.E., and Gott, J.R. 1972, Ap.J., **176**, 1.

Holmberg, E. 1958, Lund. Medd. Ser. II Nr. 136.

Hubble, E. 1926, Ap.J., **64**, 321.

Hubble, E. 1936, The Realm of the Nebulae, (New Haven: Yale University Press), p.36.
Huchtmeier, W.K., and Richter, O.-G. 1989, Astr.Ap., **210**, 1.
Kormendy, J. 1985, Ap.J., **295**, 73.
Kormendy, J., and Djorgovski, S. 1989, Ann. Rev. Astr. Ap., **27**, in press.
Kotanyi, C.G. et al. 198x,
Larson, R.B., Tinsley, B.M., and Caldwell, C.N. 1980, Ap.J., **237**, 692.
Lundmark, K. 1926, Uppsala Medd., **19B**, No.8.
Malumuth, E.M., and Richstone, D.O. 1984, Ap.J., **276**, 413.
Matthews, T.A., Morgan, W.W., and Schmidt, M. 1964, Ap.J., **140**, 35.
Merritt, D. 1984, Ap.J., **276**, 26.
Ortolani, S., and Gratton, R.G. 1988, Pub.A.S.P., **100**, 1405.
Pierce, M.J., and van den Bergh, S. 1990, in preparation.
Prugniel, P., Nieto, J.-L., and Simien, F. 1987, Astr.Ap., **173**, 49.
Sandage, A. 1961, The Hubble Atlas of Galaxies, (Washington: Carnegie Institution).
Sandage, A. 1983, in IAU Symposium 100, Internal Kinematics and Dynamics of Galaxies, ed. E. Athanassoula (Dordrecht: Reidel), p.367.
Sandage, A., Freeman, K.C., and Stokes, R.N. 1970, Ap.J., **160**, 831.
Sandage, A., and Tammann, G.A. 1981, A Revised Shapley-Ames Catalog of Bright Galaxies, (Washington: Carnegie Institution).
Schombert, J.M. 1988, Ap.J., **328**, 475.
Shapley, H. 1938, Nature, **142**, 715.
Tinsley, B.M. 19..,
van den Bergh, S. 1960a, Ap.J., **131**, 215.
van den Bergh, S. 1960b, Ap.J., **131**, 558.
van den Bergh, S. 1960c, Pub. David Dunlap Obs., **2**, 159.
van den Bergh, S. 1976, Ap.J., **206**, 883.
van den Bergh, S. 1977, in The Evolution of Galaxies and Stellar Populations, ed. B.M. Tinsley and R.B. Larson (New Haven: Yale University Observatory), p.19.
van den Bergh, S. 1979, Ap.J., **230**, L161.
van Driel, W. 1987, Ph.D. dissertation, University of Groningen.
van Gorkom, J.H., Balkowski, C., Kotanyi, C. 1984, in Groups and Clusters of Galaxies, ed. F. Mardirossian, G. Giuricin and M. Mezzetti (Dordrecht: Reidel), p. 61.
van Gorkom, J., and Kotanyi, C. 1984, in The Virgo Cluster, ed. O.-G. Richter and B. Binggeli (Garching: European Southern Observatory), p. 61.
Whitmore, B.C., McElroy, D.B., and Tonry, J.L. 1985, Ap.J.Suppl.Ser., **59**, 1.
Wirth, A., and Gallagher, J.S. 1984, Ap.J., **282**, 85.

ELLIPTICAL GALAXIES

GARTH D. ILLINGWORTH UCO/Lick Observatory/Board of Studies in Astronomy and Astrophysics, University of California, Santa Cruz, 95064.

MARIJN FRANX Center for Astrophysics, Harvard University, 60 Garden Street, Cambridge, MA, 02138.

ABSTRACT Elliptical galaxies, and the bulges of disk galaxies, have a unique role to play in understanding how galaxies form. Their structural, kinematic and population characteristics reflect the conditions and the epoch of their formation. The results from studies in these areas, together with data gleaned from observations of galaxies at high redshift, will provide the basis for elucidating the mechanisms and the timescales at which they formed. Recent observations of the kinematics of elliptical galaxies have shown a variety of outstanding features: a significant fraction of galaxies have kinematically-distinct cores, while some even appear to rotate around their long axes. The kinematic diversity of ellipticals contrasts with a striking uniformity in their photometric properties. This regularity extends to the colors and color gradients in ellipticals. Yet it has been interesting to see that this uniformity in the photometric properties is superficial. A more careful appraisal of the data, particularly of the residuals from ellipse fitting show a wealth of interesting, if not understood, details that appear to relate to apparently disconnected global properties. Ellipticals and bulges remain a challenge for the development of a consistent picture of the formation of galaxies.

SYNOPSIS

How galaxies formed is one of the outstanding issues of contemporary astrophysics – a "Holy Grail" of astronomy. This goal has, and will continue to play a major part in observational and theoretical programs in astronomy. More definitively, we can ask:

- when did the major part of the baryonic matter agglomerate, i.e., what was t_f?
- what were the physical processes at work?
- how long did the process take before the resulting object was identifiably a galaxy in current terms, i.e., what was Δt_f?

The answers to these questions will, of course, depend upon the type of galaxy, and even can be directed to components such as the disk or the bulge, for example. The obvious approach to answering these questions is to observe the properties of objects at high redshift, and note the evolution of such properties with redshift. To some degree this has been a very successful approach. The unexpected activity and evolution of galaxies found at modest redshifts ($z <$

0.4-0.6) by Butcher and Oemler (1978), and followed up very comprehensively by Dressler and Gunn and others (Gunn 1989 – and references therein) brings one aspect of the formation/evolution process within our observational grasp.

Recently there have been dramatic discoveries at higher redshifts ($z > 1$), but it has proved very difficult to relate these objects to "normal" galaxies – they appear to be rather esoteric objects. There are clearly strong selection effects at work, and so one is faced with drawing conclusions from objects that are at the extremes of their classes. These observations are exciting, and will probably play a valuable role in defining t_f and possibly Δt_f. However, they are unlikely to give us the level of detail required to answer the questions posed earlier. To really understand galaxy formation we are going to need to put together all the data that we can get. Nearby galaxies will play a crucial role in this process. Observations of galaxies, combined with models and theory, have already, and will continue to place valuable constraints on the initial conditions under which galaxies formed, and on the processes which played a role in their evolution.

One important element of this process is the increasing sophistication and accuracy of the observations. The use of CCDs with their potential for routine photometric accuracies of <1% has resulted in systematic studies of large samples of ellipticals. Multicolor two-dimensional photometric maps are now available for more than 70 ellipticals (see, e.g., Franx, Illingworth and Heckman 1989a, and Peletier *et al.* 1990). Unfortunately, the small size of current CCDs and the resulting difficulty of deriving the background sky level on frames of nearby galaxies has resulted in systematic errors in many cases. This will continue to be a problem until large-format CCDs become routinely available.

Some years ago at the Princeton meeting on *Structure and Dynamics of Elliptical Galaxies*, Martin Schwarzschild remarked that one needed kinematical results accurate to $\sim$5 km s^{-1} for "serious" comparison with the models. Observational data are now approaching that goal. Much of the kinematical data now being obtained gives velocities and velocity dispersions accurate to 10 km s^{-1}, and in many cases to 5 km s^{-1}. Furthermore, large samples of data are now being obtained. Two-axis kinematical profiles are already available for some 40 galaxies, and will probably exceed 70 galaxies within the year with the completion of a magnitude-limited survey (Franx and Illingworth 1990a). These new data are becoming available at a time when we are seeing increasingly more sophisticated and realistic models. The N-body simulations, particularly those that include "gas", and the studies of triaxial systems (see, e.g., Carlberg 1984; de Zeeuw 1987; Barnes 1990; Franx, Illingworth and de Zeeuw 1990; Hernquist 1990) will, together with these kinematic and photometric data, enable us to get a better handle on the initial conditions in elliptical formation.

Ellipticals remain of great value for formation issues. Their apparent simplicity (a single, old, dominant population), along with the lack of extensive dust and star formation, makes them worthwhile candidates for such discussions, even though disks and disk formation remains one of the outstanding questions of galaxy formation. Furthermore, everytime it seems that there may be life beyond ellipticals, we seem to discover some new, fascinating and significant feature of ellipticals that emphasizes their value for delving into the questions surrounding galaxy formation.

We will highlight the major results of the last decade, and focus on several results of the last few years.

OBSERVATIONAL STATUS

The large body of data that has been obtained on elliptical galaxies over the last decade enables us to summarize the generic properties of ellipticals. While some of these characteristics were identified from earlier photographic work, it is valuable to see these properties quantified by the more accurate CCD data that is now available. A reasonable summary would be:

- the remarkable uniformity of the photometric properties of ellipticals – the light distributions, color gradients and line strengths, and shapes (they are elliptical with only very small deviations) bear a striking similarity from elliptical to elliptical;
- however, subtle and potentially valuable distinctions do occur in these properties – notably, shells, distortions, residual smooth 3rd, 4th, and higher-order structures, deviations from the "proto-typical" $r^{1/4}$-law profile, the wide variety of ellipticity and position angle profiles, UV "bumps" in the SED (spectral energy distribution);
- the contrast of the very diverse kinematical properties to the uniformity of the photometric properties is striking – while ellipticals may be characterized as "slow-rotators", the dispersion in their rotational properties is large, ranging from rotationally-flattened to essentially zero rotation ($< 0.02\sigma$);
- the luminosity dependence of rotation as discussed by Davies *et al.* (1983) – this feature remains an enigmatic property of elliptical galaxies;
- the striking uniformity of the global properties of elliptical galaxies, as evidenced by the "fundamental plane", i.e., by the tight distribution in the (R, σ, μ)-space – this high degree of regularity over a wide range of luminosity has been discussed extensively by Djorgovski and Davis (1987), and by Dressler *et al.* (1987);
- the derivative property, but one worth highlighting in its own right is the quite limited range of M/L in the central parts of ellipticals.

In many cases, new results help to clarify one's understanding. This has not typically been the case with ellipticals. The results listed above have not led to a clear and comprehensive picture of how ellipticals formed. However, the rather more extensive, statistically-better-defined samples of kinematic and photometric data could help us reach clarity on this question.

The last few years have been a very productive period for observational results on elliptical galaxies. We would like to highlight the observational results in three areas:

- photometric properties – deviations from ellipses;
- kinematics and kinematically-distinct nuclear components;
- color gradients.

PHOTOMETRIC PROPERTIES

CCDs have led to a resurgence of activity in galaxy surface photometry. Most studies fit ellipses to the data and derive profiles of surface brightness, ellipticity, position angle, and color as a function of radius. The residual deviations from ellipses are also usually given when significant. Kormendy and Djorgovski (1989) summarize the most recent activity in this area. The interpretation of the results of these surface photometry programs are haunted by the problem of projection. The degeneracy introduced by projection makes it very difficult to derive definitive results from the data. The isophote twists as a function of radius have long been assumed to represent the effect of projection of axial ratio changes in a triaxial system. However, they could well represent real twists in a biaxial system. For example, Gerhard (1983) produced such an apparently long-lived (on a precession time scale or longer) system in a merger simulation. Other data, typically kinematic, are needed to remove this degeneracy.

A further problem with the value of photometric data results from the effects of seeing. Unfortunately, seeing effects are significant out to 5 times the FWHM of the seeing function (Franx, Illingworth and Heckman 1989a, hereafter FIH1). Thus the gradients in ellipticity and the position angle changes near the centers of galaxies are strongly influenced by the seeing. This is unfortunate, because accurate measurements of the isophotal shapes near the centers could provide useful information. For example, supermasssive black holes can produce a significant rounding of the isophotes near the center (Gerhard 1987).

Probably the most interesting result in the arena of surface photometry over the last few years has been the realization that the deviations from ellipses appear to be potentially valuable diagnostics of other properties. These isophote deviations from ellipses are small (on the order of 0.5% typically), with some 40-50% showing deviations on the order on 1% or higher (see, e.g., Lauer 1985; Jedrzejewski 1987; Bender, Döbereiner, and Möllenhof 1988; FIH1; Peletier *et al.* 1990). Characteristic examples are given by Bender, Döbereiner, and Möllenhof and by Peletier *et al.*

These deviations are typically characterized by the residual higher-order harmonical terms in intensity along the best-fitting ellipse, and scaled to obtain an estimate of the shape of the isophote. Hence we obtain the C3, S3, C4, S4, ... terms, which are the $\cos 3\theta$, $\sin 3\theta$, $\cos 4\theta$, $\sin 4\theta$, ... deviations of the isophotes from ellipses (Jedrzejewski 1987; FIH1; Peletier *et al.* 1990). "Box-like" and "disk-like" residuals are of particular interest, and are described by the C4 terms along the best-fitting ellipse. If this term is positive, then the isophotes are disky, if it is negative, the isophotes are boxy. However, many systems show more complicated patterns which cannot be characterized by a single number. For example, NGC 1549 and NGC 636 show both disk-like and box-like residual terms at different radii (Jedrzejewski 1987; FIH1). The case of NGC 1549 is illustrated in Figure 1. The galaxy has a negative C4 term ($\cos 4\theta$) in the inner part, and a positive C4 term in the outer parts. Note that the S4 term cannot be neglected, and is actually *higher* than the C4 term for a large radial interval. It thus becomes difficult to give a simple characterization of the form of the residual distortion.

It was realized that part of the reason for this difficulty could be due to the use of the major axis as the reference for the higher-order terms. The angle used for the harmonical expansion of the residual intensity is usually taken with respect to the major axis of the galaxy, and the major axis position angle usually changes with radius. If the projection of any feature that contributes to the residual is different from that of the galaxy as a whole, the use of the varying major axis may complicate the structure of the residual. This proved to be so in some cases.

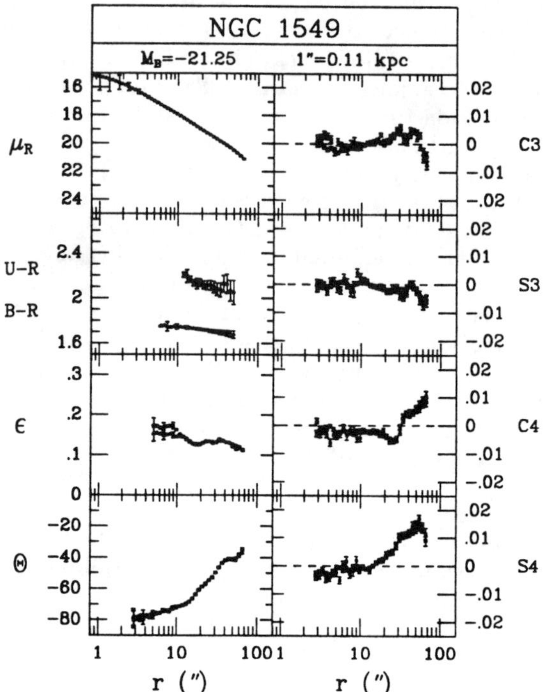

Fig. 1. Surface photometry results for NGC 1549 taken from FIH1. The radial profiles of surface brightness, color, ellipticity, position angle and the higher-order cos and sin fits to the residuals are all shown.

To see the effect of this we can take the angle with respect to a fixed position angle on the sky, say North, and then calculate the residual terms. The result for NGC 1549 is shown in Figure 2. Note that the residual terms do not change sign in this representation. Hence, the phase of the fourth order residuals is more or less constant as a function of radius. Since the position angle of the galaxy does change as a function of radius, the phase of the residuals is not constant with respect to the major axis of the galaxy.

This leads one to a preferable way of representing these terms, and that is to describe the residuals in terms of higher-order amplitudes and phases. The phases can be given, of course, with respect to both the major axis and to a fixed position angle. The same data for NGC 1549 are shown in Figure 3, but in the phase-amplitude representation, and with respect to a fixed angle (labeled NORTH) and with respect to the major axis – the constancy of the structure is striking.

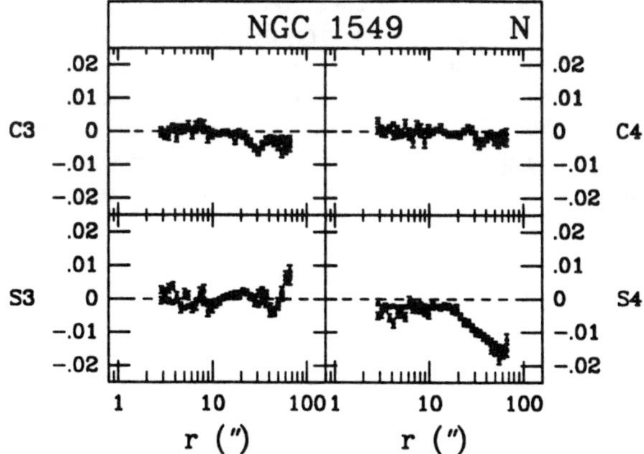

Fig. 2. The C3,4 and S3,4 terms from the surface photometry data for NGC 1549 from Figure 1, plotted with reference to a fixed position angle (North), instead of being referenced to the major axis PA as in Figure 1. Note that the phase is constant with respect to North.

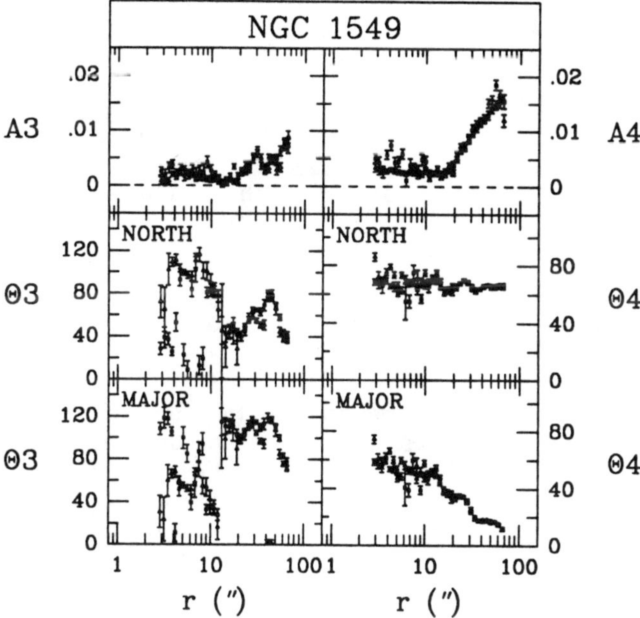

Fig. 3. The same data for NGC 1549, but shown as 3rd and 4th order amplitudes with the corresponding phases referenced to both a fixed position angle (NORTH), and to the major axis (MAJOR). The major axis position angle variation is shown in Figure 1. The value of decoupling the phases and amplitudes is clear.

A further example of the value of these terms for detecting structure was shown by FIH1 for NGC 1700. The phase-amplitude residuals for NGC 1700 in Figure 4 show a narrow feature that appears to be a somewhat asymmetric ring-like structure at ~60-70″ that is skewed 40° to the major axis. The amplitude of this feature is sufficient (5%) for it to show as a clear distortion in an image of the galaxy (FIH1), but the quantitative information along with the qualitative structure allows for a much better description of the nature of the structure.

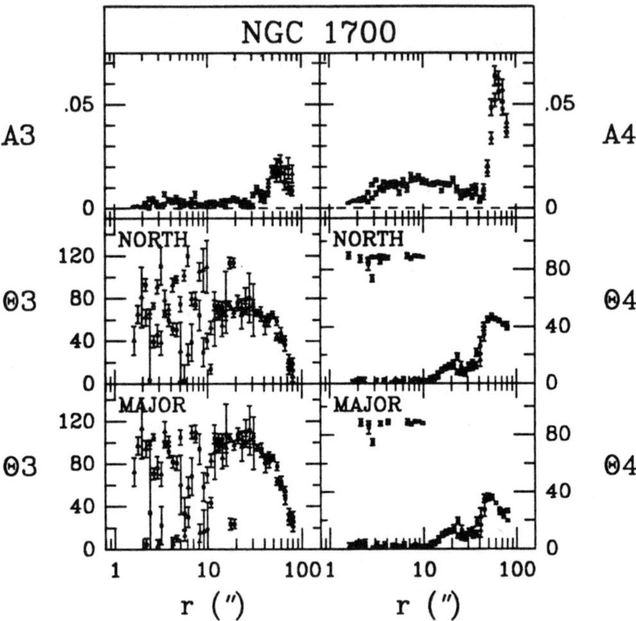

Fig. 4. As for NGC 1549 in Fig. 3, but for NGC 1700. The high-amplitude feature at 60-70″ is obvious.

While these examples indicate the utility of this approach for recognizing unusual structures in galaxies, there are more general questions raised by the existence of these terms. In particular, what is their physical significance. One question which arises is why these terms are so small; i.e., why are the isophotes of ellipticals so well described by ellipses. There is no law of stellar dynamics which states that the isophotes of galaxies should be ellipses. Given this situation, what do these terms indicate to us. It is clear that ellipticals are not the clean, simple objects indicated by typical optical images. Disk-like structures appear to be more pervasive than we thought, as is dust and gas, as the IRAS IR and Einstein X-ray surveys have shown. At first sight one might view these terms as indicative of the ubiquitous presence of disks, skewed and/or projected such as not to be aligned with the observed major axis. This might indicate, very reasonably, that the relationship of S0s to ellipticals is much more continuous than the usual classifications would indicate.

As noted by FIH1 and Peletier *et al.*, even a weak disk with its rapid rotation can disproportionally affect the measured velocity and so conclusions drawn from correlations of properties involving kinematics need to be made with care. One also needs to be cautious about the selection effects that may have arisen by excluding ellipticals with known weak disks from samples for kinematic and photometric studies by their designation as S0s. This may be quite reasonable in a qualitative morphological approach, but is risky for quantitative studies.

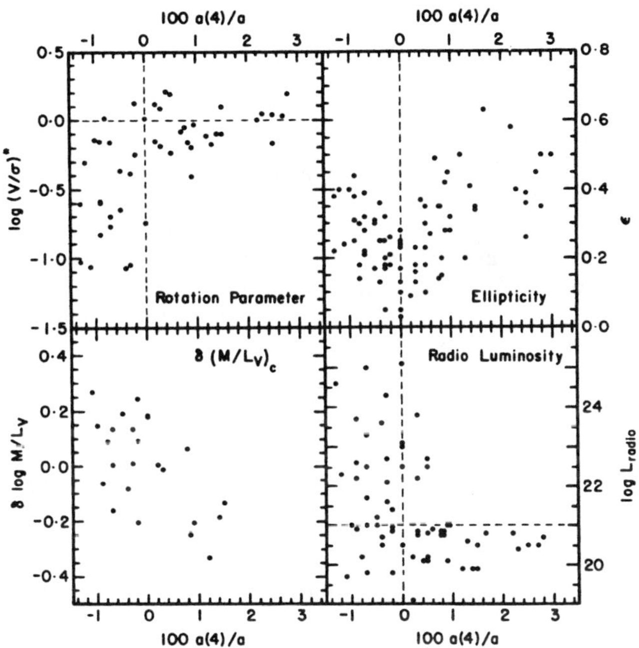

Fig. 5. Correlations of the a(4)/a (≡ C4) parameter with ellipticity, radio power, M/L, and rotational properties for the sample of data from Bender *et al.* (1989), as assembled by Kormendy and Djorgovski (1989).

However, it is these correlations of photometric, kinematic, radio, X-ray and other properties that suggest that more may be at work than just the simple inclusion of disks into ellipticals. Bender *et al.* (1989) discuss correlations between the mean residual C4 ($\cos 4\theta$) terms and a variety of parameters. Kormendy and Djorgovski (1989) also discuss these correlations in their review. Figure 5, taken from Kormendy and Djorgovski, shows some of the effects that were discussed by Bender *et al.* Systems with positive C4 terms (disk-like) rotate faster, and have lower M/L ratios than systems with negative C4 terms (box-like). Also, systems with positive C4 terms are much less likely to be radio sources (and X-ray sources – see Bender *et al.* 1989). A possible explanation is that the systems with positive C4 are interlopers among ellipticals, and are more like S0 galaxies. Another possibility is that the presence of a cold disk influences the kinematic results, as noted above, although this cannot affect the radio or X-ray luminosity. Thirdly, if the positive C4 terms are caused by the presence of a weak disk, then galaxies with high C4 terms will be those viewed at special orientations. This complicates any comparisons with the galaxies that have small or negative C4 terms.

One concern is that the use of C4 alone is not sufficient; as shown above for NGC 1549, and also for other galaxies (FIH1), the S4 term is larger than the C4 term. The use of phases and amplitudes may help. An additional step would be to derive the terms of higher order than 4th as a means of discriminating between a truly "box-like" structure or more asymmetric structures. This may allow one to identify the structures with the remnants of mergers/interactions, a suggestion that has been made for the galaxies with "box-like" isophotes – particularly for those that are radio galaxies.

A larger concern is that rotation, M/L ratio, and radio luminosity are known to correlate with the absolute magnitude of a galaxy. While some of the correlations noted by Bender *et al.* (1989) are striking, it would be valuable to elucidate the degree to which absolute magnitude is a driving factor. Larger samples may be needed to disentangle the effects of changes in absolute magnitude and of the C4 terms. It will be interesting to see whether use of distance-independent parameters like color or velocity dispersion can help to resolve some of these problems.

KINEMATICS

Again, as for the photometric results of the last few years, the availability of more sensitive and stable spectrographs and detector systems has resulted in a tremendous improvement in the quality of kinematic data and in the rate at which it could be obtained. This has allowed larger and more systematic programs to be carried out, with the ability to detect rarer characteristics, and potentially to derive results that depend on statistically significant samples. Two noteworthy kinematical results were derived.

a) Minor Axis Rotation

The first was the recognition that a significant fraction of elliptical galaxies show rotation along their minor axes. Such rotation is expected naturally if ellipticals have triaxial shapes (see, e.g., Contopoulos 1956; Kondrat'ev and Ozernoi 1979; Binney 1985). One of the most striking examples of minor axis rotation was found in a very normal-looking elliptical NGC 4261 by Davies and Birkinshaw (1986). NGC 4406 (Illingworth and Franx 1989) is another striking example, and is discussed and the data shown in *b)* below. NGC 4261 has very strong rotation along its minor axis, and almost no rotation along its major axis. Such a galaxy may be triaxial and rotating about its intrinsic short axis, or it may be triaxial-prolate, and rotating about its long axis, or it may even be rotating around an axis which is not a symmetry axis. Wagner, Bender and Möllenhof (1988), Franx, Illingworth and Heckman (1989b; hereafter FIH2), and Jedrzejewski and Schechter (1989), have found many more systems with minor axis rotation, with a wide range in v_{minor}/v_{major}. While these observations demonstrate that ellipticals are not simple oblate systems, since such systems can never show minor axis rotation, their real value lies in what they will be able to tell us about ellipticals.

Even with the limited data currently available (the 22 galaxies observed by FIH2), the kinematic misalignement proves to be a valuable tool. This should be strengthened by the results from a magnitude-limited survey of some 50+ ellipticals that is nearing completion (Franx and Illingworth 1990a). The kinematic misalignment is defined to be the angle between the minor axis of a galaxy and its rotation axis. A histogram of the kinematic misalignment is shown in Figure 6 for the 22 galaxies observed by FIH2. We see that the histogram of misalignment angles peaks near zero, declines rather rapidly at larger values, and might have a secondary peak at large misalignments. The existence of non-zero values is strong evidence for generically triaxial figures for ellipticals.

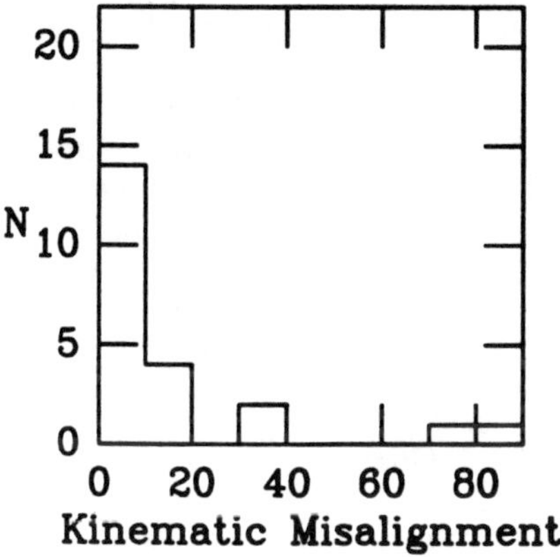

Fig. 6. Histogram of kinematic misalignment angles (see text) taken from FIH2.

There are some interesting ramifications of this result. If ellipticals are triaxial, then the streaming within these galaxies may be around the long axis and/or around the short axis of the galaxy, ignoring figure rotation (FIH2). Thus the net angular momentum of an elliptical can lie anywhere in the plane of the long and short axis of the galaxy. We might not expect to see such a strong correlation between the rotation axis and the short axis. The strong correlation is either due to stellar dynamical effects which are not reflected in the current models, or due to constraints imposed during the formation process. It is possible that the figure rotation is much more important than we think, and/or that the shapes of ellipticals are very near to oblate.

Projection effects complicate the interpretation of the kinematic misalignments for triaxial systems. Even if the angular momenta of all galaxies are aligned with the short axis of the figure, minor axis rotation will still be observed for some galaxies for a certain range of viewing angles. The existence of a few galaxies with very large kinematic misalignments poses an interesting question. Are these just the result of projection, albeit a low probability event, of essentially short axis rotators, or are they evidence for a small fraction of ellipticals being long-axis rotators? Some additional data (to be discussed below in b)) suggests that at least one of these galaxies (NGC 4406) may well be a long-axis rotator.

The high fraction of small kinematic misalignments implies that the direction of the intrinsic angular momentum is strongly coupled to the intrinsic shape of the galaxy, or more specifically, to the direction of the intrinsic short axis. Although such coupling is not required for triaxial galaxies, we might hypothesize that the angular momenta of all ellipticals are exactly aligned with the intrinsic short axes. The observed minor axis rotation is then simply due to the projection effects of the triaxial systems. Binney (1985) has shown that under this hypothesis the observed minor axis rotation may be used to constrain the intrinsic shapes of elliptical galaxies.

Probability distributions for the kinematic misalignment can be derived using the observed ellipticity distribution and assumptions about the triaxiality. These distributions can be compared with the data. A preliminary analysis of recent data showed that a scenario in which all ellipticals are triaxial, including those without minor-axis rotation, is consistent with the data (Franx, Illingworth and de Zeeuw 1990). These models show large minor axis rotations only for special viewing angles. The larger set of data from Franx and Illingworth (1990a) should give stronger constraints on the intrinsic shapes. It is worthwhile keeping in mind a potentially serious limitation of this approach. The assumption that the angular momenta of ellipticals lie along the intrinsic short axis may not be valid, though if it is the norm except for a small percentage of long-axis rotators, the approach may still be viable.

b) Kinematically-distinct Cores

The second noteworthy result of recent times has been the discovery of kinematically-distinct cores (Franx and Illingworth 1988, hereafter FI; Jedrzejewski and Schechter 1988; Bender 1988). Some 10 ellipticals have been found so far with cores that have angular momenta anti-parallel, or orthogonal to the angular momentum of the main galaxy. In addition, in some systems the core rotates in the same sense as the outer parts, but with a distinct peak in the rotation in the core. The fraction of ellipticals with kinematically distinct cores of any flavor is $\approx 25\%$ of the sample.

One of the most extreme examples of counter-rotation is IC 1459 (FI). The rotation curve is shown in Figure 7, both for the minor and for the major axis. The rotational velocity peaks near the core at a value of 170 km s^{-1}, decreases, reverses sign at about $10''$ from the center, and flattens out at 45 km s^{-1} at larger radii. Almost no rotation is observed along the minor axis, neither in the center nor in the outer parts. The core angular momentum projects essentially exactly anti-parallel to the angular momentum of the outer parts of the galaxy.

Interestingly enough, it appears that the bulk of the stars in the core *are not rotating rapidly.* Analysis of the line profiles (the velocity distribution function) in the core (at $\approx 2''$) suggests that we are dealing with a two-component system, one of which is rapidly-rotating and of low velocity dispersion ("disk-like"?), while the other is slowly rotating and has a large velocity dispersion characteristic of the galaxy as a whole. This latter component dominates, containing some 75% of the light (see FI, and Illingworth and Franx 1989).

The line profiles (i.e., the observed distribution of stars as a function of radial velocity) near the core are shown in Figure 8. The line profiles 2″ each side of the nucleus along the major axis are asymmetric and are mirror images. The dotted model fit based on the above assumption of a simple two-component system gives a dominant hot component with $v \approx 0$ km s^{-1} and $\sigma \approx 350$ km s^{-1}, and a cold component with $v \approx 300$ km s^{-1} and $\sigma \approx$ 100-150 km s^{-1}. The mean velocity at these radii is only ≈ 60 km s^{-1}, but the well-known biases in the usual Fourier and Cross-correlation methods towards the velocity of the narrower component result in the velocity of the composite system being nearer to that of the narrow peak (170 km s^{-1} in Figure 7).

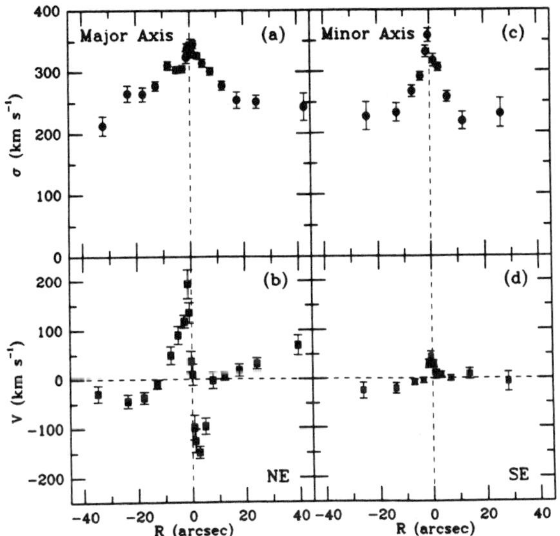

Fig. 7. The rotation and dispersion profiles for IC 1459, for both the major and the minor axis. Note the counter-rotation of the core with respect to the outer parts of the major axis, and the absence of rotation on the minor axis.

It is remarkable that IC 1459 is normal in most other aspects. The surface brightness profile does not show any obvious feature near the center (see FI for these other data). The color profile looks normal, as does the profile of "Mgb" line strength. The galaxy is an IRAS source, some dust has been found in optical images, and it has a compact radio source. The feature most indicative of an interaction is the ionized gas in the center, but interestingly, this rotates in the same sense as the outer part of the galaxy (see FI).

Bender (1988), Jedrzejewski and Schechter (1988), and Franx, Illingworth and Heckman (1989b) have found several other systems with kinematically-distinct cores. Most of these systems also look reasonably normal in other respects. NGC 5322 has a counter-rotating core similar to IC 1459. The surface photometry shows disky isophotes near the center, and boxy isophotes in the outer parts (Bender, Döbereiner, and Möllenhof 1988). NGC 4406, a large galaxy in Virgo, has a core with its angular momentum projected along the apparent short axis. The angular momentum of the outer parts projects along the apparent long axis of the galaxy (like NGC 4261). Thus this galaxy is outstanding in two distinct ways; it has minor-axis rotation in the outer parts and a misaligned core. Yet this galaxy appears to be otherwise quite normal. NGC 4494 has a rapidly-rotating core which rotates in the same sense as the outer parts. The rotation curve along the long axis shows a peak at a radius smaller than 2″, has a minimum around 7″, and rises slowly at larger radii. No rotation is found along the minor axis.

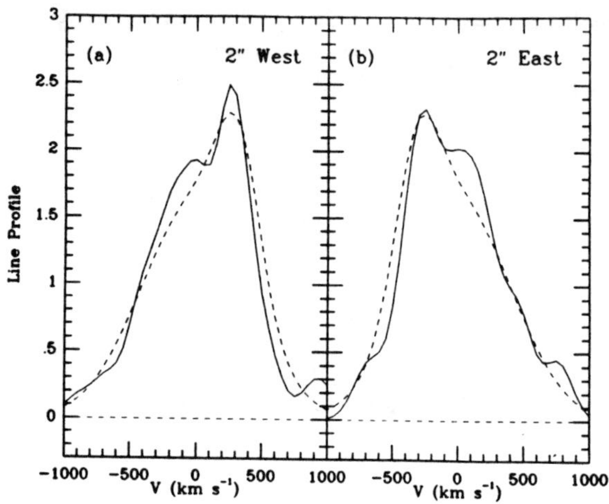

Fig. 8. The line profiles for the absorption lines in IC 1459 2″ either side of the nucleus. The full lines are the observed line profiles, while the dashed lines are the two-component model fits to the data. One component turns out to be cold and rapidly-rotating, while the other is hot and is not rotating. The line profiles are noisy on scales of a few hundred km s^{-1}.

These systems are characteristic of the kinematically-distinct cores found to date. A compilation is given by Illingworth and Franx (1989). The fraction of galaxies with obvious such cores is 20%, while an additional 10% may have weaker effects. The typical scale of the peak of the velocity in the core has not been determined, primarily because it is usually at radii of 1-2″ where the effect of seeing is large. The velocity maxima could be at, or extend to, very small radii. Such cold "disk-like" components could well be ubiquitous in elliptical nuclei, remaining undetected because their contrast against a dominant hot component is low due to inadequate spatial resolution, velocity resolution and S/N in the data.

Such reversals could well occur on larger radius scales as well. While the surveys typically map the region from $\approx 0.1 r_e$ to $\approx 1 r_e$, little is known about the form of the rotation at larger radii. Illingworth and Franx (1989) noted that NGC 4278, from data by Davies and Birkinshaw

(1988), appeared to have a velocity reversal at 20″ or larger. This is shown in Figure 9. It will be of interest to confirm the reversal in NGC 4278, and to see if subsequent surveys unearth similar cases.

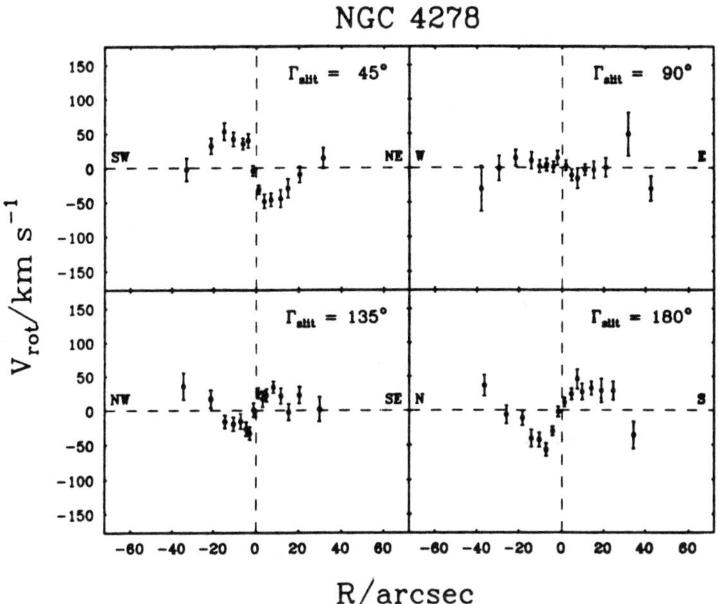

Fig. 9. The rotational velocities at four different position angles for the elliptical galaxy NGC 4278 (from Davies and Birkinshaw 1988). Note the apparent reversal in the sense of rotation at large radii, i.e., at 20-30″.

NGC 4406 is a particularly interesting case for reasons beyond its core characteristics. As noted above, the outer parts of this galaxy have an angular momentum vector which projects along the major axis of the galaxy, while that in the inner regions is orthogonal and projects along the short axis. The data are shown in Figure 10. If the alignment of the rotation axis with the major axis in the outer parts is a projection effect, then the line-of-sight lies in the plane of the short and the long axis. The core of this galaxy has an angular momentum which projects 90° perpendicular to that of the outer parts. The implication is that the rotation axis of the core is the intermediate axis. In a triaxial potential, the closed orbits around the intermediate axis are generally unstable, and hence no tube-orbits exist which can produce streaming around the intermediate axis (Heiligman and Schwarzschild 1979). Therefore, the hypothesis that alignment of the rotation axis with the long-axis is a projection effect is incorrect. The conclusion is that the outer parts of this galaxy really rotates around the intrinsic long axis! The only uncertain assumption is that the rapidly-rotating core component is not self-gravitating. Inclusion of self-gravity might stabilize a disk rotating around the intermediate length axis; although it would still be very difficult to form it in that configuration.

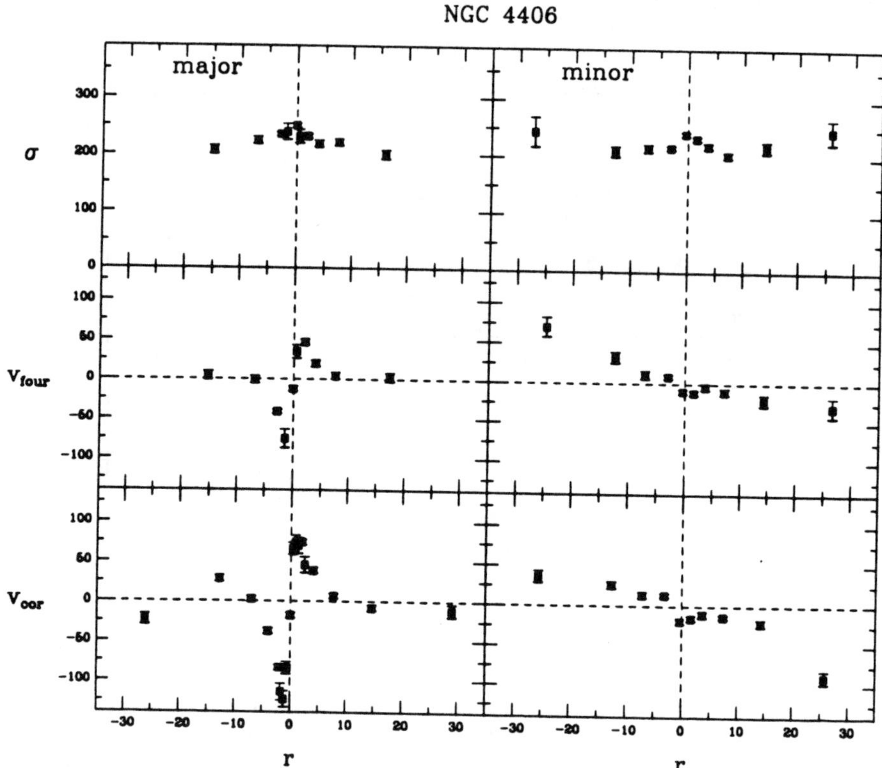

Fig. 10. The rotation and dispersion profiles for NGC 4406, for both the major and the minor axes. Note the rapidly-rotating core along the major axis, and the absence of rotation on the major axis at larger radii. Contrast this with the rotation along the minor axis. The velocities were determined using two different line widths, labelled v_{four} for the broad-lined fit and v_{cor} for the narrow-lined fit.

Three (maybe 4) ellipticals, some 10% of the sample, have $v_{min} >> v_{maj}$, which seems large given the strong bias towards zero minor axis rotation in the sample to date. How many of these really are long axis rotators? Cases like NGC 4406, interesting as they are by themselves, will greatly complicate the use of the minor-axis rotation data for determining shapes. That at least one long-axis rotator appears to exist suggests an interesting quirk in the formation process for ellipticals.

How did the kinematically-distinct cores form? Kormendy (1984) suggested that the accretion by a large, low density galaxy of a small galaxy with its high density core might leave the small galaxy intact while it sinks to the core of the large galaxy. Balcells and Quinn (1990) have verified such a scenario with N-body simulations. The end result would be a core consisting mainly of the stars from the small galaxy immersed in the large system. The small galaxy would dominate the light from the core and the kinematics of the inner parts can be very different from the kinematics of the outer parts, although the merger process by itself appears to install some order. In such a scenario, it is expected that the properties of the core of the final product reflects the properties of the original weeny galaxy, i.e., the color, velocity dispersion, and metal line strengths are those of a low-luminosity galaxy. This is not the case with the kinematically-distinct cores noted above, and suggests that this is not the mechanism by which such cores formed, although accretion or mergers are almost certainly involved.

The asymmetrical line profiles of IC 1459 suggest another formation mechanism (FI). If the two-component structure of the core is real, and not an artifact from fitting Gaussians, then the "intruding" remnant is rapidly-rotating with a v/σ of larger than 2, very atypical for either an elliptical or the bulge of a spiral. The rotational velocity of the rapidly-rotating component is near to the circular velocity of the host galaxy. This suggests that we may be seeing a disk-like system, which formed from accreted gas which was counter-rotating with respect to the stars. Alternatively, we could be seeing the stellar remnants of a small stellar system that was accreted and disrupted near the core of the large galaxy, where the tidal forces are highest. The high rotational velocity reflects the orbital angular momentum of the small galaxy, and not its internal velocities. Even for galaxies on low angular momentum orbits, the remnants could have high circular velocities so deep in the potential well. The small fractional contribution of the counter-rotating component (20%) to the light may explain why the colors and line strengths do not decrease in the core. It seems to us, however, that the former gaseous disk plus star formation picture is probably the more likely scenario, but detailed models are needed to clarify the processes. The mass of the counter-rotating component in IC 1459 was estimated by FI to be 10^{10} $M_\odot$. This is comparable to the mass of the gas involved in star burst events in the extreme IRAS galaxies (e.g., Scoville *et al.* 1986). Thus the kinematically-distinct structures could be the signatures of past star bursts.

It is not clear at what epoch these structures formed. They are most likely in dynamical equilibrium, because the dynamical time is so short near the cores (An exception may be NGC 7626, which has a very irregular velocity field – Jedrzejewski and Schechter 1989). Thus they may have formed at the time of galaxy formation, when galaxies were being assembled from smaller clumps. Further studies of these systems are clearly needed to resolve these issues.

COLOR GRADIENTS

The correlation of color with luminosity is a well-established relation for elliptical galaxies, with the more luminous galaxies being redder, and more metal rich (see, e.g., Sandage and Visvanathan 1978). While aperture photometry measures have shown that the colors of galaxies become more blue with radius (e.g., de Vaucouleurs and de Vaucouleurs 1972; Sandage and Visvanathan 1978; Persson, Frogel and Aaronson 1979), systematic mapping of such gradients has only become possible in recent years with the availability of CCDs — and even these are quite limited currently because of their small format (FIH1; Peletier *et al.* 1990).

The interpretation of the color changes as metallicity changes has become customary. It is based upon the spectroscopic line strength measurements in a variety of studies (e.g., Faber 1977; Burstein *et al.* 1984; Efstathiou and Gorgas 1985; Peletier 1989). They show that more luminous galaxies have higher line strengths, indicating higher metallicity, consistent with the redder colors of such galaxies. The data are not as clear for line strength gradients (e.g., Gorgas and Efstathiou 1987; Davies and Sadler 1987; Peletier 1989; Baum 1990), but appear to be consistent with the hypothesis that the changes within ellipticals are also due to metallicity changes in the sense that the outer parts have lower metallicity. The remainder of this discussion will assume that the color gradients reflect, in the mean, metallicity changes.

The consistency and similarity of the color gradients within ellipticals are quite striking. The $B-R$ and $U-R$ color gradients from FIH1 are shown in Figure 11. Note that these are absolutely calibrated. The mean gradients are $\Delta(B-R)/\Delta \log r = -0.07$, and $\Delta(U-R)/\Delta \log r = -0.23$. These results are consistent with the older aperture photometry results. A metallicity gradient of $\Delta[Fe/H]/\Delta \log r \approx -0.3$ produces the observed color gradients (FIH1). It is well to keep in mind that age differences could also play a role in the color gradients. More line-strength gradient data and analyses will be needed to clarify the role of age-vs-metallicity effects.

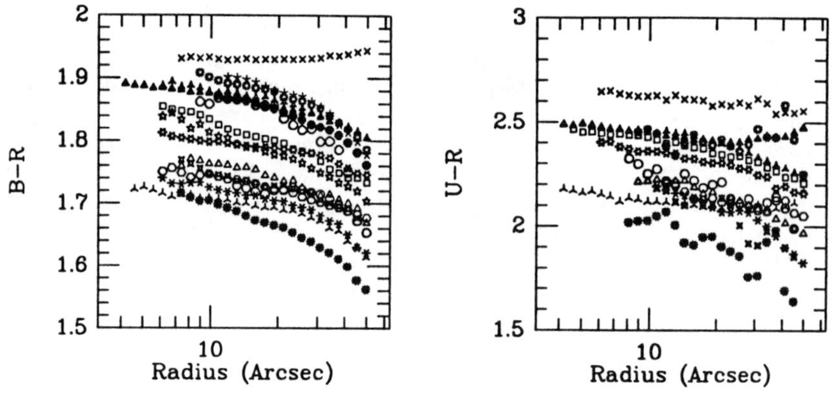

Fig. 11. The $B-R$ and $U-R$ color gradients in the FIH1 sample of galaxies.

The uniformity of the colors within and between galaxies is also quite striking. This is illustrated in Figure 12a, where the $B-R$ and $U-R$ color profiles are plotted against one another for the galaxies in the FIH1 sample. These data are consistent with the hypothesis that the same mechanism is responsible for the color gradients within galaxies as is responsible for the color differences between galaxies.

Kormendy and Djorgovski (1989) and Peletier *et al.* (1990) have compiled the available data on color gradients. It is striking that there is no correlation between absolute magnitude and mean color gradient for galaxies brighter than

$M_B \approx -19$. The Peletier *et al.* data suggest that the lowest luminosity galaxies may show smaller (zero?) color gradients, but the sample is small. It is important to keep in mind that the uncertainty on the measured gradients is large, ±0.1 mag, comparable to the $B-R$ gradients (see, e.g., FIH1). This is primarily a result of the difficulty of deriving the sky background level on the rather small CCD detectors. Hence much of the scatter in the observed color gradients may be due to measurement errors, especially for the $B-R$ gradient. It is clear that still more accurate measurements are needed; large format CCDs should provide the improved results, allowing a detailed comparison with the line strength gradient results.

While a variety of approaches can be made towards understanding the important factors that affect color gradients, it proved useful to take a simple approach towards the data by asking if there is a single relation of the form *local color = f(structural and dynamical parameters)* that explains both the color gradients within galaxies and the color differences between galaxies. It is clear that such an approach is simplistic, and has to be used with care, but it may give some valuable insight into the physical processes at work. This is developed more fully in Franx and Illingworth (1990b).

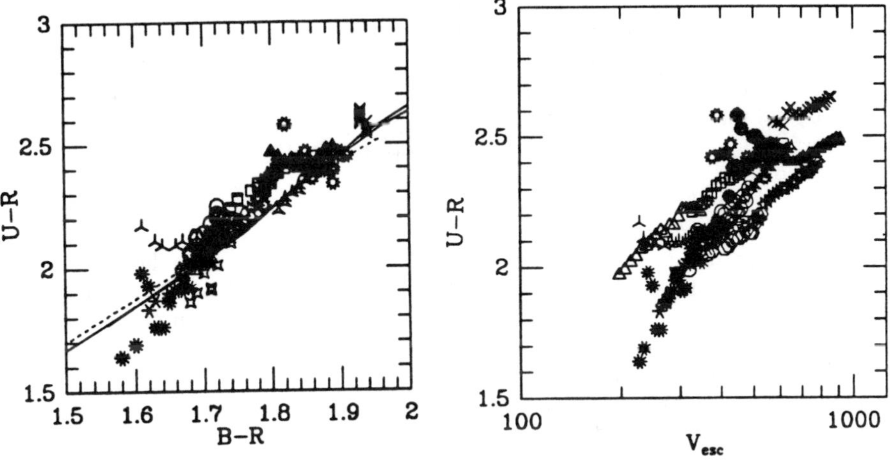

Fig. 12. a) The $U-R$ and $B-R$ color gradient data of Figure 11 are plotted against one another. b) Local color as a function of local escape velocity for a constant M/L galaxy. Compare with Figure 11, where the individual galaxy profiles are plotted versus radii.

The straightforward color-vs-density or surface brightness relations are not convincing. Taking a more sophisticated view, and building upon the idea that the color-luminosity relation is produced by a relation between binding energy and mean metallicity (e.g., Larson 1974), we have derived local color as a function of local binding energy. We have expressed the binding energy as the local escape velocity. The resulting relation for the galaxies in the sample of FIH1 is shown in Figure 12b, where we plot the color gradient data of Figure 11 as a function of escape velocity. No zero point shifts were applied, i.e., the data are as calibrated. The escape velocity was calculated assuming an $r^{1/4}$ profile for all the galaxies. The measured central velocity dispersion was used to calculate the escape velocity at an effective radius. The effective radius and velocity dispersion were taken from from Faber *et al.* (1989).

The relation in Figure 12b is surprisingly tight. The useful range in $U - R$ is about 0.7 mag, while the spread about the relation is approximately 0.1 mag. While a larger sample is needed to verify this result, it is of some interest to think about how such a relation may arise, and about its implications.

The important factors are (1) the $U - R$ color gradients of the galaxies in the sample are roughly constant at −0.23 mag per decade in radius, (2) the uniformity of the surface brightness profiles and structure of elliptical galaxies, with only small variations in the shape of the velocity dispersion profile (typically slightly negative logarithmic gradient), (3) the good correlation between the mean color and the velocity dispersion (e.g., Wirth and Gallagher 1984). Together these give a qualitative rationale for the existence of the relation seen in Figure 12b. This will be explored in more detail by Franx and Illingworth (1990b).

The escape velocities shown in Figure 12b are probably under-estimated because we did not include dark matter. However, the inclusion of dark matter does not necessarily change the result in a significant way. Franx and Illingworth (1990b) obtain a similar relation for models which do incorporate dark matter; they assume that the density profiles of halos are similar, and that the total M/L is constant.

What are the physical processes that would result in a relation between escape velocity and mean metallicity? A purely local process could be regulation of star formation by a hot (X-ray) galactic medium. Such a hot X-ray gas component has a pressure that is a function of the local escape velocity. A plausible scenario is one in which star formation proceeds until the cold gas is heated to a temperature and pressure higher than the X-ray component, and hence escapes from the locus of star formation (and from the galaxy?). For regions where the pressure of the X-ray gas is lower (lower escape velocity), star formation will be stopped after fewer generations of stars, and hence the self-enrichment will be lower. This simple-minded scenario needs to be explored more quantitatively, since it ignores the effects of cooling and conduction. Another possible explanation might be that the IMF and the yield of the star formation is a function of the local pressure of the hot gas. More global mechanisms involving, for example, the coupling of dissipation and metal enrichment, are harder to envisage, but should not be excluded given our limited understanding of galaxy formation.

SUMMARY

The contrast between the uniformity of the structural properties of ellipticals, particularly the uniformity of their color gradients and the existence of a global color-luminosity relation, and the diversity of their kinematic properties is striking. If larger samples show that the color relations noted here, particularly that of local color with escape velocity, are generic to ellipticals, theories of the formation of such galaxies will have to encompass their structural and population uniformity with the contrasting kinematic diversity.

The reconciliation of these diverse properties is a challenging task for theoretical studies, but these data and that obtained in the future will provide the foundation upon which we will build a picture of how such galaxies formed.

REFERENCES

Balcells, M., and Quinn, P. J., 1990, *Ap. J.*, submitted.

Barnes, J. 1990, in *Dynamics and Interactions of Galaxies*, ed. R. Wielen (Heidelberg: Springer-Verlag), in press.

Baum, W. A. 1990, *this volume.*

Bender, R. 1988, *Astr. Ap. (Letters)*, **202**, L5.

Bender, R., Döbereiner, S., and Möllenhof, C. 1988, *Astr. Ap. Suppl.*, **74**, 385.

Bender, R., Surma, P., Döbereiner, S., Möllenhof, C., and Madejski, R. 1989, *Astr. Ap.*, **217**, 35.

Binney, J. J. 1985, *M.N.R.A.S.*, **212**, 767.

Burstein, D., Faber, S. M., Gaskell, C. M., and Krumm, N. 1984, *Ap. J.*, **287**, 586.

Butcher, H. and Oemler, A. 1978 *Ap. J.*, **219**, 18.

Carlberg, R. G. 1984, *Ap. J.*, **286**, 403.

Contopoulos, G. 1956, *Z. Astrophys.*, **39**, 126.

Davies, R. L., Efstatiou, G., Fall, S. M., Illingworth, G. D., Schechter, P. L. 1983, *Ap. J.*, **266**, 41.

Davies, R. L., and Birkinshaw, M. 1986, *Ap. J. (Letters)*, **303**, L45.

Davies, R. L., and Birkinshaw, M. 1988, *Ap. J. Suppl.*, **68**, 409.

Davies, R. L., and Sadler, E. M. 1987, in *IAU Symposium 127, Structure and Dynamics of Elliptical Galaxies,* ed. T. de Zeeuw, (Dordrecht:Reidel), p441.

Djorgovski, S. and Davis, M. 1987, *Ap. J.*, **313**, 59.

Dressler, A. , Lynden-Bell, D., Burstein, D., Davies, R. L., Faber, S. M., Terlevich, R. J., and Wegner, G. 1987, **313**, 42.

Efstathiou, G., and Gorgas, J. 1985, *M.N.R.A.S.*, **215**, 37p.

Faber, S. M. 1977, *Evolution of Galaxies and Stellar Populations* eds. B.M. Tinsley and R.B. Larson (New Haven: Yale Univ. Obs.), p157.

Faber, S. M., Wegner, G., Burstein, D., Davies, R. L., Dressler, A., Lynden-Bell, D., and Terlevich, R. J. 1989, *Ap. J. Suppl.*, **69**, 763.

Franx, M., and Illingworth, G. D. 1988, *Ap. J. (Letters)*, **327**, L55.

Franx, M., Illingworth, G. D., and Heckman, T.M. 1989a, *A. J.*, **98**, 538 (FIH1).

Franx, M., Illingworth, G. D., and Heckman, T.M. 1989b, *Ap. J.*, **343**, 617 (FIH2).

Franx, M., and Illingworth, G. D., 1990a, in preparation.

Franx, M., and Illingworth, G. D. 1990b, preprint.

Franx, M., Illingworth, G. D., and de Zeeuw, P. T., 1990, in preparation.
Gerhard, O. E. 1983, *M.N.R.A.S.*, **202**, 1159.
Gerhard, O. E. 1987, in *IAU Symposium 127, Structure and Dynamics of Elliptical Galaxies,* ed. T. de Zeeuw, (Dordrecht:Reidel), p241.
Gorgas, J., and Efstathiou, G. 1987, in *IAU Symposium 127, Structure and Dynamics of Elliptical Galaxies,* ed. T. de Zeeuw, (Dordrecht:Reidel), p189.
Gunn, J. E. 1989, in *The Epoch of Galaxy Formation*, eds. C. S. Frenk, R. S. Ellis, T. Shanks, A. F. Heavens and J. A. Peacock, (Dordrecht: Kluwer), p. 167.
Heiligman, G., and Schwarzschild, M. 1979. *Ap. J.*, **233**, 872.
Hernquist, L. 1990, in *Dynamics and Interactions of Galaxies*, ed. R. Wielen (Heidelberg: Springer-Verlag), in press.
Illingworth, G. D. and Franx, M. 1988, in *Dynamics of Dense Stellar Systems*, ed. D. Merritt (Cambridge: Cambridge University Press) p13.
Jedrzejewski, R. I. 1987, *M.N.R.A.S.*, **226**, 747.
Jedrzejewski, R. I., and Schechter, P. L. 1988, *Ap. J. (Letters)*, **330**, L87.
Jedrzejewski, R. I., and Schechter, P. L. 1989, *A. J.*, **98**, 147.
Kondrat'ev, B. P., and Ozernoi, L. M. 1979, Sov. Astron. Lett., **5**, 37.
Kormendy, J. 1984, *Ap. J.*, **287**, 577.
Kormendy, J., and Djorgovski, S. 1989, *Ann. Rev. Astr. Astrophys.*, **27**, in press.
Larson, R. 1974, *M.N.R.A.S.*, **166**, 585.
Lauer, T. R. 1985, *M.N.R.A.S.*, **216**, 429.
Peletier, R. F. 1989, PhD Thesis, University of Groningen.
Peletier, R. F., Illingworth, G. D., Davies, R. L., Davis, L. E., Cawson, M. C., 1990, *A. J.*, in press.
Persson, S. E., Frogel, J. A., and Aaronson, M. 1979, *Ap. J. Suppl.*, **39**, 61.
Sandage, A., and Visnavathan, N. 1978, *Ap. J.*, **223**, 707.
Scoville, N. Z., Sanders, D. B., Sargent, A. I., Soiffer, B. T., Scott, S. L., Lo, K. Y. 1986 *Ap. J. (Letters)*, **311**, L47.
Wagner, S. J., Bender, R. and Möllenhof, C. 1988, *Astr. Ap. (Letters)*, **195**, L5.
Wirth, A., and Gallagher, J. S. 1984, *Ap. J.*, **282**, 85.
de Vaucouleurs, G., and de Vaucouleurs, A. 1972, Mem. R.A.S. **77**, 1.
de Zeeuw, P. T. 1987, in *IAU Symposium 127, Structure and Dynamics of Elliptical Galaxies,* ed. T. de Zeeuw, (Dordrecht:Reidel), p271.

CCD SURFACE PHOTOMETRY OF CORONA BOREALIS CLUSTER GALAXIES

DONALD H. GUDEHUS
Department of Astronomy, University of Michigan, Ann Arbor, MI

ABSTRACT The photometric properties of cluster galaxies are useful for a number of cosmological tests. Data are presented here for cluster Abell 2065 in two broad bands (V and 8000 Å), with the goal of understanding the complicating processes of luminosity evolution and dynamical evolution The data were analyzed to remove the effects of overlapping images and blurring from seeing. The radius parameter vs. magnitude plot for this cluster shows that a few of the brightest galaxies have undergone some merging, increasing their diameters by up to a factor of 2.3 times that expected for a galaxy of the same magnitude. Seeing-free color gradients are generally quite small. While merging of the brightest galaxies might be expected to wipe out color gradients, the gradients are also generally small in galaxies which are at least 1.5 magnitudes fainter than the magnitude where merging appears to have operated.

INTRODUCTION

Certain of the photometric and structural properties of cluster galaxies comprise the basic distance indicators of observational cosmology. To make full use of these indicators we must understand cluster galaxy luminosity and dynamical evolution. Galaxian surface photometry, however, is complicated by the facts that galaxies overlap with one another and with stars, and that their images are blurred by the earth's atmosphere. In addition, observations through a fixed wavelength bandpass introduce errors depending on redshift and the uncertain nature of their spectra. These issues are addressed here by 1) selecting filters to match the redshift of each cluster so as to always observe at the same rest wavelength, 2) iteratively modeling each overlapping galaxy's image to eliminate contamination, and 3) fitting surface brightness formulae convolved with the measured atmospheric point spread function with nonlinear least squares. For the central region of the Cor Bor cluster, filters were chosen in the V band and the 8000 Å band, giving effective wavelengths in the cluster's restframe of 5080 Å and 7437 Å, respectively. All observations were carried out with a CCD system (Gudehus and Hegyi 1985) on the McGraw-Hill 1.3-m telescope.

DISCUSSION

Profiles

An image is modeled by a set of elliptical isophotes, each of which has an independent x and y location, inclination, and eccentricity. About 20 to 200 parameters can describe any object whose model is improved by successive iterations as neighboring overlapping images are modeled and subtracted. After modeling most of the galaxies in two frames which were exposed for 45 minutes at each wavelength in the center of the cluster and subtracting the models from the original frames, the fields become very smooth. This process allows any individual galaxy to be analyzed free of contamination. Surface brightness as a function of distance along the semimajor axis of the modeled elliptical galaxies is fitted by nonlinear least squares (Gudehus 1987) with de Vaucouleurs formulas (1948, 1953) or modified Hubble formulas (Abell and Mihalas 1966), convolved by the actual stellar point spread function. Two galaxies which exhibit slight halos are more isolated in the field.

Colors and Color Gradients

Integrated colors give information on the role of the cluster environment in controlling star formation. An excess of faint blue galaxies in distant clusters may indicate that spirals evolve into S0's (Butcher and Oemler 1978) or that star burst activity is occuring. If merging involves galaxies of different masses, the merged galaxies should take on the color of intermediate mass galaxies. However, colors of 57 galaxies show no excess of faint blue members and no trend with magnitude. If there has been color evolution in this cluster, that process appears to be complete. If merging has taken place it is not evidenced by a variation of color with magnitude.

Color gradients can in principle give additional evidence for dynamical evolution, since the gradients should be wiped out by merging. Plots of color vs. distance from the center for the brightest ellipticals show minimal trends, but on the average indicate slightly redder centers. A seeing-independent color gradient, or color delta, is the difference between the color at the center and at one radius parameter away. Color deltas are given in Figure 1 for the brightest 15 galaxies in the frame. The values are small for the faint galaxies as well as the bright ones. While merging could explain the small color gradients in the massive galaxies, it is not clear how the lower mass galaxies could have small color gradients unless they formed that way.

Radius Parameters

Figure 2 shows Hubble radius parameter plotted against absolute magnitude for 75 ellipticals (and S0's) in A2065. I have drawn in Holmberg's (1969) field galaxy and Gudehus's (1973) faint cluster galaxy slope and fitted the theoretical merging line (slope = 1/2.5) to the brightest galaxy data. There is some evidence that a few of the most luminous galaxies have increased their size. The surface brightnesses are also fainter than the average of the less luminous galaxies, consistent with merging.

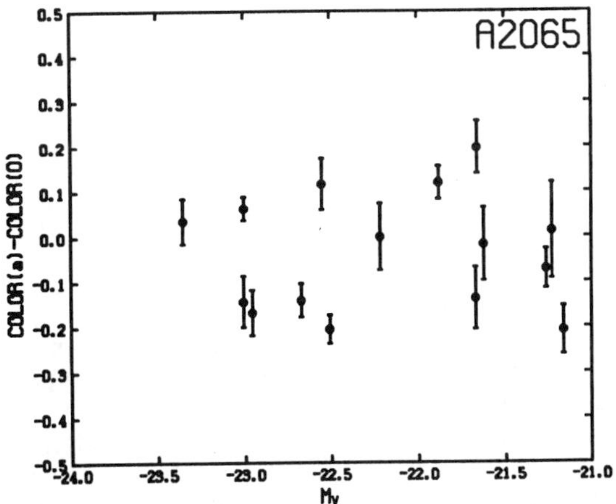

Fig. 1. Seeing-independent color gradients for the brightest 15 galaxies in the frame.

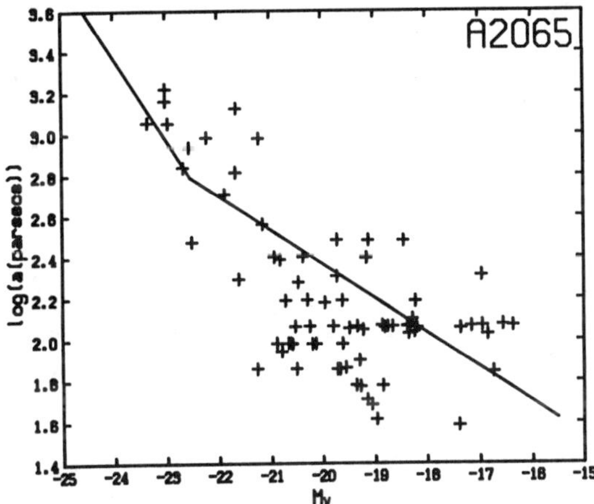

Fig. 2. Log radius parameter vs. absolute V magnitude for 75 galaxies in the frame. Because a few of the very brightest galaxies lie along the steep line, they may have undergone merging.

REFERENCES

Abell, G. O., and Mihalas, D. M. 1966, *A. J.*, **71**, 635.
Butcher, H. and Oemler, A., Jr. 1978, *Ap. J.*, **219**, 18.
Gudehus, D. H. 1973, *A. J.*, **78**, 583.
Gudehus, D. H. 1987, *Bull. A.A.S.*, **19**, 744.
Gudehus, D. H. and Hegyi, D. J. 1985, *A. J.*, **90**, 130.
Holmberg, E. 1969, *Ark. Astron.* **5**, 305.
Vaucouleurs, G. de 1948, *Ann. d'Astrophys.* **11**, 247.
Vaucouleurs, G. de 1953, *M.N.R.A.S.*, **113**, 134.

THE MASS – METALLICITY – LUMINOSITY DENSITY RELATION FOR ELLIPTICAL GALAXIES

R. DE CARVALHO and S. DJORGOVSKI
Division of Physics, Mathematics and Astronomy, California Institute of Technology, Pasadena, CA 91125

ABSTRACT Global properties of elliptical galaxies form a two-dimensional family. One aspect of this "fundamental plane" of ellipticals is a bivariate relation between the mass or luminosity, surface brightness or luminosity density, and a metallicity indicator (e.g., the Mg line strength index, or a color). In other words, there is a second parameter in the mass-metallicity relation, identified here as the luminosity density. The existence of this relation can be understood in the context of dissipative galaxy formation, reflecting both the effect of galactic winds, and an analog of the Schmidt law for star formation for proto-elliptical galaxies. Small, but systematic differences are found in the solutions for the field and for the cluster ellipticals, suggesting a systematic difference in their formative histories.

The global parameters of elliptical galaxies (e.g., mass, luminosity, density, metallicity, etc., but not the shape parameters, such as ellipticity) form a two-parameter statistical family. This can be represented by an inclined plane in the space defined, e.g., by radius, velocity dispersion, and mean surface brightness (Dressler *et al.* 1987; Djorgovski and Davis 1987, hereafter DD; Djorgovski 1987). The physical explanation behind this "fundamental plane" of early-type galaxies is that: (*a*) galaxies are bound by the Newtonian gravity; (*b*) they have a similar dynamical structure; and (*c*) the (M/L) ratio is a power-law function of the fundamental plane variables only (Faber *et al.* 1987; Djorgovski 1988; Djorgovski, de Carvalho, and Han 1988). The existence of this tight correlation implies an important regularity in the process of formation of elliptical galaxies.

We discovered a new family of distance indicator relations for elliptical galaxies, in which a distance-dependent quantity (radius or luminosity) is correlated with a linear combination of the mean surface brightness, and a metallicity term, e.g., a metallicity-sensitive color or the Mg index (de Carvalho & Djorgovski 1989). The new radius – metallicity indicator – surface brightness relations can be understood as a generalization of the fundamental plane: both the velocity dispersion and the final, average metallicity of an elliptical galaxy can be related to the escape velocity, and thus one can substitute a metallicity indicator in place of the velocity dispersion. A direct implication of the radius (or luminosity) – surface brightness – metallicity relation is that there is a "second parameter" in the mass-metallicity relation for elliptical galaxies, which we identify as the luminosity density.

Here we give explicit solutions for the bivariate relations between the virial mass (computed simply as $\sigma^2 R_e$, thus implicitely assuming that all galaxies have the same dynamical structure), mean luminosity density (computed as surface brightness / R_e), and metallicity indicators, viz., $(B-V)$ color and the Mg index. We use the data from the surface photometry by Djorgovski (1985; and DD), and the "7 Samurai" collaboration (Burstein *et al.* 1987). For galaxy distances we use the "Great Attractor" model by Lynden-Bell *et al.* (1988) and assume $H_0 = 100$ km s^{-1} Mpc^{-1} throughout.

We obtain the following scaling solutions for the DD sample, and the luminosity density in the Lick red (r_G) band:

All galaxies:	$\log M = 4.65\,(B-V) - 0.68\,\log\rho_L + \text{const.}$
Clusters:	$\log M = 10.1\,(B-V) - 0.32\,\log\rho_L + \text{const.}$
Field:	$\log M = 2.9\,(B-V) - 0.81\,\log\rho_L + \text{const.}$

All galaxies:	$\log M = 8.4\,Mg - 0.61\,\log\rho_L + \text{const.}$
Clusters:	$\log M = 13.3\,Mg - 0.32\,\log\rho_L + \text{const.}$
Field:	$\log M = 6.8\,Mg - 0.81\,\log\rho_L + \text{const.}$

For the 7S sample, and luminosity density in the B band, we obtain:

All galaxies:	$\log M = 7.5\,(B-V) - 0.75\,\log\rho_L + \text{const.}$
Clusters:	$\log M = 8.5\,(B-V) - 0.9\,\log\rho_L + \text{const.}$
Field:	$\log M = 5.5\,(B-V) - 0.55\,\log\rho_L + \text{const.}$

All galaxies:	$\log M = 10\,Mg - 0.7\,\log\rho_L + \text{const.}$
Clusters:	$\log M = 11\,Mg - 0.75\,\log\rho_L + \text{const.}$
Field:	$\log M = 8\,Mg - 0.6\,\log\rho_L + \text{const.}$

Note that some differences must exist on account of different bandpasses used, and the fact that the $(B-r)$ color should correlate with the metallicity. The errors of coefficients in these relations are of the order of 10%. In order to translate the metallicity indicators, the Mg index and the colors, into the metallicities, we compiled all available calibrations and population synthesis and analysis from the literature. The correspondence between these observables and the metallicity Z (or the [Fe/H]) is rather poorly understood. We adopt the following mean scaling relations:

$$\log Z \sim [\mathrm{Fe/H}] \sim (3.5 \pm 0.5)\,Mg$$

$$\log Z \sim [\mathrm{Fe/H}] \sim (3 \pm 1)\,(B-V)$$

Applying these conversions to the relations listed above, we derive the average scaling relations, with the higher weight given to the Mg index solutions. For the DD sample, in the Lick red (r_G) band we obtain:

All galaxies:	$M \sim Z^{2.2\pm0.6}\rho_L^{-0.55\pm0.1}$, or $Z \sim M^{0.5\pm0.15}\rho_L^{0.3\pm0.1}$
Clusters:	$M \sim Z^{3.7\pm0.6}\rho_L^{-0.38\pm0.06}$, or $Z \sim M^{0.27\pm0.05}\rho_L^{0.11\pm0.03}$

Field: $M \sim Z^{1.7\pm0.6}\rho_L^{-0.65\pm0.15}$, or $Z \sim M^{0.65\pm0.15}\rho_L^{0.45\pm0.15}$

For the 7S sample, in the B band, we obtain:

All galaxies: $M \sim Z^{2.8\pm1.0}\rho_L^{-0.7\pm0.1}$, or $Z \sim M^{0.36\pm0.13}\rho_L^{0.25\pm0.12}$
Clusters: $M \sim Z^{3.0\pm1.0}\rho_L^{-0.8\pm0.1}$, or $Z \sim M^{0.33\pm0.05}\rho_L^{0.27\pm0.12}$
Field: $M \sim Z^{2.1\pm1.0}\rho_L^{-0.6\pm0.1}$, or $Z \sim M^{0.48\pm0.23}\rho_L^{0.28\pm0.16}$

There is a fairly good agreement between the DD and 7S samples, even though they do not fully overlap, the data were obtained in different bands, using different instruments and techniques, and reduced independently. On the other hand, there are persistent systematic differences between the cluster and the field ellipticals, in the sense that the dependence of metallicity on the mass is steeper for the field galaxies. We are inclined to believe that this discrepancy is real, and that it reflects some systematic difference in the formative histories of field and cluster ellipticals, possibly the age effects.

These scaling laws should be explained by theoretical models of galaxy formation, such as those by Carlberg (1984). If the initial enrichment is regulated by galactic winds (energy input by the supernovæ), the final metallicity will depend on the escape velocity and thus the mass. The new term, luminosity density, implies an additional functional dependence at a fixed mass, similar to the Schmidt law for star formation. We note that the relations (especially for the cluster ellipticals) are fairly tight, and the residual scatter may be completely accounted for by the cumulative measurement errors.

The results presented in this paper are preliminary, and a full account of this work will be submitted for publication in the near future. RDC is on leave from Observatorio Nacional, CNPq, Rio de Janeiro, Brasil. SD is an Alfred P. Sloan Foundation fellow. We acknowledge a partial support from California Institute of Technology.

REFERENCES

Burstein, D., *et al.* 1987, Ap. J. Suppl., 64, 601.
Carlberg, R. 1984, Ap. J., 286, 416.
de Carvalho, R., and Djorgovski, S. 1989, Ap. J. Letters 341, L37.
Djorgovski, S. 1985, Ph.D. thesis, University of California, Berkeley.
Djorgovski, S., and Davis, M. 1987, Ap. J., 313, 59 (DD).
Djorgovski, S. 1988, in Starbursts and Galaxy Evolution, eds. Thuan, T. X. *et al.* , Moriond Workshop, p. 549. Gif sur Yvette: Editions Frontières.
Djorgovski, S. 1987, in IAU Symposium #127, Structure and Dynamics of Elliptical Galaxies, ed. T. de Zeeuw, p. 79. Dordrect: D. Reidel.
Djorgovski, S., de Carvalho, R., and Han, M.-S. 1988, in Extragalactic Distance Scale, eds. S. van den Bergh and C. Pritchet, A.S.P.Conf.Ser. 4, 329.
Dressler, A., *et al.* 1987, Ap. J., 313, 42.
Faber, S., *et al.* 1987, in Nearly Normal Galaxies, ed. S. Faber, p. 175. New York: Springer.
Lynden-Bell, D., *et al.* 1988, Ap. J., 326, 16.

DARK MATTER IN ELLIPTICAL GALAXIES

STEPHEN M. KENT
Harvard/Smithsonian Center for Astrophysics, Mail Stop 20, 60 Garden St., Cambridge, MA 02138

ABSTRACT Methods for measuring the mass distribution in elliptical galaxies are reviewed and an attempt is made to determine the dependence of the cumulative mass/light (M/L) ratio as a function of radius in several well-studied objects. In spite of problems with determining the mass as a function of radius in ellipticals, it appears that ellipticals show a much wider range of behavior of variation in M/L ratio with radius than do spiral galaxies of comparable luminosity. Methods of using higher order moments in velocity distributions to provide additional constraints on galaxy mass distributions are discussed briefly.

I. INTRODUCTION

Canonical wisdom holds that all galaxies harbor a halo of dark matter. Although the key physical processes that determine the morphological type of a galaxy are still not known, current theories of galaxy formation agree that the segregation of luminous from dark matter should be higher in elliptical galaxies than in spiral galaxies. In the classical picture of hierarchical clustering and dissipation (White and Rees 1978; Faber 1982), the collapse of protogalaxies by gravitational instability leads to sinking of baryonic matter to the center of a dark matter halo due to the radiation of energy by the baryons. Spiral galaxies are presumed to have acquired enough angular momentum that their collapse is halted by rotation; elliptical galaxies have a much lower specific angular momentum, and they shrink much farther until their collapse is halted by the conversion of gas to stars. Qualitatively, we would expect the ratio of dark to luminous matter inside some characteristic *optical* radius to be higher in spirals than in ellipticals. In the alternative merger picture, elliptical galaxies are formed from the collision of two or more spiral galaxies (*e.g.*, Toomre 1977; Schweizer 1982). Detailed n-body simulations (Barnes 1988) show that the luminous matter in the remnant is again more centrally concentrated relative to the halo than in the precursor.

In spiral galaxies, circular rotation curves measured from neutral and ionized gas provide a fairly direct measure of the mass distribution, and the observation that the cumulative mass/light (M/L) ratio increases monotonically with radius provides strong evidence for the existence of dark halos with a larger characteristic scale length than that of the luminous (stellar) material (*e.g.*, Rubin 1987). The mass distribution in elliptical galaxies must be derived by less direct methods using measurements such as absorption-line velocity dispersions, and the dependence of M/L ratio on radius is much more difficult to determine unambiguously. However, with the accumulation of new data on velocity dispersion profiles, HI gas disk kinematics, globular cluster velocities, and X-ray-emitting gas, the situation for measuring elliptical galaxy masses is gradually improving, and we can now begin to make quantitative statements about dependence of mass on radius for several objects.

The primary objective of this work is to collect together the best available data on elliptical galaxy masses and compare them with a comparable set of data for spiral galaxies. Only galaxies for which useful kinematic data at more than one radius have been published are used. Section II will review the methods used to convert these data to masses and M/L ratios. Section III will present the main results, along with some new results on how 4th moments may be used to resolve ambiguities in modeling the mass distribution in galaxies when discrete velocities of objects such as globular clusters are available.

II. METHODOLOGY

Table I is an attempt to collect together the best available kinematic data for a select sample elliptical galaxies. Only galaxies are included which have photometric data, stellar velocity dispersion measurements, and some other kinematic observations that permit measuring the galaxy mass beyond at least twice its half-light radius. The methods that are used to convert these data to M/L ratios are summarized here. The corresponding M/L ratios are plotted in Fig. 1.

Stellar Kinematics

If the M/L ratio of a spherically symmetric elliptical galaxy is constant, then by the virial theorem, the total mass M is given by

$$M = \frac{8.93 r_e \sigma_{obs}^2}{G f^2}, \tag{1}$$

where r_e is the $\frac{1}{2}$-light radius, σ_{obs} is some observed velocity dispersion, $f = \sigma_{obs}/\overline{\sigma}$, and $\overline{\sigma}$ is the global mean velocity dispersion (Tonry 1983). The factor f corrects for the fact that one usually observes only some average central velocity dispersion rather than the mean total dispersion. If the velocity distribution function is isotropic, then the correction factor f can be calculated as a function of aperture sizes and seeing (Tonry 1983); a value of 1.45 is used here.

With spatially-resolved velocity dispersion measurements, one can test the assumptions of velocity isotropy and constant M/L ratio by using the stellar equation of hydrostatic equilibrium. In most cases that have been studied in detail, at least one of these assumptions must be violated (Binney and Mamon 1982; Davies and Illingworth, 1986; Bicknell *et al.*, 1989). Either the velocity distribution is anisotropic (typically being dominated by radial

TABLE I Elliptical Galaxy Data

Object	D (Mpc)	M_V	r_e (kpc)	σ_* (km s^{-1})	References
Stellar Velocity Dispersions					
M87	15.7	-22.5	7.2	335	2, 9
N1052	13.4	-20.3	1.9	230	1, 2
N2974	22.5	-21.2	3.5	222	2, 6
N4278	16.4	-21.0	2.4	243	2, 13
N4472	15.7	-22.8	7.5	315	2, 13
N5128	2.5	-20.1	2.8:	150	2, 14
I2006	13.0	-19.1	1.5	128	10
HI rings			r_{HI}	V_{rot}	
N1052			16	200	12
N2974			9.8	328	6
N4278			13	255	8
N5128			4.4	240	11
I2006			9.4	202	10
Globular Clusters			r_{GC}	σ_{GC}	
M87			17.8	370	7
N5128			13.	154	5
X-ray Masses			r_X	Mass(X-ray)	
M87			45	3.7×10^{12}	4
N4472			61	3.5×10^{12}	4

References:

[1]Davies and Illingworth, 1986
[2]De Vaucouleurs, de Vaucouleurs, and Corwin, 1976
[3]Fabricant and Gorenstein, 1983
[4]Forman, Jones, and Tucker, 1985
[5]Hesser, Harris, and Harris, 1986
[6]Kim *et al.*, 1988
[7]Mould, Oke, and Nemec, 1987
[8]Raimond *et al.*, 1981
[9]Sargent *et al.*, 1978
[10]Schweizer, van Gorkom, and Seitzer, 1989
[11]van Gorkom, 1987
[12]van Gorkom *et al.*, 1986
[13]Whitmore, McElroy, and Tonry 1985
[14]Wilkinson *et al.*, 1986

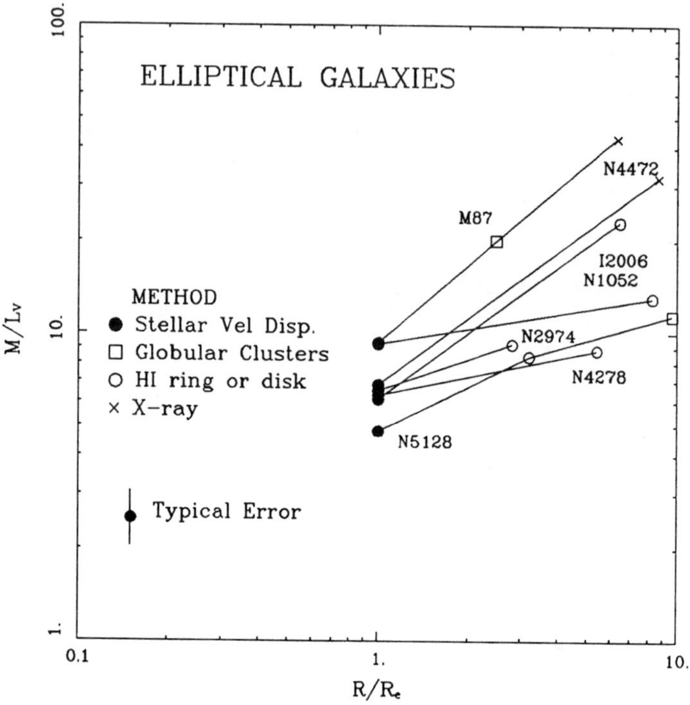

Fig. 1. Dependence of cumulative M/L ratio on radius for a sample of elliptical galaxies. Sources of data are given in Table I. Because of possible inconsistencies in the assumed distances, the curves for individual galaxies can be shifted vertically relative to each other.

motions in the center and tangential motions at large radii), or the M/L ratio varies with radius (typically declining away from the center, reaching a minimum, and then increasing again). The factor f used above was chosen to give rough agreement between the M/L ratio from Eq. (1) and the M/L ratio determined from fitting anisotropic models. If the M/L ratio varies with radius, then the value given by Eq. (1) will be correct only for a single radius; somewhat arbitrarily, this radius was picked to be r_e. Clearly, the error in the M/L ratio derived in this fashion will be large, being a combination of statistical measurement errors and systematic modeling errors.

HI Disks and Rings

HI gas disks have been found around a number of elliptical galaxies, often extending to several r_e. If the gas is on circular orbits, then one can immediately measure the galaxy mass on a much larger scale than is usually covered by stellar velocity dispersion measurements. Unfortunately, the actual picture is not so simple in many cases. For example, in NGC 4278 (Raimond *et al.*, 1981), the gas density major axis, gas kinematic major axis, and photometric major axis are all misaligned by up to 60°, and the gas kinematic major and minor axes are not perpendicular. These observations can be understood only if the gas disk is not uniformly filled and/or warped and if the gas orbits are not circular due to the underlying mass distribution being triaxial. Converting observed rotation velocities to masses is clearly quite hazardous.

With these problems in mind, five HI-determined masses are plotted in Fig. 1. The ring in IC 2006 is quite regular, and the disk in NGC 5128 is viewed edge-on, so these points should be reasonably reliable; the points for NGC 1052, NGC 2974, and NGC 4278 are less so. In some cases, more conservative values are taken for the maximum radius and rotation velocity than the published values to avoid possible problems with beam smearing.

HI Globular Cluster Kinematics

From measurements of radial velocities of a large number of globular clusters around a galaxy, one can find the mass m at any radius r by again using the equation of hydrostatic equilibrium:

$$m(r) = -\frac{r\sigma^2}{g}\left[\frac{d\ln\nu}{d\ln r} + 2\frac{\ln\sigma}{\ln r}\right], \tag{2}$$

where σ is the cluster velocity dispersion and ν is the cluster space density. This equation again assumes that the cluster velocity distribution is isotropic. Since the second term in brackets is expected to be small compared with the first (and is poorly measured, in any case), it is usually dropped. Useful numbers of globular cluster velocities have been measured for two galaxies: M87 and NGC 5128, covering a range in radius of $r/r_e = 2.5 - 6$.

X-ray Emitting Gas

The *Einstein* X-ray satellite detected low-luminosity X-ray emission from a number of elliptical galaxies (Forman, Jones, and Tucker 1985; hereafter FJT). The emission is often well in excess of what one would expect from discrete point sources, and is presumed to arise from a hot, intragalactic medium. From measurement of the gas temperature T and space density ν, one can use hydrostatic equilibrium to derive the total galaxy mass m inside radius r:

$$m(r) = -\frac{rkT}{\mu}\left[\frac{d\ln\nu}{d\ln r} + \frac{d\ln T}{d\ln r}\right], \tag{3}$$

where μ is the mean molecular mass. The gas density profiles are quite similar: $\nu \propto [(1+(r/a)^2]^{-3/4}$ (FJT), with a=0.75h^{-1} kpc. The emission can be detected to quite large radii: $r/r_e > 10$. Unfortunately, gas temperatures could be measured only for a few of the brightest galaxies, and were typically $T \approx 10^7$ K. Furthermore, a temperature gradient could be measured for only two galaxies (M87and NGC 4472), and only poorly at that (see below). Only two X-ray-determined masses are plotted in Fig. 1 for galaxies where the temperature was measured.

Knapp (1987) pointed out that the X-ray-determined masses found by FJT are often several times larger than masses determined by other methods (*e.g.*, from HI rotation curves). Part of this discrepancy probably arises because FJT assumed a fixed X-ray temperature for all galaxies

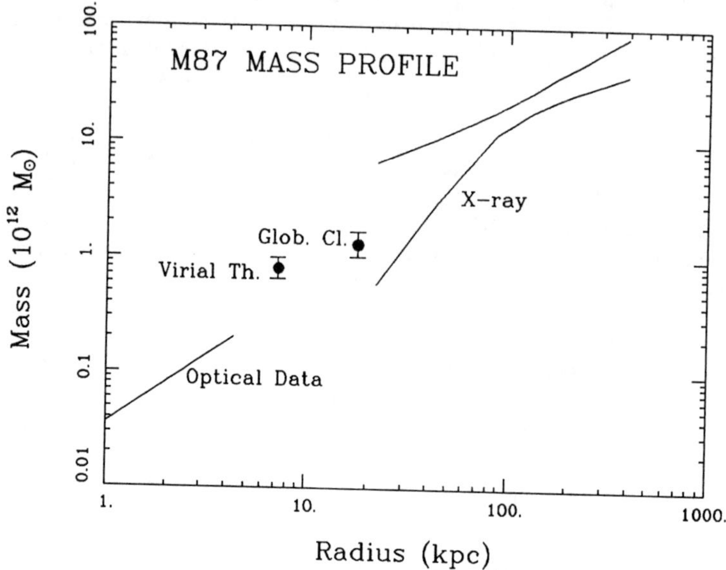

Fig. 2. Cumulative mass profile in M87 measured by four different methods. The "virial theorem" mass is determined using Eq. (1) and the data listed in Table I. The "globular cluster point" is also taken from data in Table I. The optical data are taken from Sargent *et al.* (1978). The X-ray data come from Fabricant and Gorenstein (1983).

that, in most cases, is too large; also, the X-ray determined masses refer to a radius about twice that of the other methods. M87 provides one of the best objects to compare masses measured by different methods. Figure 2 shows the cumulative mass as a function of radius determined by three separate methods. Although none of the methods overlap in radius, the $m(r)$ curve does increase smoothly, roughly as r. The limits on the X-ray masses are taken from Fabricant and Gorenstein (1983) who considered a wide range of possible models for the temperature profile consistent with all the *Einstein* data. Interestingly, the mass at 11 kpc measured from the globular cluster kinematics joins on most smoothly with models which have a positive temperature gradient; a positive temperature gradient was also inferred to exist from the *Einstein* spectral data. In such a case the mass actually increases with radius more rapidly than r. It is not yet clear if M87 is typical or exceptional in this regard, so it is not obvious if these results can be extended to other galaxies.

Other Mass Estimators

Several other mass estimators have been tried for ellipticals which are not as useful as the four discussed above, due to intrinsic difficulties with the method or lack of data. These estimators include:

1. HII rotation curves: For reasons that are not yet fully understood, HII emission-line rotations curves in spheroidal systems (including spiral galxy

bulges) fall well below the expected circular rotation velocity (Schweizer *et al.* 1989; Kormendy and Westpfal 1989). Consequently, emission-line rotation curves are not as useful in ellipticals as they are in spiral galaxies.

2. Shells: Hernquist and Quinn (1987) proposed that shells around elliptical galaxies can be used to probe the shape of the gravitational potential; however, Hernquist and Quinn (1988) point out that the results are highly model-dependent.
3. Satellite galaxy kinematics: One can measure elliptical galaxy masses at large radii by applying either the virial theorem or one of its variants to the kinematics of satellite galaxies. This method is usually limited by small number statistics and by the uncertainty of whether a given satellite is actually bound to a parent galaxy.
4. Gravitational lenses: Again, mass estimates of elliptical galaxies in lens systems are quite model dependent; furthermore, most known lenses are at high redshift and so are difficult to study.
5. Planetary nebula kinematics. One can use planetary nebulae in much the same way as globular clusters to measure galaxy masses. At present, this method has been applied only to local group galaxies (Nothenius and Ford 1986).

III. RESULTS

A fairly simple quantitative measure of the concentration of luminous to dark matter can be obtained by measuring a characteristic radius r_2 inside which the cumulative M/L ratio is twice its value at r_e: $(M/L)_{r_2}=2(M/L)_{r_e}$. This radius is relatively straightforward to determine observationally since it does not rely on ambiguous procedures such as modeling the distribution of different mass components in a galaxy or estimating the M/L ratio of the luminous component alone from population synthesis models. The dissipation scenario gives no quantitative prediction of how r_2/r_e should vary with morphological type, but if we use the observation that r_e for elliptical galaxies is roughly $\frac{1}{2}$ its value for late-type spiral galaxies of comparable luminosity (Kent 1985), then we might expect r_2/r_e to be roughly twice as large for the ellipticals as for late-type spirals. The merger scenario predicts a lesser difference: *e.g.*, in one of Barnes' (1988) simulations, r_2/r_e for the remnant is only 1.5 times the value for the precursor galaxies. However, Barnes' simulation also shows that the mean luminosity density of the remnant is about the same as the precursor galaxy, so to get the observed higher densities in ellipticals, one must still appeal to some dissipational infall of gas during the merging process that subsequently forms stars and ultimately increasing r_2/r_e.

How do current observations compare with these simple scenarios? The major results are summarized in Figs. 1 and 3. Figure 3 shows the cumulative M/L ratio as a function of radius for a representative sample of spiral galaxies with extended HI rotation curves; references for the photometry and kinematics may be found in Kent (1987). All values of M/L are scaled to a Hubble constant $H_0 = 100$ km s^{-1} Mpc^{-1}. Low-luminosity spirals have been excluded. Masses are simply taken to be $V_{rot}^2 r/G$, and the luminosities are computed by integrating the light in elliptical apertures. The galaxies all show a similar behavior in the behavior of the M/L ratio with radius. The

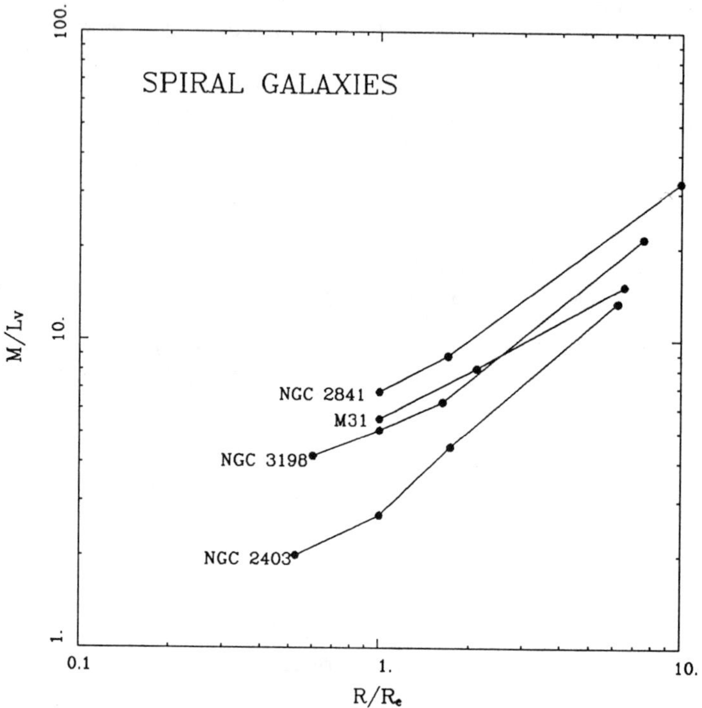

Fig. 3. Dependence of cumulative M/L ratio on radius for a sample of spiral galaxies. Luminosities are in the V-band. Radii are in units of the half-light radius R_e. Data are taken from Kent (1987) and references therein.

parameter r_2/r_e is lies in the narrow range 3-4. The corresponding data for ellipticals have been plotted in Fig. 1. Here, the observational situation is not as satisfactory, since none of the methods used to measure masses are as accurate as using spiral galaxy rotation curves. Nonetheless, the data imply that the parameter r_2/r_e varies between 2.5 and 10 for ellipticals. While values at the high end of the range are just what one would have expected from the simple galaxy formation scenarios, values at the low end are unexpected, and would imply that in some ellipticals, dark matter is as well mixed with the luminous matter as it is in spirals. Two of the galaxies with high r_2/r_e, M87 and NGC 4472, lie in the centers of groups within the Virgo cluster, and so might be exceptional because of their environment; IC 2006, however, is an isolated low-luminosity elliptical.

Clearly, more data will be needed before one can make general statements about the extent and masses of dark matter halos around ellipticals and how they vary with galaxy mass and environment. It is important to understand these dependences, not only for the problem of galaxy formation, but also to understand how they might affect the calibration of the Faber-Jackson relation by which elliptical galaxies are used for measuring distances. One might be suspicious of the curves plotted in Fig. 1, since the scatter in the M/L ratios derived from stellar velocity dispersions show a much smaller scatter than the M/L ratios at larger radii. X-ray observations of hot gas in ellipticals are potentially the most useful probe of dark matter, since they extend out to the largest radii and can be used to study a large number of galaxies. Problems with temperature measurements and temperature gradients must be resolved first, however, and may require the capabilities of AXAF.

On the theoretical side, Merrifield and Kent (1989) have been exploring ways of using higher order moments of velocity distribution functions to provide additional information in modeling the mass distribution in galaxies. If one measures the projected velocity dispersion profile in a galaxy and then solves for the intrinsic dispersions σ_r and $\beta = 1 - \sigma_t^2/\sigma_r^2$ assuming some mass distribution and using the usual method of Binney and Mamon (1982), then Merrifield and Kent show that the projected fourth moment $\langle v_p^4 \rangle$ must satisfy

$$\int_0^\infty 2\pi\mu\langle v_p^4\rangle R dR = \int_0^\infty 4\pi r^3 \nu\sigma_r^2\left[1 - \frac{2\beta}{5}\right]\frac{\partial\Phi}{\partial r}dr,$$

where μ is the projected density and ν is the space density of any suitable set of test particles, such as planetary nebulae or globular clusters, and Φ is the gravitational potential. The measured fourth moment, therefore, provides a check on how good one's guess is as to the assumed mass distribution, and it should be possible to distinguish between such cases as constant M/L ratio versus $m \propto r$. Measuring the fourth moment to sufficient accuracy (say, 10%) will require velocities for several hundred objects. Such data sets do not yet exist, but should be feasible to obtain using multi-object spectrographs.

ACKNOWLEDGEMENTS

This work is supported by NSF grant AST-8451724, the A. P. Sloan Foundation, and the Digital Equipment Corporation.

REFERENCES

Barnes, J. E. 1988, *Ap. J.*, **331**, 699.
Bicknell, G. V., Carter, D., Killeen, N. E. B., and Bruce, T. E. G. 1989, *Ap. J.*, **336**, 639.
Binney, J., and Mamon, G. 1982, *M. N. R. A. S.*, **200**, 361.
Davies, R. L., and Illingworth, G. D. 1986, *Ap. J.*, **302**, 234.
de Vaucouleurs, G., de Vaucouleurs, A., and Corwin, H. G. 1976, *Second Reference Catalog of Bright Galaxies* (Austin: University of Texas).
Faber, S. M. 1982, in *Astrophysical Cosmology*, ed. H. A. Bruck, G. V. Coyne, and M. S. Longair (Vatican: Pontifica Academia Scientarium), p. 191
Fabricant, D., and Gorenstein, P. 1983, *Ap. J.*, **267**, 535.
Forman, W., Jones, C., and Tucker, W. 1985, *Ap. J.*, **293**, 102.
Hernquist, L., and Quinn, P. 1987, *Ap. J.*, **312**, 1.
Hernquist, L., and Quinn, P. 1988, *Ap. J.*, **331**, 682.
Hesser, J. E., Harris, H. C., and Harris, G. L. H. 1986, *Ap. J. Lett.*, **303**, L51.
Kent, S. M. 1985, *Ap. J. Suppl.*, **59**, 115.

Kent, S. M. 1987, *A. J.*, **93**, 816.
Kim, D.-W., Guhathakurta, P., van Gorkom, J. H., Jura, M., and Knapp, G. R. 1988, *Ap. J.*, **330**, 684.
Knapp, G. R. 1987, in *IAU Symposium 127, Structure and Dynamics of Elliptical Galaxeis*, ed. T. de Zeeuw (Dordrecht: Reidel), p. 145.
Kormendy, J., and Westpfahl, D. J. 1989, *Ap. J.*, **338**, 752.
Merrifield, M. R., and Kent, S. M. 1989, in preparation.
Mould, J. R., Oke, J. B., and Nemec, J. M. 1987, *A. J.*, **93**, 53.
Nolthenius, R., and Ford, H. C. 1986, *Ap. J.*, **305**, 600.
Raimond, E., Faber, S. M., Gallagher, J. S., and Knapp, G. R. 1981, *Ap. J.*, **246**, 708.
Rubin, V. C. 1987, in *IAU Symposium 117, Dark Matter in the Universe*, ed. J. Kormendy and G. R. Knapp (Dordrecht: Reidel), p. 51.
Sargent, W. L. W., Young, P. J., Boksenberg, A., Shortridge, K., Lynds, C. R., and Hartwick, F. D. A. 1978, *Ap. J.*, **221**, 731.
Schweizer, F. 1982, *Ap. J.*, **252**, 455.
Schweizer, F., van Gorkom, J. H., and Seitzer, P. 1989, *Ap. J.*, **338**, 770.
Tonry, J. 1983, *Ap. J.*, **266**, 56.
Toomre, A. 1977, in *The Evolution of Galaxies and Stellar Populations*, ed. B. M. Tinsley and R. B. Larson (New Haven: Yale University Press), p. 401.
van Gorkom, J. H. 1987, *IAU Symposium 127, Structure and Dynamics of Elliptical Galaxeis*, ed. T. de Zeeuw (Dordrecht: Reidel), p. 421.
van Gorkom, J. H., Knapp, G. R., Raimond, E., Faber, S. M., and Gallagher, J. S. 1986, *A. J.*, **91**, 791.
White, S. M., and Rees, M. J. 1978, *M. N. R. A. S.*, **183**, 341.
Whitmore, B. C., McElroy, D. B., and Tonry, J. L. 1985, *Ap. J. Suppl.*, **59**, 1.
Wilkinson, A., Sharples, R. M., Fosbury, R. A. E., and Wallace, P. T. 1986, *M. N. R. A. S.*, **218**, 297.

STELLAR POPULATIONS IN ELLIPTICAL GALAXIES

WILLIAM A. BAUM
Lowell Observatory, Flagstaff, Arizona 86001, USA

ABSTRACT On the basis of integrated color indices and the strengths of spectral features, bright elliptical galaxies are believed to have high metallicities, even in their halos. Progressive metal enrichment of the stellar population almost certainly required an extended period of time to build up. That scenario is supported by evidence for a wide dispersion in metallicity and age. This paper calls attention to some early inferences about stellar populations in ellipticals, describes recent findings about their metallicity gradients, and summarizes the observed properties that theoretical models of elliptical galaxy formation must take into account. Preparations have been made to use a pixel–statistics method, which is essentially a variant of the old count–brightness method, for interpreting CCD observations of some bright ellipticals with the Hubble Space Telescope.

REMARKS ON HISTORY

The key to understanding the origin and evolution of galaxies is to investigate the stellar populations and interstellar environments within them, but no *bright* ellipticals are near enough to permit observations of individual stars. As a result, our knowledge of the stellar population in bright ellipticals is based on integrated properties such as color indices, spectral energy distributions, and spectral features.

The development of ideas about stellar populations in ellipticals has been reviewed in previous symposia by Burstein (1985), O'Connell (1986), and Pickles (1987), but I would like to add a few comments on early history. In particular, the important concept of *time–distributed star formation and progressive metal enrichment* dates back much earlier than current reviews suggest.

As Sandage (1986) points out in an excellent historical review, the idea of stellar populations as introduced by Baade (1944) was a grand unifying theme that provided a framework for investigating stellar evolution in various environments. But ideas about the stellar populations in *ellipticals* took a wrong turn when Baade proposed *the same H-R diagram* for ellipticals as for globular clusters, based on his incipient resolution of red stars in M32 and the bulge of M31. He assigned both to "pure Population II." Evidence that something was amiss with Baade's proposal came to light in the 1950s (Stebbins and Whitford 1952, Pettit 1954, Baum 1955, Baum and Schwarzschild 1955, Morgan 1956, Baum 1959, deVaucouleurs 1961).

This is illustrated by Figure 1, which is reproduced from a paper presented at a symposium in 1958. It was apparent that bright ellipticals are much redder than globular clusters. Enough was already known about evolutionary tracks to infer from these color data that the H-R diagrams could not possibly be the same unless the main–sequence luminosity functions were so different as to call for an unreasonable mass–luminosity ratio in the ellipticals. By allowing ellipticals and globulars to have *differing H-R diagrams*, the same 1958 paper went on to show by population synthesis that 8–color data for bright ellipticals could be fitted within 1% by a model made up 95% (bolometrically) of a somewhat old disk-type population (M67 plus M giants) plus only 5% of a globular–cluster–type population. That finding led to the proposal reproduced in the bottom half of Figure 1. It made use of the then–new concept of progressive metal enrichment of a population by successive generations of stars (Fowler and Greenstein 1956, Struve 1956). Thus, the metallicity should have built up in ellipticals over an extended time period, the same as it has in the Milky Way, instead of being primordial in origin. There was even a good case (Morgan 1956) for age differences among globular clusters. It is only recently, however, that the concept of temporally distributed star formation in ellipticals is gaining general acceptance.

Important contributions in the intervening years have been made by many investigators. I started to compile a list but quit when I was going to have to mention more than forty people! Fortunately, most of that work is covered by the excellent review papers already cited (Burstein 1985, O'Connell 1986, Sandage 1986, Pickles 1987).

TODAY'S VIEW

Let me now try to summarize the party line concerning stellar populations in bright ellipticals. The brightest are not only the reddest (as illustrated in Figure 1), but they were also found by McClure and van den Bergh (1968), by Faber (1972, 1973), and by others to have the strongest

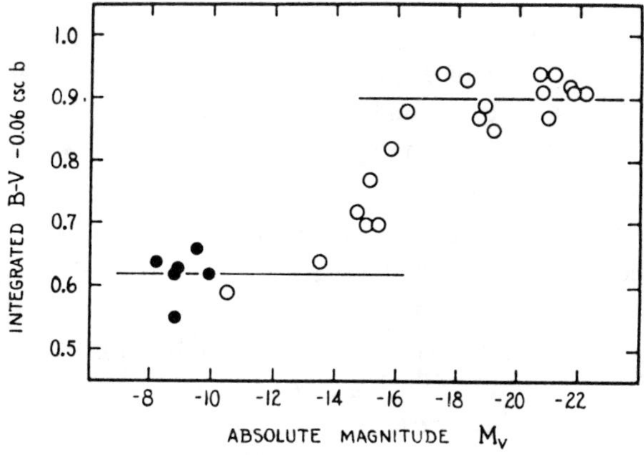

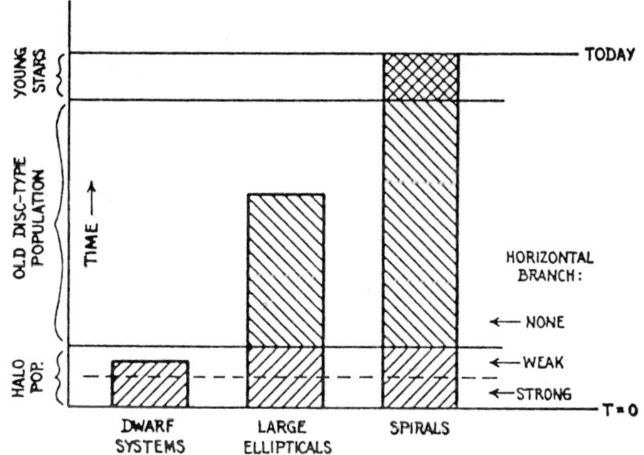

Fig. 1 *Top*: Early data on the color indices of old stellar populations, showing that elliptical galaxies (*open circles*) differ greatly from globular clusters (*filled circles*). The open circle near $M_V = -10$ is the Leo II dwarf spheroidal. *Bottom*: Early proposal for time–distributed star formation in ellipticals. This was based on the eight–color population synthesis described in the text. Both figures are from Baum (1959).

absorption features. All bright ellipticals are strong lined; in constructing a population synthesis, super–metal–rich stars have to be included.

In addition, bright ellipticals individually appear to contain a wide *dispersion* in metallicity and age. Pickles (1987) and others have found that the integrated spectrum of an individual galaxy cannot be matched by the spectra of stars of any single metallicity or age. Owing to the spectrophotometric and spectroscopic similarity of bright ellipticals to the bulges of spirals, one can look for further clues in the bulge of the Milky Way. The wide dispersion in metallicity (and presumably age) in the bulge is beautifully illustrated by Figure 6 of Whitford and Rich (1983). They found a stunning spread of [Fe/H] among 21 K giants, with values ranging from −1.5 to +1.0.

In the integrated spectra of old populations, Burstein finds a failure of CNO-based absorption–line features to correlate well with one another, with broad–band colors, or with iron absorption. Even within a homogeneous population sample such as a single globular cluster, Hesser, Carbon, and others find markedly differing strengths of NH, CH, and CN features from star to star. These results cast some doubt on the reliability of ages and metallicities inferred from integrated spectra of bright ellipticals. Burstein estimates that present uncertainties in stellar syntheses only restrict the ages of stars in bright ellipticals to lie between 5 and 15 billion years, but there is not full agreement on that point. Part of the difficulty with age dating is the lack of color–magnitude diagrams for pure super–metal–rich populations.

OUR CONTRIBUTION

Let me now report on our own current work. My principal collaborator is Bjarne Thomsen of Astronomisk Observatoriet, Aarhus Universitet.

Much of the abundance data for elliptical galaxies pertain to their central regions, because most spectroscopic features are difficult to measure at faint light levels. One notable exception is the broad magnesium absorption feature near 5100Å, which indents the spectrum over a 200Å–wide region, as nicely illustrated by the tracing of the spectrum of the M31 nucleus in Figure 7 of Spinrad and Wood (1965). The depth of that feature, relative to adjacent regions, can therefore be measured at very low light levels by low–dispersion spectroscopy or by differential photometry with medium–bandwidth filters. Magnesium–band observers, in addition to ourselves, have included Spinrad and Wood (1965), Wood (1966), Faber (1972, 1973, 1977), O'Connell (1976), Faber, Burstein, and Dressler (1977), Dressler (1984), Efstathiou and Gorgas (1985), Davies *et al.* (1987), and Couture and Hardy (1988).

Using theoretical spectrum synthesis, Mould (1978) showed that the magnesium feature is strongly sensitive to the metallicity of stellar populations and is only weakly sensitive to the initial mass function, confirming evidence found earlier by Faber (1972). Mould's modeling pertained to ages of 13 Gyr and 6 Gyr, and it was judged to be valid for a metallicity range from about twice solar down to one–third solar. At 6 Gyr, Mould's magnesium index was about 13% lower than at 13 Gyr, the giants of the younger population being hotter and having lower surface gravity. Thus, the magnesium index is not a *pure* measure of metallicity; the *age* and age–dispersion of the population also play a role. The fact remains, however, that strong magnesium is accompanied by other indicators of metal richness, so the central regions of bright elliptical galaxies are indeed metal rich.

Thomsen and I set out to explore the faint *outskirts* of bright ellipticals. Are they, like the Milky Way, super–metal–rich in the center but metal–poor in the halo? In order to reach very faint surface brightness levels, we used three interference filters of medium bandwidth, one being centered on the magnesium feature and the other two flanking it. Our observations were made with the Gunn–Westphal CCD camera attached to the Cassegrain focus of the U.S. Naval Observatory 1.5–meter reflector. The CCD was a Texas Instruments 800×800–pixel chip with a readout noise of about 15 electrons RMS.

To enable our CCD to cover each galaxy completely, we had to work on ellipticals at least as far away as the Coma Cluster. Our filters were tailored to that redshift. By working in Coma, we also had plenty of ellipticals to choose from, all at about the same distance and redshift. Each galaxy became a special project, both in observational procedure and in data reduction. The details are described in papers by Baum, Thomsen, and Morgan (1986), and by Thomsen and Baum (1987, 1989).

High–quality observations were obtained for NGC 4874, 4839, and 4881. The surface brightness profile of the least luminous one, NGC 4881, is shown in Figure 2a. It is an E0 galaxy located 18 arcminutes north of the Coma Cluster center. Each point is the *mode* of pixel values within a thin annular ring at the radius indicated. No smoothing of any kind has been done, but a "seeing" correction has been applied by an iterative process described in our 1987 paper. By the use of "super flats" and by tests made on "empty fields," one can establish that errors are below 0.2% of the night–sky level. The outermost point on the profile corresponds to 80 kiloparsecs if $H_0 = 50$ km s^{-1} Mpc^{-1}.

Magnesium indices for NGC 4881, expressed on the Mg_2 scale of Faber and collaborators, are shown in Figure 2b. Data from two sets of CCD frames are plotted separately. They were each obtained by ratioing data for the magnesium passband to similar data for the flanking passbands. The magnesium index is plotted against the one–quarter power of the radius, but by relating the index to the surface brightness scale at the very bottom, the magnesium *gradient* can be expressed as a dimension-

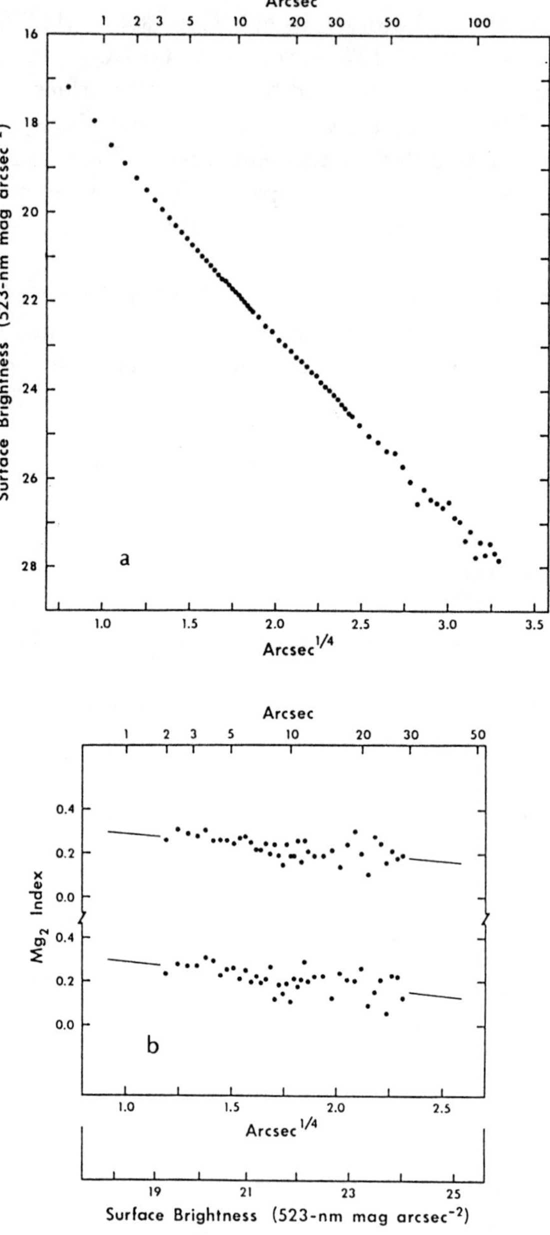

Fig. 2 *a*: Surface brightness profile of the Coma Cluster E0 galaxy NGC 4881 at a wavelength of 523 nm. Individual points represent mode values for annular rings at the radii indicated. *b*: Magnesium index profile for the Coma Cluster E0 galaxy NGC 4881. Two sets of observations are plotted.

less quantity, because the index and and the surface brightness are both expressed on magnitude scales. The use of dimensionless line–strength gradients has the virtue of permitting gradients in galaxies at various distances (and even at unknown distances) to be compared directly with one another on a common scale. Another rationale for dimensionless gradients is that, to a first approximation, surface brightness represents the column density of the stellar population, and in $r^{1/4}$–like galaxies, column densities should correlate with spatial densities.

Our results for the three bright Coma ellipticals are summarized in Table I. For magnesium–index zero points, we have adopted Dressler's (1984) nucleus values. Except for the vicinity of the NGC 4839 nucleus, we find *low magnesium gradients.* To the degree that magnesium is a tracer of metallicity, low *metallicity* gradients are therefore indicated. That finding is consistent with the low iron gradients measured directly in the inner regions of bright ellipticals by Davies and Sadler (1987).

TABLE I Magnesium Gradients in Coma Ellipticals

Galaxy	Mg_2 Nucleus	[Fe/H] Nucleus	Gradient $\delta Mg_2/\delta\mu$	Mg_2 at μ=24	[Fe/H] at μ=24
NGC 4874	0.33	-	–0.032	0.20	-
NGC 4881	0.29	-	–0.023	0.17	-
NGC 4839 bulge	0.30	-	–0.055	-	-
NGC 4839 halo	-	-	–0.020	0.16	-
Average	0.31	+0.31	–0.025*	0.18	–0.20

* Average of *halo* gradients.

If we use the relationship of the Mg_2 index to the metallicity [Fe/H] given in 1981 by Terlevich *et al.* (1981), our results imply nucleus metallicities about twice the solar value, while halo metallicities at a surface brightness of 24th mag arcsec^{-2} average about two–thirds of solar. Thus, at that location, the halo populations of these Coma Cluster ellipticals appear to be several times more metal rich than globular clusters that inhabit similar environments, such as the globulars in giant ellipticals investigated by Strom *et al.* (1981), Huchra (1988), Cohen (1988), and Harris (1988).

Combining our results for the three Coma ellipticals with various findings of others, we call attention to the following observed properties of galaxies that any successful theory of elliptical galaxy formation needs to address: **(1)** Giant elliptical galaxies have high metallicities but low metallicity gradients, resulting in relatively *metal–rich halos.* **(2)** Giant ellipticals are inferred to have triaxial figures supported by anisotropic velocity dispersion, rather than by rotation. **(3)** Globular clusters in giant ellipticals have lower metallicities than halo stars in the same regions. **(4)** The distribution of globular clusters in a giant elliptical tends to be less centrally concentrated than the main body of stars is. **(5)** Halo stars in the Milky Way (a disk galaxy) are not so metal rich as those in giant ellipticals.

No single existing model can explain all these observed properties. The comparatively high metallicities of elliptical galaxies, particularly in their inner regions, argues (as we did 30 years ago) for a history with many generations of star formation in order to build up the observed amount of metal enrichment. That process may be difficult to quantify because of the probable dependence of the initial mass function on metallicity and the associated rate of high–Z production. In any case, *the process of metal enrichment will evidently have required an extended period of time.* That remains true even if mergers are contributing, as discussed by Fall (1979).

The extended time scale of elliptical galaxy formation is supported by other evidence. Based on a review of spectral syntheses of ellipticals, O'Connell (1986) estimates that the last epoch of vigorous star formation in ellipticals probably occurred only 5 to 10 billion years ago, long after their initial collapse and the formation of metal–poor globular clusters. The extended time scale is also consistent with the 1985 theory of Fall and Rees in which metal–poor globular clusters should form during protogalactic collapse, but star formation in the main body of a galaxy should be delayed until the metallicity of the gas exceeds [Fe/H] ~ -1.

PLANS FOR THE FUTURE

Future work on the stellar population in bright ellipticals could usefully proceed along several fronts, but I would particularly like to describe a technique that the Hubble Space Telescope Wide–Field/Planetary Camera (WF/PC) team plans to use. Figure 3 shows a simulation of a portion of NGC 3379 we made in 1984 to explore how *pixel statistics* can be used to characterize the upper end of the H–R diagram. It would be difficult to do conventional CCD photometry on individual stars in this field, but there is a lot of statistical information present.

Several ellipticals with M_V between -16 and -22 are scheduled for deep WF/PC exposures with wide–V (F555W) and wide–I (F785LP)

Fig. 3 *Left:* 1984 simulation of a region in the E0 galaxy NGC 3379 where the surface brightness is V = 22 mag arcsec^{-2}, imaged with the HST Wide Field Camera. The V modulus was assumed to be 29.5 mag., and the dominant population was assumed to be old but metal rich. ***Right:*** Expected pixel–to–pixel noise for a featureless surface of the same brightness.

filters. After the WF/PC image of an elliptical has been flat–fielded, brightness gradients can be subtracted and uncluttered areas can be selected for statistical analysis. Figure 4 illustrates by example what sort of pixel brightness variations might be expected, based on the simulation in Figure 3. In Figure 4a, a histogram of pixel values in the simulated image is compared with the taller and narrower histogram of pixel values that would have been produced on the CCD by a featureless area of the same surface brightness. If one histogram is subtracted from the other, the differential histogram looks like Figure 4b; its shape and amplitude are quite sensitive to different assumed populations and distances. The plan therefore is to create a library of simulations in two colors for various assumed populations and distances, and later to look for the best match to actual Space Telescope images using all available statistical methods. Our simulations indicate that this pixel–statistics technique should be effective at least as far as the Virgo Cluster (modulus ~31 magnitudes), where we have scheduled WF/PC observations of three ellipticals: NGC 4552, NGC 4649, and (jointly with John Bahcall) NGC 4472.

The WF/PC frames will, of course, also contain valuable data concerning globular clusters associated with those ellipticals. The Virgo globulars can then be compared, better than before, with globulars associated with ellipticals in the Coma Cluster which are scheduled for similar

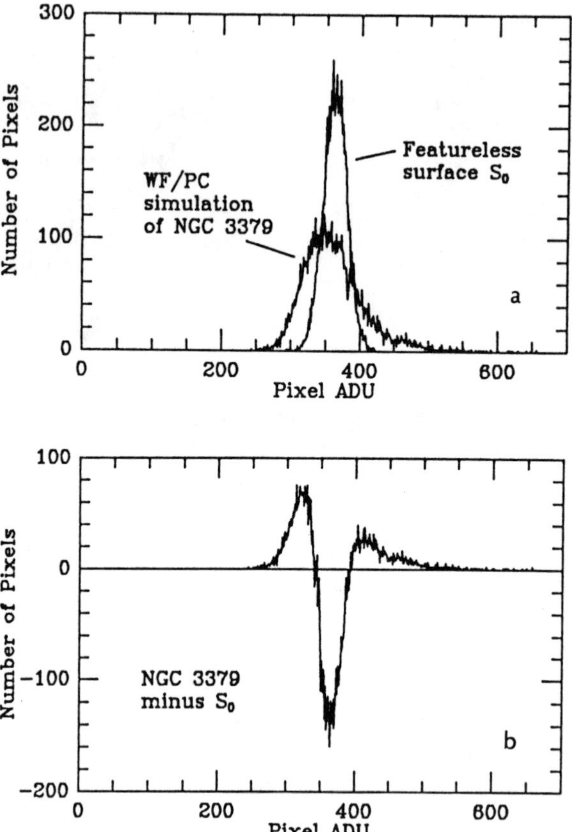

Fig. 4 *a*: Histograms of pixel DN values in the simulated HST–WF/PC images in Figure 3. *b*: Difference between those two histograms. Cases for various assumed distances and populations can be reduced to statistical parameters and compared with data from actual HST-WF/PC frames.

WF/PC observations. The inferred distance of Coma should provide an estimate of H_o that is relatively insensitive to local departures from a smooth Hubble flow.

The expectation of using the pixel–statistics technique was incorporated in 1985 in our team's "GTO" program. In principle, it is an extension of the count–brightness technique explored by Baum and Schwarzschild (1955). It has also been assessed very recently by Tonry and Schneider (1988) as a method for obtaining absolute distances of ellipticals by *assuming* that the absolute magnitude of the stars dominating the pixel statistics is already known. But that assumption may be rather dicey. It may be better *initially* to use other criteria for estimating the distances of ellipticals (such as those in the Virgo Cluster), and then see what pixel statistics tell us about their stellar populations. In that way, Space Telescope WF/PC observations of the ellipticals in Virgo should enable us to *calibrate* pixel statistics as a distance tool, in addition to reducing our ignorance about stellar populations.

REFERENCES

Baade, W. 1944, *Ap. J.*, **100**, 137.

Baum, W. A. 1955, *Astron. J.*, **60**, 153.

Baum, W. A. 1959, *Pub. Astron. Soc. Pac.*, **71**, 106. Also see *Suppl. Ann. d'Ap.*, **8**, 23.

Baum, W. A., and Schwarzschild, M. 1955, *Astron. J.*, **60**, 247.

Baum, W. A., Thomsen, B., and Morgan, B. L. 1986, *Ap. J.*, **301**, 83.

Burstein, D. 1985, *Pub. Astron. Soc. Pac.*, **97**, 89.

Cohen, J. G. 1988, *Astron. J.*, **95**, 682.

Couture, J., and Hardy, E. 1988, *Astron. J.*, **96**, 867.

Davies, R. L., Burstein, D., Dressler, A., Faber, S. M., Lynden-Bell, D., Terlevich, R. J., Wegner, G. 1987, *Ap. J. Suppl.*, **64**, 581.

Davies, R. L., and Sadler, E. M. 1987, in *IAU Symposium 27, Structure and Dynamics of Elliptical Galaxies*, ed. T. de Zeeuw (Dordrecht: Reidel), p. 441.

deVaucouleurs, G. 1961, *Ap. J. Suppl.*, **5**, 233, 1961.

Dressler, A. 1984, *Ap. J.*, **271**, 512.

Efstathiou, G., and Gorgas, J. 1985, *Mon. Not. Roy. Astron. Soc.*, **215**, 37P.

Faber, S. M. 1972, *Astron. Astrophys.*, **20**, 361.

Faber, S. M. 1973, *Ap. J.*, **179**, 731.

Faber, S. M. 1977, in *The Evolution of Galaxies and Stellar Populations*, eds. R. Larsen and B. Tinsley (New Haven: Yale University Observatory), p. 157.

Faber, S. M., Burstein, D., and Dressler, A. 1977, *Astron. J.*, **82**, 941.

Fall, S. M. 1979, *Nature*, **281**, 200.

Fall, S. M., and Rees, M. J. 1985, *Ap. J.*, **298**, 18.

Fowler, W. A., and Greenstein, J. L. 1956, *Proc. Nat. Acad. Sci.*, **42**, 173.

Harris, W. E. 1988, in *IAU Symposium 126, Globular Cluster Systems in Galaxies*, eds. J. E. Grindlay and A. G. D. Philip (Dordrecht:

Reidel), p. 237.

Huchra, J. 1988, in *IAU Symposium 126, Globular Cluster Systems in Galaxies*, eds. J. E. Grindlay and A. G. D. Philip (Dordrecht: Reidel), p. 255.

McClure, R. D., and van den Bergh, S. 1968, *Astron. J.*, **73**, 313.

Morgan, W. W. 1956, *Pub. Astron. Soc. Pac.*, **68**, 509.

Mould, J. R. 1978, *Ap. J.*, **220**, 434.

O'Connell, R. W. 1976, *Ap. J.*, **206**, 370.

O'Connell, R. W. 1986, in *Spectral Evolution of Galaxies*, eds. C. Chiosi and A. Renzini (Dordrecht: Reidel), p. 321.

Pettit, E. 1954, *Ap. J.*, **120**, 413.

Pickles, A. 1987, in *IAU Symposium 127, Structure and Dynamics of Elliptical Galaxies*, ed. T. de Zeeuw (Dordrecht: Reidel), p. 203.

Sandage, A. 1986, *Ann. Rev. Astron. Astrophys.*, **24**, 421.

Spinrad, H., and Wood, D. B. 1965, *Ap. J.*, **141**, 109.

Stebbins, J., and Whitford, A. E. 1952, *Ap. J.*, **115**, 284.

Strom, S. E., Forte, J. C., Harris, W. E., Strom, K. M., Wells, D. C., and Smith, M. G. 1981, *Ap. J.*, **245**, 416.

Struve, O. 1956, *Sky and Telescope*, **15**, 391 and 449.

Terlevich, R., Davies, R. L., Faber, S. M., and Burstein, D. 1981, *Mon. Not. Roy. Astron. Soc.*, **196**, 381.

Thomsen, B., and Baum, W. A. 1987, *Ap. J.*, **315**, 460.

Thomsen, B., and Baum, W. A. 1989, *Ap. J.*, **347**, in press.

Tonry, J., and Schneider, D. P. 1988, *Astron. J.*, **96**, 807.

Whitford, A. E., and Rich, R. M. 1983, *Ap. J.*, **274**, 723.

Wood, D. B., 1966, *Ap. J.*, **145**, 36.

STELLAR CONTENT OF NEARBY GALAXIES: II. THE LOCAL GROUP DWARF ELLIPTICAL GALAXY M32

Wendy L. Freedman
The Observatories of the Carnegie Institution of Washington
813 Santa Barbara St., Pasadena, CA, 91101.

ABSTRACT A field in the elliptical galaxy M32 has been resolved into stars using a CCD camera on the Canada-France-Hawaii telescope (CFHT). From these observations a color-magnitude diagram and luminosity function have been obtained for the brightest stars in this companion galaxy to M31.

V, *R* and *I* CCD frames of a field offset two arcminutes south of the nucleus of M32 were obtained at the prime focus of the CFHT in 0.6 arcsec seeing. The frames were reduced using DOPHOT, a two-dimensional stellar photometry program written by P. Schechter. Details of the reduction process can be found in Freedman (1989).

The *I*-band luminosity function for the M32 data is plotted in Fig. 1. These data have been corrected for both Galactic and M31 field contamination. A sharp increase in numbers is visible at I = 20.0 mag; while a second apparent break is seen at I = 20.5 ± 0.1 mag. In order to proceed in estimating the distance to M32, an assumption about the type of stellar population being observed is required. Conversely, in order to interpret the features in the luminosity function and color-magnitude diagrams (Fig. 2 below), an *a priori* knowledge of the distance is needed. Since an independent determination of the distance modulus is as yet unavailable, the distance modulus to M32, and subsequent interpretation of the stellar populations, have been estimated for *both* discontinuities in the luminosity function, one at I = 20.1 mag, and the other at I = 20.5 mag.

Interpreting the brightest discontinuity in the luminosity function as being the tip of the first red giant branch, and adopting the method outlined by Mould, Kristian and da Costa (1983) and Freedman (1989), yields a distance modulus estimate of 24.0 ±0.2 mag, corresponding to a distance of 630 ± 60 kpc. The adoption of the bright discontinuity in the luminosity function as the tip of the first red giant branch in M32 thus yields a distance modulus which is 0.3 mag (or about 100 kpc) closer than M31, consistent with the conclusion of Ford, Jacoby and Jenner (1978) that M32 is situated in front of M31 rather than behind it. However, given the uncertainties in the method applied here, the best that can be said is that these data are consistent with previous studies rather than providing a firm value for the relative modulus of the two galaxies.

If the fainter discontinuity in the luminosity function at I = 20.5 mag is interpreted as being the onset of the first red giant branch, and adopting the procedure described above, a distance modulus of 24.3 ±0.2 mag is obtained. This value is now comparable to the distance modulus of M31 (e.g., see the review by van den Bergh 1989).

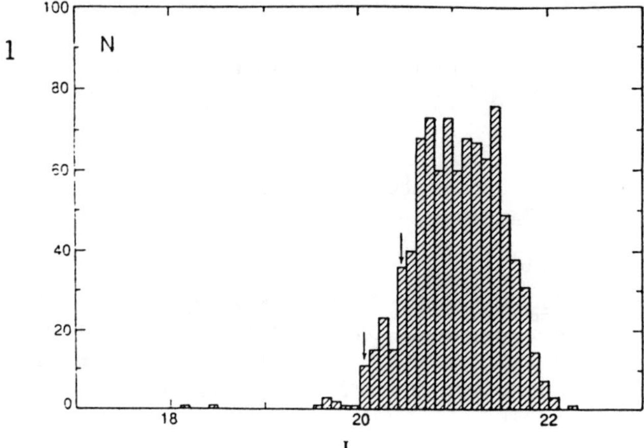

Fig. 1. Histogram *I*-band luminosity functions for stars with $1 < (V\text{-}I) < 2.5$ in the field-corrected CCD Field of M32. Arrows indicate the two discontinuities.

In Figure 2a, a field-corrected color-magnitude *I* versus *V-I* is plotted, and theoretical isochrones for giant stars from Green, Demarque and King (1987) are superposed, with a range of metallicities from -1.7 to -0.3 dex, and an age of 17 Gyr. The sequences have been shifted to the distance modulus of 24.0 mag obtained above, and to account for a color excess of $E(V\text{-}I)= 0.12$ mag. In Fig. 2b, the isochrones are shifted by 24.3 mag. A plausible fit is also obtained with this further modulus, but in this case the number of stars above the tip of the first red giant branch is considerably greater, as discussed further below.

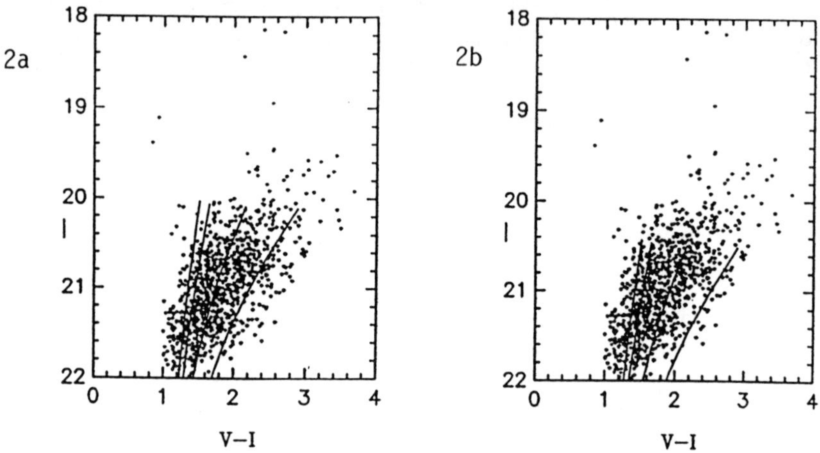

Fig. 2 Field-corrected *I* versus *(V-I)* color-magnitude diagrams with the theoretical isochrones shifted to a distance modulus of (a) 24.0 mag and (b) 24.3 mag.

DOES M32 HAVE AN INTERMEDIATE-AGE COMPONENT?

Stars of intermediate-age (4-5Gyr) will extend approximately one bolometric magnitude above the tip of the first red giant branch (Mould and da Costa 1988).

In Figs. 2a and b, the stars left above the tips of the isochrones are plotted as open circles for emphasis. Fig. 2a, the ratio of stars above and below the tip of the adopted first red giant branch (allowing for the presence of AGB stars below the tip) is 0.06, a factor of four lower than that predicted for a population with an intermediate age of 5 Gyr. However, adopting an apparent first red giant branch tip luminosity of 20.4 mag, yields a corresponding ratio of 0.22. This ratio is a factor of four greater than the previous estimate. In this interpretation, a significant fraction of the stars in this population may have ages in the range of 5-8 Gyr. This result is consistent with spectroscopic studies of the nucleus of M32.

Despite the uncertainties in estimating the numbers of stars in the age range 3-8 Gyr, the lack of stars brighter than I = 19.5 mag, does provide a firm constraint, indicating the absence of a younger 1 to 2 Gyr population in M32.

SUMMARY

If it is accepted that the observed stars in M32 are on the red giant branch, the relative numbers of stars above and below the adopted giant branch tip suggest that a significant population of asymptotic giant branch stars may be present which could have formed as recently as 5 Gyr ago. However, interpretation of the luminosity function for these red stars is not unique; if a second (brighter) discontinuity seen at $I = 20.15 \pm 0.10$ mag is interpreted as the tip of the first red giant branch, it suggests that the age of most of the M32 stars is comparable to those in Galactic globular clusters. A firm constraint on the existence and/or importance of a younger component requires an independent distance estimate to M32 (which will likely be forthcoming with the launch of Space Telescope).

This work was supported in part by NSF grant AST 87-13889.

REFERENCES

Baade, W. (1944). Astrophys. J., **100**, 137.

Ford, H. C., Jacoby, G. H., and Jenner, D. C. (1978). Astrophys. J., **223**, 94.

Freedman, W. L. (1989). Astron. J., October, in press.

Green, E. M., Demarque, P., and King, C. R. (1987). *The Revised Yale Isochrones and Luminosity Functions*, (New Haven: Yale).

Mould, J. R. and Da Costa, G. S. (1988). A.S.P. Conf. Series, **1**, *Progress and Opportunities in Southern Hemisphere Optical Astronomy*, eds., V. M. Blanco and M. M. Phillips, p. 197.

Mould, J., Kristian, J., and da Costa, G. (1983). Astrophys. J., **270**, 471.

O'Connell, R. W. (1980). Astrophys. J., **236**, 430.

van den Bergh (1989). Astr. and Astr. Rev., in press.

DWARF CEPHEIDS IN THE CARINA DWARF GALAXY

JAMES NEMEC
Department of Geophysics and Astronomy, University of British Columbia, Vancouver, B.C. Canada, V6T 1W5

MARIO MATEO
The Observatories of the Carnegie Institution of Washington, 813 Santa Barbara Street, Pasadena, CA 91101

ABSTRACT

We report the discovery of ~15 dwarf Cepheids in a 4′×4′ field in the Carina dwarf galaxy. Light curves are presented for the nine stars that have known periods. The periods range from $0^{d}_{.}051$ to $0^{d}_{.}072$, the visual amplitudes are between 0.10 to 0.60 mag, and the mean visual magnitudes are in the interval 21.3 to 23.2 mag. If the nine variables are 15 Gyr old and are similar to the SX Phe stars found recently in several globular clusters, then seven of the stars appear to obey the fundamental-mode *P-L* relationship. Two of the stars are unusually luminous and may be long-period first-overtone pulsators. On the other hand, since 70% or more of the stars in Carina are ~7 Gyr old, with [Fe/H]~−1.9, it is also possible that some of the stars are of intermediate age. In a deep c-m diagram all of the variables appear to be luminous main-sequence stars. The fact that there is only a small number of faint subgiants at the base of the red giant branch supports the Mould and Aaronson conclusion that the fraction of 15 Gyr old stars in Carina is very low.

DISCUSSION

For the past two years we have been conducting a program to identify, and study in detail, Pop II dwarf Cepheids (*i.e.* SX Phe stars) and eclipsing binaries among the blue stragglers (BSs) in several globular clusters (Mateo *et al.* 1988; Nemec 1989). These studies have shown that ~30% of all Pop II BSs are photometrically variable. Of these, ~80% are SX Phe stars. The other 20% are either close contact binaries or wider eclipsing systems. The discovery of three contact binaries (with orbital periods ~8 hrs) among the BSs in N5466 (Mateo *et al.* 1990), and the discovery that the eclipsing binary blue-straggler NJL 5 in ω Cen is a certain cluster member (see Jensen and Jorgensen 1985, and Margon and Cannon 1989), provides direct evidence that at least some BSs are binary systems (see Eggen and Iben 1989).

The SX Phe stars in globular clusters generally have periods between 50 and 120 min, and visual amplitudes in the range 0.2 to 0.8 mag. The stars appear to obey two well-defined *P-L* relationships, one for fundamental-mode pulsators, and one for first-overtone pulsators. These relationships are useful for distance estimation. Since the masses of SX Phe stars are relatively large (~1.2 $M_{\odot}$), and because there has been

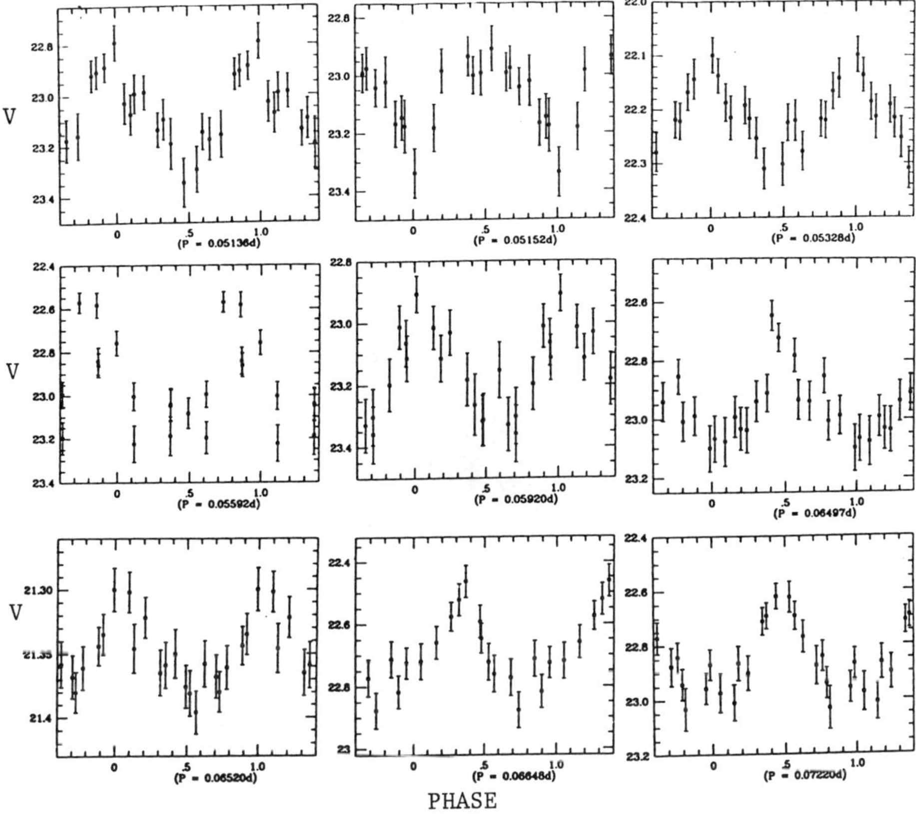

Fig. 1. *V* light curves for nine of the dwarf Cepheids.

no recent star-formation in globular clusters, the currently favoured explanation for SX Phe stars is that they are coalesced binary systems in the faint extension of the Cepheid instability strip.

The goals of our program are: (1) to establish the frequencies of occurrence of SX Phe stars, contact binaries, and wider eclipsing systems in globular clusters; (2) to establish reliable *P-L* relationships for SX Phe stars so that distances can be accurately estimated; (3) to derive masses for the SX Phe stars (relative to the RR Lyrae stars, or double-mode masses), and for the binary systems; (4) to map out the red and blue edges of the faint instability strip; (5) to test if period change rates of the SX Phe stars give information on changing mean densities (at present, the three best studied SX Phe stars in the field of our Galaxy all have decreasing periods, which may suggest that their mean densities are increasing); and (6) to improve our general understanding of the formation and evolution of binary systems in globular clusters.

In this paper we report preliminary results of a major extension of our program to the nearby Carina dwarf galaxy. Carina is a remarkable hybrid system containing both intermediate-age (i.e. ~7 Gyr old) and old (~15 Gyr) stars. Among its members are at least nine carbon stars, seven of which are of high luminosity and intermediate age, and two of which are of lower luminosity and which appear to be similar to the

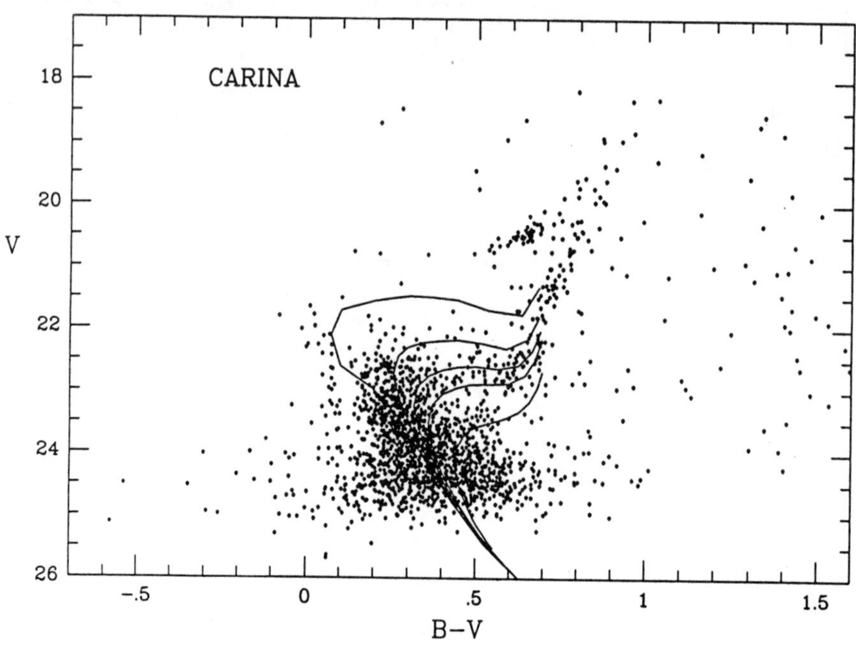

Fig. 2. Color-magnitude diagram for the Carina dwarf galaxy, based on photometry of the averaged frames. The isochrones are for ages of 2, 4, 6, 8 and 15 Gyr, and [Fe/H]=−1.0 (VandenBerg 1985). The diagram contains ~2000 stars. Clearly, most of the subgiants that have turned off the main-sequence have an age near 7 Gyr.

old binary CH-stars in ω Cen (Mould *et al.* 1982). Mould and Aaronson's (1983) c-m diagram shows that, despite having a low mean metal abundance, [Fe/H] ~ −1.9, as many as ~70% of the Carina stars appear to be 7.5±1.5 Gyr old, and may have formed in a major burst of star formation. The ~100 RR Lyrae stars in Carina discovered by Saha, Monet and Seitzer (1986) seem to provide evidence for an underlying population of ~15 Gyr old stars, but their age is highly uncertain.

On two nights in March 1989 the prime-focus CCD camera (No.1 TI chip) on the CTIO 4-m telescope was used to search a 4′× 4′ field in Carina for faint photometric variables. The field that was observed contains three RR Lyrae stars, and overlaps the two fields observed by Mould and Aaronson (1983). Exposure times were 500-s in B and V, and consecutive exposures were taken over a total of seven hours. A control field located 1°south of Carina was also observed. From DOPHOT photometry of the CCD frames (see Mateo and Schecter 1989), ~15 dwarf Cepheids were discovered in Carina. Fig.1 shows V light curves for nine of the variable stars. The preliminary periods, which range from $0\overset{d}{.}051$ to $0\overset{d}{.}072$, are recorded below the light curves of the respective stars. The mean V magnitudes of the stars are between 21.3 to 23.2, and visual amplitudes are in the interval 0.1 and 0.6 mag.

If the variables are SX Phe stars similar to those found in the globular clusters ω Cen (see Da Costa, Norris and Villumsen 1986), NGC 5466 (Nemec and Harris 1987; Mateo *et al.* 1988, 1990 in preparation), and NGC 5053 (Nemec 1989; Nemec *et al.* 1990, in preparation), then: (1) they are among the longest period SX Phe stars that are known; (2) the *P-L* diagram for SX Phe stars (see Nemec 1989) suggests that seven of the stars are fundamental-mode pulsators, and that the two remaining variables, which are unusually bright, are first-overtone pulsators; and (3) they are probably relatively massive, with $M{\sim}1.2\ M_{\odot}$. On the other hand, since 70% or more of the stars in Carina are ~7 Gyr old (see below), it is also possible that some fraction of the variables are intermediate-age dwarf Cepheids. Until further analysis is done we cannot distinguish between these two possibilities.

In addition to searching for faint variables, we have combined the best 18 V frames and all 12 B frames, and have made photometric measurements of all the stars on the averaged frames. Fig.2 shows the resulting c-m diagram. The reddest stars in the diagram are field stars in the line of sight to Carina, and the faint variables are among the upper-main-sequence stars. Also plotted are theoretical isochrones for the ages 2, 4, 6, 8 and 15 Gyr, and [Fe/H]=−1 (from VandenBerg 1985). The bulk of the main-sequence turnoff stars are found only ~2 mags fainter than the red horizontal branch clump, and not at the canonical globular cluster turnoff-luminosity, which occurs ~3.5 mag fainter than the RHB. This diagram also shows that almost all of the Carina stars have ages between 4 and 8 Gyr, and that at the base of the giant branch (*i.e.,* near V=23.5 and B–V=0.65), where one might expect to see 15 Gyr old subgiants, there are very few stars. These observations support Mould and Aaronson's (1983) conclusion that the proportion of 15 Gyr old stars in Carina is very small, perhaps as little as 10% of the visible stars.

REFERENCES

DaCosta, G., Norris, J. and Villumsen, J. 1986, *Ap. J.*, **308**, 743.

Eggen, O. and Iben, I.Jr. 1989, *A.J.*, **97**, 431.

Jensen, K.S, and Jorgensen, H.E. 1985, *Astr. Ap. Suppl.*, **60**, 229.

Margon, B. and Cannon, R. 1989, *Observatory*, **109**, 82.

Mateo, M., Harris, H., Nemec, J., Olszewski, E. 1988, *Bull. A.A.S.*, **20**, 717.

Mateo, M. and Schecter, P. 1989, in *The First ESO Data Analysis Workshop*, in press.

Mateo, M., Harris, H, Nemec, J. and Olszewski, E. 1990, *Ap. J.*, *in press.*

Mould, J. and Aaronson, M. 1983, *Ap. J.*, **273**, 530.

Mould, J., Cannon, R., Aaronson, M. and Frogel, J. 1982, *Ap. J.*, **254**, 500.

Nemec, J. and Harris, H. 1987, *Ap. J.*, **316**, 172.

Nemec 1989, in *The Use of Pulsating Stars in Fundamental Problems of Astronomy*, IAU Colloq.111, *ed.* E. Schmidt, Cambridge University Press.

Saha, A., Monet, D. and Seitzer, P. 1986, *A.J.*, **92**, 32.

VandenBerg, D.A. 1985, *Ap. J. Suppl.*, **58**, 711.

THE GLOBULAR CLUSTER SYSTEM OF THE EDGE-ON Sb GALAXY NGC 5170

Philippe Fischer
Dept. of Physics, University of Victoria, Victoria, B.C., V8W 2Y2

James E. Hesser
Dominion Astrophysical Observatory, 5071 W. Saanich Rd., Victoria, B.C., V8X 4M6

Hugh C. Harris
United States Naval Observatory, P.O. Box 1149, Flagstaff, AZ, 86002

Gregory D. Bothun
Dept. of Astronomy, University of Michigan, Ann Arbor, MI 48109-1090

ABSTRACT A CCD study of the edge-on Sb galaxy, NGC 5170, was carried out to search for its globular cluster system. A total of $\sim 130 \pm 20$ presumed clusters were detected within 2′ of the galaxy center. Distance moduli of 31.2 ± 0.3 and 32.3 ± 0.3 were derived for the galaxy using differing assumptions about the globular cluster luminosity function. Using the second value, and making all the necessary corrections, we estimate a total cluster population of $\sim 815 \pm 320$; and a specific frequency, $S=1.3 \pm 0.5$, or $S_{spheroidal} = 3.9 \pm 1.5$. A power-law fit to the radial density distribution of the cluster candidates yields an index of -1.69 ± 0.26.

I. INTRODUCTION

Globular cluster systems (GCSs) provide a powerful probe of the formation epoch of individual galaxies, and may offer a pure Population II distance indicator. To understand which physical elements of the galaxy halo formation process are universal, and which are peculiar to a particular morphological type, we need to carry out extensive comparisons between the GCSs of spirals and ellipticals. This paper describes the analysis of one Sb galaxy, NGC 5170. NGC 5170 was chosen for observation because it is seen very near edge-on ($i \approx 90°$), and it has a high luminosity ($M_V = -22.0 \pm 0.25$).

II. OBSERVATIONS

A total of three galaxy frames and two background frames were taken. In addition a total of 34 standard stars were measured throughout the observing run. All of the frames were taken with the CTIO 4^m telescope using the Johnson V filter and the RCA CCD (300×508 $0.6''$ pixels).

III. REDUCTION

The reduction procedure for the NGC 5170 frames consisted of five steps: A) subtraction of the galaxy light and profile-fitting photometry of the remaining objects; B) culling of non-stellar objects; C) calibration of the photometry; D) derivation of completeness limits; and E) construction of the luminosity function. Fig. 1 shows the background-subtracted luminosity function for the globular clusters of NGC 5170. The fainter bins have been corrected for incompleteness.

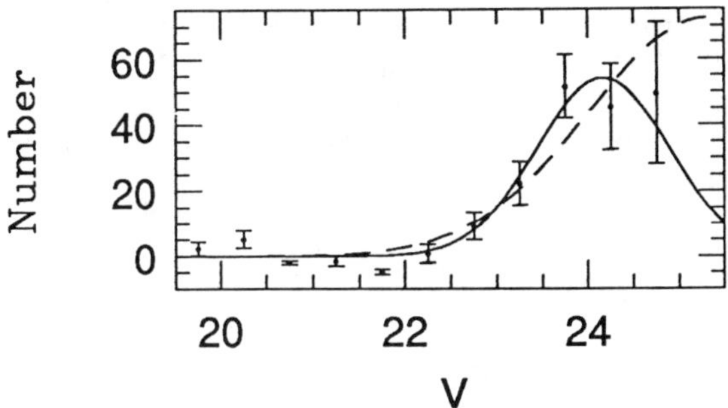

Fig 1.– The GCLF for NGC 5170. The solid curve represents the three-parameter fit, while the dashed curve has $\sigma = 1.3$.

IV. RADIAL NUMBER DENSITY

Weighted least-squares solutions of the form,

$$Log(\sigma) = C_1 + C_2 Log(R), \quad \text{and}$$
$$Log(\sigma) = C_3 + C_4 R^{\frac{1}{4}},$$

were performed. The coefficients thus derived are $C_1 = 1.28 \pm 0.07$ $C_2 = -1.69 \pm 0.24$, $C_3 = 4.36 \pm 0.42$, and $C_4 = -3.06 \pm 0.44$ ($R_e = 1.4'$). Preliminary comparisons of the cluster number density distribution with the galaxy halo light seem to indicate a close agreement, as is sometimes seen in other galaxies.

V. DISTANCE DETERMINATION

To use the globular cluster luminosity function (GCLF) as a distance indicator, it is necessary to assume that the luminosity function is, to first order, a gaussian distribution with a known intrinsic peak luminosity. In the case of NGC 5170, the large error bars in the faintest bins of Fig. 1 render it difficult to determine whether the peak has been reached. Therefore there are two ways to approach the interpretation of our LF.

The first solution, using a weighted least-squares fit to the gaussian for the faintest six bins in Fig. 1 and allowing all three gaussian parameters to vary, yielded a peak luminosity of $V_o = 24.16 \pm 0.23$, and a dispersion of $\sigma = 0.73 \pm 0.21$. The σ derived here is small compared to those typically found for other galaxies [generally $1.0 \leq \sigma \leq 1.5$ is obtained for local group galaxies (W.E. Harris 1988)]. The small σ inferred from our data may well be an artifact of the larger uncertainties in the faintest bins of the luminosity function, leading us to prefer an alternate approach.

Hanes and Whittaker (1987) have demonstrated that reliable estimates of the peak luminosity can be obtained, even when one is considerably short of the peak magnitude, if the correct value of σ is adopted. If values of $M_{V_o} = -7.20 \pm 0.04$, $\sigma = 1.3 \pm 0.05$ are adopted (Fischer 1989) a distance modulus of 32.3 ± 0.3 is inferred (the first technique yields 31.2 ± 0.3).

The Fisher-Tully (FT) relationship in H yields $(m-M)_H = 32.5 \pm 0.14$. This is within one standard deviation of the GCLF distance derived earlier adopting $\sigma = 1.30$. *This is the first direct comparison between these two techniques for the distance to a specific galaxy.*

VI. SPECIFIC FREQUENCY

In order to compare GCSs of galaxies differing greatly in intrinsic luminosity, the concept of specific frequency, S, was developed (W.E. Harris and van den Bergh 1981). Before S can be estimated for NGC 5170, it is necessary to correct for the number of clusters which were missed in the analysis.

This can be accomplished through the use of the $R^{\frac{1}{4}}$ law version of the radial number density function (to fill in the areas that could not be measured) and the GCLF (to fill in the clusters too faint to measure), yielding a total of 815 ± 320 clusters. Using $M_V = -22.0$ for the galaxy yields $S = 1.3 \pm 0.5$.

VII. CONCLUSIONS

1) The GCS has been detected as a significant excess of counts above "background" levels within $2'$ of the galaxy center. 2) The radial density distribution for NGC 5170 clusters exhibits a power-law index (-1.69 ± 0.24) consistent with its luminosity. There is relatively good agreement with the halo light. 3) The inferred distance moduli range from $(m-M) = 31.2 \pm 0.3$ mag to $(m-M) = 32.3 \pm 0.3$, mag depending on what assumptions are made regarding the GCLF parameters. 4) Adopting $(m-M) = 32.5$, the total size of the GCS is estimated to be 815 ± 320, yielding $S = 1.3 \pm 0.5$ (for a bulge/disk ratio of 0.5, $S_{sph} = 3.9 \pm 1.5$).

REFERENCES

Fischer, P. 1989, M.Sc. Thesis, University of Victoria, Victoria, B.C., Canada.

Hanes, D.A., and Whittaker, D.G. 1987, *A.J.*, **94**, 906.

Harris, W.E., and van den Bergh, S. 1981, *A.J.*, **86**, 1627.

Harris, W.E. 1988, in *The Extra-Galactic Distance Scale*, eds. S. van den Bergh and C.J. Pritchet, (Provo: BYU Press) p. 231.

THE EVOLUTION OF DISK GALAXIES AT Z=0

ROBERT C. KENNICUTT, JR.
Steward Observatory, University of Arizona, Tucson, Arizona 85721

ABSTRACT Observations of the star formation properties of nearby disk galaxies provide important clues to their past evolution. I review recent progress in the measurement of global star formation rates in galaxies, with special emphasis on those techniques which may be applied to distant galaxies, and discuss what these data reveal about the past and future stellar birthrates in disks, the physical processes which regulate disk evolution, and the role of evolution in the Hubble sequence.

INTRODUCTION

The development of observational techniques for determining reliable star formation rates (SFRs) in galaxies is providing information on the evolution of disks which in many ways complements what is learned from observations of distant galaxies at large lookback times. I will first give a brief progress report on the available observational methods for measuring SFRs, with special emphasis on those techniques which may be extended to high redshift galaxies, and follow with a summary of what these observations tell us about the present, past, and future evolution rates of disks. I will then discuss two long standing problems in the field, the physical regulation of the stellar birthrate in disks, and the role of evolution in the Hubble sequence.

QUANTITATIVE DETERMINATIONS OF STAR FORMATION RATES

One of the main motivations for observing distant galaxies is to eventually measure the evolution in the stellar birthrate with cosmological lookback time, and nearby galaxies are the testing ground for such measurements of global birthrates. Ten years ago the only quantitative method for measuring SFRs in galaxies was color evolution modelling (e.g., Searle, Sargent, and Bagnuolo 1973; Larson and Tinsley 1978). Application of this method led to much of our current understanding of disk evolution, but the large uncertainties inherent in the technique severely limited the application of the derived SFRs. During the past decade several more direct techniques for measuring integrated SFRs in galaxies have been developed, and I briefly summarize them below. A more detailed discussion is given in Kennicutt (1990).

Lyman Continuum Photon Counting

This is currently the most widely applied technique. For an ionization bounded HII region the number of photons emitted in a recombination line or in the nebular continuum is proportional to the number of Lyman continuum ionizing photons emitted by the exciting stars, and thus for large HII regions or galaxies, which involve the combined photoionization of hundreds or thousands of OB stars, the total nebular luminosity can be used to directly infer the massive SFR. The most frequently applied tracer is the Hα emission line (e.g., Kennicutt and Kent 1983, Kennicutt 1983, Gallagher, Hunter, and Tutukov 1984), but Brackett lines (Telesco 1988 and references therein) and thermal radio continuum measurements (e.g., Klein and Gräve 1986, Duric, Bourneuf, and Gregory 1988) have also been applied, especially in heavily obscured regions.

The primary limitation of this technique is its strong dependence on the form of the stellar initial mass function (IMF). The nebular emission only directly traces stars above 10–15 $M_\odot$, a population which comprises approximately 10–20% of the total stellar mass for a conventional IMF (e.g. Scalo 1986), and thus the bulk of the total SFR must be extrapolated or measured using other techniques. By combining the equivalent width of the Hα emission line with integrated UBV colors it is possible to constrain the IMF slope down to roughly a solar mass (Kennicutt 1983), and reduce the error in the total SFR to less than 50%. SFRs derived from the Hα line must also be corrected for extinction, which is typically 0.5–2 magnitudes in normal spirals and irregulars, but can be much higher in IR-luminous galaxies.

Ultraviolet Continuum Fluxes

The UV continuum radiation longward of the Lyman break can be measured directly, and surveys of nearby galaxies in the 1500-3200 Å range have been conducted by groups at Marseille (e.g., Donas *et al.* 1987) and at Goddard (e.g., Smith and Cornett 1982). The expected launch of several UV-sensitive telescopes in the next 2 years (e.g., HST, ASTRO) will add considerably to this database.

Compared to the Lyman photon counting techniques this method has several advantages; all of the massive star formation is measured, not just the fraction occurring in HII regions, and the flux at these longer wavelengths traces a somewhat broader range of stellar masses. It is also the technique which is most easily extended to high redshifts, since this spectral region is accessible in the optical window to $z \simeq 8$. The principle disadvantage of the technique is extinction, which can amount to several magnitudes at these wavelengths, and is difficult to measure, due to the patchiness of the obscuration.

Far Infrared Continuum

Considerable effort is currently being directed at calibrating the far infrared continuum emission (20–400 microns) of galaxies as quantitative SFR tracers. At these wavelengths the predominant radiation is thermal emission of the interstellar dust, and in normal spirals this represents a major contributor to the bolometric luminosity.

Unfortunately the extraction of quantitative SFRs from the IRAS broadband fluxes has proven to be more problematic than initially anticipated. In normal galaxies the dust is heated by stars with a large range of masses and ages, and the fraction of FIR emission which is directly tied to recent star formation varies from nearly 100% in starburst regions down to a few percent or less in early-type galaxies (e.g., Lonsdale and Helou 1987, Walterbos and Schwering 1987, Buat and Deharveng 1988). If it is known a priori that the the UV radiation field is dominated by young stars, then the FIR emission can be used as a quantitative SFR tracer, but for most normal spirals an accurate multi-component model for the FIR radiation is needed before one can hope to extract meaningful SFRs. The potential reward from such an effort is enormous, because the IRAS survey covered the entire sky, and the database is free of many of the selection biases which plague the existing optical and UV surveys.

Resolved Stars

The most direct possible approach to measuring the SFR in a galaxy would be to resolve the individual stars, and use stellar photometry and spectroscopy to derive the SFRs and IMFs directly. Photometric surveys, most notably by Freedman (1985), have provided data on the luminosity functions and spatial distributions of massive stars in the nearest galaxies, but as discussed by Massey (1985), spectral types for complete samples are required if accurate absolute SFRs are to be determined. Such a survey is being undertaken for the Magellanic Clouds by Massey *et al.* (1989a, b), and this enormous undertaking will eventually provide a fundamental calibration and test for the other techniques discussed here.

Consistency Tests and Comparisons of Techniques

Until recently most of these techniques were applied to independent and largely disjoint samples of galaxies, so it was difficult to evaluate the accuracy of any individual method. A few preliminary comparisons of the different tracers have recently been made, however, which show gratifying agreement in the SFR scales, at least for the massive stars. For example Buat, Donas, and Deharveng (1987) compared SFRs derived from UV continuum fluxes and Hα measurements (cf. Fig. 4 below), and found good agreement, provided that care was taken to correct both sets of measurements for extinction. These SFRs are also consistent with those derived from modelling of the broadband colors of galaxies (Kennicutt 1983), implying that the integrated energy distributions of disk galaxies from the Lyman continuum to the near infrared can be understood with a common set of SFRs and an IMF with approximately a Salpeter slope between 1–60 $M_{\odot}$.

Massey *et al.* (1989a, b) have tested the Hα derived SFRs in two giant HII regions directly, by comparing the predicted ionizing fluxes with their actual stellar contents. In one object the fluxes agree to within the errors, while in the other the Hα flux underestimates the actual OB stellar content by 40%, possibly due to density bounding effects. A larger sample is needed to evaluate the overall reliability of the Lyman photon based SFR scale, but the initial results are most encouraging.

An obvious weakness in these comparisons is that they only test the reliability of the SFRs derived for the massive end of the stellar population, at best for stars above 5–10 $M_\odot$. Uncertainty in the form and variation of the IMF at lower masses is easily the most important source of systematic error in all current measurements of SFRs in external galaxies. As discussed earlier, modelling of the stellar continuum provides a constraint on the IMF slope down to about 1 $M_\odot$, and reduces the uncertainty in the total SFRs to the 50% level, but this only applies to the average properties of large samples of normal disk galaxies. Virtually nothing is known about the systematic variation of the IMF as a function of abundance, galaxy type, or star formation environment, and in at least one type of object, starburst galaxies, there are strong hints of abnormalities in the IMF (see Scalo 1986 and references therein). Since any such dependence of the IMF would likely correlate with galaxy age and distance, sorting out these effects is essential before we can quantitatively compare the spectra of nearby and distant star forming galaxies.

Application to Distant Galaxies

It should be possible to apply these techniques to objects spanning a range of redshifts, in order to directly measure the past evolution of the disk SFR. Unfortunately the most sensitive and well studied tracer, the Hα emission line, redshifts out of the optical window at relatively small lookback times, and hence other emission line or continuum indices in the blue and near-UV region have been explored as alternatives.

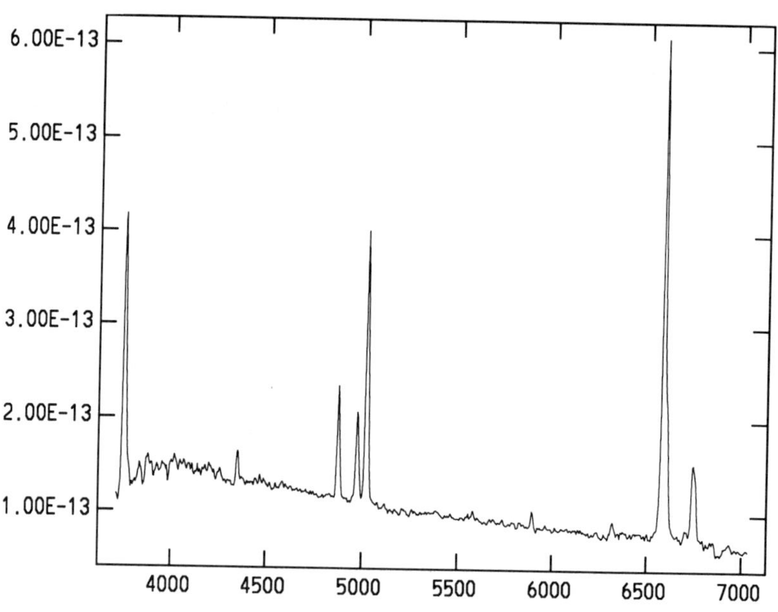

Fig. 1. Integrated spectrum (central 45") of NGC 3310, a nearby blue Sbc galaxy with an unusually high SFR.

One possibility is to use the strong nebular emission lines shortward of Hα as Lyman photon tracers. In active star forming blue galaxies, such as the example illustrated in Figure 1, NGC 3310, the higher order Balmer lines and the forbidden lines of [OII]λ3727 and [OIII]λ4959, 5007 are strong, and should be detectable by ground-based telescopes out to redshifts of 0.5–1. Gallagher, Bushouse, and Hunter (1989) compiled spectrophotometric data for a large sample of low-redshift blue galaxies, and found a mean [OII]/Hβ ratio (corrected for stellar absorption) of 3.2, with a scatter of only a factor of 3. They conclude that either line, but especially [OII]λ3727, could serve as a quantitative SFR tracer, confirming the suggestion of Dressler and Gunn (1982).

These conclusions may not apply, however, to distant galaxies with SFRs which are comparable to those in normal present-day spirals. Figure 2 illustrates the spectrum of a typical Sc galaxy, NGC 6181, observed by Kennicutt and Congdon (1990) as part of a survey of integrated spectra of nearby galaxies. Despite the relatively strong appearance of the Hα+[NII] emission lines, [OIII]λ5007 is virtually absent, Hβ appears in absorption, and only the [OII]λ3727 line is detectable. NGC 6181 is a relatively active galaxy; in Sa-Sb galaxies the blue emission lines would be virtually undetectable at high redshift.

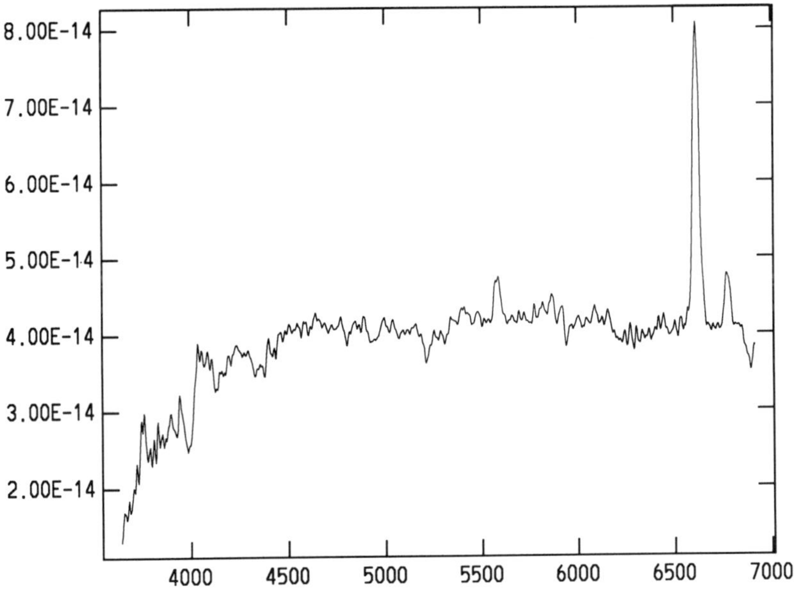

Fig. 2. Integrated spectrum (central 45") of the Sc galaxy NGC 6181.

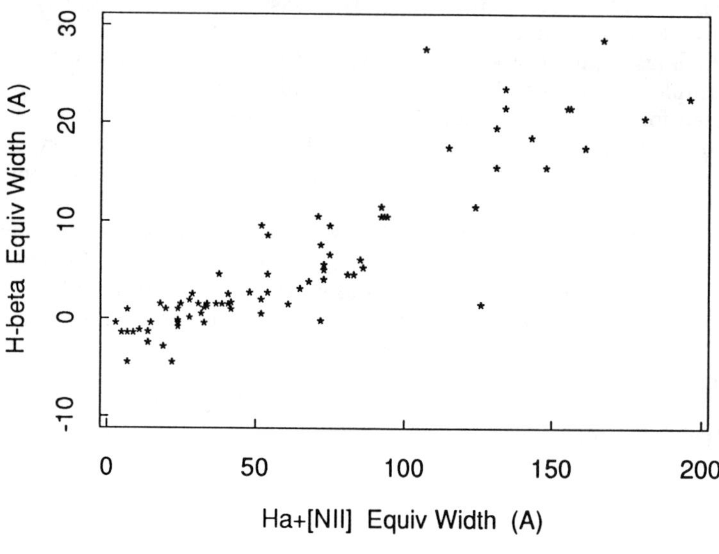

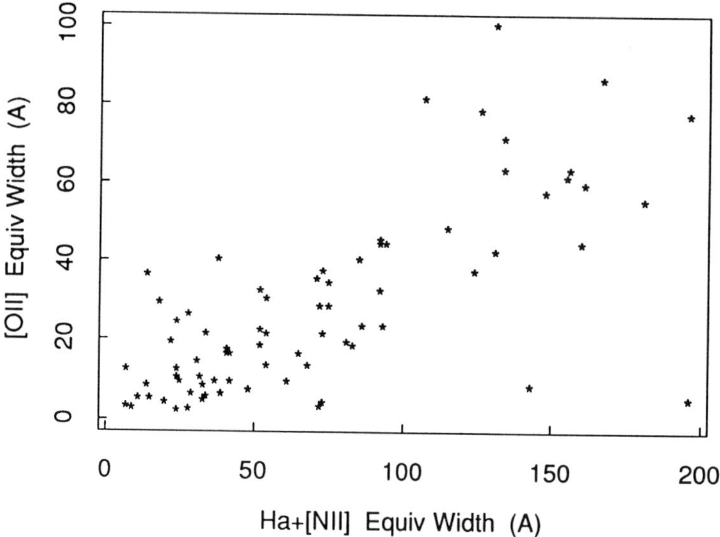

Fig. 3. Correlation between the equivalent widths of the Hβ (upper) and [OII] (lower) emission lines with Hα+[NII] in the integrated spectra of nearby galaxies in the survey of Kennicutt and Congdon (1990). The Balmer equivalent widths have not been corrected for the effects of stellar absorption.

Figure 3 shows the relations between the equivalent widths of Hβ, [OII], and Hα for the Kennicutt and Congdon sample. While the line strengths are well correlated overall, most normal spiral galaxies possess Hα equivalent widths of less than 50 Å, and over this range the scatter in the correlations is large. Hβ is dominated by stellar absorption in most of the objects, and hence is not very useful as a SFR tracer, even if an approximate absorption correction is applied. [OII] is more promising. The line is stronger and is less influenced by stellar absorption problems, but the Hα/[OII] ratio still shows a factor of ten dispersion, probably caused by variations in reddening, nuclear emission, and nebular excitation. Nevertheless these results are encouraging, because they suggest that the large body of [OII] emission line data being obtained for intermediate redshift galaxies can be applied as quantitative SFR indices, in much the way that Hα is applied locally.

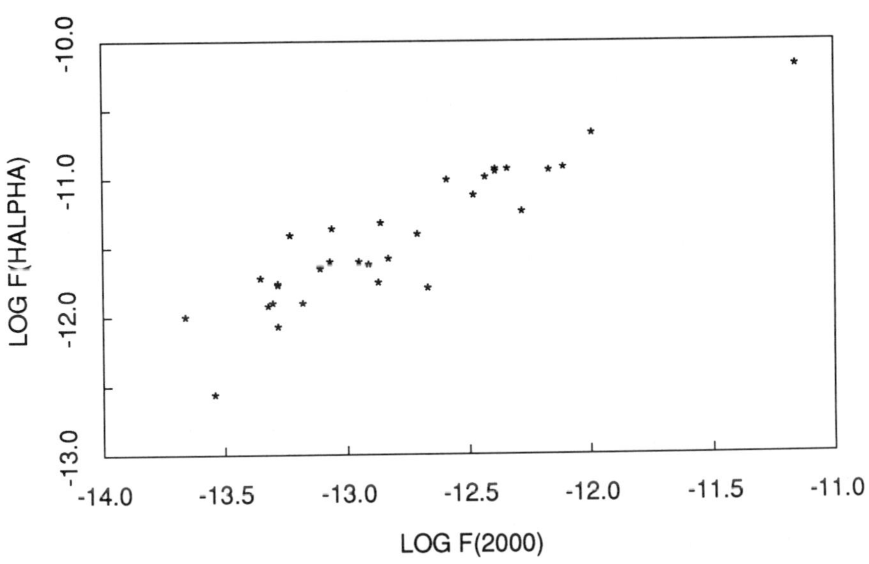

Fig. 4. Correlation between integrated UV continuum and Hα fluxes for nearby spiral galaxies. Replotted version of a figure in Buat *et al.* (1987).

For $z > 1$ even the [OII] line is difficult to observe, and for objects at these higher redshifts the UV continuum flux may provide the most reliable measure of the SFR. Figure 4 shows the correlation between 2000 Å continuum flux and Hα flux for a sample of spirals studied by Buat, Donas, and Deharveng (1987). If this result holds for a larger and more diverse sample of galaxies, it should be possible to derive quantitative estimates of SFRs from integrated photometry or low-dispersion spectra of very distant ($z > 3$) galaxies.

CURRENT, PAST, AND FUTURE EVOLUTION RATES OF DISKS

Total SFRs have been measured for a few hundred galaxies using one or more of the techniques described in the previous section, ranging in Hubble type from spirals and irregulars (Kennicutt 1983, Gallagher, Hunter, and Tutukov 1984) to S0-Sa galaxies (Pogge and Eskridge 1987; Dressel, O'Connell, and Telesco 1989, Thronson *et al.* 1989), and in luminosity from giant spirals to dwarfs (e.g., Hunter and Gallagher 1985, Klein and Gräve 1986, Thronson and Telesco 1986, Hodge, Lee, and Kennicutt 1989). The data provide a self-consistent picture of the mean star formation properties of galaxies along the Hubble sequence.

The current star formation rates, integrated over all stellar masses and normalized to the mass of the Milky Way Galaxy, range from 0.1–1 $M_{\odot}yr^{-1}$ in the earliest disk galaxies to 1–100 $M_{\odot}yr^{-1}$ in the blue Sc and irregular galaxies. The mean rates are strongly correlated with Hubble type, but vary by 1-2 orders of magnitude within each type. The mechanisms responsible for this variation are not well understood (see next section), but both systematic effects (e.g., gas content, interactions) and stochastic variations (e.g., star formation bursts) appear to contribute.

The mean SFRs can also be compared with the total masses of stars and gas in the galaxies, in order to derive characteristic time scales for the past and future star formation in disks. The birthrate histories of disks have been discussed by several authors (Kennicutt 1983, 1986; Gallagher *et al.* 1984, Sandage 1986). A useful parameter is the ratio of the current SFR to the past rate averaged over the age of the disk, which we will denote as "b" following Scalo (1986). The average value of b ranges from of order 0.1 in Sab-Sb disks, the earliest types for which reliable data are available, to near unity in Sbc-Sc disks. In many late-type spirals and irregulars the current rate is higher than the average past rate ($b > 1$). These results confirm conclusions drawn a decade earlier by Searle, Sargent, and Bagnuolo (1973), and Larson and Tinsley (1978), based on modelling of galaxy colors, and the idea that disks evolve relatively steadily with time no longer appears to be a controversial result.

Strong independent confirmation for this picture has come from studies of the age distribution of stars in the solar neighborhood, most recently from chromospheric age dating of GK stars by Barry (1988), and from modelling of the luminosity function of white dwarfs by Noh and Scalo (1989) and Iben and Laughlin (1989). These studies suggest that the birthrate in the Galactic disk, averaged over intervals of several hundred Myr, has remained constant to within a factor of 2–3 since star formation began, confirming earlier work of Twarog (1980), based on isochrone fitting.

By combining the current SFRs with the total masses of gas, one can derive a characteristic time scale for gas consumption and future star formation in disks. For spirals the consumption times range from less than 1 Gyr in starbursting systems to more than a Hubble time in low surface brightness and gas rich red spirals, with a median (for Shapley-Ames or RC2 selected galaxies) of 4–5 Gyr (Larson, Tinsley, and Caldwell 1980, Kennicutt 1983). These estimates include both atomic and molecular gas and a modest correction for recycling of the remaining gas. Similar calculations applied to dwarf irregular galaxies generally yield longer consumption times (Sandage 1986).

The consumption times for spirals are considerably shorter than derived previously (Roberts 1963), and imply the "burnout" of a significant fraction of the current-day spirals in a fraction of a Hubble time, unless the gas supply is replenished from outside. The effective consumption times are even shorter when one takes into account that much of the gas lies outside of the star forming disk, and thus that star formation will probably be curtailed long before all of the gas is consumed. The interpretation of these time scales remains controversial; the short times are often used as supporting evidence for copious amounts of gaseous infall, but the possibility of rapid evolution in the gas supply and stellar birthrate in disks cannot be excluded.

WHAT REGULATES THE EVOLUTION OF DISKS?

A fundamental problem in galaxy evolution is the relationship between the large scale SFR and the properties of the interstellar medium. Parametric relations between the SFR and the density or other properties of the interstellar gas form the basis of virtually all contemporary galactic evolution models, but our understanding of these relations, from both an observational and a theoretical point of view, is still very primitive. Here I briefly summarize some of the recent progress in this area. For a more detailed discussion see Kennicutt (1990).

Most galaxy evolution models are still based on the parametrization suggested by Schmidt (1959), in which the SFR is assumed to scale as a simple power of the gas density:

$$R = a\rho^n$$

Since the Schmidt model was proposed 30 years ago several dozen papers have been written which attempt to quantify the relation and measure the index n (see Berkhuijsen 1977 and Freedman 1984 for references), but they have not converged on a consistent form for the law. Part of this inconsistency is caused by selection effects in the data used to define the law, but large variations ($0 < n < 4$) are often found within the same galaxies, raising the question of whether a universal Schmidt law exists. Other observations are fundamentally inconsistent with a single power-law SFR vs gas density relation. For example, the star formation properties of starburst galaxies, spiral arms, and gas-rich S0 galaxies all require a highly nonlinear response of the SFR to gas density, whereas the radial distributions and temporal behavior of the SFRs in late-type disk galaxies are consistent with a very mild (linear or shallower) density dependence of the SFR.

These problems have stimulated several observational and theoretical efforts to re-examine the dependence of the large scale SFR on gas density and other disk properties. Although much of this work is at a preliminary stage, encouraging progress has been made toward a more physical formulation of the star formation law.

Observations

Several recent surveys of HI, CO, and Hα emission in nearby galaxies have made it possible to define the empirical SFR vs gas density relations from first principles. These results can be conveniently divided into comparisons of the total SFRs and gas masses of galaxies, and spatially resolved studies of SFRs and gas surface densities within specific galaxies.

Correlations between the total SFRs and gas contents of galaxies are of particular interest for galaxy evolution modelling, because they test whether the global evolution properties of disks can be reliably described by simple single-zone models. Several papers have called attention to the tight, roughly linear correlations between the total Hα or far-infrared luminosities of galaxies and their total gas masses, especially in molecular form, and these have been used to argue for a global Schmidt law with $n = 1$, or a constant 'star formation efficiency' in galaxies (e.g., Young *et al.* 1986). As pointed out by Stark *et al.* (1986) and Kennicutt (1989, 1990), these linear correlations between extrinsic properties of galaxies are largely an artifact of scaling, the fact that "bigger galaxies have more of everything" (Stark *et al.* 1986). When care is taken to remove the effects of scaling by normalizing the total SFRs and gas masses, either by scaling them to galaxy luminosity or to disk area (yielding average surface densities), the character of the correlations changes markedly (Buat *et al.* 1989, Kennicutt 1989). The mean SFR is roughly correlated with total (molecular plus atomic) and mean HI surface density, but is only weakly correlated with the mean surface density of molecular gas (as inferred from CO).

The correlation between the disk-averaged SFRs and total gas densities confirms that there is some empirical foundation for single-zone disk evolution models, but the interpretation of the other correlations is less clear. For example, the Hα vs HI correlation could reflect a causal relation between atomic gas density and the SFR, akin to a Schmidt law, or it could result from the formation of HI from the photodissociation of molecular gas, in rough proportion to the SFR (Shaya and Federman 1987, Tilanus and Allen 1989). Likewise there are several possible explanations for the poor SFR-CO correlation, including variations in extinction and the CO/H2 conversion factor, or simply nonlinearities in the star formation law which would tend to smear out any Schmidt law on global scales.

More direct information on the form of the star formation law can be gleaned from comparisons of spatially resolved SFRs and gas densities within individual galaxies. (Since volume densities of stars and gas are virtually impossible to measure in most external galaxies, these studies probe the relation between the projected SFR and gas surface densities.) Again, recent surveys of the star formation, HI, and CO surface density distributions in nearby galaxies have led to new insights into the behavior of the star formation law.

Perhaps the most significant development has been the recognition of the importance of star formation thresholds. Several studies have shown that molecular cloud formation and star formation tends to be abruptly curtailed below a critical threshold gas density, usually in the range $10^{20} - 10^{21}$ H cm^{-2} (e.g., Davies, Elliott, and Meaburn 1976, Skillman 1987, Guiderdoni 1987, Ohta, Sasaki, and Saito 1989). Kennicutt (1989) has extended this analysis by compiling radial profiles of Hα, HI, and CO emission for a large sample of field spirals with active star forming disks. All show radial cutoffs in the SFR, but with a large range of threshold densities, $3 - 50 \times 10^{20}$ H cm^{-2}, or 2-40 $M_{\odot}pc^{-2}$. Most of the thresholds occur at radii where the gas is predominantly atomic, but in at least two cases the thresholds occur well within the optical disk, where the gas is predominantly molecular.

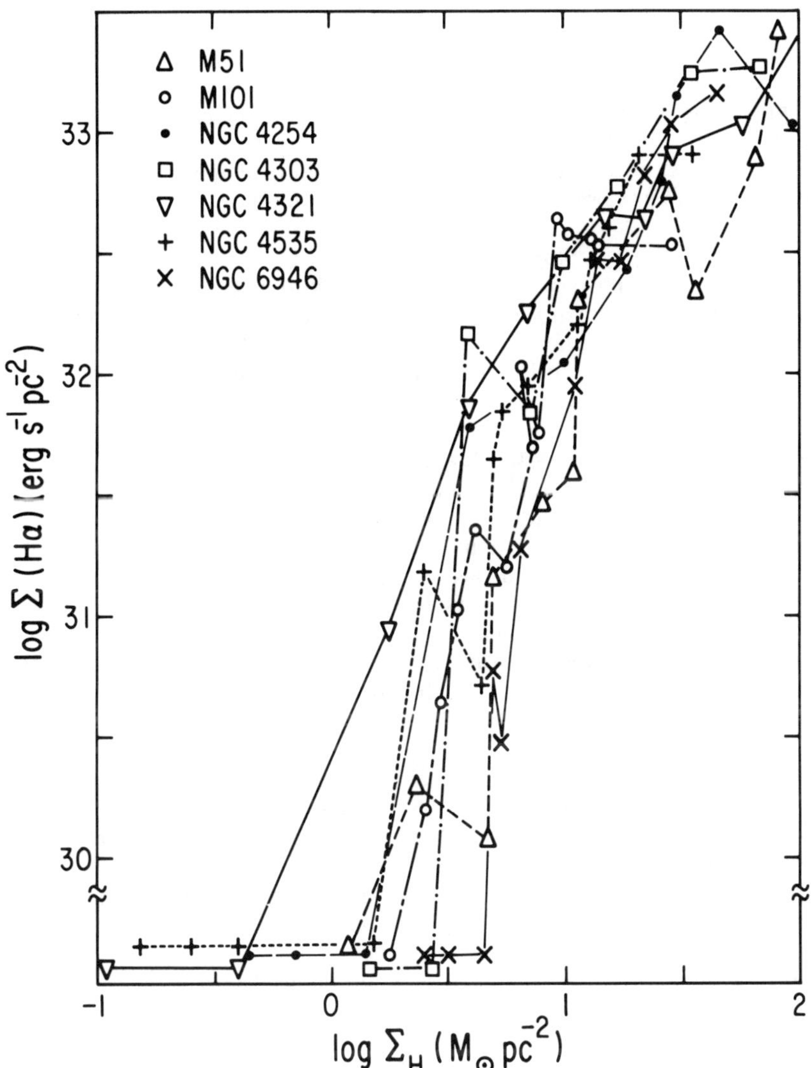

Fig. 5. Correlation of Hα surface brightness and total gas surface density in 7 Sc galaxies. Points are azimuthal averages at constant galactocentric radius, so other parameters may be influencing the SFR.

Figure 5 shows the SFR-density relations for 7 nearby Sc galaxies. At high gas densities, well above the threshold, the relation is well represented by a Schmidt law with slope $n \simeq 1.3$, and nearly constant zeropoint. At low densities, well below the thresholds, star formation is almost completely suppressed, while near the thresholds the slope of the SFR-density relation is much steeper than a normal Schmidt law. Figure 5 only shows late-type galaxies with active star forming disks; when the same data are compared for early type, less active star forming disks, the upper, linear part of the star formation law is absent, but the other two regimes are present (Kennicutt 1989).

The existence of distinct star formation regimes, separated by a threshold density which may vary within or between galaxies, may help to resolve many of the observational inconsistencies with a Schmidt law which were discussed earlier. The large variation in the observed threshold densities may also provide clues to the physical basis of the star formation law.

Theoretical Models for the Star Formation Law

Recently there has been a resurgence of theoretical interest in understanding the basic features of the star formation law, and in developing physically-based alternatives to the Schmidt law. A thorough discussion of this area is beyond the scope of this review, but it is useful to describe a few examples which illustrate the direction of this work.

There is substantial evidence associating the global star formation thresholds with large scale gravitational instabilities in the gas disks, as originally suggested by Quirk (1972), and subsequently discussed by a number of workers, including Elmegreen (1979), Fall and Efstathiou (1980), Cowie (1981) and Larson (1988). Guiderdoni (1987) and Kennicutt (1989) have shown that the observed thresholds are similar to the critical densities for gravitational stability (Toomre 1964, Goldreich and Lynden-Bell 1965). The same type of analysis reveals that gas density distributions themselves closely follow the gravitational stability limit expected for an isothermal Toomre disk (Zasov and Simikov 1988, Kennicutt 1989, Oey and Kennicutt 1989).

This is illustrated in Figure 6, which shows the radial dependence of the gas density normalized to the critical surface density for gravitational stability (Q), using the Toomre (1964) stability criterion, and a constant disk velocity dispersion of 6 km s^{-1}, for 15 galaxies with active star forming disks (see Kennicutt 1989 for details). The radial coordinate in each galaxy is normalized to the observed star formation threshold radius. Most of the disks lie remarkably close to the stability limit; Q rarely varies by more than a factor of 5 within individual disks, even though the absolute surface densities vary by more than two orders of magnitude.

Figure 6 also shows that there is a constant value of Q (roughly 0.7) which reproduces the observed star formation thresholds in all of the galaxies. Galaxies with low disk SFRs (usually types S0 to Sb) are usually characterized by gas densities which lie near, but below the $Q \simeq 0.7$ threshold (Kennicutt 1989). If this interpretation is correct, then the observed shape of the star formation law is due to large scale gravitational stability processes.

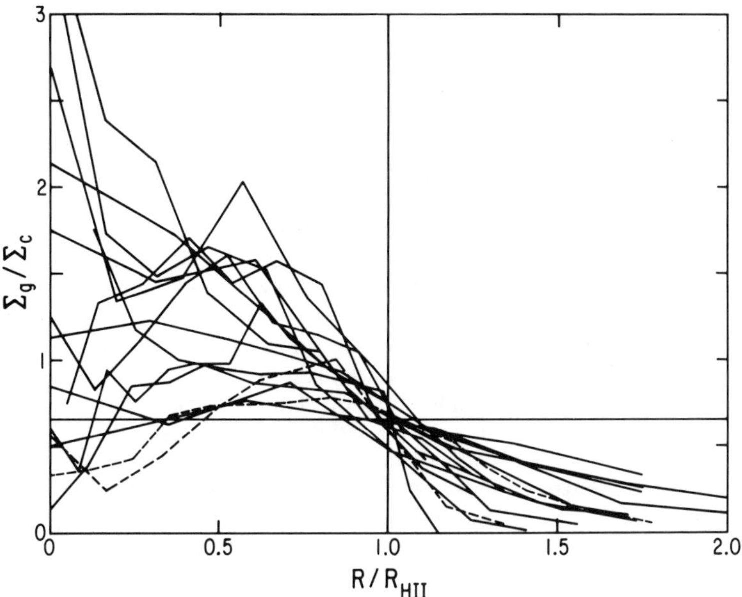

Fig. 6. Radial dependence of the ratio of gas surface density to the critical density for gravitational stability (Toomre criterion), for 15 galaxies with active star formation. The radial coordinate is normalized to the observed threshold radius.

Other physical mechanisms are needed to understand the form of the star formation law above the threshold. Unfortunately the roughly power-law shape at high gas densities is consistent with any number of physical processes. For example, an $n = 2$ Schmidt law, which appears to describe the small scale star formation in the Galactic disk, can be explained nicely by a cloud-cloud collision model (e.g., Scoville and Good 1987), but as pointed out by Larson (1988) several other mechanisms can produce Schmidt laws with similar indices. Other workers have proposed modified versions of a Schmidt law, based on the expected dynamical behavior of the interstellar gas (e.g., Wyse 1986, Wyse and Silk 1989).

Several workers have investigated the possibility that large scale star formation is a self-regulated process, where pressure input into the interstellar medium from supernovae, stellar winds, or radiation reaches equilibrium with the gravitational binding pressure of the gas (e.g., Franco and Cox 1983, Dopita 1985, 1990; Silk 1987). When applied on galactic scales these models often make relatively simple predictions about how the SFR should scale, and these may be subject to observational tests.

EVOLUTION AND THE HUBBLE SEQUENCE

I conclude by discussing (somewhat speculatively) another fundamental question which can be addressed by these observations: To what extent is the Hubble sequence, especially the S0 and spiral sequence, an evolutionary sequence? It is well known that Hubble's own thinking on this subject was heavily influenced by Jeans (1929), who postulated that galaxies evolve from ellipticals to spirals and then irregular galaxies. Forty years later the reigning paradigm was that any evolution along the Hubble sequence, if it occurred at all, was restricted to the epoch immediately following galaxy formation. Perhaps even more deeply rooted in much current thinking is the assumption that Hubble's morphological sequence is uniquely defined by an underlying "manifold" of physical properties, and that the basic structure of this one or two-parameter sequence was defined at the time of galaxy formation.

While it is clear that these contemporary paradigms lie closer to the truth than those of Hubble's time, many of the observations discussed earlier suggest that evolution may play a very significant role in the Hubble sequence, especially among the spiral and lenticular galaxies. Perhaps the strongest circumstantial evidence is the relatively short gas consumption lifetimes of many spiral galaxies. Sc disks, which have been forming stars at a steady rate over a large fraction of the Hubble time, will exhaust their present gas supplies in the inner disks in a small fraction of the Hubble time; indeed the gas densities will be driven below the levels required for global star formation in even less time. Subsequently these Sc galaxies will quickly evolve to systems more closely resembling small bulge Sa-Sb galaxies, or perhaps the low surface brightness galaxies described by Bothun at this meeting. Perhaps these latter galaxies represent formerly active star forming disk systems which exhausted their gas supplies in the recent past, as has been suggested for many S0 disks by Gregg (1989) and Bothun and Gregg (1989).

One implication of such evolution would be that there is more than one evolutionary path to a particular galaxy type. For example, S0 galaxies may form as the result of at least three distinct evolutionary processes, rapid star formation and gas depletion immediately after disk formation (the conventional view– see Sandage 1986), as a result of ram pressure stripping of the gas disk of a spiral entering a rich cluster (Dressler 1984 and references therein), or from gas depletion by star formation in what was previously a spiral galaxy, the scenario described above. Conversely, one can envision many scenarios which would lead two identical protogalaxies to evolve into very different types of galaxies today. The distinctions are of more than academic interest, because they could seriously influence the interpretation of observations of disk galaxies at large lookback times.

ACKNOWLEDGEMENTS

I acknowledge the support of the National Science Foundation, through Grant AST-8613257.

REFERENCES

Barry, D. C. 1988, *Ap. J.*, **334**, 446.
Berkhuijsen, E. M. 1977, *Astr. Ap.*, **57**, 9.
Bothun, G. D., and Gregg, M. D. 1989, *A. J.*, in press.
Buat, V., and Deharveng, J. M. 1988, *Astr. Ap.*, **195**, 60.
Buat, V., Deharveng, J. M., and Donas, J. 1989, *Astr. Ap.*, in press.
Buat, V., Donas, J., and Deharveng, J. M. 1987, *Astr. Ap.*, **185**, 33.
Cowie, L. L. 1981, *Ap. J.*, **245**, 66.
Davies, R. D., Elliott, K. H., and Meaburn, J. 1976, *Mem. R.A.S.*, **81**, 89.
Donas, J., Deharveng, J. M., Laget, M., Milliard, B., and Huguenin, D. 1987, *Astr. Ap.*, **180**, 12.
Dopita, M. A. 1985, *Ap. J. (Letters)*, **295**, L5.
Dopita, M. A. 1990, in *The Interstellar Medium in External Galaxies*, ed. J. M. Shull and H. A. Thronson (Dordrecht: Kluwer), in press.
Dressel, L. L., O'Connell, R. W., and Telesco, C. M. 1989, in *The Interstellar Medium in External Galaxies*, ed. D. Hollenbach and H. Thronson, NASA Conference Publications, in press.
Dressler, A. 1984, *Ann. Rev. Astr. Ap.*, **22**, 185.
Dressler, A., and Gunn, J. E. 1982, *Ap. J.*, **263**, 533.
Duric, N., Bourneuf, E., and Gregory, P. C. 1988, *A. J.*, **96**, 81.
Elmegreen, B. G. 1979, *Ap. J.*, **231**, 372.
Fall, S. M., and Efstathiou, G. 1980, *M.N.R.A.S.*, **193**, 189.
Franco, J., and Cox, D. P. 1983, *Ap. J.*, **273**, 243.
Freedman, W. L. 1984, Ph.D. thesis, University of Toronto.
Freedman, W. L. 1985, *Ap. J.*, **299**, 74.
Gallagher, J. S., Bushouse, H., and Hunter, D. A. 1989, *A. J.*, **97**, 700.
Gallagher, J. S., Hunter, D. A., and Tutukov, A. V. 1984, *Ap. J.*, **284**, 544.
Goldreich, P., and Lynden-Bell, D. 1965, *M.N.R.A.S.*, **130**, 97.
Gregg, M. D. 1989, *Ap. J.*, **337**, 45.
Guiderdoni, B. 1987, *Astr. Ap.*, **172**, 27.
Hodge, P., Lee, M. G., and Kennicutt, R. C. 1989, *Pub. A.S.P.*, **101**, 640.
Hunter, D. A., and Gallagher, J. S. 1985, *Ap. J. Suppl.*, **58**, 533.
Hunter, D. A., and Gallagher, J. S. 1986, *Pub. A.S.P.*, **98**, 5.
Iben, I., and Laughlin, G. 1989, *Ap. J.*, **341**, 312.
Jeans, J. L. 1929, *Astronomy and Cosmogony*, (Cambridge: Cambridge University Press).
Kennicutt, R. C. 1983, *Ap. J.*, **272**, 54.
Kennicutt, R. C. 1986, in *Stellar Populations*, ed. C. Norman, A. Renzini, and M. Tosi, (Cambridge: Cambridge University Press), p. 125.
Kennicutt, R. C. 1989, *Ap. J.*, in press.
Kennicutt, R. C. 1990, in *The Interstellar Medium in External Galaxies*, ed. J. M. Shull and H. A. Thronson, (Dordrecht: Kluwer), in press.
Kennicutt, R. C., and Congdon, C. W. 1990, in preparation.
Kennicutt, R. C., and Kent, S. M. 1983, *A. J.*, **88**, 1094.
Kennicutt, R. C., Edgar, B. K., and Hodge, P. W. 1989, *Ap. J.*, **337**, 761.
Klein, U., and Gräve, R. 1986, *Astr. Ap.*, **161**, 155.

Larson, R. B. 1988, in *Galactic and Extragalactic Star Formation*, ed. R. E. Pudritz and M. Fich (Dordrecht: Kluwer), p. 435.
Larson, R. B., and Tinsley, B. M. 1978, *Ap. J.*, **219**, 46.
Larson, R. B., Tinsley, B. M., and Caldwell, C. N. 1980, *Ap. J.*, **237**, 692.
Lonsdale, C. J., and Helou, G. 1987, *Ap. J.*, **314**, 513.
Massey, P. 1985, *Pub. A.S.P.*, **97**, 5.
Massey, P., Garmany, C. D., Silkey, M., and DeGioia-Eastwood, K. 1989, *A. J.*, **97**, 107.
Massey, P., Parker, J. W., and Garmany, C. D. 1989, *A. J.*, in press.
Noh, H.-R., and Scalo, J. 1989, *Ap. J.*, in press.
Oey, M. S., and Kennicutt, R. C. 1989, in *The Interstellar Medium in External Galaxies*, ed. D. Hollenbach and H. A. Thronson, NASA Conference Publication Series, in press.
Ohta, K., Sasaki, M., and Saito, M. 1989, *Pub. A.S. Japan*, **40**, 653.
Pogge, R. W., and Eskridge, P. B. 1987, *A. J.*, **93**, 291.
Quirk, W. J. 1972, *Ap. J. (Letters)*, **176**, L9.
Roberts, M. S. 1963, *Ann. Rev. Astr. Ap.*, **1**, 149.
Sandage, A. 1986, *Astr. Ap.*, **161**, 89.
Searle, L., Sargent, W. L. W., and Bagnuolo, W. G. 1973, *Ap. J.*, **179**, 427.
Scalo, J. M. 1986, *Fund. Cos. Phys.*, **11**, 1.
Schmidt, M. 1959, *Ap. J.*, **129**, 243.
Scoville, N. Z., and Good, J. C. 1986, in *Star Formation in Galaxies*, ed. C. J. Lonsdale Persson, NASA Conference Publication 2466, p. 3.
Shaya, E. J., and Federman, S. R. 1987, *Ap. J.*, **319**, 76.
Silk, J. 1987, in *Star Forming Regions*, ed. M. Peimbert and J. Jugaku (Dordrecht: Reidel), p. 663.
Skillman, E. D. 1987, in *Star Formation in Galaxies*, ed. C. J. Lonsdale Persson, NASA Conference Publication 2466, p. 263.
Smith, A. M., and Cornett, R. H. 1982, *Ap. J.*, **261**, 1.
Stark, A. A., Knapp, G. R., Bally, J., Wilson, R. W., Penzias, A. A., and Rowe, H. E. 1986, *Ap. J.*, **310**, 660.
Telesco, C. M. 1988, *Ann. Rev. Astr. Ap.*, **26**, 343.
Thronson, H. A., Tacconi, L., Kenney, J., Greenhouse, M., Margulis, M., Tacconi-Garman, L., and Young, J. 1989, *Ap. J.*, in press.
Thronson, H. A., and Telesco, C. M. 1986, *Ap. J.*, **311**, 98.
Tilanus, R. P. J., and Allen, R. J. 1989, *Ap. J. (Letters)*, **339**, L57.
Toomre, A. 1964, *Ap. J.*, **139**, 1217.
Twarog, B. A. 1980, *Ap. J.*, **242**, 242.
Walterbos, R. A. M., and Schwering, P. B. W. 1987, *Astr. Ap.*, **180**, 27.
Wyse, R. F. G. 1986, *Ap. J. (Letters)*, **311**, L41.
Wyse, R. F. G., and Silk, J. 1989, *Ap. J.*, **339**, 700.
Young, J. S., Schloerb, F. P., Kenney, J. D., and Lord, S. D. 1986, *Ap. J.*, **304**, 443.
Zasov, A., and Simikov, S. 1988, *Astrophysica*, **29**, 190.

NEARBY LUMINOUS BLUE GALAXIES

John S. Gallagher III
AURA, Inc. and Space Telescope Science Institute, Suite 701, 1625 Massachusetts Ave. N.W., Washington, DC 20036

ABSTRACT Some galaxies are comparable to spirals in terms of blue luminosities but show optical evidence (blue colors, strong emission lines) for large Pop I stellar components. These luminous blue galaxies provide examples of rapidly evolving large galaxies in the nearby universe. This paper discusses the evolutionary status of luminous blue galaxies and their possible relationships to the excess numbers of faint blue galaxies that are being found by deep surveys of galaxies.

INTRODUCTION

Galaxies with very blue optical colors are of special interest since they are likely to contain either large populations of young OB stars or AGN. In this paper I will briefly review optical properties of luminous galaxies that fall in the former category, which I refer to as luminous blue galaxies or LBGs. LBGs are defined as having blue optical colors ($U-B < -0.1$), higher luminosities than the normally blue irregular galaxies ($L_B > 0.1L^*$, where L^* is the characteristic luminosity of the Schecter fit to the local galactic luminosity function; e.g. de Lapparent et al. 1989), and lack evidence for significant optical light contributions from AGNs.

LBGs are thus those systems which optically appear to have the least evolved stellar populations among galaxies with spiral-like luminosities in the nearby universe. Galaxies with dominant young stellar populations could belong to one of three evolutionary categories (cf. Sandage 1963, Burbidge et al. 1963): young galaxies where a minimal old stellar population is present, slowly evolving galaxies where the star formation rate (SFR) has not declined with time so that young stars overwhelm the optical light from older stars, and starburst systems where the star formation rate has recently increased thereby producing an excess of young, blue stars. Our data suggest that a large fraction of LBGs (but not all!) can be understood in terms of recent, collisionally-induced star bursts.

Some LBGs are readily detected by surveys for blue galaxies that can be carried out with Schmidt telescopes either spectroscopically (e.g. the Markarian or Michigan-Tololo surveys) or by color (e.g. the Haro, Kiso, or Case surveys). Many LBGs were also included in Zwicky's famous surveys for galaxies with unusual morphologies (see Sargent 1970). These techniques, and especially objective prism surveys, are biased towards galaxies with high surface brightnesses. Low surface brightness LBGs also exist; e.g. in the form of luminous DDO "dwarf" galaxies (Thuan and Seitzer 1979; Hunter and Gallagher 1985). The remainder of this paper concerns only the more traditional high surface brightness LBGs (HS-LBGs).

I review the properties of this class of galaxy, mainly on the basis of optical observations of a sample of HS-LBGs obtained in collaboration with D. Hunter and H. Bushouse which is tabulated in Gallagher et al. (1989). Far infrared characteristics derived from the IRAS data base are given for a few members of this sample in Hunter et al. (1989). These data provide the basis for an exploration of the evolutionary status of local HS-LBGs.

It is also interesting to compare characteristics of nearby HS-LBGs with those of blue galaxies which are prominent in faint galaxy surveys (e.g. Kron 1980, Broadhurst et al. 1988; Tyson 1988; Cowie et al. 1988). I will then turn to the question of whether selection effects associated with the properties of HS-LBGs could partially explain the overrepresentation of this class of galaxy in faint galaxy samples.

PROPERTIES OF HIGH SURFACE BRIGHTNESS LBGs

Observed Characteristics

Photometric parameters for a sample of typical HS-LBGs selected from the Markarian catalogs have been beautifully summarized by Huchra (1977): they have $-0.1 > U-B > -0.8$ with a median of about $U-B = -0.3$, blue surface brightnesses that are up to 30-50% higher than normal galaxies, and make up about 5-10% of the nearby galaxies. Huchra also demonstrated that HS-LBGs are distinctive in having almost any morphology except that of a normal spiral galaxy. In particular, LBGs exist with E-like, de Vaucouleurs law brightness profiles (e.g. Loose and Thuan 1986), S0 forms, as well as with the expected peculiar structures (e.g. the clumpy irregulars, Heidmann 1983; see also Sargent 1970; Salzer et al. 1989a). Examples of HS-LBGs are shown in Figures 1 and 2. Among the more actively star-forming luminous galaxies, the correlation between stellar population and morphology that exists for most normal galaxies breaks down (cf. Morgan and Osterbrock 1969).

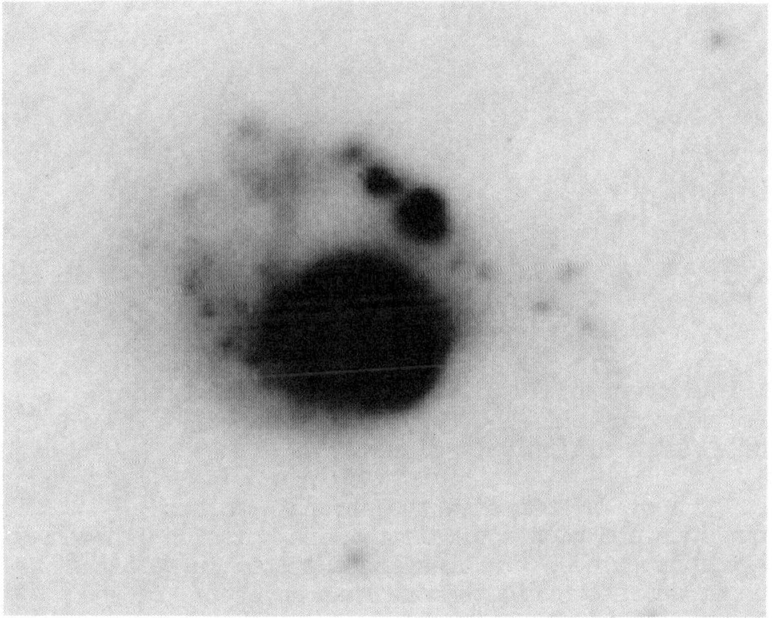

Fig. 1. R-band image of III Zw 33 obtained under good seeing condition with the KPNO 2.1-m telescope. The structure of this galaxy is suggestive of effects from a recent interaction.

HI observations were obtained for a large sample HS-LBGs by Gordon and Gottesman (1981). They showed that HS-LBGs have abnormal HI profiles relative to their luminosities in that "double-horned" HI line profiles commonly seen in spirals are rare. Hydrogen mass to blue light ratios are typical of those seen in late type spirals, but some of the most luminous HS-LBGs have very high HI contents ($M_{HI}>10^{10}$ solar masses). No HS-LBGs have been detected in CO (Gordon et al. 1982); so the molecular gas content of HS-LBGs is still uncertain. Even so, it is clear that as a class HS-LBGs are gas-rich on the basis of their HI gas contents alone.

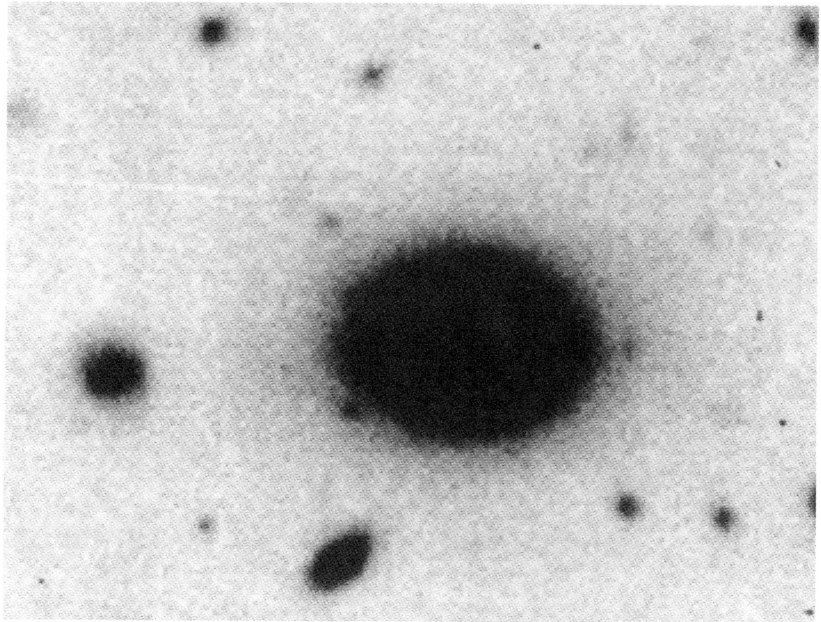

Fig. 2. R-band image of IV Zw 93, also taken with the KPNO 2.1-m telescope and TI CCD system. This galaxy is an example of a "blue elliptical" class of HS-LBG.

Optical spectra of HS-LBGs show the strong H-Balmer absorption lines and high equivalent width emission lines from HII regions. These features are expected in galaxies where younger stars contribute more than about 1/3 of the optical light. A typical HS-LBG blue region spectrum is shown in Figure 3. Our group has analyzed these spectra via stellar population synthesis techniques. We find that the mix of stellar temperatures that is required to reproduce the optical spectrum is between that expected for constant star formation rates (SFRs) and star bursts where the SFR has increased by more than a factor of three in less than 0.5 Gyr. Empirical emission line abundances further suggest oxygen abundances between roughly solar and LMC values (see also Salzer et al. 1989b).

Evolutionary States

The galactic evolutionary phases responsible for HS-LBGs has yet to be firmly established. Some progress has been made by measuring current SFRs and estimating the Roberts time (time scale to deplete the current gas content; M_{HI}/SFR). We calculated SFRs via standard models for a few LBGs where we have obtained global H-alpha flux measurements (Hunter and Gallagher 1986). The SFRs that we derived were based on a Salpeter initial mass function and standard assumptions about the Lyman continuum luminosities of O stars (Gallagher and Hunter 1987). No correction was made for internal absorption of H-alpha luminosity by galactic dust; these SFRs are lower limits.

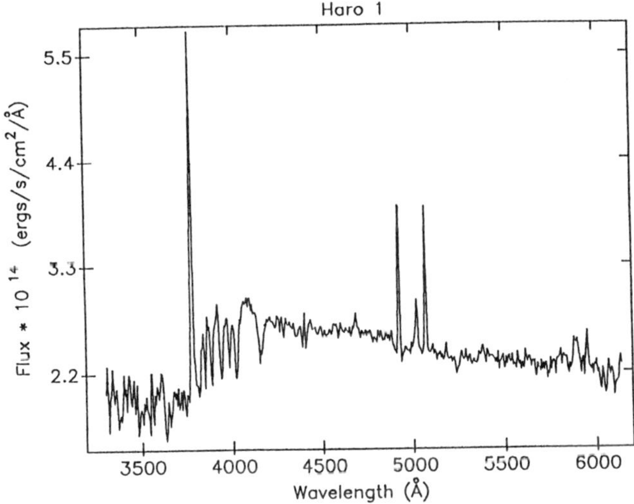

Fig. 3. This blue region spectrum of Haro 1 was obtained with a 22 arsec aperture spectrophotometer on a KPNO 0.9-m telescope. The strong emission and Balmer absorption lines are typical of HS-LBGs.

Far infrared observations presented in Hunter et al. (1989; see also Maehara et al. 1988) show that many HS-LBGs are likely to have considerable dust absorption optical depths in regions of active star formation. The SFRs based on optical data therefore might be too low by factors of 2 or more. Interestingly, optical spectra do not show strong evidence for internal reddening in either the decrements of higher Balmer series emission lines or in the slopes of the continua, even in galaxies with pronounced FIR excesses. Optical spectra alone do not provide a complete evolutionary picture for HS-LBGs.

Global SFRs per unit area based on observed H-alpha fluxes are similar to those seen in spiral galaxies. However, the star formation process is often centrally concentrated in HS-LBGs, and central SFRs per area that we derive from optical spectra are very high, often amounting to the production of the total local Galactic disk stellar surface density in only 1 Gyr. When corrections are made for effects of internal absorption using the IRAS far infrared data, global SFRs could be several times those found in spirals, but even so the HS-LBGs do not stand out as being truly extraordinary in terms of estimated current SFRs (as compared, e.g. with M82-type systems where the SFRs are well above those of normal galaxies).

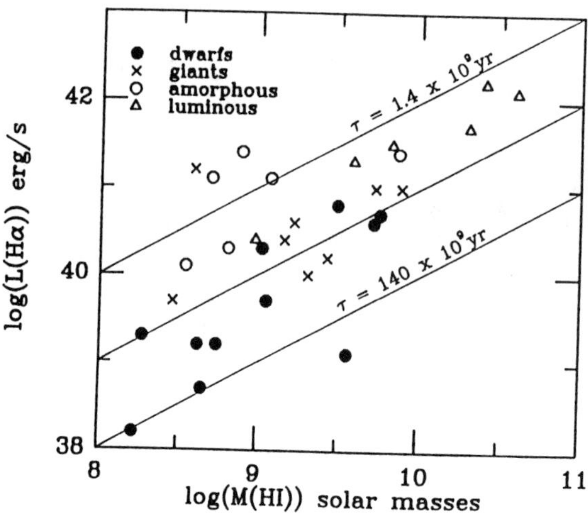

Fig. 4. Roberts times are plotted for blue galaxies using only optically derived SFRs and gas masses derived from HI contents. LBGs are evolving on time scales that are short as compared with a Hubble time.

Figure 4 illustrates Roberts times for HS-LBGs based on optical SFRs and HI gas content corrected for a normal He abundance. Also shown are other classes of less luminous blue galaxies. While HS-LBGs do have short Roberts' times (typically several Gyr), they are not candidates for extreme star bursts where the Roberts times are less than 1 Gyr. Unfortunately this diagram is subject to major uncertainties that work in opposite directions. The FIR data suggest reductions in evolutionary time scales by factors of 2-5, which require that many HS-LBGs be in rapidly changing evolutionary states.

On the other hand, the molecular gas content is unknown, but if comparable to that in spirals, might be a significant fraction of the HI mass thereby increasing the Roberts' time by a factor of 2. Finally, much of the HI in galaxies is often located at large galactic radii where it is not readily available for star formation; this also leads to an overestimate of the Roberts times. In summary, characteristic time scales for gas depletion in HS-LBGs are most likley in the range of 1-3 Gyr, or about a factor of 2 shorter than Roberts times derived in a similar way for normal spiral galaxies.

HS-LBGs are a relatively rapidly evolving class of galaxy, and therefore either must be young or experiencing a transitory evolutionary phase of enhanced SFR (see Maehara et al. 1988 and references therein). The properties of most HS-LBGs, however, are inconsistent with a very young galaxy model. The mix of stellar population temperature classes is wrong and metal abundances are too high for very young galaxies. Moderate youth (age since onset of major star formation of 1 Gyr or so) or slow formation are both workable models for HS-LBGs (e.g. the study of II Zw 23 by Keel 1988). Collisionally-induced star bursts are feasible explanations for 1/2-2/3 of HS-LBGs where disturbed neighbors or other morphological evidence exist for a recent external perturbation. The mechanisms that account for the other types of HS-LBGs remain to be discovered.

RELATIONSHIPS TO FAINT BLUE GALAXIES

Faint galaxies tend to have bluer colors with increasing optical magnitude (Kron 1980, Tyson 1988, Cowie et al. 1988). Even in moderately faint samples a disquieting fraction of all galaxies have blue galaxy characteristics. For example, the Durham/ Anglo-Australian Telescope deep redshift survey sample detects galaxies that preferentially have [O II] emission line equivalent widths that are larger than those found in nearby spiral galaxies (Broadhurst et al. 1988). The typical D/AAT survey galaxy resembles an LBG in terms of emission line properties (Gallagher et al. 1989). The two examples of faint blue galaxies examined by Gallagher and Hamilton (1988) further suggests that the LBG parallel probably holds for blue galaxies seen to redshifts of z=0.5-0.7.

This analogy immediately tells us that the moderately faint blue galaxies are not necessarily in unique evolutionary phases; galaxies that could produce the observed characteristics are seen as HS-LBGs in the nearby universe. Note, however, that HS-LBGs are rare in normal catalogs of galaxies because of selection effects. Arp (1965) emphasized that objects with "compact" structures will tend to be missed by galaxy catalogs that select on the basis of angular size (see also the discussion of observational selection effects by G. Ellis 1980). Galaxy morphology is also a problem for LBGs: low surface brightness LBGs are confused with dwarf irregulars while HS-LBGs are easily misclassified (e.g. as planetary nebulae, such as Abell 76, Talent et al. 1982).

The preponderance of galaxies with LBG-like spectral characteristics at even moderate redshifts, however, is a puzzle. Such galaxies make up <10% of the local luminous galaxy population. Based on measured evolutionary time scales, pronounced evolution in the fraction of galaxies with HS-LBG characteristics would only be expected for look back times of more than 5 Gyr, which correspond to z>0.5. Either galaxies evolve more rapidly than we think, or there is a problem with sample selection in the mid-level surveys.

Given the current level of understanding of galaxy evolution, the latter conclusion would hardly be a surprise! However, the analysis of number counts by Broadhurst et al. appears to exclude models in which the excess of blue galaxies arises from smoothly declining SFRs in as the field galaxy population ages.

As an alternative Broadhurst et al. suggest that the excess of faint blue galaxies is due to the presence of a star bursting population of galaxies, a conclusion that would also be consistent with the observed parallels between local LBGs and faint blue galaxies. The cosmololgically interesting question is then whether the fraction of the galaxy population experiencing such bursts has changed significantly over only the past few Gyr.

This issue may prove difficult to resolve because of selection effects. The HS-LBGs are especially likely to be included in surveys for galaxies at moderate-to-high redshifts. Their ultraviolet surface brightnesses over

10-20 kpc scales may run from 30% to 100% higher than those of spirals in the U band to more than 10 times higher in the 100-200 nm rocket ultraviolet spectral region (based on the limited OAO and ANS ultraviolet photometry of galaxies). Preferential selection of galaxies with HS-LBG characteristics at z>0.2 is therefore a serious problem that must be examined in detail before we can conclude that the spectral properties of field galaxies have evolved significantly during the past few Gyr.

Unfortunately, the question of rate of evolution of the field galaxy population thus remains as a challenge for the hopefully not too distant future. No doubt such matters will be summarized in the historical introductions of the Astronomical Society of the Pacific's bi-centennial meeting.

ACKNOWLEDGEMENTS

Much of the research reported here on LBGs has been done in collaboration with Howard Bushouse and Deidre Hunter. Support for multi-wavelength investigations from the NASA ADP and O/IR observations from NOAO are gratefully acknowledged. I also thank the Organizing Committe for the opportunity to participate in this meeting.

REFERENCES

Arp, H. C. 1965, Ap.J., 142, 402.

Broadhurst, T. J., Ellis, R. S., and Shanks, T. 1988, M.N.R.A.S., 235, 827.

Burbidge, E. M., Burbidge, G. R., and Hoyle, F. 1963, Ap.J., 138, 873.

Cowie, L. L., Lilly, S. J., Gardner, J. and McLean, I. S. 1988, Ap.J. Lett., 332, L29.

Ellis, G. F. R. 1980, Ann. N.Y. Acad. Sci., 336, 130.

Gallagher, J. S. and Hunter, D. A. 1987, in Star Formation in Galaxies, ed. J. L. Persson, NASA Conference Publication 2466, p.259.

Gallagher, J. S. and Hamilton, D. 1988, Ap.J., 330, 661.

Gallagher, J. S., Bushouse, H., and Hunter, D. A. 1989, A.J., 97, 700.

Gordon, D. and Gottesman, S. J. 1981, A.J., 86, 161.

Gordon, M. A., Heidmann, J., and Epstein, E.E. 1982, P.A.S.P., 94, 415.

Heidmann, J. 1983, Highlights in Astronomy, 6, 611.

Hunter, D. A. and Gallagher, J. S., 1985, Ap. J. Suppl., 58, 533.

Hunter, D. A. and Gallagher, J. S. 1986, P.A.S.P., 98, 5.

Hunter, D. A., Gallagher, J. S., Rice, W. L., and Gillett, F. C. 1989, Ap.J., 336, 152.

Huchra, J. P. 1977, Ap.J. Suppl., 35, 171.

Keel, W. C. 1988, Astron. Ap., 202, 41.

Kron, R. 1980, Ap.J. Suppl. 43, 305.

de Lapparent, V., Geller, M. J., and Huchra, J. P. 1989, Ap.J., 343, 1.

Loose, H. H. and Thuan, T. X. 1986, Ap.J., 309, 59.

Maehara, H., Hamabe, M., Bottinelli, L., Gougenheim, L., Heidmann, J., and Takase, B. 1988, P.A.S.J., 40, 47.

Morgan, W. W. and Osterborck, D. E. 1969, A.J., 74, 515.

Salzer, J. J., MacAlpine, G. M., and Borosn, T. A. 1989a, Ap.J. Suppl., 70, 447.

Salzer, J. A., MacAlpine, G. M., and Boroson, T. A. 1989b, Ap.J. Suppl., 70, 479.

Sandage, A. R. 1963, Ap.J., 138, 863.

Sargent, W. L. W., 1970, Ap.J., 160, 405.

Thuan, T. X. and Seitzer, P. 1979, Ap.J., 231, 680.

Tyson, J. A. 1988, A.J., 96, 1.

A SEARCH FOR WOLF-RAYET STARS AT THE GALACTIC CENTER

M.W.WERNER* & J.R.STAUFFER* 245-6, NASA-Ames Research Center, Moffett Field, CA 94035

E.E.BECKLIN, Institute for Astronomy, University of Hawaii, Honolulu, HI 96822

Understanding the stellar population at the Galactic Center is important for evaluating the history of star formation in this region and also for understanding the luminosity and ionization sources which are known to be concentrated in the central parsec of the galaxy (Werner and Davidson, 1989). Such studies of the nearest galactic nucleus also offer the hope of illuminating the processes occurring in more distant galactic nuclei. Infrared images of the core of the Galaxy by Rieke et al (1989) show that the components of the IRS16 complex, which is close to the dynamical center of the Galaxy, stand out as blue (T>10,000K) objects against the background of cool red stars which dominate the stellar population in this region. Allen et al (1989) and Werner and Stauffer (1988) have suggested that these and other hot objects within the central few arecminutes of the Galaxy might be Wolf-Rayet stars. In particular, Allen et al call attention to the similarity of the 2um spectrum of one of the Galactic Center objects to those of the WN9/Ofpe stars in the Large Magellanic Cloud, while Werner and Davidson detail other similarities and point out that a handful of such stars could provide the luminosity and ionization of the central parsec of the Galaxy.

We have explored this hypothesis with infrared spectroscopy in the 3um window, using the Cooled Grating Spectrometer at the IRTF with a spectral resolution of 300 km/s and an angular resolution of 2.5 arcsec. We chose to observe the HeII (7-6 + 11-8) complex at 3.092um, which is strong in the spectra of both WN and WC stars in the Galaxy (Hillier, 1982; Williams, 1982) but is not expected to be strongly excited in the weakly-ionized diffuse plasms at the Galactic Center.

The preliminary data obtained to date are shown in the Figure. The spectra have been divided by the spectrum of a reference star to remove atmospheric effects and are plotted on arbitrary linear scales so that equivalent widths can be compared. The lowest spectrum shows the HeII complex together with the nearby Pfund-epsilon line of HI in the high excitation planetary nebula NGC7027, in which the lines are unresolved.

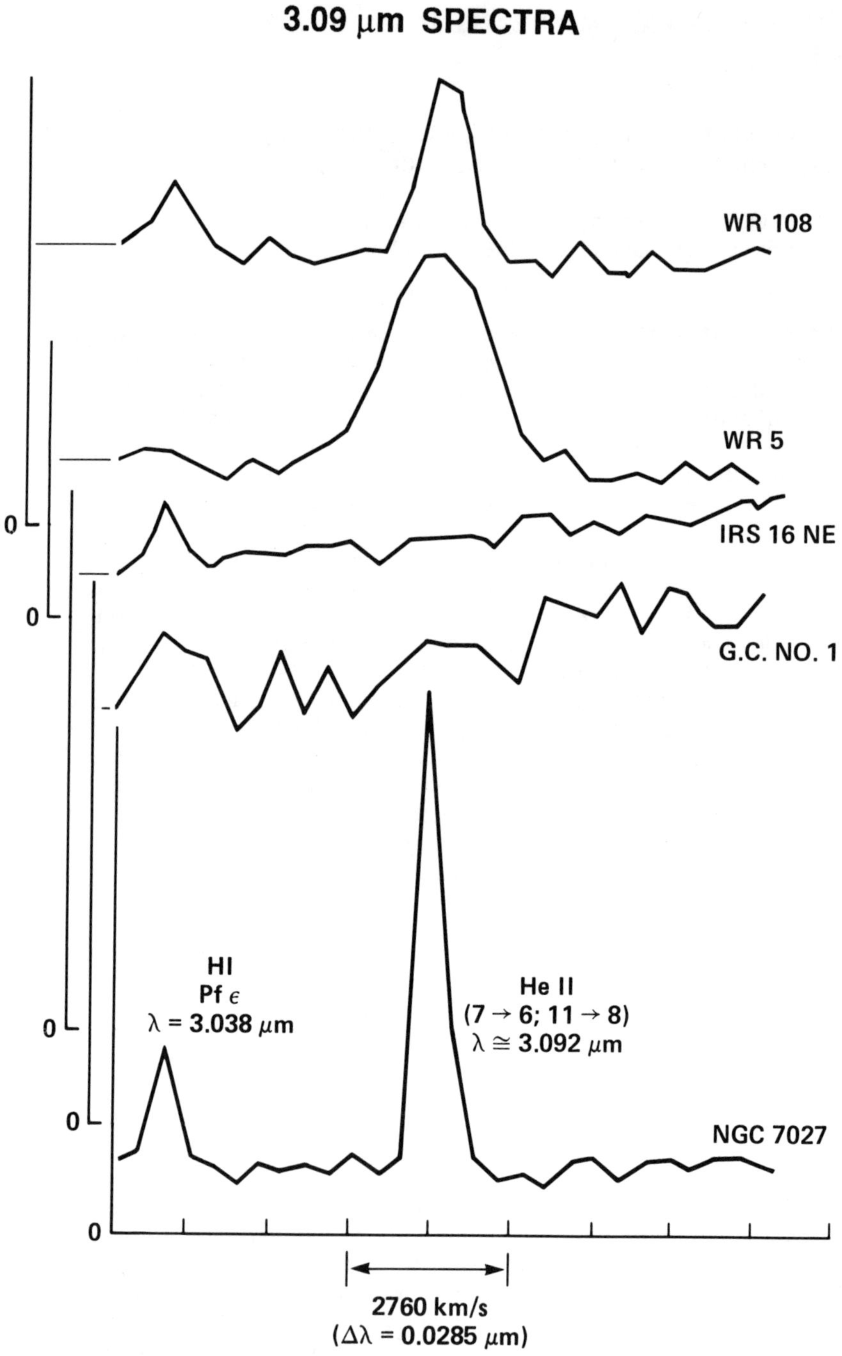

3.09 μm SPECTRA
WR 108
WR 5
IRS 16 NE
G.C. NO. 1
NGC 7027
0
0
0
0
0
HI
Pf ϵ
λ = 3.038 μm
He II
(7 → 6; 11 → 8)
λ ≅ 3.092 μm
2760 km/s
(Δλ = 0.0285 μm)

The upper two spectra show the 3.092um feature in two galactic WR stars: WR108 is the only WN9/Of star known in the Galaxy and is thus the closest analogue to the LMC stars cited by Allen et al. WR5 is a WC6 star and has a remarkably strong and broad HeII line. The line is resolved in WR108 as well.

The corresponding spectra of two objects in the Galactic Center, IRS16NE and GC#1, are also shown. GC#1 is the object singled out by Allen et al as a possible WR star. Neither galactic center object shows evidence for the 3.092um line. Three-sigma upper limits on the line/continuum ratio in the Galactic Center are ~0.5 times the ratio observed in WR108 and ~0.1 times that seen in WR5. Thus these incomplete preliminary observations provide no additional evidence for the presence of Wolf-Rayet stars in the core of the Galaxy.

---

*-Visting Astronomer, IRTF

REFERENCES

Allen, D.A., et al. 1989, in IAU Symposium #136, The Galactic Center, ed. M. Morris

Hllier, J.D. 1982, in IAU Symposium #99, Wolf-Rayet Stars, ed. C.W.H. de Loore and A.J.Willis, p. 225.

Rieke, G.H., Rieke, M.J., and Paul, A.E. 1989, Ap.J., 336, 752.

Werner, M.W., and Davidson, J.A. 1989, in IAU Symposium #136, The Galactic Center, ed. M. Morris, p. 423.

Werner, M.W., and Stauffer, J.R. 1988, BAAS, 20 1097.

Williams, P.M. 1982, in IAU Symposium #99, Wolf-Rayet Stars, ed. C.W.H. de Loore and A.J.Willis, p. 73.

PROFUSE CENTRAL STAR FORMATION IN 'SO' GALAXIES

L. L. DRESSEL
Dyer Observatory, Vanderbilt University, Nashville, TN 37235

R. W. O'CONNELL
Leander McCormick Observatory, University of Virginia, Charlottesville, VA 22903-0818

C. M. TELESCO and R. DECHER
Space Science Laboratory, ES-63, NASA/Marshall Space Flight Center, Huntsville, AL 35812

ABSTRACT Radio continuum flux densities, far infrared flux densities, far infrared colors, and ultraviolet and optical spectra indicate that high rates of star formation are occurring in some SO galaxies, especially barred galaxies. Radio continuum and infrared images show that the star forming region is typically concentrated within the central few kpc of the galaxy. The high incidence of central star formation in barred galaxies suggests that bar-driven inflow of gas is fueling the star-forming regions in these galaxies.

INTRODUCTION

As a class, SO galaxies are characterized by a lack of resolved bright stars in the disk. However, several lines of evidence support the hypothesis that a high rate of star formation is occurring at the centers of some SO galaxies.

DATA AND INTERPRETATION

Far Infrared Colors and Powers

In star-forming regions, dust absorbs much of the ultraviolet emission of the hot massive young stars and re-emits the energy as infrared radiation. This radiation is warmer than that produced by dust in the general interstellar medium. Powerful warm infrared sources can thus indicate sites of intense star formation. Globally warm powerful sources have been found to occur most frequently in SO and early-type spiral galaxies in a study of IRAS data (Lonsdale et al. 1985) for a complete sample of 738 galaxies brighter than $14^{m}.5$ mag (Dressel 1988). The detection rate above F(60μ) = 1.5 Jy is much higher for SBO and SBa galaxies (27%) than for SO and Sa galaxies (7%). Bar-driven inflow of gas may thus be fueling the star-forming regions in many cases.

Radio-Infrared Flux Correlation

As shown by Helou, Soifer, and Rowan-Robinson (1985), radio continuum flux density and far infrared flux are extremely well correlated for both ordinary spiral disks and for starburst galaxies. This correlation presumably results from a relationship between the numbers of luminous massive young stars (which heat the dust) and of recent supernovae (which contribute to the radio emission). The magnitude-limited SO + SBO sample shows a similar correlation, consistent with the suggestion that the far infrared flux is produced by starlight-heated dust.

Radio Maps

VLA radio continuum observations of a few SO galaxies with warm, bright infrared sources show that the radio emission is centrally concentrated with a variety of morphologies. For example, UGC 12618 has a diffuse source with a central peak; UGC 03265 has a ring-shaped source. These sources have total diameters of about 6 kpc and 4 kpc, respectively for H_0 = 50 km s^{-1} Mpc^{-1}. The major blue diameters of the galaxies are 52 kpc and 43 kpc, respectively (Nilson 1973). The radio sources thus occupy the central 1% of the surface area of the optical disks.

Infrared Images

We have imaged six early-type galaxies at 10.8 microns at the NASA Infrared Telescope Facility with a 5 x 4 array of 4.3" x 4.3" FWHM pixels. All of the infrared sources are centered on the optical galaxies and are unresolved or marginally resolved. The sources in these galaxies are thus smaller in FWHM than the approximately 2 kpc FWHM pixels.

Ultraviolet Evidence of Hot Stars

Despite extinction by dust, enough ultraviolet light escapes many of the infrared-bright SO galaxies to make them unusually bright in the ultraviolet. Eight of the twelve SO + SBO galaxies in the magnitude-limited sample with warm, bright infrared sources (F(100μ)/F(60μ) < 2.0, F(60μ) > 1.5 Jy) are Markarian galaxies. Only one of these Markarian galaxies is known to be a Seyfert galaxy. The colors of non-Seyfert Markarian galaxies are consistent with aging bursts of star formation (Huchra 1977).

Optical Evidence of Hot Stars

Several SO galaxies with warm bright infrared sources were included in the spectrophotometric survey of early-type galaxies by O'Connell and Dressel (1978). The centers of the galaxies were observed with the KPNO 2.1m telescope + IIDS through an 8.4 arcsec aperture with a resolution of 11Å. The spectra of the infrared-bright SO galaxies generally show a strong A star component (deep Balmer absorption lines) and usually show emission lines typical of HII regions.

CONCLUDING REMARKS

A key question to our understanding of these galaxies is whether they "really" are SO galaxies, or at least would have been recognized as SO galaxies before the episode of central star formation began. Some of Nilson's classifications (used here) have been confirmed by Sandage or de Vaucouleurs and collaborators from better plates; some of the galaxies may be misclassified Sa galaxies; some are apparently difficult to classify because of mixed characteristics, faint "non-SO" features, or peculiarities (due to the central star-formation process?). More optical imaging is needed to characterize the host galaxies and to study the evolution of their star-forming regions.

ACKNOWLEDGEMENTS

This research was supported at V.U. by grant AST-8818606 from the National Science Foundation and by funds from NASA administered by the American Astronomical Society.

REFERENCES

Dressel, L. L. 1988, Ap. J. (Letters), 329, L69.
Dressel, L. L., and Condon, J. J. 1978, Ap. J. Suppl., 36, 53.
Helou, G., Soifer, B. T., and Rowan-Robinson, M. 1985, Ap. J. (Letters), 298, L7.
Huchra, J. P. 1977, Ap. J., 217, 928.
Lonsdale, C. J., Helou, G., Good, J. C., and Rice, W. 1985, Cataloged Galaxies and Quasars Observed in the IRAS Survey, (Jet Propulsion Laboratory).
Nilson, P. 1973, Uppsala General Catalogue of Galaxies, Nova Acta, Sec. V:A, Vol. 1.
O'Connell, R. W., and Dressel, L. L. 1978, Nature, 276, 374.

A NON-STEADY COOLING FLOW MODEL FOR NGC 1275 (PERSEUS)

AVERY MEIKSIN

Space Telescope Science Institute

3700 San Martin Drive, Baltimore, MD 21218 and

The Johns Hopkins University

ABSTRACT The evolution of a cooling flow onto a central dominant galaxy in a galaxy cluster is studied using numerical spherically symmetric hydrodynamic computations. It is argued that the flows may be non-steady, with a short central cooling time and X-ray emitting central temperature maintained by gravitational heating due to the evolution of the cluster potential well. A model for NGC 1275 in Perseus is presented as an example. The model offers the possibility of reconciling the high X-ray inferred mass accretion rates with the severe constraints imposed by optical and radio measurements.

INTRODUCTION

In ~30 clusters of galaxies, X-ray measurements have revealed cooling times less than a Hubble time. In such clusters, a quasi-static inflow of gas may arise. The accretion rates estimated from X-ray observations range from a few up to $1000 M_\odot\,\mathrm{yr}^{-1}$, amounting to $\sim 10^{10} - 10^{13} M_\odot$ over the cluster lifetime. The fate of the accreted material, however, is currently unknown. Possibly the flows give rise to cD's: a dominant galaxy is found at the center of every cooling flow. (See Fabian, Nulsen, and Canizares 1984 for a review.)

Several characteristic properties of cooling flows which any good theory must explain include: (1) a *short central cooling time*, often $\lesssim 10^9$ yr; (2) an *inwardly decreasing temperature profile*; and (3) an *inwardly decreasing mass accretion rate.* The fundamental cooling flow question is: Since the cooling time is short, why is the ICM still hot? The simplest answer is that the flows are in a steady state. To maintain a steady state, the accreted material must somewhere be removed. The inwardly decreasing mass accretion rate suggests material is condensing out of the flow. Optical and radio observations, however, impose severe constraints on the amount of accreted material, which further restrict cooling flow models: (4) a *low star formation rate* with a normal disk IMF is inferred from optical color measurements, with a SFR one-tenth or less the X-ray estimated value, and (5) *small amounts of HI and*

H_2 are inferred from radio measurements, amounting to $< 1\%$ of the X-ray estimated accreted mass. One suggestion has been that the accreted material is converted into low mass stars ($\langle M \rangle \sim 0.5 M_\odot$) with $> 99\%$ efficiency.

An alternative is that the flows are not in a steady state. In this case, an inwardly decreasing accretion rate arises from the non-steady nature of the flow rather than from condensation. A non-steady flow could result from evolution of the cluster. The deepening of the cluster potential acts as a heat source that can compensate for the radiative losses. One such model is presented here for NGC 1275.

A TIME-DEPENDENT MODEL

The model includes two sources of external gravity, a static central dominant galaxy and a cluster with a King density profile, the central value of which increases with time. The self-gravity of the gas is included. The gas cools radiatively with half-solar abundances. Cosmological outer boundary conditions are imposed, and $\Omega = 1$ and $H_0 = 50\,\mathrm{km\,s^{-1}Mpc^{-1}}$ are assumed. The computations are started at $z = 1.5$ ($t = 3 \times 10^9$ yr), and terminated at $z = 0$ ($t = 13 \times 10^9$ yr). A mass sink is included according to $\rho_s = f q_s \rho / t_{\rm cool}$, where ρ is the local ICM gas density, $t_{\rm cool}$ is the local gas cooling time, f is the fractional deviation from hydrostatic equilibrium, and q_s is an efficiency factor, set to 0.5 below. The sink acts only to remove gas.

The fluid equations are solved using a spherically symmetric finite difference scheme, kindly provided by E. Bertschinger.

The results of a model for Perseus, assuming a cluster core radius of 350 kpc, are presented in Figure 1. Despite a cooling time less than 2×10^9 yr within the inner 15 kpc at $t = 11 \times 10^9$ yr, the flow has not reached a steady state by $t = 13 \times 10^9$ yr (Fig. 1*d*). Condensates form at the rate $\sim 60\ M_\odot\,\mathrm{yr}^{-1}$ by $z = 0$, and amount to $6 \times 10^{10}\ M_\odot$ by this time. The accretion rate at the cooling radius (210 kpc) is 1100 $M_\odot\,\mathrm{yr}^{-1}$. The accumulated condensate mass, which would presumably end as stars and cold gas, is <1% of the accretion rate at the cooling radius times 10^{10} yr. The remainder of the accreting material is still settling. The heating by cluster evolution permits a non-steady flow in which the local cooling-induced decrease in gas entropy arises primarily from an increasing gas density rather than from a decreasing gas temperature. Not until $\sim 18 \times 10^9$ yr does the flow reach a steady state. The accretion rate at this time is 1400 $M_\odot\,\mathrm{yr}^{-1}$, and $1.7 \times 10^{12}\ M_\odot$ of stars have accumulated.

CONCLUSIONS

A non-steady model is presented for the cooling flow onto NGC 1275 in which the potential well of Perseus is allowed to deepen with time. The inflowing gas settles within the central regions of the cluster, with a high central gas temperature ($>10^7$ K) and a short, though decreasing, central cooling time sustained by gravitational heating due to the evolution of the cluster. Although the flow will eventually reach a steady state in this model, the time

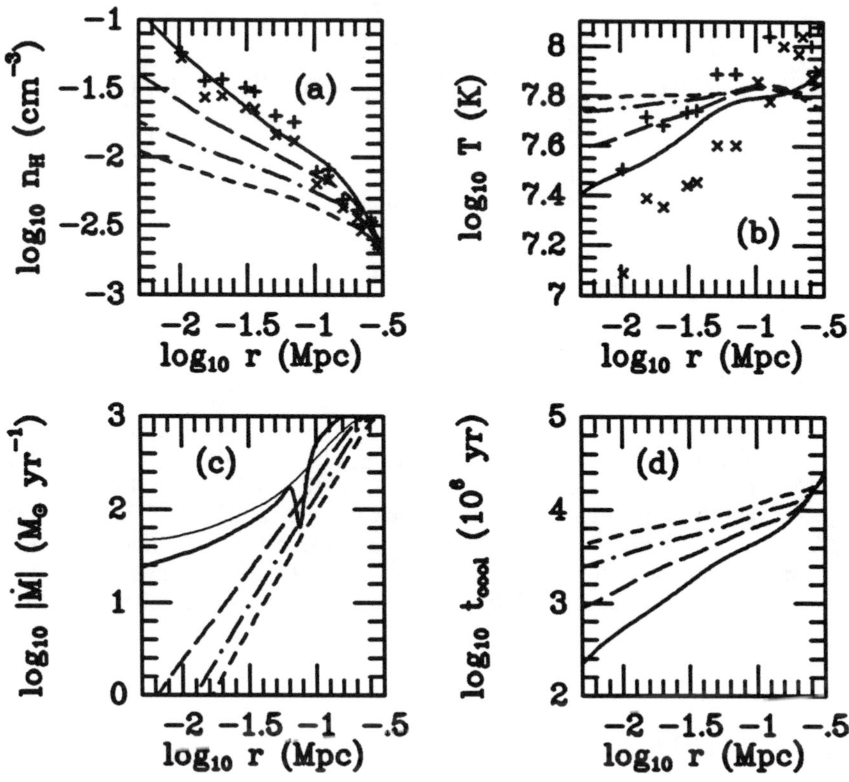

Fig. 1 A non-steady model for the cooling flow in NGC 1275. A core radius of 350 kpc is assumed for Perseus. Shown are the hydrogen number density, temperature, mass accretion rate, and cooling time at 7 (- - -), 9 (— · — · —), 11 (— — —), and 13×10^9 yr (——). The thin line in the $\dot{M}$ plot represents an estimate for $\dot{M}$ at $t = 13 \times 10^9$ yr, computed as the ratio of integrated luminosity to local gas enthalpy. The data (×'s and +'s) are from Fabian, Hu, Cowie, and Grindlay (1981), for two of their model potentials.

required is much longer than a central cooling time. The flow in Perseus may be non-steady and remain non-steady for another several $\times 10^9$ yr.

Less than 1% of the accreted material is found to condense out of the flows, presumably into stars and cold gas. Such a small amount is consistent with the X-ray, optical, and radio measurements of Perseus (NGC 1275) and a standard disk IMF. This may offer the possibility of reconciling the extremely high accretion rates inferred from X-ray measurements with the severe constraints imposed by observations in the other wavebands. Nonetheless, even at 1 – 10% of the X-ray estimated values, Perseus and possibly other cooling flows may still be sites of prodigious star formation, surpassed in present day galaxies only by starbursts.

HIGH RESOLUTION CO IMAGES OF SEYFERT GALAXIES

Margaret Meixner and Melvyn Wright
Astronomy Dept., Univ. of California, Berkeley, CA 94720

Rich Puchalsky and Leo Blitz
Astronomy Program, Univ. of Maryland, College Park, MD 20742

ABSTRACT The Hat Creek millimeter array[1] has been used to image the CO (J=1-0) emission of three Seyfert galaxies, NGC 3227, NGC 7469 and NGC5033, chosen for their strong single dish CO detections. Extensive u-v coverage was obtained for all three galaxies resulting in 2″-3″resolution. The CO emission in NGC 3227 and NGC 7469 appear as compact structures with respective major axes of 1.2 and 5 kpc, centered on the active nuclei, containing substantial fractions (~80% and ~ 50%) of the single dish flux. The CO emission in NGC 5033 is not detected at such high resolution, implying a CO structure size of 20″-60″(1.2-3.6 kpc).

INTRODUCTION

One of the current problems in active galaxy research is to determine to what degree one can separate Seyfert and starburst activity. Several composite Seyfert/starburst objects are known to exist, including NGC 1068 (Balick and Heckman 1985) and NGC 7469 (Heckman *et al.* 1986). It has been suggested that these two phenomena are related, in an evolutionary sense (Norman and Scoville 1988). However, the mechanisms and the nature of the fuels for these phenomena are not well understood. In a recent review, Hernquist (1989) combined the various hypothesized mechanisms into a hierarchical framework in which starburst and Seyfert activity follow sequentially. In this model, large scale events, such as tidal interaction between companion galaxies, drive gas into the inner ~ 1 kpc by non-axisymmetric perturbations to the potential. Self-gravity then takes over, compressing the gas and causing a starburst. From this point there are several ways to create a central black hole (Rees 1984) and hence cause Seyfert activity. In this paper, we investigate the role of the molecular gas in fueling the activity in the nucleus by observing CO in three relatively nearby (12-64 Mpc) Seyfert 1 galaxies, NGC 3227, NGC 7469, NGC 5033, chosen for their strong single dish CO fluxes (Heckman *et al.* 1989).

[1] Operated by the University of California, Berkeley, the University of Illinois and the University of Maryland, with support from the National Science Foundation.

RESULTS AND CONCLUSIONS

1 . Using the interferometer, we have detected 80% (470 Jy km s^{-1}) of the single dish flux observed in NGC 3227 and 50% (320 Jy km s^{-1}) of that observed in NGC 7469. We did not detect any CO emission in NGC 5033 (<40 Jy km s^{-1} /3″beam), although the integrated flux measured with a single-dish is similar to that in NGC 3227, which is at a comparable distance. We obtained upper limits of ~0.06 Jy/3″beam for the 2.7mm continuum emission in all three galaxies.

2 . The high resolution of the observations and the close proximity of NGC 3227 permit us to observe the molecular gas structure at the linear scale of the narrow line region (NLR) for the first time. The two peaks of CO emission (Fig. 1) appear to straddle the nucleus of the galaxy and to be spatially coincident with the NLR. Since the physical conditions in the NLR would destroy the molecular gas, Seyfert/starburst activity must be episodic or the gas must be replenished.

3 . Blue-shifted and red-shifted CO emission maps, created for both NGC 3227 and NGC 7469, demonstrate that the gas moves in the same sense as rotation curves determined for the galaxies supporting the notion that the molecular gas has rotational motion about the nucleus. In the case of NGC 7469 , these maps confirm the interaction of the molecular gas with the circumnuclear starburst.

4 . H_2 mass estimates derived from the CO luminosity employing the Galactic $N(H_2)/I(CO)$ conversion factor (Bloeman *et al.* 1986) seem improbably high when compared to dynamical masses. The discrepancy might result from an overestimate of the molecular mass, caused by a high gas temperature (40-50K) or by non-virial motion of the gas.

5 . The results of this study are consistent with the idea that interacting galaxies cause gas to concentrate in the nucleus thereby feeding starburst and Seyfert activity. Both Seyferts with high central concentrations of CO emission, NGC 3227 and NGC 7469, are interacting galaxies listed in Arp's catalogue (Arp 1966) and have circumnuclear star formation or starburst. The third Seyfert, NGC 5033 has no detectable centrally concentrated gas emission implying either an alternate mechanism of feeding the central black hole or perhaps the end stage of this process.

ACKNOWLEDGEMENTS

We would like to thank Tim Heckman for donating his CCD images, and useful discussion and Andrew Wilson, Jon Carlstrom, Heidi Kirkpatrick, Alex Filippenko , and Joe Shields for their constructive comments. Zonta International supported Margaret Meixner throughout her research for this paper.

REFERENCES

Arp,H. 1966,*Ap. J. Suppl.*,**14**,1.
Balick, B. and Heckman, T.M. 1985,*A.J.*,**90**,197.

Bloemen, J.B.G.M., Strong, A.W., Blitz, L., Cohen, R.S., Dame, T.M., Grabelsky, D.A., Hermsen, W., Lebrun, F., Mayer-Hasselwander, H.A., Thaddeus, P. 1986, *Astr. Ap.*, **154**,25.

Heckman, T.M., Beckwith, S., Blitz, L., Strutskie, M., and Wilson, A.S. 1986, *Ap. J.*, **305**, 157.

Heckman, T.M., Blitz, L., Wilson, A.S., Armus, L. and Miley, G.K. 1989, *Ap. J., in press.*

Hernquist, L. 1989, *Annals of the New York Academy of Sciences, in press.*

Norman, C. and Scoville, N. 1988, *Ap. J.*, **332**,124.

Rees, M.J. 1984, *Ann. Rev. Ast. Ap.*, **22**, 471.

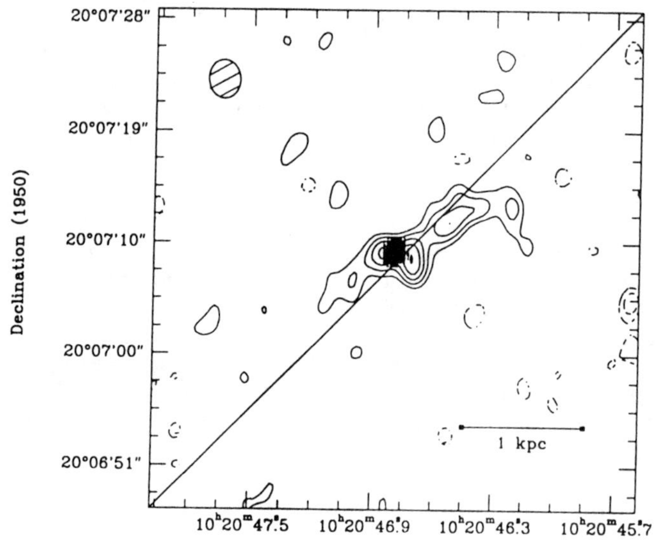

Fig. 1 NGC 3227: Total CO intensity contours ($\pm 24 = 2\sigma, \pm 36, 48, 60, 72$ Jy km s^{-1} / 2.″8 ×2.″3 beam) overlaid on Hα (halftone)

CCD PHOTOMETRY OF THE RING GALAXY ARP 146

L.D. Spight, A.B. Schultz and P.T. Colegrove
Department of Physics, University of Nevada

M.A. DiSanti and U. Fink
Lunar and Planetary Laboratory, University of Arizona

ABSTRACT We report new broad band B, V, R, and I CCD observations of Arp 146. The images were obtained with the 61" Mt. Bigelow telescope. The B-V color indices for the intruder and ring galaxy are respectively 0.80 and 0.52 . Knots in the ring galaxy are evidence of new star generation resulting from the encounter. The integrated BVRI and color indices indicate that the encounter has well mixed the dust between the two galaxies.

INTRODUCTION

Within the general category of collisional galaxies, ring galaxies are a particularly interesting morphological subclass resulting from special collisional parameters (Lynds & Toomre, 1976). As examples of non-nuclear coherent starburst activity over extended regions, they provide fertile ground for observational investigation and a natural laboratory for testing sophisticated computational and theoretical models of galactic shock wave propagation (Appleton & Struck-Marcel, 1987). Of the several dozen known ring galaxies (Few & Madore, 1986; Theys & Spiegel, 1976; Thompson, 1977), only a handful show the purity of an uncomplicated structure, featuring a clearly empty ring with a smooth or knotty composition and an obvious nearby companion. These and similar ring galaxies have been surprisingly little studied.

Arp 146 is a classic example of an empty ring galaxy. Discovered by Dewhirst and catalogued by Arp and by Vorontsov-Velyaminov (VV 790), it is a well-defined, 18"x11" elliptical ring approximately 20 kpc in diameter, separated from its companion by about 13", with a heliocentric velocity of about 22615 km/s at galactic coordinates 1 = 93.30, b = -66.98 . Freeman and de Vaucouleurs (1974) reported photographic magnitudes of 15.85 and 15.66 respectively for the ring and its companion. Appleton and Struck-Marcell (1987), using IRAS data, determined an integrated far infrared flux below the detection threshold. Majewski and Hereld (1987) have made near-infrared and UBVR measurements but have not yet published their results. Beyond this information, little data or analysis is available.

Spatial and photometric information for many ring systems is needed to clarify and codify the observational and theoretical work on starburst phenomena in presumably simpler systems. We have begun a systematic study of selected rings in an effort to obtain this information, using Arp 146 with its crisp ring as the structure with which to compare others.

OBSERVATIONS AND DATA REDUCTION

A CCD camera based on a TI 800 x 800 three phase chip and off-axis reducing optics were used to obtain 900 sec exposures through B, V, R, and I filters. The B and V filters are Johnson filters, while the R and I filters are tied to the Cousins system. Landolt stars were imaged to obtain the nightly photometric zero point. IRAF implemented on a SUN system was used for the image processing. After bias and dark frames were subtracted, each data frame was ratioed with the respective dome flats. All frames were registered to a common coordinate system. The PSF (point spread function) was determined for each frame and each was convolved with a gaussian function to obtain equivalent PSF's. One set of apertures was selected for all the photometric measurements.

MORPHOLOGY AND PHOTOMETRY

Figure 1 is a contour of the blue image of Arp 146, showing four well-defined knots of starburst activity. The remnant of the ring's pre-encounter nucleus is apparently imbedded in knot A; the B-V and B-R colors are brighter in A than in the other knots. The B-R and R-I colors for all these regions indicate that considerable dust overlies the region. Knot A apparently has a larger old stellar population fraction than the other regions. In the B-V and B-R, it and the bridge region between the ring and the companion are closer in character to the companion than to the other knots. This and the distribution of the dust in the bridge and in the region around knot A and downward toward knot B argue that the companion lies in the foreground of the projected image and that the northwest corner of the ring is tipped toward us in the line of sight. The north side of the ring is extremely dim in both B and V but has V-R and R-I colors virtually the same as in the anti-tidal region of the companion, suggesting an older population of stars, dust and gas, perhaps that more characteristic of the pre-collision epoch. The center of the ring is well evacuated of material, with magnitude and color values very close to nearby clear sky regions. Table I gives a representative sample of the integrated magnitude and color data.

TABLE I Visual Magnitudes and Color Indices

REGION	V	B-V	B-R	R-I
Whole System	15.84	0.59	1.03	0.54
Companion	17.14	0.80	1.32	0.62
Ring	16.35	0.52	0.90	0.48
Knot A	19.69	0.83	1.15	0.71
Knot B	20.16	0.57	0.81	0.48
Knot C	20.30	0.67	0.85	0.50
Knot D	20.36	0.48	0.63	0.37

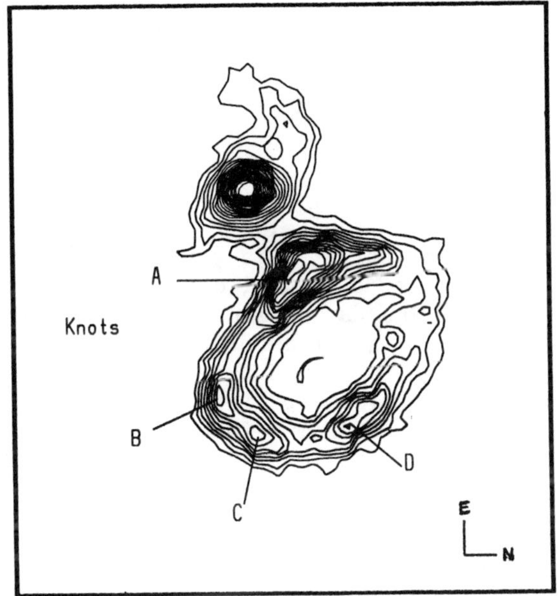

Fig. 1 Contour plot of the blue image of Arp 146.

REFERENCES

Appleton, P. & C. Struck-Marcell 1987, Astrophys. J., 323, 480.
Few, J. & Madore, B. 1986, Mon. Not. R. Astr. Soc., 222, 673.
Freeman, K. & G.de Vaucouleurs 1974, Astrophys. J., 194, 569.
Ghigo, F., Astron. J. 1980, 85, #3, 215.
Lynds, R. & A. Toomre 1976, Astrophys. J., 209, 382.
Majewski, S. & M. Hereld 1987, AAS Abstracts, 19, #4, 1032.
Theys, J. & E. Spiegel 1976, Astrophys. J., 208, 650.
Thompson, L. 1977, Astrophys. J., 211, 684.

COLOR MAPS OF ARP 146

A.B. Schultz, L.D. Spight, and P.T. Colegrove
Department of Physics, University of Nevada

M.A. DiSanti and U. Fink
Lunar and Planetary Laboratory, University of Arizona

ABSTRACT We present four color maps of Arp 146. The structure and color of the ring galaxy and its companion show evidence of a bridge of material between the companion and the remnant nucleus of the original galaxy now forming the ring. Broad band spatial coverage clearly defines regions of starburst occurrence.

INTRODUCTION

Arp 146 is a well defined ring galaxy system with the pre-collision nucleus of the target galaxy apparently imbedded in the ring itself. At least four well defined knots in the ring are evident in the CCD images, regions of new star formation precipitated by the shock front generated by the collision between the intruder and target galaxies.

Color index photometry as applied to such a galaxy yields spatial information about the distribution of the stars, gas and dust in that galaxy. CCD imaging reduces spatial averaging as performed with aperture photometers, and spatial resolution is enhanced. Stellar populations can be delineated, as well as formations of knots and filaments caused by dynamic processes. Color index variations in the images of the two galaxies can provide important information about the evolution of the collisional system.

IMAGE PROCESSING

CCD images were obtained using the LPL CCD camera and off axis guider at the UAO 154 cm Catalina reflector. Exposures of 900 seconds duration each were taken using B, V, R, and I filters. Stars chosen from Landolt's (1983) list of standard UBVRI photometric stars were imaged through the same filters. The B and V filters were Johnson (1955), and the R and I filters were Cousins (1976).

Each image was processed by first subtracting a bias frame followed by ratioing with the respective dome flat. The imaging technique reduces spatial averaging as performed with photoelectric photometers and consequently enhances the spatial resolution of knots and filaments. All data frames were then shifted to a common coordinate system specified by the V filter image in order to determine the color difference frames. The alignment of the frames is good to 0.1 of a pixel. The individual data frames were then convolved with a gaussian function to assure equivalent point spread functions in all frames. The final step before creating the scaled magnitude frames was to subtract the average sky from each frame. Consequently, the residual sky values in the color maps are not representative of the true sky color.

The known magnitudes and extracted counts/sec of the Landolt standard stars were used to calculate the mag/pixel for each frame. The color maps were created by subtracting two color magnitude frames. No second order color transformations to the standard UBVRI system were made. The image processing was performed using IRAF.

CONCLUSIONS

Figure 1 is a contour plot of the red image of Arp 146 and contours of the color maps. It is apparent that the B-R color map is redder that the B-V color map, indicating that there is a large amount of well mixed, distributed dust in ARP 146. The B-V and B-R color maps show an extended bridge of material overlying the old nucleus of the ring galaxy and the proximate part of the companion. This suggests that the remnant nucleus of the ring galaxy is behind a tidal arm of the foreground galaxy.

Problems in the interpretation of the color maps are caused by the usual guiding, focusing, and centering errors as well as seeing conditions. These potential problems can result in differences in the PSF (point spread function) between the different filter images. Consequently, color maps may contain spurious color differences caused by the subtraction process itself. Thus, only broad collective behavior of the stars, gas and dust in a system is reliably revealed.

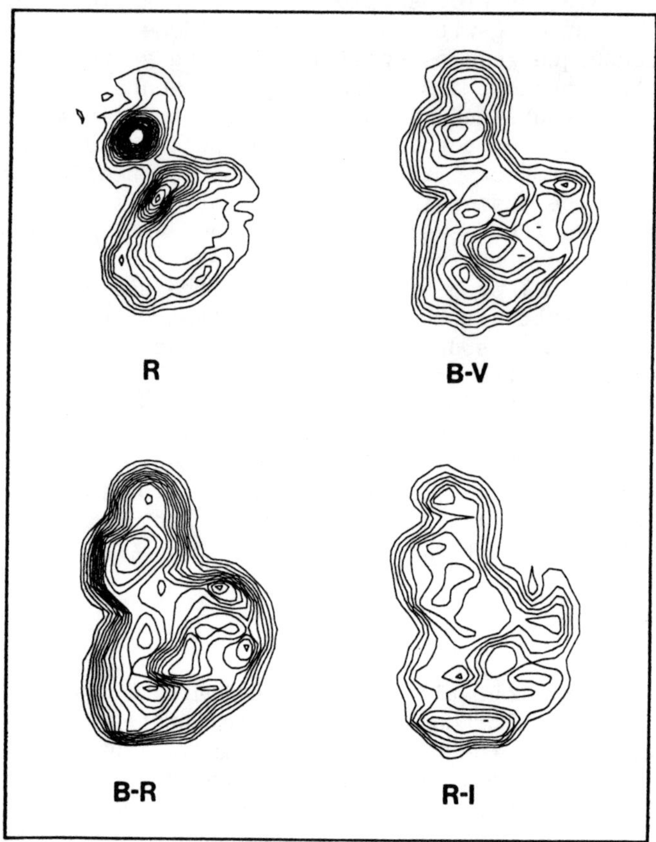

Fig. 1 Contour plots for Arp 146: (R) surface brightness contour of the red image; (B–V), (B–R) and (R–I) color contour maps. Color contour intervals are drawn at 0.1 intervals.

ACKNOWLEDGMENTS

The CCD camera was operated under NASA grant NSG 7070. Special thanks goes to SUN Microsystems, INC. for donating the computer workstation to our astronomy program, without which this research would not have been possible.

REFERENCES

Cousins, A.W.J. 1976, Mem. R. Astron. Soc., 81, 25.
Johnson, H.L. 1955, Ann. Astrophys., 18, 292.
Landolt, A.U. 1983, Astron. J., 88, 439.

EXTRACTING COSMOLOGICAL AND EVOLUTIONARY PARAMETERS FROM DISTANT GALAXY OBSERVATIONS

GUSTAVO BRUZUAL A.

Centro de Investigaciones de Astronomía (CIDA), Apartado Postal 264, Mérida 5101-A, Venezuela.

ABSTRACT Models for the galaxy number counts and color distributions in SA 68 (Koo 1986) built according to a Non-Negative Least Squares (NNLS) algorithm (Lawson and Hanson 1974) are discussed. The value of χ^2_{red} that characterizes each model can be used as a goodness of fit indicator to decide in favor of a model over another, providing an objective and unbiased criterion to extract cosmological and evolutionary parameters from a data set.

INTRODUCTION

The problem of extracting evolutionary parameters from a set of observations can be posed as follows: given a set of observations, extract from these data as much quantitative information as possible related both to the sample being studied and to the assumed cosmological model. Any algorithm deviced to perform this task must be objective and unbiased and, in the optimal case, avoid any subjective considerations. The solution to this problem is not easy. In a typical case, one has to derive the set of evolutionary and cosmological parameters that describes a data set such as CAT 68 (Koo 1986) shown in Fig. 1.

MAGNITUDE- AND COLOR-REDSHIFT DIAGRAMS

The magnitude-redshift diagram (Hubble diagram) and the color-redshift diagram are often used to infer values of cosmological and evolutionary parameters. This procedure can be very uncertain if the properties of the sample are not well known or not modeled appropriately. As an illustration of the problems that arise consider Fig 2. The top frame in Fig. 2 shows the predicted behavior of the color index $U+J-F-N$ (Koo 1986) versus redshift (z) for a set of μ models from Bruzual (1983), improved with the addition of a new stellar library that includes the Gunn and Stryker (1983) stellar spectral energy distributions (sed's) in the optical range. These lines define the locus where galaxies of most morphological classes and apparent magnitudes should fall in this plane for $H_o = 50$, $q_o = 0$, and galaxy age $t_g = 16$ Gyr. The μ parameter used in the exponential star formation rate (SFR) to build these models ranges from $\mu = 0.95$ to $\mu = 0.01$ in steps $\Delta\mu = 0.05$ for the top 20 lines, for which the Salpeter ($x = 1.35$) initial mass function (IMF) was used. The 7 bluest models were built with $\mu = 0.01$ and the following IMF's $x =$ 1.20, 1.10, 0.85, 0.67, 0.50, 0.30, and 0.05.

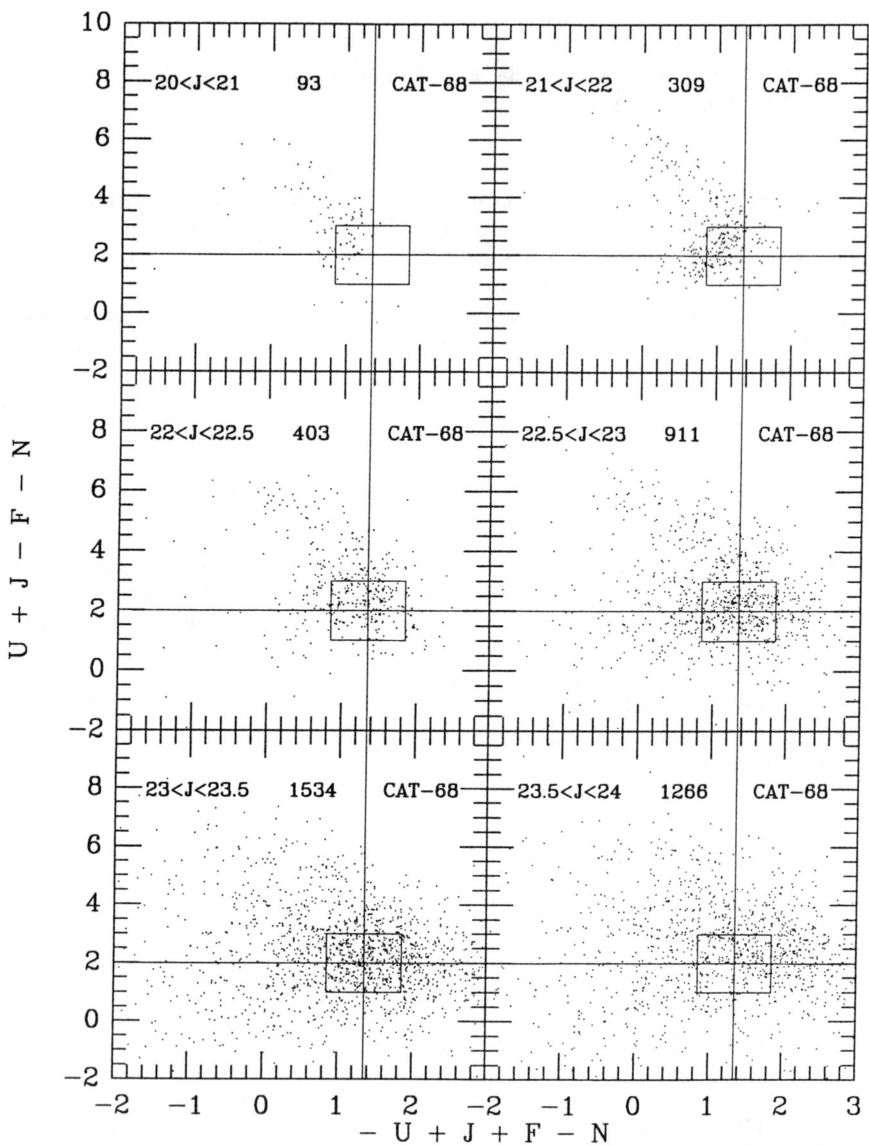

Figure 1. The $U+J-F-N$ versus $-U+J+F-N$ distribution of galaxies from CAT 68 in the indicated ranges of J magnitude (Koo 1986). The lines which divide each plot into four rectangles intersect near the centroid of the observed distribution. The inner square serves as a reference when comparing with model predictions. The number at the top indicates the number of dots in the frame.

The other four frames in Fig. 2 show the loci predicted to be occupied in this plane by galaxies of different optical J magnitudes. Each dot in these plots represents a galaxy. To build these plots galaxies of all types have been assumed to be distributed in absolute magnitude according to a color dependent Schechter (1976) luminosity function. The sample incompleteness and the random errors in the measured fluxes have been modeled for CAT 68 using the parameters indicated by Koo (1986). The nearby bluest galaxies are intrinsically faint and are thus excluded from entering into the bright J frames. At $z > 0.5$ the red galaxies are too faint to be included in any of our frames. Thus most of the galaxies observed in CAT 68 at $J > 22$ are predicted to have a relatively narrow range in SFR.

This example shows that empty regions in the observed color-z diagrams when compared to the predicted equivalent diagrams do not mean that the models and the evolutionary parameters used are inappropriate. Models should not be forced to agree with the data before the sample incompleteness and the random errors in the photometry have been built into the models. The simple procedure of shifting observed and predicted diagrams up and down until reasonable overlap is obtained may lead to inaccurate determinations of evolutionary and cosmological parameters.

STANDARD MODELS FOR GALAXY NUMBER COUNTS

Several authors have followed what may be called the standard approach to modeling galaxy number counts. Models that follow the time evolution of the spectrophotometric properties of galaxies have been used by Bruzual and Kron (1980), Tinsley (1980), Bruzual (1981), Koo (1981), Shanks et al. (1984), King and Ellis (1985), Yoshii and Takahara (1988), Guiderdoni and Rocca-Volmerange (1987), and Broadhurst, Ellis, and Shanks (1988) to predict the galaxy number counts. The amount of spectral evolution for each type of galaxy follows directly from the evolving spectra input into the count model. In all these models the following quantities must be specified in order to build the model:

1) Morphological type and color class of galaxies to include in the model.

2) Spectral energy distribution corresponding to each class, as well as its time evolution.

3) Luminosity function (M^*,α) for each class, including any time (i.e., z) dependence.

4) Total number of galaxies per unit co-moving volume.

5) Relative fraction of each class at a given apparent magnitude or co-moving volume.

6) Cosmological model: H_o, q_o, Λ.

7) Detailed knowledge of the sample incompleteness and of the random errors in the photometry as a function of apparent magnitude for the data set being studied.

Evolutionary and cosmological parameters are not extracted from these models. On the contrary, values for these parameters must be known or assumed in advance in order to be input into the models. Thus the standard technique for modeling galaxy number counts tests the degree of consistency of our assumptions

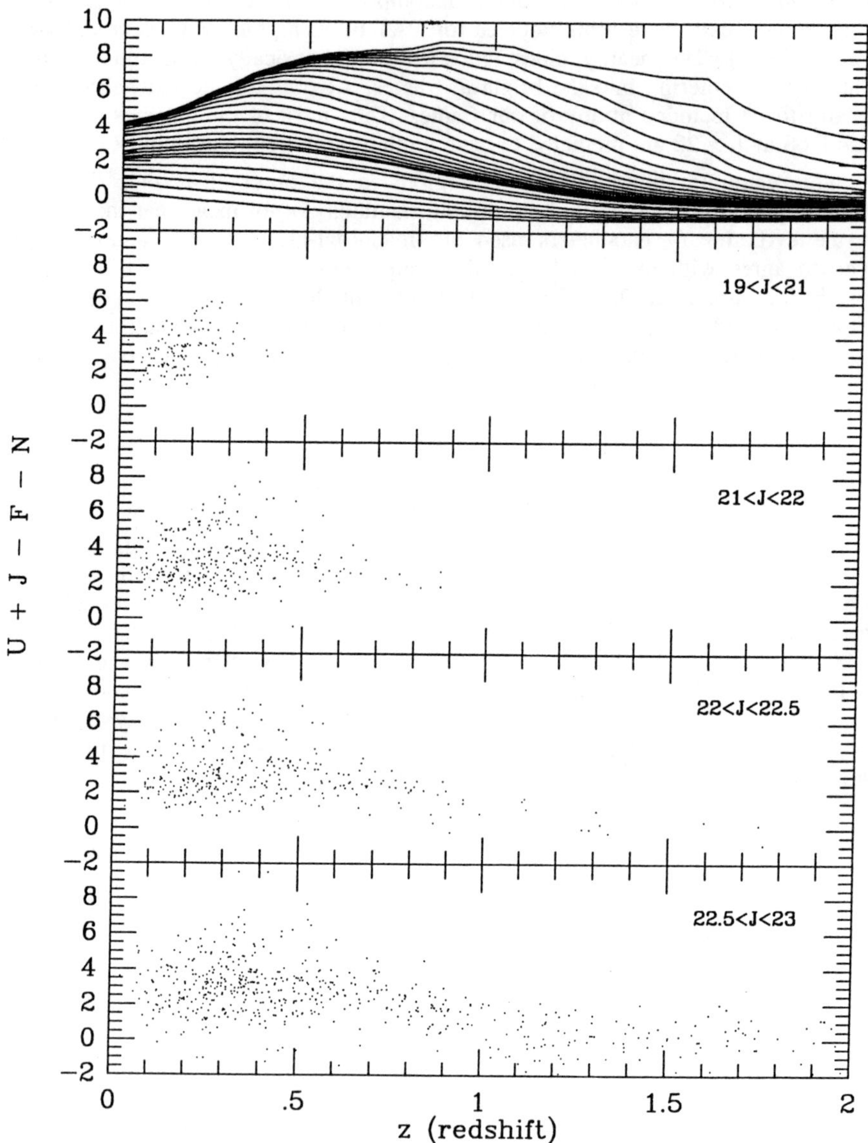

Figure 2. Top frame: Behavior of the $U+J-F-N$ color index with redshift for the sed's used in Model 1 (see Tables 1 and 2). Lower frames: expected distribution as a function of J magnitude when the sample incompleteness and the random errors in the photometry for CAT 68 (Koo 1986) are included in the models.

and the data, but do not provide us with the best set of values for the evolutionary parameters. Only a very small region of parameter space has been explored.

The following example illustrates the type of problems that arise from these models. All of the authors mentioned above (except Broadhurst, Ellis, and Shanks) interpreted the steepening of the number counts around $J = 23$ as due to a general increase in the amount of UV light seen in the evolving early type galaxies for $z > 1$, during the high star formation phases of their evolution. This interpretation prompted the Durham/AAT and the KPNO redshift surveys (Broadhurst, Ellis and Shanks 1988, Koo and Kron 1987) both of which have failed to reveal the expected numbers of high z galaxies.

It is clear then that we must build better models, based on an objective technique which allows us to explore a large volume of parameter space at once. The following describes a general technique which not only takes advantage of diverse forms of data simultaneously, but also provides objective selection of the best models.

OBJECTIVE APPROACH TO MODELING GALAXY COUNTS AND COLOR DISTRIBUTIONS

The following approach to modeling galaxy number counts and color distributions is described in detail by Bruzual and Koo (1989). Only a brief summary is included here. The distributions of observed properties that we want to reproduce with our model will be used to build an m-vector D (data vector) as explained below. Each component of this vector is a data point. For each of n color classes we predict a value for each point in the data vector and build an identical m-vector C_i with the predictions of our model for the *ith*-class. The *ith*-column of the matrix P (prediction matrix) is then filled with the vector C_i.

The Non-Negative Least Squares (NNLS) algorithm (Lawson and Hanson 1974) provides an effective means to solve the following problem. Given an $m \times n$ matrix, P, and an m-vector, D, compute the n-vector, X, which solves the least squares problem

$$PX = D \tag{1}$$

subject to $X_i \geq 0$ for all i.

The *ith*-component of the solution vector, X_i, represents the weight assigned to the *ith*-class in the NNLS solution.

If we want to reproduce with our model a $j \times k$ table that contains j lines of k columns each, we build the data vector D of dimension $j \times k$ (the number of bins in the table) by filling elements $D(1+(l-1)\times j)$ through $D(l \times j)$ with the data in the *lth*-column of the table ($1 \leq l \leq k$). If another table is available, we continue filling D at element $D(j \times k+1)$. Any number of tables can be used and the tables do not need to have equal number of lines and/or columns. Each C_i vector is built in identical fashion from the model predictions for the *ith*-class and then used to fill the *ith*-column of the prediction matrix P.

The reduced χ^2 corresponding to the NNLS solution is given by

$$\chi^2_{red} = \frac{\sum \frac{1}{p_i}(p_i - d_i)^2}{N_f}, \tag{2}$$

where p_i and d_i are the predicted and the observed number of galaxies inside the *ith*-bin, respectively. N_f is the number of degrees of freedom. The summation extends over the total number of bins used in the fit. Data points with zero galaxies are not included in the NNLS fit. The contribution to χ^2_{red} of bins with $d_i <$ 100 must be given special treatment (e.g. Gehrels 1986).

Some of the advantages of the proposed method include:.

a) All the data available can be explored at once.

b) There is no need to introduce a normalization based in most cases on a different set of data. The solution vector X contains information about the absolute number of galaxies of each class that are present in the data vector.

c) No subjective judgment is needed to derive the solution vector X, as long as the matrix P covers a sufficiently large region of parameter space.

d) χ^2_{red} can be used as a goodness of fit indicator to decide in favor of a model over another.

RESULTS

The models listed in Table 1 were computed in order to explore the technique described in the previous section. The quantities d_i were derived from the $U-J$ *vs.* J, $J-F$ *vs.* J, $F-N$ *vs.* J, $U+J-F-N$ *vs.* J, $2U-2F$ *vs.* J, and $(U+J-2F)/2$ *vs.* J observed distributions for CAT 68 (Koo 1986). For $J < 19$ the $J-F$ *vs.* J distribution for field galaxies of Butcher and Oemler (1985) was used. The total number of resulting bins N_{bins} is given in Table 1. The quantities p_i were derived from the model sed's listed in Tables 2 and 3, assuming the cosmological and evolutionary models indicated in Table 1. The Schechter (1976) galaxy luminosity function was used in all models.

Table 1
Characteristics of Models for CAT 68

Model	1	2	3	4
H_o	50	50	50	50
q_o	0	0.5	0	0
t_g (Gyr)	16	13	16	16
Evolution	Yes	Yes	No	Yes
N_{bins}	896	896	896	694
N_{cc}	14	13	13	8
N_f	882	883	883	686
χ^2_{red}	4.2	6.9	9.9	37.2

The input model sed's used in Models 1 to 3 are listed in Table 2, together with the weight χ_i assigned to the ith class by the NNLS algorithm. Classes without an entry in Table 2 were not chosen by the algorithm. Each color class is identified by the value of the μ parameter appearing in the Bruzual (1983) exponentially decaying SFR and by the value of the exponent x describing the

IMF. The sed's used in Model 4 were computed by adding a second burst of star formation to the standard c-model of Bruzual (1983). The second burst starts at age t_2, lasts for 0.25 Gyr, and transforms into stars a quarter of the galaxy mass (assumed to be $10^{11}\ M_\odot$). A total of 30 model sed's were input into Model 4. t_2 was varied from 1.5 Gyr to 15.5 Gyr in 0.5 Gyr steps (29 models). The standard c-model was also included. For all the sed's in Model 4 the Salpeter (x = 1.35) IMF was assumed. Table 3 lists only those sed's selected by the NNLS fit.

N_{bins} represents the number of bins used in the fit. All bins with $d_i \geq 1$ were included in Models 1 to 3. Only bins with $d_i \geq 3$ were used in Model 4. No reasonable fit was possible for the latter model if bins with less than 3 galaxies were included. The number of degrees of freedom N_f was defined as

$$N_f = N_{bins} - N_{cc} \tag{3}$$

where N_{cc} is the number of color classes (or galaxy sed's) selected by the NNLS algorithm.

Table 2
Sed's Used in Models 1 to 3

		Mod-1	Mod-2	Mod-3
x	μ		$100X_i$	
0.05	0.01			37.80
0.30	0.01			
0.50	0.01			5.94
0.67	0.01	3.07	2.82	2.92
0.85	0.01	4.14	3.21	10.20
1.10	0.01	6.26	5.79	17.30
1.20	0.01	2.45	2.62	
1.35	0.01		0.43	0.06
1.35	0.05	3.85	1.26	3.38
1.35	0.10	1.56	5.53	3.92
1.35	0.15	2.43	0.04	1.08
1.35	0.20	0.72	2.59	1.67
1.35	0.25	1.04		2.35
1.35	0.30	0.62	1.10	
1.35	0.35	0.97		1.59
1.35	0.40			
1.35	0.45	0.79		
1.35	0.50	0.32		
1.35	0.55			
1.35	0.60			
1.35	0.65	1.63	0.60	
1.35	0.70		2.41	
1.35	0.75		1.54	
1.35	0.80			
1.35	0.85			
1.35	0.90			
1.35	0.95			0.73

The sampling incompleteness and random errors in the photometry were introduced into the models using the parameters derived by Koo (1986) for CAT 68. The predictions of these models are compared to the observations of CAT 68 in the following pages.

Table 3
Sed's Used in Models 4

t_2(Gyr)	1.5	4	8	9.5	11.5	13	13.5	15.5
$100X_i$	7.71	1.27	0.05	0.51	2.37	0.65	0.38	4.58

DISCUSSION

The larger residuals observed in Fig. 3 for Models 2, 3, and 4 than for Model 1 are consistent with the value of χ^2_{red} listed in Table 1. The number counts (Fig. 4) are reproduced equally well by Models 1 and 2. There is an excess in the predicted number of galaxies in the J and F bands past the completeness limit. The similarity between the predicted counts for Models 1 and 2 shows that galaxy counts are not sensitive to cosmology. Models 3 and 4 predict fewer galaxies than observed at magnitudes fainter than 21 in all bands. This is expected from the residuals of the fit shown in Fig. 3.

The color distributions shown in Fig. 5 are similar for Models 1, 2, and 3, and are close to the observed distributions, while Model 4 deviates badly from the observations. The color-color diagrams of Fig. 6 show very similar distributions to that of Fig. 1 for Models 1 and 2, even though the differences in the number of points in each frame of Figs. 6a and 6b with respect to Fig. 1 are larger than anticipated. The diagram for Model 4 (Fig. 6d) shows a much poorer fit to the data, with voids and excesses of galaxies in various color ranges. By visual inspection the diagram for Model 3 (Fig. 6c) is not that different from Model 1 (Fig. 6a) except for the smaller number of galaxies in each frame. This is consistent with the deficiency in the number counts noticed in Fig. 4 and shows that the brightening effect of spectral evolution is important in increasing the number counts at a given apparent magnitude.

Table 4
Median z for Different Models

J range	**Mod-1**	**Mod-2**	**Mod-3**	**Mod-4**
$19.0 < J < 21.0$	0.15	0.16	0.12	0.18
$21.0 < J < 22.0$	0.24	0.22	0.19	0.42
$22.0 < J < 22.5$	0.34	0.32	0.24	2.63
$22.5 < J < 23.0$	0.54	0.49	0.28	2.66
$23.0 < J < 23.5$	0.83	0.81	0.36	2.66

The redshift distributions of Fig. 7 are very similar for Models 1 and 2. As expected, Model 3 with non evolving sed's predicts that galaxies are seen up to much lower z's than in Models 1 and 2. Fig. 7d shows that in Model 4 a very large fraction of distant galaxies is present at all J's. The color ranges covered by the sed's used in Model 4 were examined and found to be equivalent to those of Models 1 to 3, with a somewhat redder blue envelope. Galaxies in Model 4 are fainter at any $z < 2$ than in the other Models, and the NNLS algorithm was forced to pick up the most distant galaxies at their brightest to minimize χ^2. This is clearly shown by the large weight assigned to the sed with $t_2 = 1.5$ Gyr with respect to the others (Table 3). This is not an intrinsic limitation of Model 4, but of the particular choice of the sed's input into it. By using second bursts of higher relative strength, galaxies can be made brighter and the fit improved. However, by using stronger second bursts we approach the sed's of Models 1 to 3, and hence the physical distinction between these Models and Model 4 tends to disappear.

Table 4 lists the median z of the distributions shown in Fig. 7. These numbers should be compared with the observational values listed in Table 5. The redshift distributions of Models 1 and 2 are equally consistent with the observations. Model 3 is marginally consistent up the magnitude level for which the z distribution is known. This model is likely to be ruled out when the redshift surveys reach deeper magnitude levels.

Table 5
Median z for Observed Distributions

J range	ζ	Source
$20.0 < b_J < 21.5$	0.22	Broadhurst et al. (1988)
$21.0 < b_J < 22.5$	0.30	Colless et al. (1988), Ellis (this volume)
$21.0 < J < 22.0$	0.24	Koo and Kron (1987)

Visual inspection of the previous figures alone does not allow us to favor Model 1 or Model 2 (or even Model 3). The χ^2_{red} listed in Table 1 shows that Model 1 provides a better fit to the data (2 sigma on the average). Model 2 provides a 2.7 sigma fit, whereas Model 3 is above the 3 sigma level.

The models presented in this paper are only for illustration of the NNLS technique and are far from covering all of parameter space. Other choices of the cosmological and evolutionary parameters should be tried before a claim can be made in favor of a given set of values. We have shown how χ^2_{red} can be used as a goodness of fit indicator, as well as a pointer towards those regions of parameter space that should be explored further.

REFERENCES

Broadhurst, T. J., Ellis, R. S., and Shanks, T. 1988, *M.N.R.A.S.*, **235**, 827.

Bruzual A., G. 1981, *Ph.D. thesis,* University of California, Berkeley.

Bruzual A., G. 1983, *Ap. J.*, **273**, 105.

Bruzual A., and Koo, D. C. 1989 (in preparation).

Bruzual A., and Kron, R. G. 1980, *Ap. J.*, **241**, 25.

Butcher, H. R., and Oemler. Jr., A. 1985, *Ap. J. Suppl.,* **57**, 665.

Colless, M., Ellis, R. S., and Taylor, K. 1988, in *The Epoch of Galaxy Formation, (NATO ASI Series Vol. 264),* eds. C. S. Frenk, R. S. Ellis, T. Shanks, A. F. Heavens, and J. A. Peacock (Dordrecht: Kluwer), p. 359.

Gehrels, N. 1986, *Ap. J.,* **303**, 336.

Guiderdoni, B., and Rocca-Volmerange, B. 1987, *Astron. Astrophys.,* **186**, 1.

Gunn, J. E., and Stryker, L. L. 1983, *Ap. J. Suppl.,* **52**, 121.

King, C. R., and Ellis, R. S. 1985, *Ap. J.,* **288**, 456.

Koo, D. C. 1981, *Ph.D. thesis,* University of California, Berkeley.

Koo, D. C. 1986, *Ap. J.,* **311**, 651.

Koo, D. C., and Kron, R. G. 1986, in *Observational Cosmology (IAU Symposium 124),* eds. A. Hewitt, G. Burbidge, and L. Z. Fang (Dordrecht: Reidel), p. 383.

Lawnson, C. L., and Hanson, R. J. 1974, in *Solving Least Squares Problems,* Prentice Hall.

Schechter, P. 1976, *Ap. J.,* **203**, 297.

Shanks, T., Stevenson, P. R. F., Fong, R., MacGillvray, H. T. 1984, *M.N.R.A.S.,* **206**, 767.

Tinsley, B. M. 1980, *Ap. J.,* **241**, 41.

Yoshii, Y., and Takahara, F. 1988, *Ap. J.,* **326**, 1.

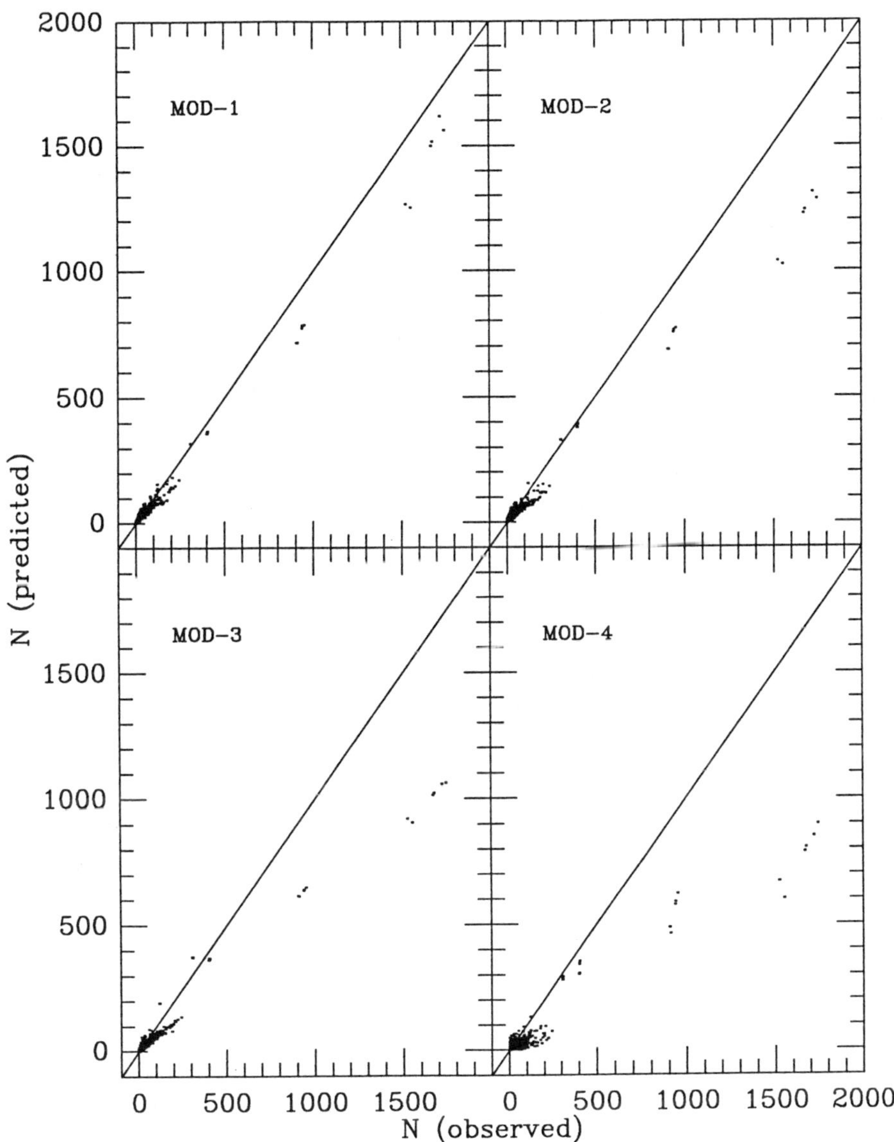

Figure 3. Comparison of the number of predicted and observed galaxies in each bin for the different models. The number of bins used in the NNLS fit for each model is indicated in Table 1. The slope 1 line is included for reference.

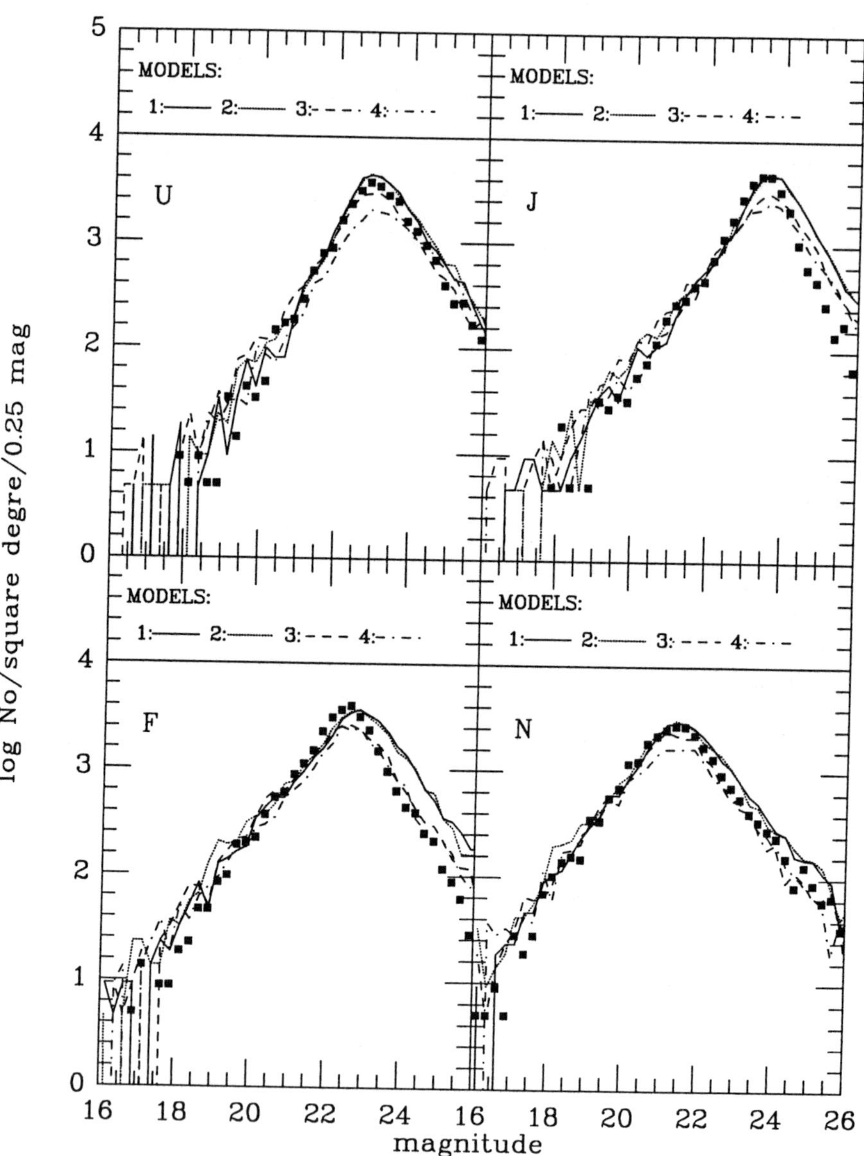

Figure 4. Differential number counts per square degree in the indicated bands plotted every 0.25 mag. The solid squares correspond to CAT 68 from Koo (1986). The lines correspond to the predicted counts from Models 1 to 4.

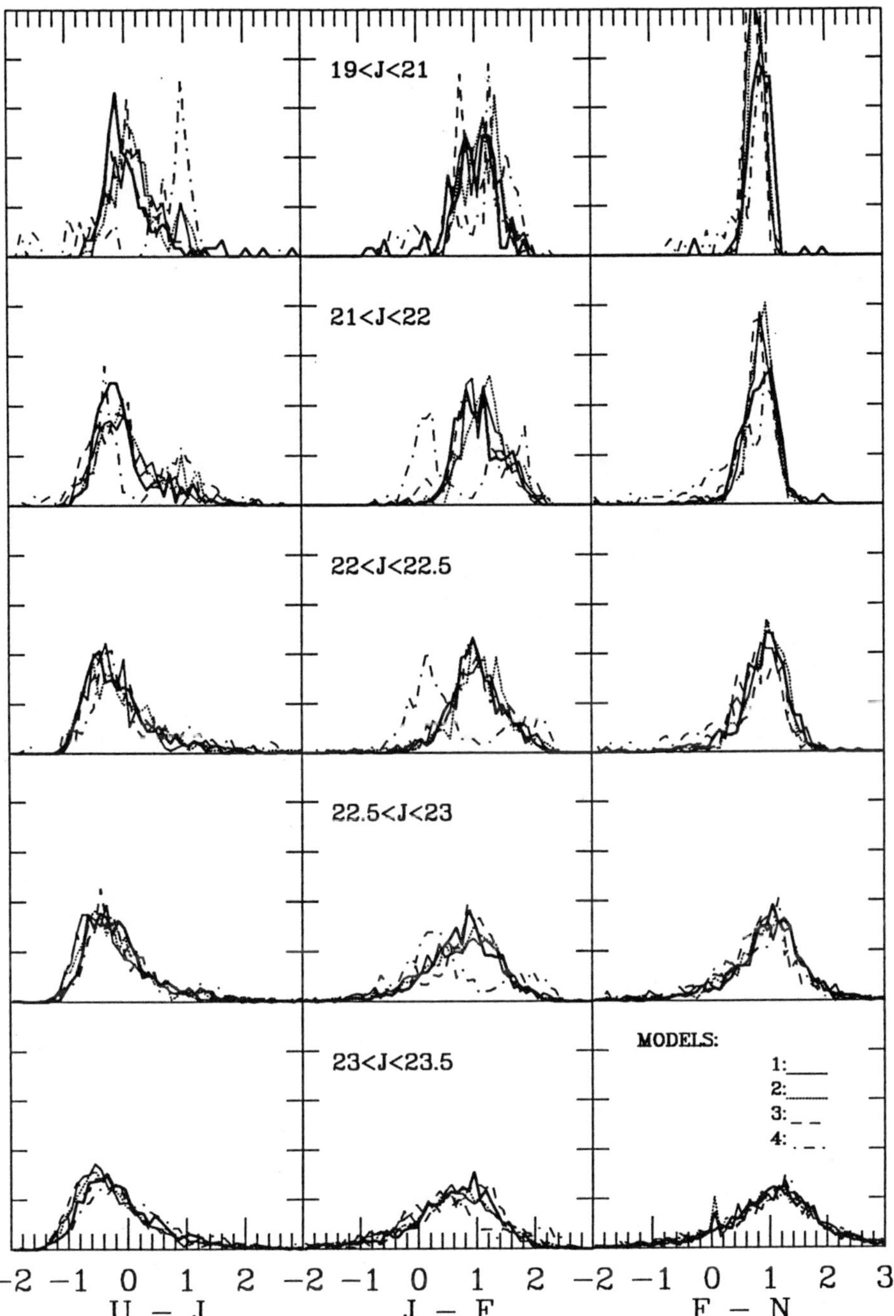

Figure 5. $U-J$, $J-F$, and $F-N$ normalized color distributions of galaxies in the indicated J ranges. The heavy line corresponds to CAT 68 (Koo 1986). The other lines represent the predicted distributions from Models 1 to 4. The vertical scale varies from 0 to 0.25 and indicates the fraction of the total number of galaxies in each 0.1 magnitude bin in color.

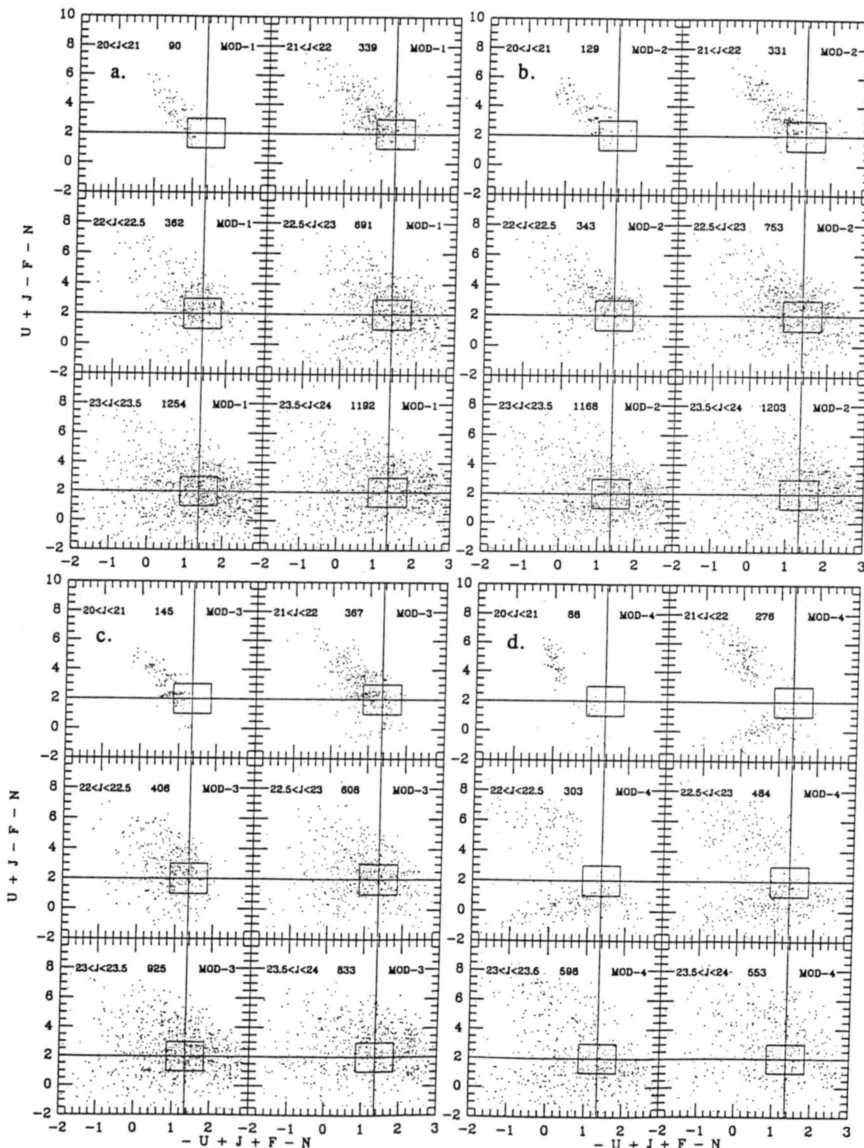

Figure 6a,b,c,d. The predicted *U+J-F-N* versus *-U+J+F-N* distribution for galaxies. At the top of each frame the intervals of *J* magnitude, the number of dots in the frame, and the model number are indicated. The lines intersect near the centroid of the observed distribution, and the rectangles serve as a reference when compared to Fig. 1.

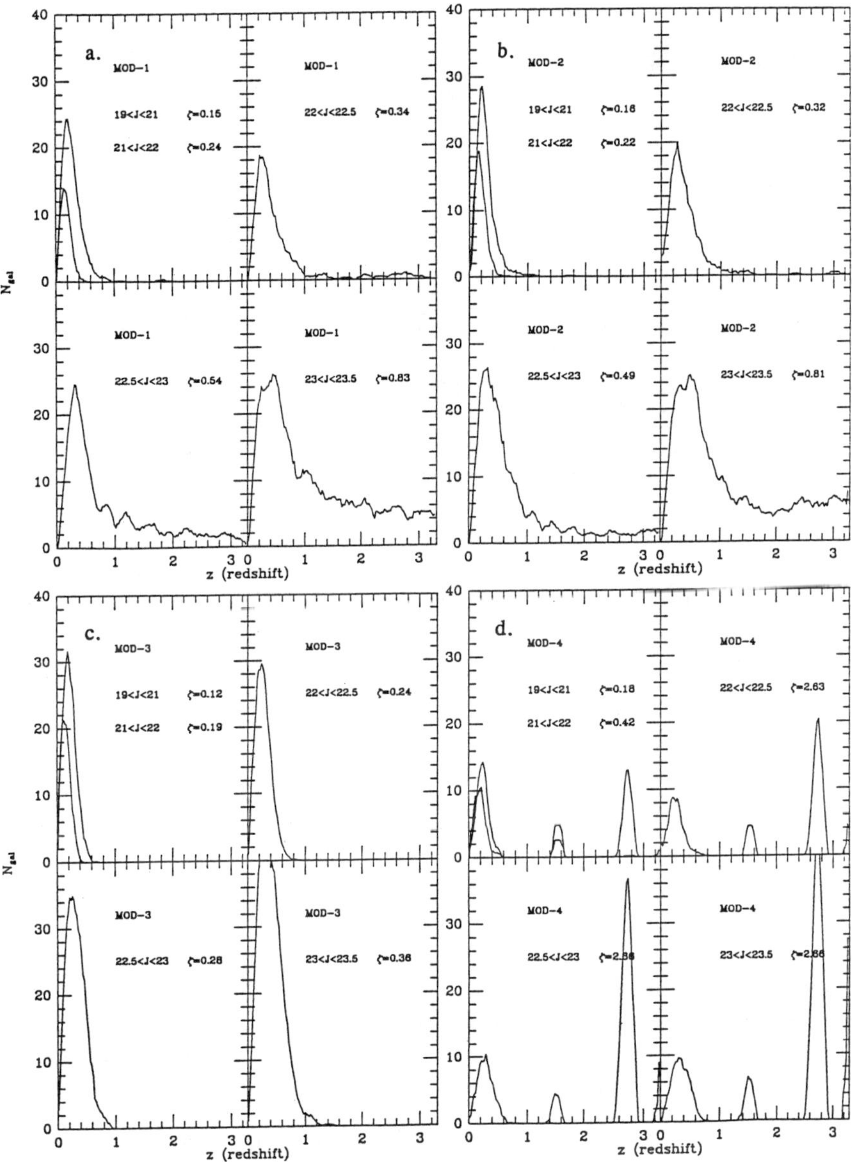

Figure 7a,b,c,d. Predicted redshift distributions. At the top of each frame the model number and the intervals of *J* magnitude are indicated. ζ gives the median redshift of the distribution in the frame. The vertical scale represents the number of galaxies per 0.214 square degrees (the size of CAT 68) in each 0.025 bin in redshift.

SPECTRAL EVOLUTION OF CLUSTER GALAXIES

ALAN DRESSLER
The Observatories of the Carnegie Institution of Washington
813 Santa Barbara St., Pasadena, CA 91101

JAMES E. GUNN
Princeton University, Peyton Hall, Princeton, NJ 08544

ABSTRACT There is strong evidence for spectral evolution of galaxies in rich clusters since $z \sim 0.5$. Studies of 14 clusters at $0.20 < z < 0.55$ by several different groups have consistently revealed a higher incidence of galaxies active in star formation compared to similar low-z clusters. Detection of spectral characteristics rare in present-epoch galaxies suggest that this enhanced star formation may have occurred in bursts, perhaps triggered by the first interaction of a gas-rich galaxy with the hot intracluster gas. As such this type of activity may be mainly the result of environmental influence rather than pure galactic evolution with cosmic time. However, new observations of the amplitude of the 4000 Å break obtained by Dressler and Gunn indicate that by $z \gtrsim 0.7$ even the reddest (presumably oldest) cluster galaxies show evidence of evolution. This places a limit of the bulk of star formation $2 < z < 5$ for these galaxies.

INTRODUCTION

As this symposium is appropriately dedicated to Edwin Hubble, the father of extragalactic astronomy, we would like to tell you what Hubble had to say about the subject of galaxy evolution. We would like to, but we can't, because Hubble didn't have a lot to say on this subject. Given the way he felt about a theoretical approach towards understanding the universe, which Simon White recalled when he recited to us the conclusion of "Realm of the Nebulae", it may not be surprising that Hubble didn't join into a discussion that was, in the early 1950's, more ripe with speculation than observational data.

Of course, Hubble's seminal work on the redshift-magnitude relation eventually led to an appreciation of how vital a knowledge of galaxy evolution is in the study of cosmology, a point reemphasized by Simon Lilly in his talk at this symposium. It is fitting here at Berkeley to acknowledge the heroic effort in the 1970's by Hyron Spinrad to push the Hubble diagram to record distances. Dressler remembers a night during his thesis work at Lick Observatory when Hy confided that he was really studying galaxy evolution, not cosmology.

The first attempts at realistic galaxy evolution models, for example Beatrice Tinsley's work in the 1970's, suggested only slight changes in spectrophotometric properties of galaxies over lookback times $z \lesssim 0.5$. This expectation applied especially to early-type systems - ellipticals, S0's, and the bulges of spirals - in which no significant star formation is found at the present epoch. Because rich clusters of galaxies are dominated by these types, the claim of the pioneering study by Butcher and Oemler (1978, 1984) of strong spectral evolution of cluster galaxies since $z = 0.5$ came as a great surprise.

However, since that time there have been challenges to the notion that such recent evolution is unexpected. Population synthesis models of early-type galaxies have consistently led to the conclusion that star formation has been substantial as recently as $z \sim 1$ (O'Connell 1985, Pickles 1988, Rose 1985). In addition, the role of mergers in producing early-type galaxies (Toomre 1977), which undoubtedly leads to late fireworks, has been championed by Schweizer (1986). Related to this are numerical simulations of universes dominated by cold dark matter (Davis *et al.* 1985) in which galaxy building by merging appears to continue to relatively late epochs.

Improvements in the quantum efficiency of detectors have led to a direct approach to the study of spectral evolution by providing observations of galaxies at large lookback times. The challenge is to assemble a representative sample of galaxies at a remote epoch for comparison with low-z galaxies which can reasonably be assumed to be descended from similar galaxies. Observations of galaxies in clusters provide a practical though imperfect solution to the problem of assembling a representative sample. Because of their high densities and predominance of early type galaxies, clusters do offer us an assortment of average galaxies in typical environments, but they do offer a volume-limited sample of not-rare galaxies under similar conditions for a range of epoch $z \lesssim 1$. Though there may have been some question at first as to whether distant clusters were, by selection, all extraordinarily rich or dense, more complete surveys like that done by Gunn, Hoessel, and Oke (1986) have identified clusters that appear, based on their space densities and luminosity function, to be ancestors of at least the richer present-epoch clusters like Coma or Hercules.

Butcher and Oemler (BO) measured the color distribution of galaxies in high-z clusters and found a significant blue tail, which they correctly speculated was due to a population of star forming galaxies nearly absent from today's rich clusters. Spectroscopic follow-up confirmed that many of these star-forming galaxies were, indeed, cluster members, not merely interlopers along the line of sight. In 1981 Gunn and Dressler began a program of low-resolution spectroscopy of distant cluster galaxies, starting with the two original clusters studied by Butcher and Oemler. We confirmed cluster membership for a significant number of the blue galaxies, thus confirming the BO effect (Dressler and Gunn 1983, Dressler, Gunn, and Schneider 1985). However, their spectral characteristics contradicted BO's interpretation that these bluer galaxies were normal star-forming spirals and perhaps the ancestors of the abundant S0's seen in today's clusters. Many of our spectra showed strong Balmer absorption with little or emission, or a very high rate of ongoing star formation. Such spectra are more typical of galaxies with bursts of star formation and less like the smooth, steady star formation seen in present-epoch spirals.

In the years since these early results, many more clusters have been added, including our own sample of 5 additional clusters which we review below. Lavery and Henry (1986, 1988) confirmed a weak BO effect for three clusters at $z = 0.2$, also emphasizing the importance of starbursts which they suggest are triggered by mergers. The rich cluster A370 at $z = 0.36$ has been studied spectroscopically by Lavery and Henry (1987) and by Mellier *et al.* (1988), who also find a significant population of star-forming galaxies, many of which appear to be in a starburst phase. MacLaren *et al.* (1988) have found far-UV excesses for galaxies in A370 with otherwise normal (redder) colors, consistent with the starburst interpretation. The most detailed analysis supporting the starburst model is included in a study of 3 clusters at $z = 0.31$ by Couch and Sharples (1988), who again find a good deal more star-forming activity than is typical in today's rich clusters.

There has been, then, good agreement among those who have studied the populations of high-z clusters. Therefore, our own completed study of 7 clusters at $0.35 < z < 0.55$ with uniform sampling characteristics, should provide representative statistical results. In addition, we also have made some progress on collecting data for 4 clusters $0.70 < z < 0.95$, which can be used to add to the data base.

Our data consist of observations made mainly with 4-Shooter camera/spectrograph on the the Palomar Hale 200-inch telescope. Typically we have photometry in three colors for 100-300 galaxies in 5' x 5' fields (~1-2 Mpc on a side) centered on each cluster. Photometry for wider areas around some clusters has been obtained, but has not yet been analyzed. Our low-resolution (~20 Å) spectra of 25-50 objects per field are used to determine redshifts and as a crude measure of star formation rate. Photographic multislits have been used to obtain ~10 of these spectra per exposure. The faintest objects, reaching $m_r \sim 23$, require up to 10 hours of integration to reach $S/N \sim 10$.

The results from the $z \sim 0.45$ sample are summarized by Gunn (1988), and we will only briefly review and update them here. Spectra have been obtained for 249 galaxies, of which 172 are members of clusters and 77 are "field" galaxies (which can themselves be in low-density foreground or background groups). Shown as spectral composities published by Gunn, the galaxies with strong Balmer absorption and little or no Balmer or forbidden emission lines, what we have called $E + A$ galaxies, remain a significant fraction of detected cluster members, 23/172. Galaxies with strong emission lines account for a slightly bigger fraction 37/172. We take these together under the label *active galaxies.* After correcting for incompleteness based on biases in sampling as a function of color, we derive an *active fraction* of 30 +/− 4 percent, with variations from cluster to cluster that are, at most, marginally significant. The remaining spectra, which we call *passive galaxies* have early-type spectra virtually indistinguishable from those of present-epoch ellipticals and S0's.

An *active fraction* of 30% is much higher than the fraction of such galaxies in low-z clusters. For example, based on the statistics of more than 1000 cluster galaxies at z ∼ 0.04 by Dressler, Thompson, and Shectman (1985), the frequency of emission-line galaxies is enhanced by a factor ∼3. *E + A* galaxies, practically absent from the low-z sample, are enhanced by an order-of-magnitude.

As discussed in Gunn (1988), it is the strong Balmer lines of the *E + A* galaxies and the high equivalent widths $W_{[OII]} \sim 20$ Å of many emission-line galaxies that has led us and others to identify many of these as starburst galaxies, or galaxies in a post-starburst phase. Detailed discussions can be found in Dressler and Gunn (1983) and Couch and Sharples (1988).

Lavery and Henry (1988) have suggested galaxy collisions and mergers as a mechanism for the starbursts, but our visual inspections of the fields we have studied showed no increase in the number of near-neighbors for *active galaxies*, not even for galaxies caught in-the-act of a burst. We prefer the idea that interaction with the intracluster medium (ICM) triggers the increase in star formation. The gas pressure at the center of a cluster is 10-100 times greater than the pressure in the interstellar medium (ISM) of today's typical spiral galaxy. We have suggested that the sweeping of the warm, intercloud medium by ram pressure and its replacement by a much higher-density gas may have triggered the harder-to-remove molecular clouds to collapse in a torrent of star formation.

The biggest challenge for a model of starburst activity in high-z clusters is to explain why it was so prevalent at z ∼ 0.50 but nearly absent today. Hydrodynamical models by Evrard (1989a,b) for the evolution of the ICM in clusters in a cold-dark-matter universe may help explain why our proposed mechanism loses much of its punch between z = 0.5 and z = 0. In his models Evrard finds a steeply rising pressure in the ICM that flattens within the cluster core. At z ∼ 0.5 the transition zone is abrupt, a "shock" which occurs at ∼0.5 Mpc at pressures well above typical ISM values. The model predicts that as a cluster evolves the shock moves outward, becoming more gradual and at a pressure comparable to or below typical ISM pressures. If correct, this model would then help explain the decline in starburst activity. It is likely, however, that an additional factor is required, which we identify as the probable decline in gas fraction in all galaxies since z = 0.5

Some circumstantial evidence supports our model. *Active* galaxies have a velocity dispersion ∼70% higher than *passive* galaxies. Furthermore, they seem to avoid the cluster core, appearing first at a radius of ∼0.5 Mpc. There are insufficient statistics to decide if their numbers decline rapidly beyond a few megaparsecs, as the model predicts, but it is known that few examples of strong $E + A$ galaxies are found in studies of high-z field galaxies. In addition to better statistics and better models, the validity of this idea will be tested by HST images which should show if *active* galaxies have a chaotic distribution of star formation instead of the regular pattern found in normal spirals.

EVIDENCE FOR EVOLUTION WITH COSMIC TIME

The evolution of at least the *active* population in rich clusters since z $\sim$ 0.5 is now well established, but it seems quite possible that this could be closely related to the cluster environment and *its* evolution. As such, the relevence of this observation to the evolution of more typical galaxies like our own may be limited.

To search for evidence of evolution with cosmic time, it is reasonable to start with *passive* galaxies, since their spectra are not confused by recent star formation. If *passive* galaxies contain only aging stars from a long-finished burst, models of spectral evolution (*e.g.,* Bruzual 1983, Rocca-Volmerange 1987) predict detectable evidence by z = 0.7 of a hotter main sequence turnoff. With far-UV flux, Balmer (absorption) line strengths, and the amplitude of the 4000 Å discontinuity as thermometers, colors and spectra of galaxies at z $\sim$ 0.5 turn up at least some galaxies which, as expected, are indistinguishable from the oldest galaxies we see today (Hamilton 1986, Dressler 1987, MacLaren *et al.* 1988).

We have been extending our sample to higher redshift $0.70 \lesssim z \lesssim 1.0$ where we expect measurable changes. We have focussed on 4 clusters at z = 0.70, 0.74, 0.76, and one which has turned out to be two clusters at z = 0.90 and 0.94. So far we find that contamination by other clusters or groups is more serious than in our z $\sim$ 0.5 sample, as expected. This results in an unfortunately small yield in cluster members when studying these fields, but an interesting and probably favorable consequence is that these clusters are more normal - not as high-density or rich as we might have thought (and feared).

At present we have obtained 86 spectra in these clusters, almost half in Cl1322+3027, a very rich Coma-like cluster at z = 0.76. After three years of perseverance we have accumulated 25 members of this cluster 13 of which are *active*! For the whole sample to date, 26 of the confirmed 47 cluster members are *active*. Most have strong emission lines but there are 5 $E + A$ galaxies as well. This *active* fraction of 55% is startling, but it is important to remember that we are selecting these galaxies from their brightness in the rest frame UV, probably prejudicing the sample somewhat. Regardless, it seems likely that these clusters have an *active* fraction at least as great as the ~30% found for the z $\sim$ 0.5 clusters.

We have measured the amplitude of the 4000 Å discontinuity for galaxies in this sample, taking it as a sensitive measure of the temperature of the main sequence turnoff. As shown by Hamilton (1986), Dressler and Shectman (1988), and Kimble, Davidson, and Sandage (1988), any dependence of D(4000) with absolute luminosity, presumably correlated with metal abundance, is weak for luminous galaxies like those in this sample. Thus, we expect the 4000 Å discontinuity to be reduced by either a hotter turnoff for an old population or contamination by younger, hotter stars, but not by variations in metal abundance from galaxy to galaxy or as a function of lookback time.

The results are shown in histogram form in Fig. 1. The bottom panel shows the distribution of 4000 Å discontinuity amplitude D(4000) for 76 galaxies in A548 and A1644, two typical low-z clusters representative of the much larger sample by Dressler and Shectman (1988). The 47 galaxies in clusters with $z = 0.35$-0.40 show a shift in the centroid of D(4000), but the upper boundary is little changed. The shift in the centroid is basically the BO-effect, seen as a lowering in many cluster members of D(4000) by ongoing or recent star formation. By $z = 0.54$, the 28 cluster galaxies appear to show a slight decrease in the upper boundary as well, but the effect is still not convincing. By $z = 0.70$-0.76, the 34 galaxies in the sample show strong evidence for a shift in both centroid and upper boundary. The 9 galaxies at $z = 0.90$-0.94 continue the trend.

It would appear that there is strong evidence here for the evolution of even the passive galaxies since $z = 0.7$. What are the major sources of uncertainty? One could worry about aperture effects, since we are sampling a larger area at high-z than at low-z. If the cause of this aperture effect is abundance gradients, we would expect the sign of the effect to be in the same sense as observed. On the other hand, D(4000) is insensitve to metal abundance, and the trend in the data develops from $z = 0.4$ and *above*, over which redshift interval the metric size changes very little. Alternatively, have we preferentially missed the oldest, reddest cluster members by selecting in the rest-frame ultraviolet? Probably not, because we sample at least 2 magnitudes into the luminosity function and the color differences we are concerned with are only a few tenths of a magnitude. It seems certain that our sample of the brightest galaxies in the ultraviolet contains at least some of the reddest galaxies, for example, we include the two cD galaxies in the cluster Cl1322+3027. Yet we find *no* galaxies with $D(4000) > 2.0$, a result in strong contradiction with the data for low-z cluster galaxies.

Taking this result at face value, then, we can infer something about the age of these oldest galaxies in clusters. We have observed a 15-20% decrease in the upper limit of D(4000) by $z \sim 0.8$. As described by Gunn (1988), we can use stellar evolution models by Tinsley and Gunn (1976) or Bruzual (1983) to place limits on the epoch of the bulk of star formation as $2 < z < 5$ for universe with $\Omega = 1$. The lower bound is not very model dependent and thus places rather a strong limit on the amount of late star formation that could

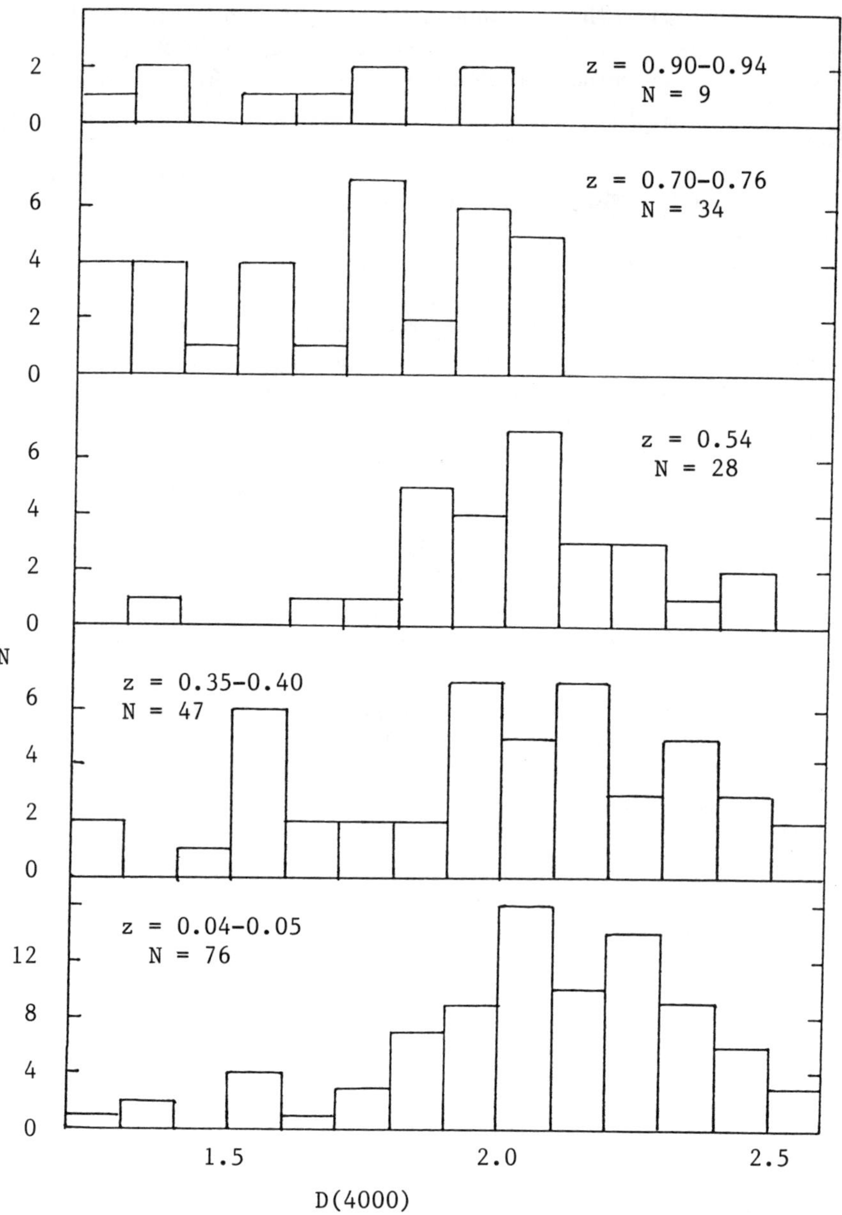

Fig. 1 – Histogram of the amplitude of the 4000 Å discontinuity D(4000) as a function of redshift for the sample. The centroid of the distribution moves steadily to lower values with increasing redshift, mostly a reflection of the BO-effect of a higher frequency of star formation. The lowering of the upper bound, reputed to be due to a hotter main sequence turnoff in even the oldest galaxies, becomes evident at $z = 0.70$, as expected for models in which the bulk of star formation occurred between $2 \lesssim z \lesssim 5$.

have occurred in these galaxies. The upper bound is less well determined because of the small time interval at higher z, but it seems to argue against the bulk of star formation happening at very high redshift $z \gtrsim 20$.

These constraints apply to ~40% of galaxies in the dense regions of rich, distant clusters. This is approximately the fraction and location of ellipticals in such clusters today, so we might, for the sake of argument at least, identify these *old* galaxies with ellipticals in clusters. Of course, ellipticals in clusters are nearly indistinguishable from ellipticals outside clusters, so why not apply this age constraint for *all* of today's ellipticals? If the preceding line of reasoning has any validity, it has an important implication for theories of galaxy formation: 10-20% of all galaxies formed the bulk of their stars before $z = 2$. This could provide a useful constraint for models of structure formation like cold dark matter, perhaps easier to evaluate than the constraint imposed by the formation of a few extremely rare objects at $z = 5$.

FUTURE PRIORITIES

The morphology of high-z galaxies in and out of clusters remains the most important missing piece of the puzzle. How does the observed evolution of high-z galaxies relate to the evolution of morphological types, and specifically, to the formation of disks and bulges? What is the role of mergers and the morphology of starbursts? These and similar questions we can soon hope to address with data from the Hubble Space Telescope.

Larger samples of cluster galaxies, including spectra and colors from the UV to the near-IR are needed to establish with certainty the trends discussed here, and to look for more subtle second-order effects like cluster-to-cluster variations. Particularly in the study of clusters at $z \sim 1$, the capabilities of future telescopes in the 8-10m class will be vital. Higher-resolution, higher-S/N spectra of the more interesting objects will be needed to better constrain models of the star formation history - another task for the new telescopes.

Finally, it will be important to assemble a more representative sample of high-z galaxies in *all environments*. Perhaps the least-biased and most straighforward way to do this is saturated coverage of an area of sky, picking up field, cluster, and supercluster galaxies where they lie. This task is just beyond present capabilities and certainly beyond the generosity of time allocation committees, but it is an idea whose time is coming.

REFERENCES

Bruzual, G. 1983, *Ap.J.*, **273**, 105.

Butcher, H. and Oemler, A. Jr. 1978, *Ap.J.*, **219**, 18.

Butcher, H. and Oemler, A. Jr. 1984, *Ap.J.*, **285**, 426.

Couch, W. J., and Sharples, R. M. 1987, *M.N.R.A.S*, **229**, 423.

Davis, M., Efstathiou, G., Frenk, C. S., and White, S. D. M. 1985, *Ap.J.*, **292**, 371.

Dressler, A., and Gunn, J. E. 1982, *Ap.J.*, **263**, 533.

Dressler, A., and Gunn, J. E. 1983, *Ap.J.*, **270**, 7.

Dressler, A., Gunn, J. E., and Schneider, D. P. 1985, *Ap.J.*, **294**, 70.

Dressler, A. 1987, in *Nearly Normal Galaxies from the Plank Time to the Present*, ed. S. M Faber, ed., Springer-Verlag, New York p. 265.

Dressler, A., and Shectman, S. A. 1988, *A.J.*, **95**, 284..

Dressler, A., Thompson, I. B., and Shectman, S. A. 1985, *Ap.J.*, **288**, 481.

Evrard, A. 1989a, submitted to *Ap. J.*, .

Evrard, A. 1989b, in *The Insterstellar Medium in External Galaxies*, eds. D. Hollenbach and H. A. Thronson, in preparation.

Guiderdoni, B., and Rocca-Volmerange, B. 1987, *A.&A.*, **186**, 1.

Gunn, J. E. 1988, in *Epoch of Galaxy Formation* eds. R. Ellis and who, who publishes.

Gunn, J. E., Hoessel, J. G., and Oke, J. B. 1986, *Ap.J.*, **306**,, 30.

Hamilton, D. 1985, *Ap.J.*, **297**, 371.

Kimble, R. A., Sandage, A., and Davidson, A. F. 1988, preprint.

Lavery, R. J., and Henry, J. P. 1986, *Ap.J.(Letters)*, **304**, L5.

Lavery, R. J., and Henry, J. P. 1987, *Ap.J.*, **323**, 473.

Lavery, R. J., and Henry, J. P. 1988, *Ap.J.*, **330**, 596.

MacLaren, I, Ellis, R. S., Couch, W. J. 1988, *M.N.R.A.S.*, **230**, 249.

Mellier, Y. Soucail, G., Fort, B., Mathez, G. 1988, *A.&A.*, **199**, 13.

O'Connell, R. W. 1985, in *The Spectral Evolution of Galaxies*, eds. C. Chiosi and A. Renzini, (Dordrecht: Reidel), p. 321.

Pickles, J. A. 1985, *Ap.J.*, **296**, 340.

Rose, J. A. 1985, *A.J.*, **90**, 1927.

F. Schweizer 1986, "Colliding and Merging Galaxies", *Science*, *231*, 227.

Tinsley, B. M., and Gunn, J. E. 1976, *Ap.J.*, **203**, 52.

Toomre, A. 1976, in *The Evolution of Galaxies and Stellar Populations*, eds. B. M. Tinsley and R. B. Larson, (New Haven: Yale University Press), p. 401.

THE BRIGHTEST GALAXIES IN CLUSTERS

SUKETU P. BHAVSAR
Department of Physics and Astronomy, University of Kentucky, Lexington, KY 40506-0055

ABSTRACT It has been shown, using extreme value theory, that first-ranked galaxies in rich clusters cannot all be statistical, the tail end of a luminosity function. The distribution of their absolute magnitudes requires that at least a fraction of them belong to a separate class of objects. This conforms with the popular (but debated) view that some of the brightest galaxies in clusters (eg the cD galaxies) are formed by a "special" process. A statistical analysis of the magnitudes of the brightest galaxies, under the general assumption that they consist of both special and normal galaxies, leads to the following "most probable" situation. About 73% of the Abell richness 0 and 1 clusters have a special galaxy. The luminosities of these special galaxies have a Gaussian distribution with a dispersion of 0.3 mag. Their metric magnitudes are on average 0.5 mag brighter than the average magnitudes of the brightest normal galaxies. Statistical analysis does not rule out the possibility that all first-ranked galaxies in rich clusters are special.

INTRODUCTION

The remarkably small dispersion (≈0.35 mag) in the magnitudes, M_1, of the first brightest galaxies in clusters, has made them indispensable as "standard candles". They have, for this reason, become the most powerful yardsticks in observational cosmology. Yet, there has existed disagreement as to the nature of these objects. The two opposing viewpoints have been: that they are a class of "special" objects (Peach 1969; Sandage 1976; Tremaine and Richstone 1977); and at the other extreme, that they are "statistical", representing the tail end of the luminosity function for galaxies (Peebles 1968, Geller and Peebles 1976).

In 1928, in a classic paper, R.A. Fisher and L.H.C. Tippett had derived the general asymptotic form that a distribution of extreme sample values should take - independent of the parent distribution from which they are drawn! This work was later expanded upon by Gumbel (1958). Bhavsar and Barrow (1985) applied these results of Fisher and Tippett (1928) and Gumbel (1958) on excellent data for M_1 obtained by Hoessel, Gunn and Thuan (1980,[HGT]), showing that the statistical hypothesis can be rejected with 95-99% confidence. This result does not necessarily imply that all brightest galaxies are special. It does demand though, that not all first ranked galaxies in clusters are statistical. The question as to their nature, implied by their distribution in magnitudes, was recently addressed by this author (Bhavsar 1989).

MODEL FOR THE BRIGHTEST GALAXIES

It was shown that a "two population model" (Bhavsar 1989) in which the first brightest galaxies are drawn from two distinct populations of objects - a class of special galaxies and a class of extremes of a statistical distribution - is needed. Such a model can explain the details of the distribution of M_1 very well. A further detailed statistical analysis by Bhavsar and Cohen (1989) of alternatives to the assumed Gaussian distribution of the magnitudes of special galaxies, along with a study of the confidence limits of the parameters determined by maximum-likelihood and other statistical methods has determined a "best" model which we shall describe here. Parameters determined purely on the basis of a statistical fit of this model with the observed distribution of the magnitudes of the brightest galaxies, are exceptionally consistent with their physically determined and observed values from other sources.

From extreme value theory, for clusters of similar size, the probability density distribution of M_1 for the "statistical" galaxies is given by

$$p = a \exp [a(M_1 - M_o) - e^{a(M_1 - M_o)}], \quad (1)$$

where a is the parameter which measures the steepness of the luminosity function at the tail end. M_o is the mode of the distribution and is a measure of the cluster mass, but with only a logarithmical dependence on the cluster mass. The distribution (1) is often referred to as the first Fisher-Tippett asymptote, or Gumbel distribution.

The probability density distribution of the quantity that measures the brightness of the special galaxies is assumed to be a Gaussian. This is the most general expression for the distribution of either the magnitudes or luminosities of these galaxies, if they arise from some process which creates a standard mold with a small scatter, arising because of intrinsic variations as well as experimental uncertainty in measurement. It turns out that the models in which the luminosities have a Gaussian distribution work marginally better. This may have been expected, luminosity being the physical quantity.

If a fraction of the rich clusters have a special galaxy which competes with the normal brightest galaxy for first-rank, then the distribution that results is derived in Bhavsar (1989). The best fit model is determined by statistical criteria (Bhavsar and Cohen 1989). This model requires 73% of the richness 0 and 1 clusters to have a special galaxy which is on average half a magnitude brighter than the average brightest normal galaxy. As a result, about 66% of all first-ranked galaxies in richness 0 and 1 clusters are special. This is because in 7% of the clusters, though a special galaxy is present, it is not the brightest. These values are somewhat different than Bhavsar (1989) because there the model assumed that the magnitudes, rather than the luminosities, of the special galaxies had a Gaussian distribution. The results here are a marginally better fit to the data (Bhavsar and Cohen 1989) than Bhavsar (1989). We show in figure 1 the excellent fit of this model to the distribution of M_1 for the Hoessel, Gunn and Thuan (HGT) rich clusters.

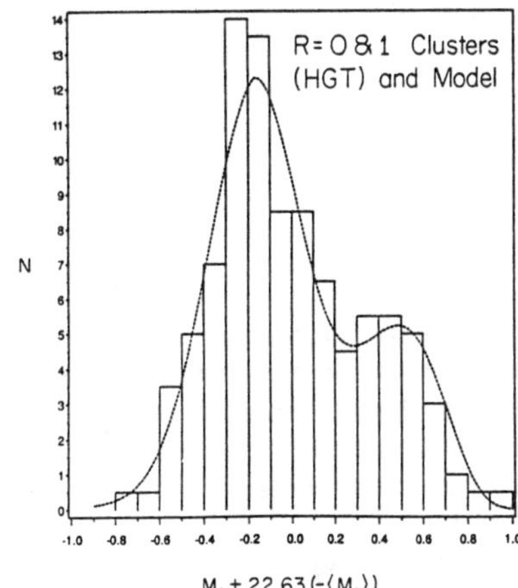

Fig. 1.

CONCLUSION

Although it is generally appreciated that some of the brightest galaxies in rich clusters are a morphologically distinct class of objects (eg cD galaxies); we have approached the problem from the viewpoint of the statistics of their distribution in M_1, and conclude that indeed some of the brightest galaxies in rich clusters are a special class of objects, distinct from the brightest normal galaxies.

ACKNOWLEDGEMENTS

I thank John Connolly, director of the Center for Computational Sciences, University of Kentucky, for providing the funds to attend this meeting.

REFERENCES

Bhavsar, S.P. 1989, Ap.J. **338** 718.
Bhavsar, S.P., and Barrow, J.D. 1985, M.N.R.A.S. **213**, 857.
Bhavsar, S.P., and Cohen, J.P. 1989, Ap.J. submitted.
Fisher, R.A., and Tippett, L.H.C. 1928, Proc. Cambridge Phil. Soc. **24**, 180
Geller, M.J., and Peebles, P.J.E. 1976, Ap. J. **206**, 939.
Gumbel, E.J. 1958, Statistics of Extremes, (New York: Columbia U.Press).
Hoessel, J.G., Gunn, J.E., and Thuan, T. X. 1980, Ap.J. **241**, 486.
Peach, J.V. 1969, Nature, **223**, 1140.
Peebles, P.J.E. 1968, Ap.J. **153**, 13.
Peterson, B.A. 1970, Ap.J. **159**, 333.
Sandage, A. 1976, Ap.J. **205**, 6.
Tremaine, S.D., and Richstone, D.O. 1977, Ap. J. **212**, 311.

GALAXY EVOLUTION AND FORMATION

LENNOX L. COWIE and SIMON J. LILLY
Institute for Astronomy, University of Hawaii
2680 Woodlawn Drive
Honolulu
Hawaii 96822, U.S.A.

ABSTRACT. Results from the Hawaii imaging and spectroscopy deep survey are discussed. Cumulative galaxy counts now exceed 10^6 deg^{-2}. To $B = 24$, the bulk of these galaxies lie at low redshift. Flat spectrum sources constitute a fraction of about 30–40% of the sources at $B = 24$. There are not enough of these to explain the high number counts, but they are an exceedingly interesting population in their own right. We argue that they are unlikely to lie at $z \sim 1$, but could be either massive starbursters at $z \sim 2 - 3$ or dwarf starbursting galaxies at very low redshift ($z \lesssim 0.1$). In either case they constitute a major episode in galaxy formation.

1. Introduction

This is an exciting period in the study of galaxy evolution and formation, when it appears we must be reaching the magnitude limits in both imaging and spectroscopy where we probe both cosmology and galaxy formation. Yet, as intense observational efforts have pushed these limits outward, it has become clear over the last year or so that there are profound conceptual problems arising, and considerable confusion as to the correct interpretation. The three key problems at this point may be stated as follows:

(1) Spectroscopic studies show that the great bulk of galaxies, even to limits as faint as $B = 24$, are quite local ($z \lesssim 0.5$). This is consistent with no evolution in the luminosity of near-L_* galaxies. Yet, at the same time, the number counts of galaxies are rising much more rapidly than no-evolution models predict, and exceed the no-evolution model counts by factors of nearly an order of magnitude at $B = 24$. This problem has been stressed by Ellis and his collaborators (Broadhurst et al. 1988; Colless et al. 1989) and might perhaps be referred to as the Durham paradox.

(2) A possibly related problem is the continued failure of the number counts to converge even as we exceed number counts of 10^6 deg^{-2}. The problem here is that, in any standard cosmology, there is insufficient volume to fit such large numbers of galaxies in. While various forms of this argument have been given (Cowie 1988; Koo 1989a,b), it has stressed most recently by Koo, and we refer to this as Koo's paradox.

(3) Is the blueing trend seen in the faint galaxy population (Kron 1980; Koo 1986; Tyson 1988) partially caused by a new blue population rising very abruptly at $B \sim 23$ or so? Or is it simply an ongoing trend, where we see more irregular galaxies as we go fainter, and in particular, some galaxies lying just beyond $z = 1$ whose flat spectrum is caused by the Balmer break lying beyond the reddest band?

Recent spectroscopy argues against the $z = 1$ interpretation, but (despite our initial interpretation of one object as having to be at $z \sim 3$ [Cowie and Lilly 1989], based on an emission line which subsequent spectra did not confirm), unfortunately does not distinguish between a local population of ultra-blue dwarves or a $z \sim 3$ population of starbursters. We refer to the nature of the very blue, or 'flat spectrum', galaxies as the flat spectrum question.

Table 1. Hawaii deep survey bandpasses and limits

	$\Delta\lambda$ (angstroms)	Pixel size (arcsec)	1σ limit seeing aper. (AB)	1σ limit 3.5″ aper. (AB)
K	22100 ± 2100	1.2	24.3	23.9
I	8340 ± 450	0.214	27.1	26.3
V	5425 ± 450	0.214	28.0	27.2
B	4470 ± 430	0.214	28.4	27.6
U′	3400 ± 150	0.214	27.6	26.8
Spectral slit	4000 – 8000	0.6×3	...	~ 28

As with all observational questions, the resolution of the three paradoxes and questions requires better data, and, in particular, added information through new tools such as additional colors. The Hawaii deep survey (summarized in Table 1) was motivated by the potential of attacking the whole problem with the much wider wavelength coverage obtained by extending the surveys into the K band and down as far into the ultraviolet as was possible from the ground (3400 Å). This depended on the UV sensitive CCDs and the new IR arrays. It was also motivated by the possibility of pushing the surveys to such extreme limits (most particularly in the I band — cf. Table 1) that we could use robust techniques (large apertures) to measure magnitudes and colors, while still maintaining high S/N in magnitude ranges of interest ($I_{AB} < 25$). (Throughout this talk we shall use the more physically based system of c.g.s. magnitudes [$AB = -48.6 - 2.5 \log f_\nu$] so that zero colors correspond to flat f_ν. They are related to normal magnitudes approximately as $[K, I, V, B]_{AB} = [K, I, V, B] + [1.99, 0.49, 0, -0.19]$.) Finally, because the IR arrays are physically small, the survey areas cover very small patches of sky (four, each of 1.3 arcmin^2) and, while there are several hundred objects identifiable, the number of bright objects ($B_{AB} < 24$) is sufficiently small (about 10 per field) that it is feasible to identify them all spectroscopically.

As we shall outline in the following sections, because of its depth, the Hawaii survey has clearly exacerbated both the Durham and the Koo paradoxes to levels where they are extremely hard to understand. The flat spectrum question unfortunately remains open and may be very hard to resolve observationally.

2. Deep imaging and spectroscopy of the small survey areas

The development of CCDs and infrared arrays, when combined with the extremely powerful observational technique of median sky-flat generation, allows one to obtain extraordinarily deep images throughout the visible and near-IR windows. The same shift and median-flat techniques can also be applied to slit spectroscopy, and prove to be extremely effective in this mode also.

The median sky flat technique consists of obtaining multiple exposures, each moved by a few arcseconds in a non-replicating pattern. The individual exposure time is optimized by making each exposure sky-background limited but otherwise as short as possible to maximize the total number of frames while not making telescope slewing overhead and CCD readout times too large a fraction of the total time. Median values of a large number of exposures on either side of an individual exposure are used to generate the sky flat for the exposure and the exposures are then registered and added to form the final frame. At this stage, cosmic rays and defects are simply filtered out.

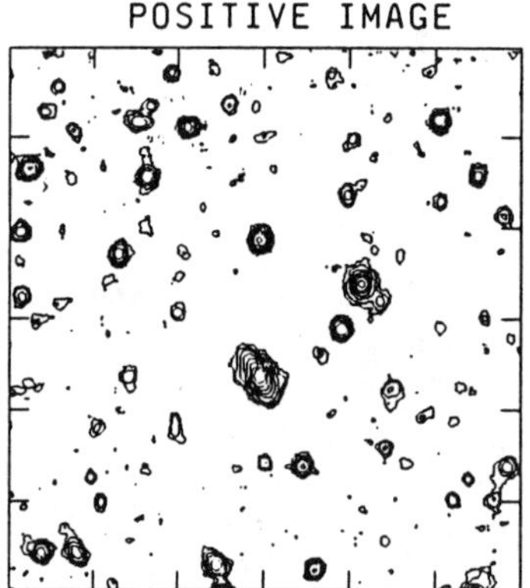

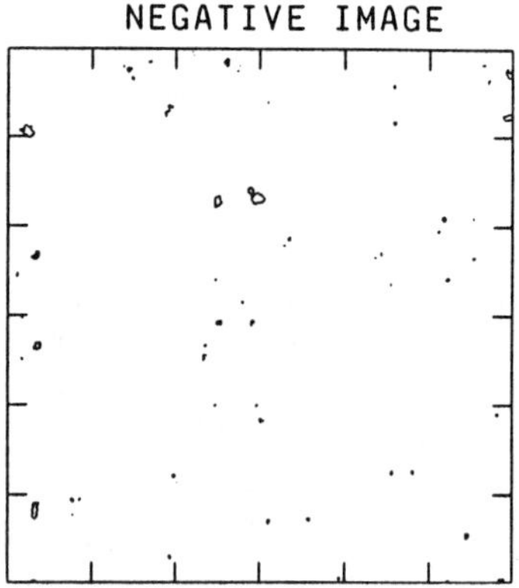

Figure 1 The central 1 arcmin^2 of the SSA22 field, contoured at equal positive and negative isophotes, such that extended objects lie at about the 5 σ level. The absence of significant numbers of objects in the negative frame shows that the flat-fielding techniques are working properly and that a high level of confidence may be placed in the objects seen in the positive image. Although about 100 objects are seen in the frame, overlap is not a major problem in distinguishing objects because of the extremely high resolution (0.8″ FWHM). At the faint limit, about 90% of galaxies added to the frame can be easily recovered.

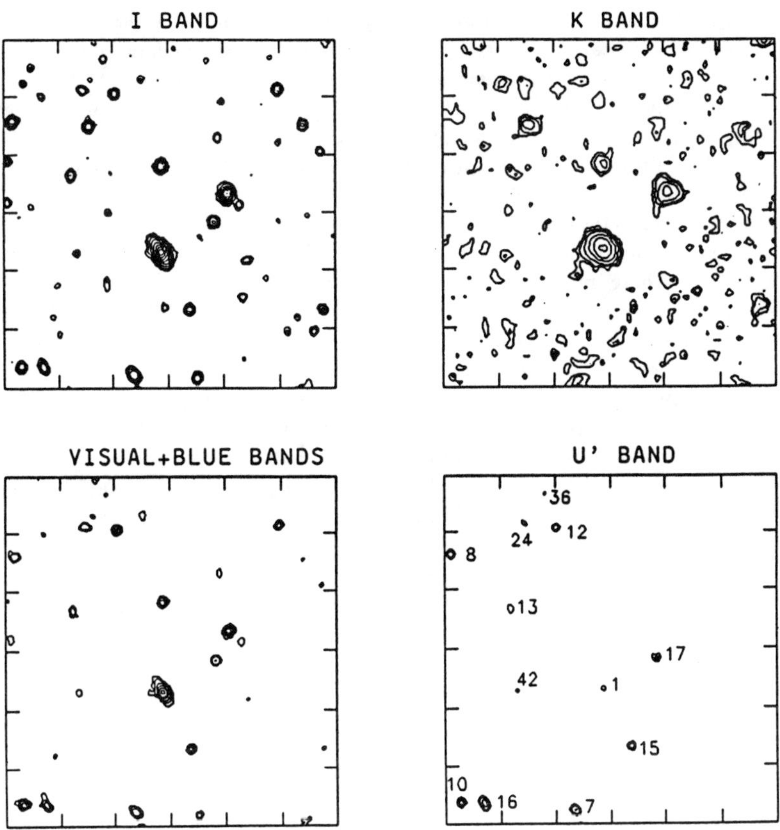

Figure 2 Central 1 arcmin^2 of SSA22 shown at constant f_ν surface brightness level, except in K, where the contour levels are a factor of 2 higher. Flat spectrum objects should appear invariant from frame to frame. Objects still visible in the U' band are labelled for reference in the text and tables.

The power of this technique is most needed in the K band where the sky background is enormously high. In IRCAM at the 3.8m UKIRT, the total background (sky + telescope + instrument) is about 14.5 mag arcsec^{-2} in K_{AB} magnitudes. To reach interesting results we must go to at least $K_{AB} = 23$ in a 14 arcsec^2 aperture, or a limiting surface brightness of about 26 mag arcsec^{-2}. That is, we must flatten to 3 parts in 10^5. The median sky flattening technique allows us to do this. Currently, a typical K band observation in our deep field region is around 12 hours of actual on-target observing time, and consists of 150 individual frames. The σ in the final image is only 50% higher than the shot-noise value. Improved observing techniques and software have recently allowed us to drop exposure times by a factor of four, which may allow us to obtain essentially perfect sky-limited images at K even for the much longer on-target observing times (36 hours) we are now obtaining. In the much easier cases of optical imaging and spectroscopy, the data is already perfect and the only limit to the depth which can be achieved is the amount of telescope time and, ultimately, the crowding problem. Current survey depths on the fields are summarized in Table 1. (Throughout this paper, we shall use 3.5″ diameter apertures, convolved with the seeing disk. While these large apertures are extremely expensive in S/N, they give extremely robust magnitudes and colors.)

The crowding problem is greatly alleviated by the high image quality on the 3.6 m CFHT where most of the optical images were obtained. This is shown in the picture of Figure 1, which is a 12-hour image with a FWHM quality of 0.8″. The exactly 1 square arcminute of sky shown contains 102 objects (400,000 objects per square degree) at a detectability level greater than 5 σ (as can be seen from the negative image). Simulations show that, even at this depth, 90% of faint-end galaxies added to the picture can be recovered as individual objects, though their outer isophotes may overlap with those of other galaxies. At these high resolutions, crowding can probably be handled straightforwardly to a galaxy surface density of about 10^6 degree^{-2} but beyond this it will be a progressively severe problem.

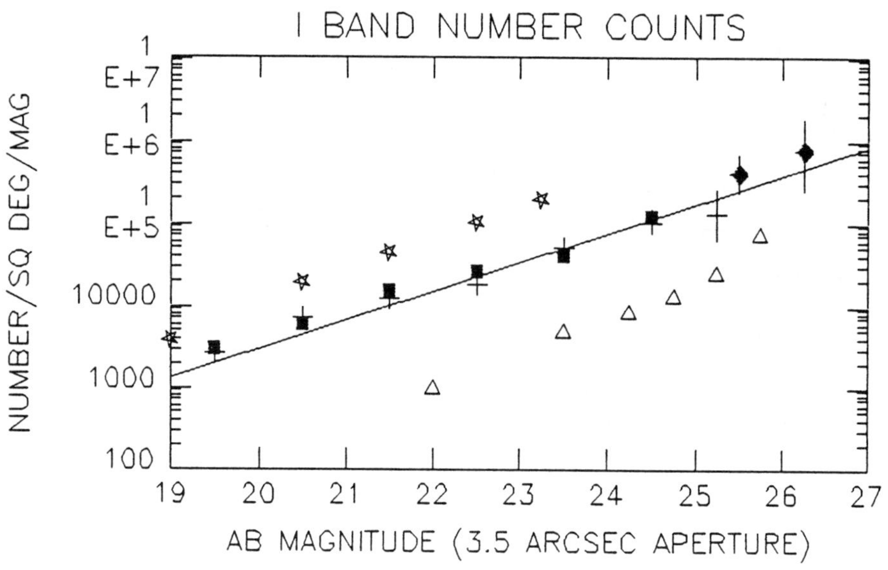

Figure 3 Number counts of I (crosses, [all fields], boxes [SSA22 fields only], diamonds [slit counts in SSA22 field], solid line fit to Tyson data at $I_{AB} < 24.5$.) The stars show the K band counts and the triangles the U' counts.

Figure 2 shows the multicolor data for this same field contoured at fixed AB magnitude surface brightnesses, showing the uniform depth of the images and also the drop in the number counts (to a fixed AB magnitude) as one moves to the bluer bandpasses. The reader should note the existence of flat f_ν objects such as SSA22–16 and also of objects such as SSA22–24 which are flat between I and B but then drop by about a factor of 2.5 in the U' band. We will return to discuss the flat-spectrum sources in Section 4.

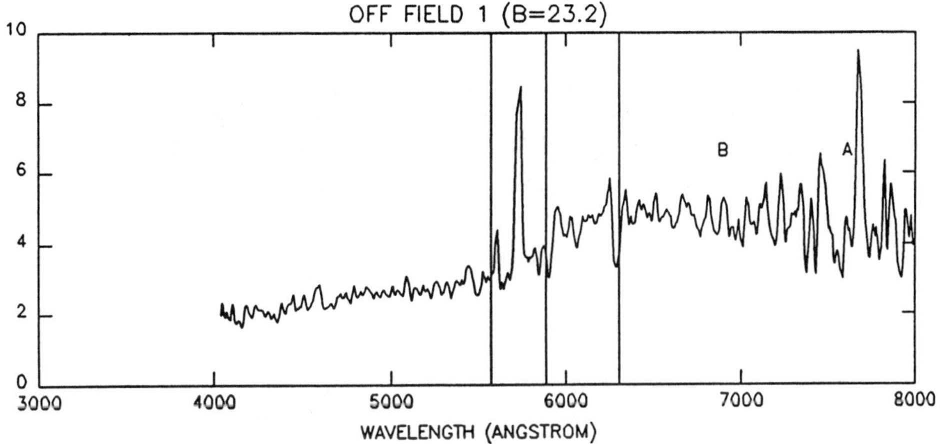

Figure 4 Spectrum of a $B = 23.2$ galaxy lying at $z = 0.53$. The solid lines mark the positions of the stronger night sky lines, and the letters denote the atmospheric bands. The strong emission lines seen are [O II] λ 3727 Å, Hβ, and [O III] λ 5007 Å.

The number counts (U'_{AB}, I_{AB}, and K_{AB}) for these fields are shown in Figure 3. The counts are automated detections corrected for incompleteness with the usual simulations. At the bright end ($I_{AB} < 24.5$) they agree extremely well with previous observations (Tyson 1988). In the I band we have extended the counts to $I_{AB} = 26.25$ using the even deeper slit images we shall discuss below. The cumulative counts probably exceed 10^6 degree^{-2} even allowing for possible errors at the fainter end. There is no sign of turnover in the counts. As was stressed by Koo in this conference, the sheer number density of galaxies at these levels is becoming extremely difficult to understand from number-count modelling (e.g. Yoshii and Takahara 1988). We shall return to this point in Section 3.

The final data that we have available on these fields is substantial deep spectroscopy on the SSA22 field. This spectroscopy uses single slits positioned on a nearby bright star and rotated to position them on one or more of the faint targets in the field. Median sky flat techniques are used and the bright star is used to spatially register the exposures, giving sky-limited spectra. Several such slit observations, ranging up to 20 hours in exposure time, have been made using the University of Hawaii wide-field grism ($3'' \times 300''$ slit, FWHM = 40 Å, $\lambda = 4000 - 8000$ Å) on CFHT and we have observed and identified most of the ten $B < 24$ objects on the 22 hour field. Of the ten $B < 24$ objects in the SSA22 area, two are clearly stellar, based on morphology. Of the remaining 8 galaxies, four are at quite unambiguous redshifts of $z < 0.4$, two (22–24 and 22–16) are flat-spectrum objects which we discuss in Section 4, and two have yet to be observed. Two additional objects with $B < 24$, but outside the deeply imaged SSA area, also fell on the slit. These also are at moderate redshift. One of these latter objects is shown in Figure 4.

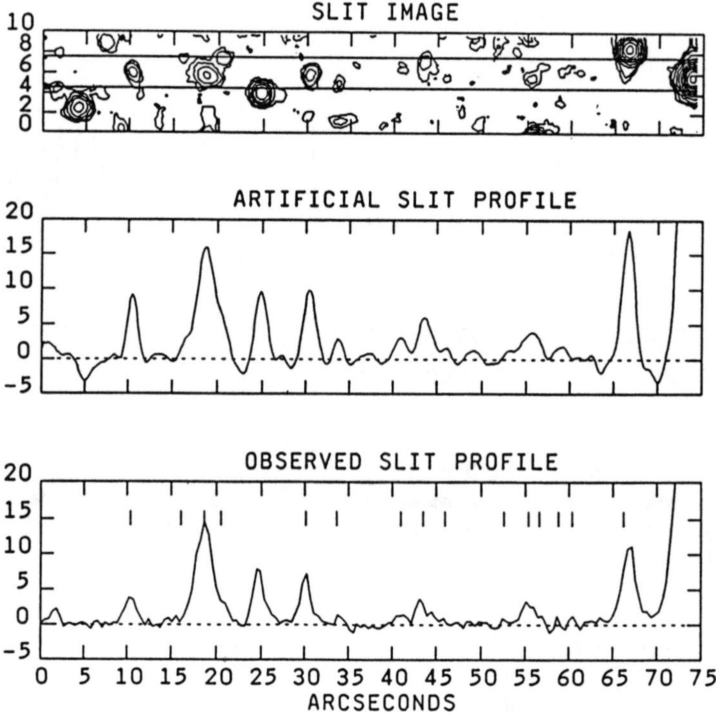

Figure 5 The observed slit profile (4000–7500 Å) in a 20 hr integration (lowest panel) is compared with the image of SSA22 ($I+B+V$), showing the slit position (top panel) and the expected surface brightness in the slit (middle panel). The slit covers 225 arcsec^2 and contains 15 marked objects with centroids lying in the slit, excluding the guide star (right hand edge) and the target SSA22–24 at $18''$.

The wide wavelength coverage and the extremely long integration times in these spectra make the wavelength-integrated slit images the deepest exposures of the sky which have ever been obtained. This is shown in Figure 5, where such a slit profile is compared with the image of SSA22. We have used these slit images to extend the I band counts by about a magnitude over the imaging number counts by identifying objects in the slit and then measuring the magnitudes on the I band image. Counts may also be made directly on the slit, but, because of the wide slit size, require substantial corrections for object overlap and slit width effects. On Figure 5, excluding the target (here SSA22–24) and the guide star, we can clearly see 15 objects whose centers lie within the 225 arcsec^2 area slit from the I band observations. Simulations show approximately 50% incompleteness at this level owing to crowding, giving a surface density of approximately 400 arcmin^{-2} or about 1.5×10^6 degree^{-2}.

3. Distribution of galaxies

There are two very curious features of the data described above. The first is the enormously high overall surface density of galaxies and the second is the very high surface density of low-z galaxies (the Koo paradox and the Durham paradox, respectively). Both of these points were discussed by speakers at this conference, based on shallower surveys. Koo argued that the high surface density of counts implies $q_0 = 0$ (see also Koo 1989), while Ellis argued that differential evolution in the luminosity function is required at $z \sim 0.2 - 0.3$ to account for the number counts in spectroscopic observations of complete samples to $B = 22.5$. The actual situation appears to be much more complex and possibly much more serious for our model cosmologies.

The simplest way to see the Koo problem is to consider the required number density of counts within a redshift-limited volume and to compare this with the local density of galaxies to some limiting magnitude (Koo 1989). This is shown in Figure 6, where we plot the required number density as a function of limiting redshift to produce a surface density of 10^6 degree^{-2}. Even in a volume to $z = 5$ in a $q_0 = 0$ universe, we find that we require $\sim 5 \times 10^{-3}$ galaxies Mpc^{-3}. As can be seen from Figure 6, the required density is several times higher for $q_0 = 0.5$. In currently favored Schechter functions, we must see all galaxies to about 5 mag below L_* *throughout the entire volume* to obtain this many galaxies. We shall argue below that seeing the fainter-end galaxies throughout the volume is not a viable possibility. Koo's interpretation of this phenomenon (i.e. $q_0 = 0$) simply depends on the fact that it is somewhat easier to get close to the number density with $q_0 = 0$, but neither $q_0 = 0.5$ or $q_0 = 0$ explains the observations without invoking some physical cause such as physical merging to increase the number density at high z (Cowie 1988).

However, the Durham problem may in fact be the same problem as the Koo problem at low redshifts where neither cosmology-dependent volume effects nor merging can be invoked so easily. Basically, to $B = 21$, the mean galaxy redshift is $z = 0.23$ and nearly all galaxies lie within $z = 0.4$. At $B = 22.5$ the sample is marginally incomplete, but the mean redshift is 0.3 and 80% of the galaxies lie within $z < 0.6$ (Broadhurst et al. 1988; Colless et al. 1989). If galaxies do not undergo luminosity evolution, then at $B = 21$ the number counts are too few by a factor of 1.3 and, at 22.5, by a factor of 2 (Yoshii and Takahara 1988).

The present results extend this to $B = 24$ where the median redshift is 0.35 and the largest conclusively observed redshift is 0.56. At this magnitude at least 50% of

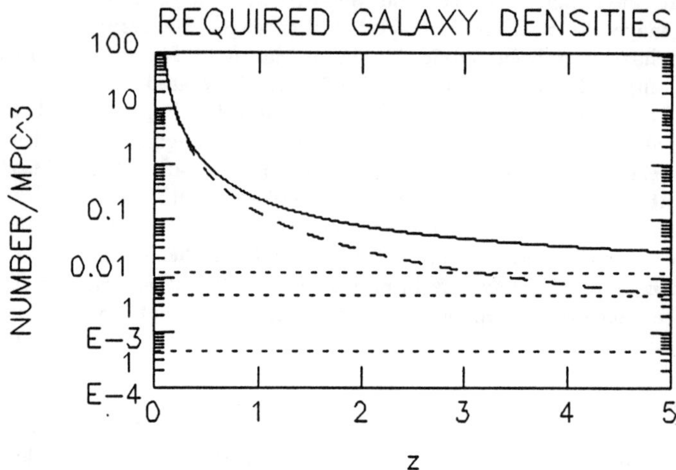

Figure 6 Required average co-moving number density of galaxies to produce 10^6 deg^{-2} objects is shown as a function of the z to which the objects can be seen. The solid line is for $q_0 = 0.5$, the dashed line for $q_0 = 0.02$ and the dotted lines show the cumulative density of galaxies brighter than L_*, 0.1 L_*, and 0.01 L_*.

the galaxies appear to lie at low redshift and to lie on the no-evolution mean track. Since the mean number counts in no-evolution models are off by a factor of $\sim 5 - 10$ at this magnitude (Yoshii and Takahara 1988; Ellis 1989) we have several times more galaxies than are expected in the available volume.

The simplest explanation of this phenomenon is that adopted by Broadhurst et al. of brightening the low end of the luminosity function to produce more near-L_* galaxies. This can work at $B = 21$ but has serious problems by $B = 24$. The average night sky surface brightness of these galaxies at low redshift ($z < 0.4$) is 3×10^{-25} ergs cm^{-2} s^{-1} arcsec^{-2} which would require formation of about 10^{-34} g cm^{-3} of metals on average (see Section 4). Massive galaxies can just account for this (Cowie 1988) but because the available mass drops as $\sim L^{-1}$ as one moves to lower luminosities, galaxies at less than 0.1 L_* simply do not contain enough mass, and produce enough metals, to sustain the required burning.

It is not clear at this point what this effect is in fact due to. It could be that the cosmology is bizarre (cosmological constant?) and the problem is caused by underestimating the avalable volume even at low redshift. Alternatively, the number density of near-L_* galaxies (i.e. the normalization of the Schechter function) could be rising as we move outward in z, reflecting large-scale cosmological inhomogeneity. Neither possibility seems attractive, and it still may be that we can slip through without such radical consequences by invoking a combination of physical merging, differential evolution, and tuning of the local galaxy parameters.

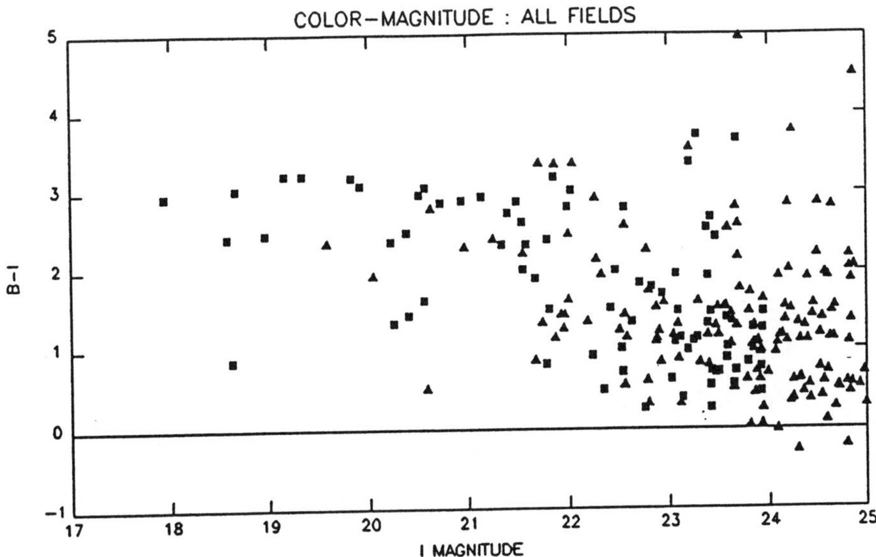

Figure 7 $(B-I)_{AB}$ vs. I_{AB} for our galaxy sample. The boxes show a wide field sample down to $I_{AB}=24$, and the triangles show the deep field data to $I_{AB}=25$.

4. Flat spectrum galaxies

As we move toward fainter magnitudes, galaxies show a sharp blueing trend (Kron 1980; Koo 1986; Tyson 1988) which is most easily seen in color-magnitude plots such as the I_{AB}, $(B-I)_{AB}$ plot shown in Figure 7. While the general blueing trend is clearly partly caused by selection toward Im and Sd galaxies owing to the K-correction, at I magnitudes between 22 and 23 a substantial population appears with colors only slightly redder than flat f_ν ($(B-I)_{AB}=0$) (Tyson 1988). We call this the 'flat-spectrum' population, as being physically clearer than referring to it as the blue galaxy population.

While, as can be seen in Figure 7, there are blue galaxies at brighter magnitudes, the onset of the flat spectrum population is abrupt and appears to signify that we are seeing in some sense a distinct population. This is shown in Figures 8a and 8b where we plot the fraction of blue color selected galaxies as a function of magnitude in our own data (Figure 8a) and in a combined sample of data from Koo (1986) and Tyson (1988) (Figure 8b). In both cases the fraction of blue selected objects rises rapidly to constitute about 40% of the counts at an I magnitude of 25 and a B magnitude of 24.5. This onset magnitude is not a function of the color criterion, as would be expected if we were simply seeing a blueing trend. As is shown in Figure 8a, a weak color selection criterion which contains galaxies at

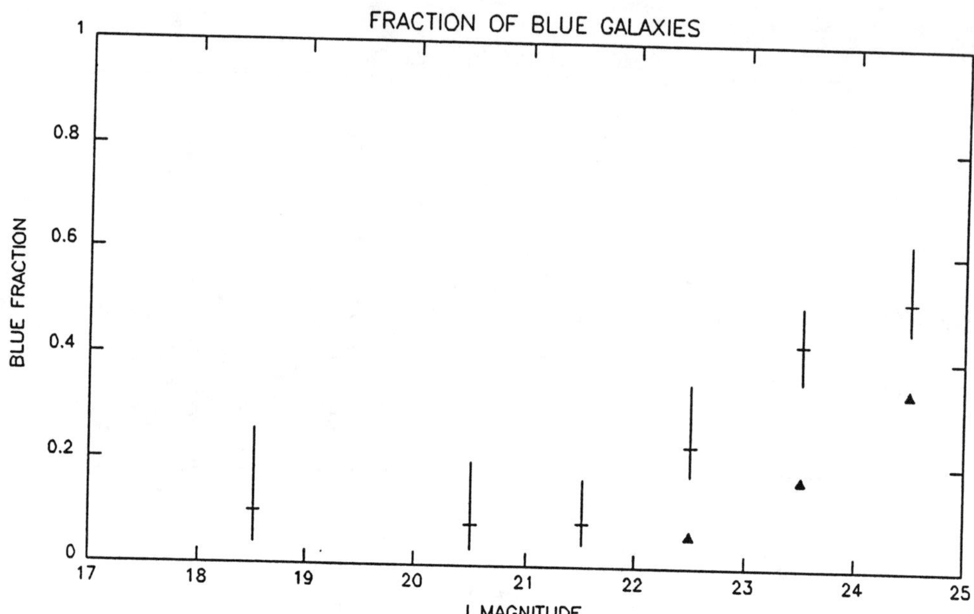

Figure 8a Fraction of counts with $(B-I)_{AB} < 1.0$ (crosses with error bars) and $(B-I)_{AB} < 0.5$ (triangles) as a function of I_{AB} magnitude. Both show an abrupt rise at $I_{AB} = 22.5$.

brighter magnitudes when compared with a stringent criterion, shows the same onset of a comparable number of new galaxies at the same magnitude. The rapidity of the rise is best seen in Figure 8b, where, in two magnitudes, the population rises by nearly three orders of magnitude. It should be emphasized that, contrary to earlier speculation, this population is too small a fraction to help explain the rise in the number counts above no-evolution discussed in Section 3. From Figure 8b we can see that, at $B < 22.5$, (the Colless et al. spectroscopic sample) we expect considerably less than 3% of the galaxies to lie in the flat spectrum population, and, though by $B_{AB} = 23 - 24$, (the present magnitude range for spectroscopic redshifts) this will have risen to a value of about 20 – 30%, this is not nearly large enough to account for the $B = 24$ discrepancy.

Defining a simple color criterion to clearly separate the flat spectrum population from low-redshift Im or Sc/Sd galaxies is not straightforward, even using a color-color plot such as the $(B-I)_{AB}$ vs $(U'-B)_{AB}$ of Figure 9. The color condition $(B-I)_{AB} < 0.61$ shown picks out most of the anomalous objects (15 galaxies out of 96 objects), but includes several galaxies that lie near the low-redshift end of the Im and Sd tracks and are most probably low-redshift galaxies. In one such case (SSA22-10, $(B-I)_{AB} = 0.59$, $(U'-B)_{AB} = 0.59$) we have obtained a spectroscopic redshift of 0.18. About 6 of the galaxies chosen according to the simple $(B-I)_{AB}$ criterion are ambiguous in this way, leaving 9 clearly anomalous galaxies, or 10% of the population

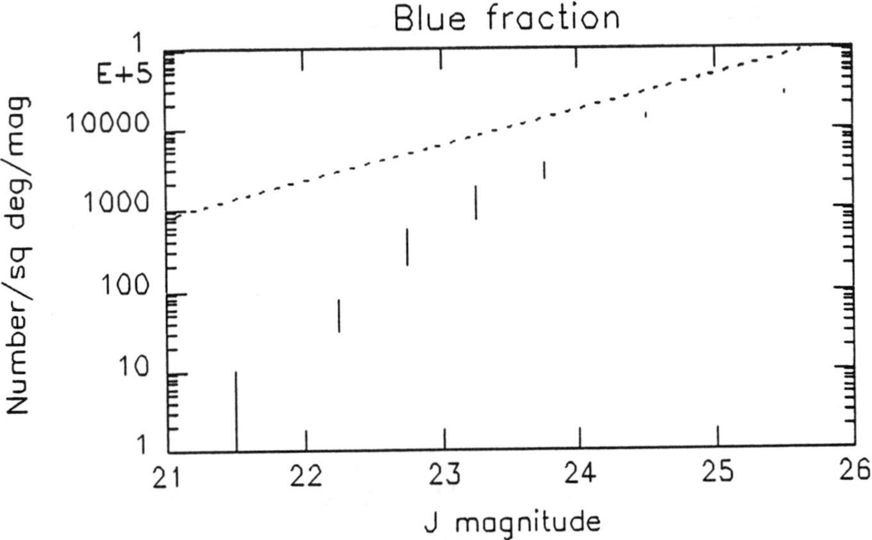

Figure 8b Blue counts ($J - R < 0.6$) shown as error bars versus J magnitude, while dotted line shows total counts.

to $I_{AB} = 24.5$. The cumulative fraction of new objects to $I_{AB} = 24.5$ required in Figure 8a is about 15%, so that it is reasonable to identify the two populations and assume that the additional counts are formed by the galaxies lying in the anomalous portion of the color-color plane.

As is well known, Im galaxies will show flat spectrum regions at wavelengths shorter than the Balmer break ($\lesssim$ 3650 Å) and longer than the Lyman continuum break ($\gtrsim$ 1000 Å) (e.g. Bruzual 1989; Songaila, Cowie and Lilly 1990) and we can produce flat spectrum galaxies in $B - I$ from this type of population provided the redshifts of the objects lie between $z \sim 1$ and $z \sim 3$. More intensely star-bursting galaxies could have flat f_ν at all wavelengths $\lambda > 1000$ Å. It is therefore possible that the onset of the flat spectrum population corresponds simply to the magnitude at which we see relatively unevolved Im galaxies beyond about $z = 1.3$ where the passage of the Balmer break through the I band moves the color from $(B-I)_{AB} \sim 0.6$ to $(B - I)_{AB} \sim 0$. However, a no-evolution galaxy redshift track makes it unlikely we would see Im galaxies to $z = 1.3$. Also, in direct observational terms, it is more likely we would have seen galaxies between $z \sim 0.6$ and $z \sim 1.3$ and we have seen none. Furthermore, in at least one case (SSA22–24, discussed below), there seems no clear way to reconcile the observed spectrum with this interpretation.

Given that this is not the explanation, we may be seeing luminous starbursting galaxies at any redshift from $z = 0$ to near 3. At $0.1 < z < 2$, we may expect an intense starburster to lie near the origin of the $(B-I)$, $(U' - B)$ plane in the absence of intrinsic reddening, but at higher redshift the passage of the Lyman forest and then the intrinsic Lyman continuum break and any possible Lyman limit systems through the U' band, and the Lyα forest into the B band, will move the track up

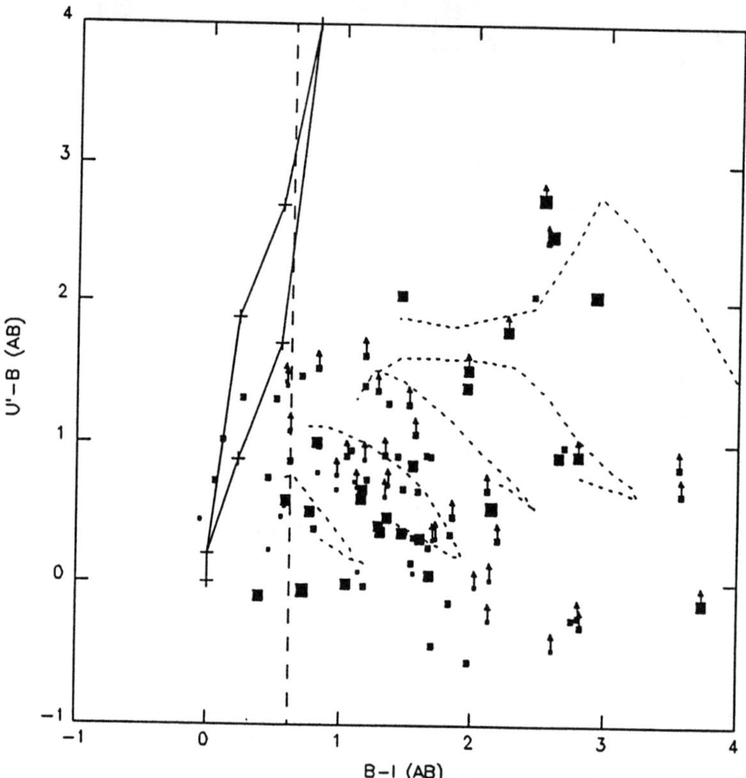

Figure 9 Color-color plot of the deep field sample with $I_{AB} < 24.5$. The box size reflects the I_{AB} magnitude of the galaxy. 19 galaxies, all with $I_{AB} > 23.5$, which were not detected in B or U', are omitted from the diagram. The short dashed lines show the tracks of normal galaxies from $z = 0$ to $z = 1.2$. Most objects lie in these zones. The solid line shows the track of a flat spectrum object with a Lyman break of 5 (left) or 2 (right). Crosses mark $z = 2$, 2.25, 2.85, 3.25. The vertical dashed line shows the $(B-I)_{AB}$ criterion used to separate flat spectrum sources from normal galaxies.

and to the right in the plane, as shown in Figure 9. (It should be emphasized that the IGM effects are statistical in nature, and that the tracks in Figure 9 are based on average effects on quasar spectra derived from the data of Wolfe et al. (1986). Individual objects can deviate significantly from the tracks.)

Now, from Figure 9, we can see that, while there are two or three objects which might be rapid starbursters at $z < 2$ if the population is at high z at all, the bulk of the population probably lies between redshifts of 2 and 3, based on the distribution of $U' - B$ colors. It should be emphasized that we would not pick out $z \gg 3$ objects using the flat spectrum criterion because of the strong Lyα forest reduction of the B magnitude. Some such objects could be present and not be too easily identified.

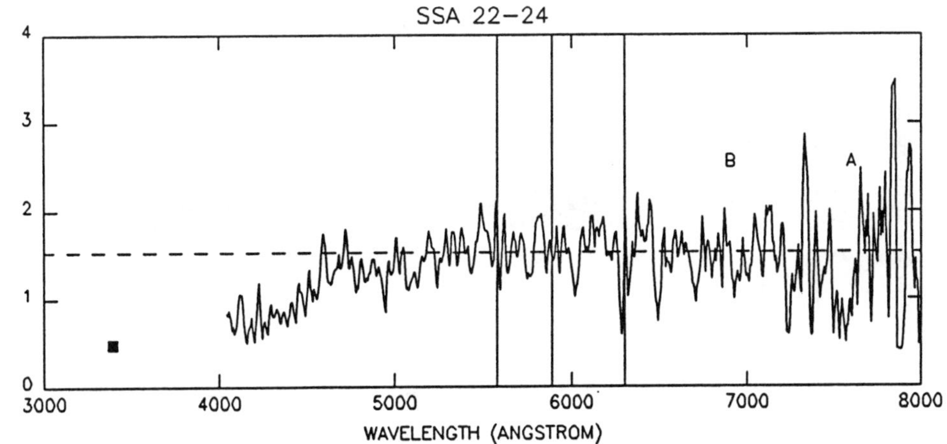

Figure 10 A 20 hr spectrum of the flat spectrum galaxy SSA22–24. The solid box shows the result of U' imaging at 3400 ang. The short-wavelength break could be either the Lyman continuum break, as we previously suggested, or the onset of the Lyman forest, placing the object at $z \gtrsim 2.7$. However, it could also be the Balmer break, placing it at $z \sim 0.1$.

We have attempted to identify two of these objects spectroscopically , the purely flat SSA22–16: $(B - I)_{AB} = 0.39$, $(U' - B)_{AB} = -0.10$, and the breaking SSA22–24: $(B-I)_{AB} = 0.12$, $(U'-B)_{AB} = 1.02$. Figure 10 shows a 20 hr spectrum of SSA22–24. On the basis of our first 6 hr spectrum, we suggested (Cowie and Lilly 1989) that this galaxy was at 3.38. The emission line on which this identification is premised is not confirmed in the second spectrum and the only major feature now present in the spectrum is the break at short wavelengths, which we had previously identified as the Lyman continuum break. In isolation, this feature could be the Ly break or the Ly forest, placing the galaxy at $z \gtrsim 2.7$, or even the Balmer break at $z \sim 0.1$ (see below).

Given the shape of the spectrum, it is very natural to ask the question whether SSA22–24 could not lie at very low redshift and the observed drop not be a Balmer break. Such a deep break seems unlikely in such a flat object but the most serious argument against this possibility lies in the high surface density of these objects (2 arcmin^{-2} to $I = 24.5$) which, if they are located within $z < 0.1$, would imply an enormous volume density. Furthermore, in this case there is no obvious way to explain the abrupt rise in the counts. Rather, we would expect to see a Euclidean shape in the flat spectrum number counts. However, this does remain an alternative viable possibility, given our failure as yet to obtain a definitive redshift for any of these objects.

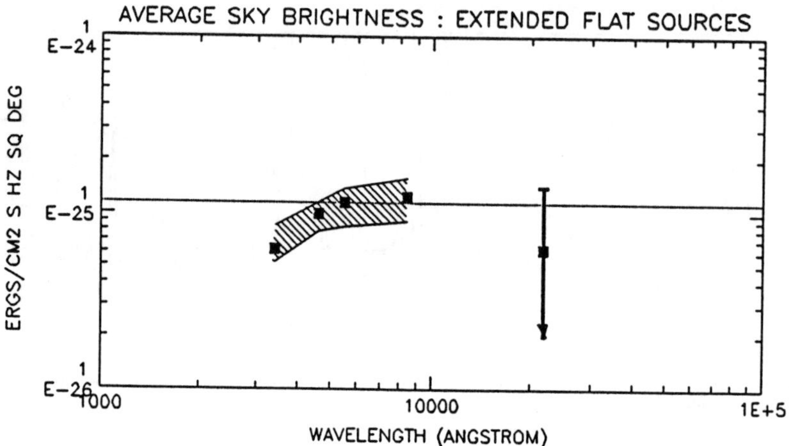

Figure 11 The average sky surface brightness of the ten clearly anomalous flat spectrum galaxies. The shading shows the range of variation from field to field. The K band is the mean value for 5 of the objects where good data is available. The bar on the K shows the 1 σ error.

Finally, we return to an argument that we have stressed continuously regarding the flat spectrum sources (Lilly and Cowie 1987; Cowie 1988; Cowie 1989; Songaila, Cowie and Lilly 1990). If the flat spectrum character of the sources is due to UV emission from massive stars, then, independently of cosmology or the redshift distribution of the sources, their average surface brightness on the sky, S_ν (i.e. their contribution to the extragalactic background light) when measured in frequency units, is given by

$$S_\nu = \frac{1}{4\pi} \varepsilon_\nu (\rho Z) \ , \tag{1}$$

where ε_ν is the energy produced per Hz for each gram of metals the massive stars release and (ρZ) is the current cosmological density of metals which was produced by the sources. The quantity ε_ν is almost independent of IMF and has a value of around 5000 ergs gm^{-1} Hz^{-1} (cf. Songaila, Cowie and Lilly 1990), giving

$$S_\nu = 3.6 \times 10^{-25} \left(\frac{\rho Z}{10^{-34} \text{ g cm}^{-3}} \right) \quad \text{ergs cm}^{-2} \text{ s}^{-1} \text{ Hz}^{-1} \text{ deg}^{-2} . \qquad (2)$$

The measured sky surface brightness of the flat spectrum sources (excluding sources such as SSA22–10 which, from their position in the color-color plane, might be normal low-redshift galaxies) is shown in Figure 11. To $I = 24.5$, these galaxies have a sky surface brightness of 1.3×10^{-25} ergs cm^{-2} s^{-1} Hz^{-1} deg^{-2}. This already implies that these galaxies have made a substantial fraction of the metals currently known in the Universe (Cowie 1988), probably at least 20%, and possibly considerably more, particularly if, as seems likely, there is significant surface brightness in such sources at fainter magnitudes.

This work was supported in part by NASA grant NAGW 949. We are grateful to A. Songaila for help in writing this paper, and to J. Gardner, I. McLean and E. Hu for making available some of the data used.

References

Broadhurst, T. J., Ellis, R. S., and Shanks, T. 1988, *M.N.R.A.S.*, **235**, 827.

Bruzual, G. 1989, quoted by S. D. M. White in *The Epoch of Galaxy Formation*, ed. C. Freanck et al. (NATO Advanced Science Institute Series), p. 15.

Colless, M., Ellis, R. S., and Taylor, K. 1989, preprint.

Cowie, L. L. 1988, in *The Post-Recombination Universe*, ed. N. Kaiser and A. Lasenby (NATO Advanced Science Institute Series), p. 1.

Cowie, L. L. 1989, in *The Epoch of Galaxy Formation*, ed. C. Frenck et al. (NATO Advanced Science Institute Series).

Cowie, L. L. and Lilly, S. J. 1989, *Ap. J. (Letters)*, **336**, L41.

Kron, R. G. 1980, *Ap. J. Suppl.*, **43**, 305.

Koo, D. 1986, *Ap. J.*, **311**, 651.

Koo, D. 1989a, in *The Epoch of Galaxy Formation*, ed. C. Frenck et al. (NATO Advanced Science Institute Series), p. 71.

Koo, D. 1989b, present volume.

Lilly, S. J. and Cowie, L. L. 1987, in *Infrared Astronomy with Arrays*, ed. C. G. Wynn-Williams and E. E. Becklin (Honolulu: U.H. Institute for Astronomy), p. 473.

Songaila, A., Cowie, L. L., and Lilly, S. J. 1990, *Ap. J.*, in press.

Tyson, A. J. 1988, *A. J.*, **96**, 1.

Wolfe, A., Turnshek, D. A., Smith, H. E., and Cohen, R. D. 1986, *Ap. J. Suppl.*, **61**, 249.

Yoshii, Y. and Takahara, F. 1988, *Ap. J.*, **326**, 1.

THE COSMIC SUBMILLIMETER BACKGROUND: A SIGNATURE OF THE INITIAL BURST OF GALAXY FORMATION?

S. DJORGOVSKI and W. N. WEIR
Division of Physics, Mathematics and Astronomy, California Institute of Technology, Pasadena, CA 91125

ABSTRACT We propose a heuristic model for the origin of the cosmic submillimeter background (SMB), reported by the Nagoya-Berkeley collaboration. The SMB is interpreted as a direct signature of an epoch of (initial) galaxy formation at $z_{gf} \sim 10 - 15$. The sources of the SMB are proposed to be dust-shrouded starburst protogalaxies, similar to the luminous *IRAS* galaxies at low redshifts. We interpret them as the progenitors of old stellar populations at low redshifts, ellipticals, bulges, and stellar components of the halos. The largest allowed time scales for the star formation in these models are in the range $FWHM \sim 0.2 - 0.6$ Gyr, for $\Omega_0 = 0.1$; for $\Omega_0 = 1$, the allowed widths are about a factor of two lower. In order not to overproduce the baryonic mass density, it is necessary that the IMF in these starbursts is biased towards high-mass stars; however, a substantial range in the IMF parameters is allowed. This postulated population of protogalaxies may be an important contributor to the diffuse soft x-ray background. The predicted surface density of protogalaxies would be in the range $\sim 10-100$ arcsec^{-2}, which is consistent with all relevant anisotropy measurements available at this time.

Recently, the Nagoya-Berkeley collaboration reported the results of a measurement of the spectrum of the cosmic microwave background (CMBR) from a rocket-borne platform (Matsumoto *et al.* 1988). They found an "excess" emission in the submillimeter wavelength region, above the level expected from the CMBR and the interstellar dust, and argued that this submillimeter background (SMB) is not caused by any instrumental effect, or any known Galactic source, and that it is most likely cosmological in origin. If this background is real, it must be very important, on account of its energetics: the energy density of SMB, $u_{SMB} \simeq (9 \pm 2) \times 10^{-14}$ erg cm^{-3} is 10 – 20% that of the CMBR, $u_{CMBR} = 4.26 \times 10^{-13}$ erg cm^{-3}, assuming $T_{CMBR} = 2.74°$K.

The explanation for the SMB which has received most attention to date involves energy generation at a high redshift, e.g., in pregalactic (population III) stars, very massive objects, accreting pregalactic black holes, exploding cosmic strings, or some other exotic process, and its reprocessing by dust at some intermediate redshifts, e.g., $z > 3$ (Carr 1987; Rowan-Robinson and Carr 1988; Hogan and Bond 1988; Hogan 1988; Silk 1988; Adams *et al.* 1989; etc.)

Other models include comptonization of the CMBR (perhaps unlikely on the account of energetics required), decay of some hypothetical relic particles, etc.

We explore a heuristic variant of the dust-reradiation model. We propose that the sources of the SMB are similar to the low-redshift, extremely luminous far-infrared galaxies, which we tentatively identify as protogalaxies, progenitors of the present-day ellipticals and bulges, and perhaps also old disks. The SMB is thus interpreted as a signature of the epoch of *initial* galaxy formation. The advantage of our approach is in using ready-made and realistic sources of the SMB, and thus obviating the need for detialed radiation transfer computations. No new process or a previously unknown kind of objects needs to be postulated. Our intent is to explore whether such a model for the SMB can be made by using reasonable parameters, and without violating any relevant astrophysical constraints.

For the template spectra used in our models, we use fits to the observed *IRAS* data on Arp 220 and M82 in the forms $\nu B_\nu(45°\mathrm{K})$, or $\nu^2 B_\nu(55°\mathrm{K})$, where $B_\nu(T)$ is the Planck function. We use $\Lambda = 0$ Friedman models, with $H_0 = 75$ km s^{-1} Mpc^{-1}, and two values of the density parameter, $\Omega_0 = 0.1$ and $\Omega_0 = 1$. We represent the time history of the luminosity generated per comoving Mpc^3 as a gaussian in the rest-frame, specified by the dispersion σ^2, and the peak epoch t_c, corresponding to the redshift z_c. The flux contribution from each redshift shell is added up and normalized by fitting the final spectrum to the observed SMB data. A grid of models is computed for a range of σ and z_c, for each of the four possible combinations of the template dust spectra and Ω_0. There is a notable parameter coupling in the models, in that the larger values of σ require larger values of z_c.

We search for the longest epoch of galaxy formation which is consistent with the data. The best fits are for the lowest values of σ. However, for the $\Omega_0 = 0.1$ models, values of σ of up to $\sim (1-2.5) \times 10^8$ yr are compatible with the data, corresponding to the $FWHM \sim 0.2 - 0.6$ Gyr. For the $\Omega_0 = 1$ models, the allowed widths are about a factor of two lower. These time scales are comparable to the free-fall times for normal galaxies, and they are constrained primarily by the observed shape of the SMB spectrum.

In order to overproduce the cosmological density of the burned-out stars, we require that the IMF in starbursts which can generate the SMB is biased towards the more massive stars, which may be the natural mode of star formation in starbursts. Given that, a large range of possible IMF parameters is allowed. A large portion of the mass processed in the starbursts, perhaps as much as 90%, should be now locked in low-luminosity and dark stellar remnants, i.e., cool white dwarfs, neutron stars, or stellar-mass black holes. These models thus predict that the dark matter within the visible parts of galaxies should be baryonic, and consist of faint or dark stellar remnants. A similar prediction was already made by Silk (1988).

The peak luminosity densities in our models can be converted to the comoving dust densities in the mid-starburst. The later cannot exceed the amount of metals generated by that time, $\Omega_* Z$. If we associate the stellar populations generated in these bursts with the present-day metal-rich old stellar populations, their final metallicities are in the range $Z \sim (1-2)Z_\odot$. Thus, the plausible mid-burst metallicities are $Z \sim 0.01$. With $\Omega_* \sim 0.1$, the dust densities requred by our models are well within the allowed range.

Finally, we can estimate the expected surface density of SMB sources. For the $\Omega_0 = 0.1$ models, and sources with luminosities $L_{SMB} = f_{11} \times 10^{11} L_{\odot}$, the surface density requred to reproduce the observed brightness of the SMB is $85/f_{11}$ arcsec^{-2} $= 1.1 \times 10^9/f_{11}$ degree^{-2}, corresponding to the r.m.s. of $0.11\sqrt{f_{11}}$ arcsec between the sources. For the $\Omega_0 = 1$ models, the surface density is $9.4/f_{11}$ arcsec^{-2} $= 1.2 \times 10^8/f_{11}$ degree^{-2}, corresponding to the r.m.s. of $0.33\sqrt{f_{11}}$ arcsec between the sources. The SMB should thus be rather smooth on scales greater than a few arcsec. Kreysa and Chini (1988) limits at 1.3 mm and on scales ~ 30 arcsec are consistent with these estimates.

It may be possible to find a spectroscopic signature of the starbursts at $z \sim 10 - 15$. Many of the commonly observed lines from the low-redshift dusty starbursts, e.g., Paα 1.875 μm, Brα 4.05 μm, or Brγ 4.05 μm, the H_2 molecular lines at 2.22 and 2.12 μm, or the PAH band at 3.3 μm may be just barely detectable with the *SIRTF*, or other future FIR/Sub-mm missions. A number of the fine structure lines of metals, e.g., [O III] 88.36 μm, [O I] 145.53 μm, or [C II] 157.74 μm would be redshifted into the atmospheric windows at millimeter wavelengths, and may be detectable with the next generation of receivers from the ground.

A population of starburst galaxies at large redshifts which can produce the SMB, could also be an important contributor to the diffuse extragalactic soft x-ray background (XRB). X-ray emission has been observed from star forming regions, and interacting galaxies (Fabbiano 1989, and references therein). The ratios of x-ray and infrared luminosities in these objects are $L_X/L_{IR} \sim (1 - 5) \times 10^{-4}$. The Harwit *et al.* (1987) model also predicts $L_X/L_{IR} \sim 5 \times 10^{-4}$. It may be more than a coincidence that the ratio of the energy density in the soft XRB, a few times 10^{-17} erg cm^{-3}, and that of the SMB, $\sim 10^{-13}$ erg cm^{-3}, is of the similar order.

The full account of this work is presented by Djorgovski and Weir (1990). We acknowledge a partial support from California Institute of Technology, Alfred P. Sloan Foundation (SD), and the NSF (WNW).

REFERENCES

Adams, F. C., *et al.* 1989, Ap. J., 344, 24.

Carr, B. J. 1987, in Comets to Cosmology, ed. A. Lawrence (Berlin: Springer-Verlag), p. 265.

Djorgovski, S., and Weir, N. 1990, Ap. J. in press (Mar. 1990).

Fabbiano, G. 1989, Ann. Rev. Astr. Ap., 27, in press.

Harwit, M., *et al.* 1987, Ap. J., 315, 28.

Hogan, C. J., and Bond, J. R. 1988, in The Post-Recombination Universe, eds. N. Kaiser, and A. Lasenby (Dordrecht: Kluwer), p. 141.

Hogan, C. J. 1988, Aph. Lett. and Comm., 27, 125.

Kreysa, E., and Chini, R. 1989, in Astronomy, Cosmology and Fundamental Physics, eds. M. Caffo *et al.* , (Dordrecht: Kluwer), p. 433.

Matsumoto, T., *et al.* 1988,Ap. J., 329, 567.

Rowan-Robinson, M., and Carr, B. 1988, in The Post-Recombination Universe, eds. N. Kaiser, and A. Lasenby (Dordrecht: Kluwer), p. 125.

Silk, J. 1988, in Dark Matter, eds. J. Audouze and J. T. T. Van (Gif sur Yvette: Editions Frontières), p. 3.

DARK MATTER, GALAXIES, AND BACKGROUND RADIATION

ERIC R. WOLLMAN
Department of Physics and Astronomy, Bates College, Lewiston, ME 04240.

ABSTRACT Fundamental properties of the system of galaxies can be understood if one assumes that the galaxies have evolved from gravitational aggregations of large grains of solid matter.

INTRODUCTION

This contribution outlines a new model of the dark matter and galaxies. The model is developed from two assumptions: 1) that the dark matter consists of large solid grains, and 2) that the apparent paucity of quasars at large redshifts is due to extinction by the grains. According to this model, the galaxies have evolved from gravitational aggregations of the grains. Within a galaxy, direct collisions among grains vaporize the grains. The vapor forms the luminous central portion of the galaxy, while residual unvaporized grains compose the galactic dark matter halo. Analysis of this evolution leads to simple expressions for the mass and luminosity of the luminous portion of the galaxy and provides an understanding of the existence and extent of hot galactic halo gas. The expression for the luminosity is roughly consistent with the observed Tully–Fisher relation. Since the grains are large, they are effective thermalizers of radiation.

THE MODEL

Consider an extended homogenous fluid of identical spherical solid grains of mass m and radius r_s. The grains are composed of matter of density ρ_s. Assume the grains to be gravitationally clumped into relaxed aggregations, which I will refer to as "condensations." For simplicity, treat each condensation as a truncated singular isothermal sphere with internal mass distribution specified by the circular orbit speed v. An isothermal sphere of grains is a galaxy of age $t = 0$.

Within a condensation, collisions among grains vaporize the grains, converting them to some combination of hot gas, dust, and grain fragments. The essential results of this process are outlined below. A more detailed analysis is presented elsewhere (Wollman 1989).

Because of the radial density gradient, the size of the vaporized region grows over time. Let $R_V(t)$ be the condensation radius at which the grain collision time equals the age t of the condensation. Then $R_V(t)$ is the radius inside of which grains have been substantially vaporized and is given by

$$R_V(t) \approx \left(\frac{v^3 t}{G\kappa}\right)^{\frac{1}{2}} \tag{1}$$

where $\kappa \equiv \rho_s r_s$. The vapor generated by grain collisions aggregates into stars and thereby forms a luminous core. The mass M_C of this luminous core and the rate of generation of vapor are

$$M_C(t) \approx \frac{v^2 R_V}{G} \approx \left(\frac{v^7 t}{G^3 \kappa}\right)^{\frac{1}{2}}. \tag{2}$$

$$\frac{d}{dt} M_C(t) \approx \left(\frac{v^7}{4G^3 \kappa t}\right)^{\frac{1}{2}} \propto v^{3.5}. \tag{3}$$

Then if stars form promptly and if the initial mass function of stars is independent of the star formation rate, the luminosity of the condensation is proportional to $v^{3.5}$. Let ϵ be the fraction of the total mass energy of newly generated vapor which is released by stars as radiation. Since much of this energy is released on a timescale short compared to the age of the condensation, the luminosity L_C of the core is

$$L_C \approx \epsilon c^2 \frac{d}{dt} M_C \approx \epsilon c^2 \left(\frac{v^7}{4G^3 \kappa t}\right)^{\frac{1}{2}}. \tag{4}$$

In order to constrain the value of κ ($\equiv \rho_s r_s$), I assume that the paucity of detectable quasars at large redshifts is due to extinction by the grains. For critical cosmic mass density ($\Omega = 1$), the result is (see Wollman 1989)

$$\kappa \approx \frac{3cH_o}{8\pi G}\left[(z_c+1)^{\frac{3}{2}} - 1\right] \tag{5}$$

where z_c is the redshift at which the extinction optical depth is unity. Equations (1) – (4) can then be written

$$R_V \approx \left(\frac{16\pi v^3}{9c}\right)^{\frac{1}{2}} \frac{1}{H_o f} \left(\frac{t}{t_o}\right)^{\frac{1}{2}}$$

$$M_C \approx \left(\frac{16\pi v^7}{9G^2 c}\right)^{\frac{1}{2}} \frac{1}{H_o f} \left(\frac{t}{t_o}\right)^{\frac{1}{2}}$$

$$\frac{d}{dt} M_C \approx \left(\frac{\pi v^7}{G^2 c}\right)^{\frac{1}{2}} \frac{1}{f} \left(\frac{t_o}{t}\right)^{\frac{1}{2}}$$

$$L_C \approx \epsilon \left(\frac{\pi v^7 c^3}{G^2}\right)^{\frac{1}{2}} \frac{1}{f} \left(\frac{t_o}{t}\right)^{\frac{1}{2}}$$

where $f \equiv \left[(z_c+1)^{\frac{3}{2}}-1\right]^{\frac{1}{2}}$ and t_o is the present age of the expansion. (For $\Omega = 1$, $t_o = \frac{2}{3H_o}$ where H_o is the Hubble constant.)

PROPERTIES OF THE GALACTIC–SCALE SYSTEM

The orbit speed v is accurately known for many galaxies. A typical value is $200\,\frac{\text{km}}{\text{s}}$. Observations of quasars indicate $z_c \sim 2$ (Heisler and Ostriker 1988). Let $h \equiv \frac{H_o}{100\,\frac{\text{km}}{\text{s Mpc}}}$ and $v_o \equiv 200\frac{\text{km}}{\text{s}}$. Then for $t \approx t_o$ and $z_c = 2$,

$$R_V \approx 60\left(\frac{v}{v_o}\right)^{\frac{3}{2}}\frac{1}{h}\ \text{kpc}$$

$$M_C \approx 5\times10^{11}\left(\frac{v}{v^o}\right)^{\frac{7}{2}}\frac{1}{h}\ \text{M}_\odot$$

$$\frac{d}{dt}M_C \approx 40\left(\frac{v}{v^o}\right)^{\frac{7}{2}}\frac{\text{M}_\odot}{\text{y}}$$

$$L_C \approx 6\times10^{14}\epsilon\left(\frac{v}{v^o}\right)^{\frac{7}{2}}\ \text{L}_\odot.$$

For hydrogen–rich grains and a standard initial mass function, ϵ is in the vicinity of 10^{-4} so that

$$L_C \sim 10^{11}\left(\frac{v}{v^o}\right)^{\frac{7}{2}}\ \text{L}_\odot.$$

From equation (5), the grain diameter is of order 1 cm. Consequently, the grains efficiently thermalize hot radiation emitted prior to the time corresponding to redshift $z = z_c \sim 2$. The thermalization should be effective out to present–epoch wavelengths of about $(z_c + 1)$ times the grain diameter, or several centimeters. Furthermore, it is easily shown that an individual galaxy is optically thin, so that radiation from any one galaxy is thermalized over a very large volume of space. Consequently, the thermalized radiation will be highly isotropic.

As is seen, the model yields simple and viable expressions for the mass and luminosity of the luminous core of a galaxy and provides an understanding of the existence and extent of hot galactic halo gas. The result for the luminosity, $L \propto v^{3.5}$, is roughly consistent with the observed Tully–Fisher relation. The conversion of hot radiation to a cool and isotropic background is a necessary consequence of the basic assumptions of the model. The general accordance with observations combined with the lack of parametric freedom justifies further investigation of the model even though the model may be incompatible with elements of standard cosmological theory.

REFERENCES

Heisler, J., and Ostriker, J.P. 1988, *Ap. J.*, **332**, 543.
Wollman, E.R. 1989, in preparation.

VOIDS IN THE LYMAN α FOREST

JILL BECHTOLD
Steward Observatory, The University of Arizona
Tucson, AZ 85721

ABSTRACT Voids in the distribution of Lyman α forest clouds are discussed. Generally, the Lyman α forest clouds at $z \approx 3$ are more uniformly distributed than present-day galaxies. Two possible voids caused by the "proximity" effect are presented.

INTRODUCTION

Quasar absorption lines can be used to trace the large-scale distribution of mass at high redshift, and to study the evolution of structure with time. They can probe very large volumes of space: for a QSO at $z_{em} = 3$, the redshift interval between emission Lyman α and Lyman β corresponds to 235 h_o^{-1} Mpc if the clouds were to survive until today ($h_o^{-1} \equiv H_o$ / 100 km sec^{-1} Mpc^{-1}; q_o=1/2). In addition, the mean separation of Lyman α forest lines at $z = 3$ is 4.4 h_o^{-1} Mpc. Thus, observations of even a handful of high redshift QSOs can adequately sample structure on scales which are comparable to the largest scales probed by galaxy redshift surveys.

In this talk, I review one aspect of the study of large scale structure, the question of whether or not significant voids exist in the distribution of Lyman α forest lines. It is well-known that large voids exist in the present-day galaxy distribution (e.g. the Boötes void with a diameter of about 62 Mpc; Kirshner, Oemler, Schechter and Shectman 1987 and references therein), and that more modest voids are quite common in general (e.g. deLapparent, Geller and Huchra 1986; Koo, Kron and Szalay 1987; da Costa et al. 1989). If the Lyman α

forest clouds are in fact proto-galaxies, it is of interest to ask whether they "fill" the voids, or show the same voided structure as present-day galaxies. In addition, one expects voids to be created by the ultraviolet radiation from intervening quasars along the line-of-sight ("the proximity effect"), if quasars are long-lived and their radiation is isotropic (Bajtlik, Duncan and Ostriker 1988; Kovner and Rees 1989). Thus, the presence or absence of voids in the distribution of Lyman α forest clouds has several potentially interesting ramifications.

In the next section I discuss one controversial candidate void in the Lyman α forest, and present new echelle data of the object in question, Q 0420-388. In section III, I discuss voids caused by the proximity effect.

THE CANDIDATE VOID TOWARD Q0420-388

Carswell and Rees (1987) first looked for voids in the Lyman-α forest, using echelle spectra of PKS 2000-330 (Carswell et al. 1987) and Q 0420-388 (Atwood, Baldwin and Carswell, 1985; hereafter ABC). They concluded that the distribution of intervals between lines was consistent with a Poisson distribution. Given that two lines of sight were sampled, they estimated that the volume filling factor of 50 Mpc voids is less than 5% for $2.59 < z < 3.75$. This may be compared to the estimated volume filling factor of ~80% for 20-50 Mpc voids in galaxies at z=0 (deLapperant, Geller and Huchra 1986). Thus, the Lyman α forest lines appear to be more uniformly distributed than local galaxies.

Crotts (1987) reanalyzed the same data, and argued that Q 0420-388 has a significant void at mean redshift $<z> = 2.575$, with extent $\Delta z = 0.037$. Ostriker, Duncan and Bajtlik (1988) questioned Crotts' analysis, and asserted that this void has an *a-priori* probability of ~20% of being found by chance. Crotts (1989) clarified his position, and Duncan, Ostriker and Bajtlik (1989) elaborated on theirs. Duncan, Ostriker and Bajtlik (1989) further analysized a quite heterogeneous sample of moderate resolution data of 18 QSOs from the literature, including the echelle spectra of PKS 2000-330 and Q0420-388. They concluded that that voids with size 10-70 Mpc have a volume filling factor of less than 20% at the 90% confidence level. Bi, Bonner and Chu (1989) also analyzed the spectrum of Q 0420-388 and argued that Crotts' void is real at the 98% confidence level. Finally, Pierre, Shaver and Iovino (1988) constructed a model of the Universe consistent with the distribution of present-day galaxies, and showed that it was inconsistent with the observed distribution of Lyman α forest clouds at high redshift.

Why do different authors come to different conclusions when analyzing the same data? The primary reason is the large uncertainty in the number density of the clouds. Clearly, the significance of any particular observed gap in the distribution of lines increases as the number density of lines increases. The situation for the Lyman-α forest is complicated by the fact that the number density of clouds is evolving with redshift, and hence is a function of wavelength.

Following Ostriker, Bajtlik and Duncan (1988), if the evolution in the number of clouds per unit redshift, dN/dz, is given by

$$\frac{dN}{dz} = A(1+z)^{\gamma} \tag{1}$$

then the *a priori* probability, P, of finding a gap bigger than a particular gap, scaled to account for evolution, is

$$P(x) = e^{-x} \tag{2}$$

where

$$x = \frac{\Delta z}{< \Delta z >} \tag{3}$$

and

$$< \Delta z >^{-1} = \frac{dN}{dz}. \tag{4}$$

The *a priori* probability, $P_>$, of finding a gap bigger than a particular gap, x_{gap}, is

$$P_> = 1 - (1 - e^{-x_{gap}})^N \tag{5}$$

Clearly, $P_>$ depends strongly on the number density of lines, N, or in other words, the value of γ and A. Unfortunately, A and γ are, for these purposes, poorly determined. The number of QSOs which have adequate spectra is small (< 20), so that the total number of Lyman-α forest lines is small (< 500). What's worse, the Lyman-α forest lines in all published spectra are obviously blended. Thus, differences in resolution and signal-to-noise effect the number of lines counted in subtle ways, even when line lists are constructed carefully above some rest equivalent width limit.

Line blending can have two effects, which bias the number of lines counted above a given equivalent width threshold in opposite ways. As the spectral resolution is increased, a single line can break up into two or more lines, *below* the threshold, thus *decreasing* the number of lines counted. On the other hand, a single line can break up into two or more lines which are all *above* the threshold, thus *increasing* the number of lines counted. Several authors have estimated the relative importance of these two effects by simulating spectra. Liu and Jones (1988) concluded that γ is systematically underestimated, because of blending and the presence of noise. Parnell and Carswell (1988) argued that the effects almost cancel, so that γ is virtually unchanged. Finally, Ostriker, Bajtlik and Duncan (1989) and Duncan, Ostriker and Bajtlik (1989) argued that for the purpose of discussing the existence of voids, the first effect dominates. They correct the observed $\gamma \approx 2.2$ to $\gamma \approx 3$, in order to account for blends.

Empirically, one can look at the question of line-blending by comparing moderate and high resolution spectra of the same object, in the small number of cases where this is possible. The results are summarized in Table I. The high resolution data for Q 2126-158 and Q 0420-388 were obtained at the 100 inch Dupont telescope at the Las Campanas Observatory with the echelle spectrograph and 2D Frutti photon counter, in collaboration with S.A. Shectman. Details of the observations and analysis are briefly described in Bechtold and Shectman (1988). The resolution is about 19 km sec^{-1} FWHM, with continuous spectral coverage from 3500 to 7000 Å.

Table I shows that the situation is complicated, and that naive corrections for line-blending are probably not appropriate. In one case, the increased resolution resulted in a *decrease* in the number of lines above the threshold, whereas in the other case, the number of lines is *unchanged*.

Returning to the issue of the significance of the candidate void in Q 0420-388, Figure 1a shows the Las Campanas spectrum of the Lyman-α forest at z=2.58 for that object. The 5-sigma equivalent width limit is plotted as a function of wavelength in Figure 1b, and is typically 0.15 Å (observed), or a rest equivalent width limit of 0.04 Å, compared to 0.25 Å in the ABC spectrum. Interestingly, the weak features seen in ABC spectrum are present in the Campanas spectrum as well, and are in fact significantly above the threshold. However, they are below ABC's threshold. Thus, although we find lines in the candidate void, they are below the equivalent width threshold in ABC's data.

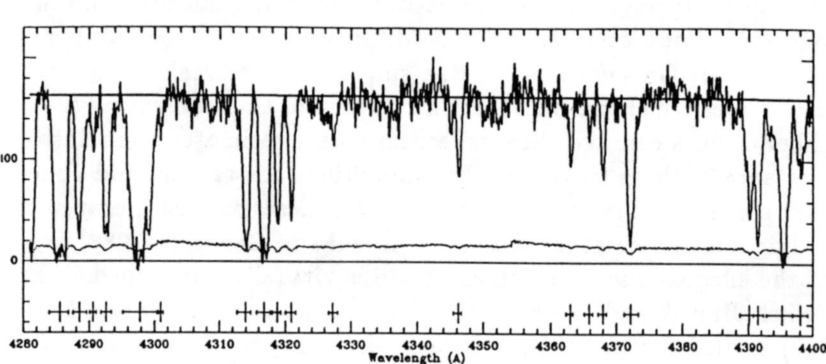

1b

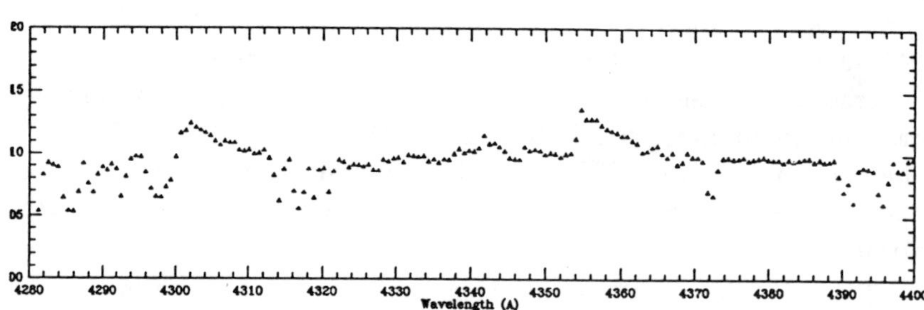

Figure 1. (a) Las Campanas Echelle spectrum of Q 0420-388, in the region of the candidate void. Lines significant above the 5σ level are indicated by crosses below the spectrum. (b) The 5-sigma equivalent width limit (observed) as a function of wavelength for the spectrum shown in (a), observed.

TABLE I. Line Blending and Statistics of the Lyman α Forest

QSO	(1)	(2)	(3)	(4)	(5)
2126-158	3.28	0.32	90 km sec^{-1}	42	*a*
		0.32	19 km sec^{-1}	32	*b*
0420-388	3.12	0.25	33 km sec^{-1}	73	*c*
		0.25	19 km sec^{-1}	73	*b*

Notes:
(1) Emission-line redshift.
(2) Rest equivalent width limit, Å.
(3) Spectral resolution (FWHM).
(4) Number of lines detected between emission Lyman α and Lyman β.
(5) Reference. (a) Sargent *et al.* 1980. (b) Bechtold and Shectman 1990.
(c) Atwood, Baldwin and Carswell 1985.

How significant are the biggest gaps in the Campanas echelle data? At present, we do not have a large enough redshift range to derive γ and A independently. Instead, we assume different values of γ, and derive the normalization, A, which reproduces the number of lines we see. The results are given in Table II. For each QSO, we give the largest gap seen, and its significance. Clearly, the gap in Q 0420-388 is the largest gap seen, but it is not significant, with $\sim$ 60% probability of being seen by chance, given a Poisson distribution of gaps.

VOIDS AND THE PROXIMITY EFFECT

Following Paczynski's suggestion, Bajtlik, Duncan and Ostriker (1988) and Kovner and Rees (1989) have discussed how the enhanced ionization of Lyman α clouds by quasars along the line-of-sight will cause "clearings" in the Lyman α forest, or voids. If one were to see a void attributable to the proximity effect at the same redshift as a known foreground quasar, one could conclude that (a) the lifetime of the foreground quasar is at least as long as the time it takes light to travel transversely between the two lines of sight, and (b) that the radiation of the foreground quasar is isotropic. (Note that in a single QSO clouds at $z_{abs} < z_{em}$ are observed at a later cosmic time than the

TABLE II. Significance of Largest Gaps

QSO	z_{em}	Δz_{gap}	z_{gap}	$P_>$	
2126-158	3.28	0.010	2.644	0.999	$\gamma = 2.4$
				0.999	$\gamma = 3.0$
0420-388	3.12	0.037	2.575	0.606	$\gamma = 2.4$
				0.666	$\gamma = 3.0$
0130-403	3.03	0.014	2.575	0.999	$\gamma = 2.4$
				0.999	$\gamma = 3.0$
All QSOs		0.037	2.575	0.969	$\gamma = 2.4$
				0.983	$\gamma = 3.0$

QSO itself, so do not set as interesting limits.) Since information on quasar lifetimes is crucial to important issues such as the interpretation of quasar luminosity functions, it is interesting to ask if examples of such voids exist.

Cristiani and Shaver (1987) have searched for bright quasars near the line of sight of Q 0420-388 with a grism imager at ESO. They find no quasar at z $\approx$ 2.58 to cause the candidate void discussed in the previous section. However, they found a V=20.8 QSO at z_{em} = 2.403, separated by 2.1 arcmin from Q00420-388. Unfortunately, the Lyman-α forest at z_{abs} = 2.403 is blueward of emission Lyman β in Q 0420-388, so it is difficult to look for a corresponding void.

Two examples of close QSO pairs with moderate resolution spectra of the Lyman α forest are available. Even though a discussion of only two examples is somewhat anecdotal, the results are interesting.

Sargent, Young and Schnieder (1982) observed Q 1623+269 and Q 1623+268 which have z_{em} = 2.61 and 2.518 respectively, and are separated by 3 arcmin on the sky. Crotts (1989) later reobserved these two objects and also observed one other nearby QSO. Q 1623+269 is very bright, V = 16.9, so one expects to see a void in the spectrum of the higher redshift object (see Bajtlik, Duncan and Ostriker (1988), Figure 9). In fact, there is a large gap just blueward of this, with Δ z = 0.067 and <z> = 2.49. (Note, one doesn't necessarily expect the void to be centered at the redshift of the foreground QSO.) However, if one adopts $\gamma = 2.4$ and $A = 3.0$ (Duncan, Ostriker and Bajtlik 1989) for this moderate dispersion data, then the probability that one would find this void by chance (having observed this particular QSO) is $P_> = 0.36$. This estimate for the probability is conservative, since it does not really reflect the probability of finding a gap is at the *known* redshift of the foreground QSO. However, this example, although suggestive, is not compelling.

The second example, shown in Figure 2, has a more significant gap. Q 0302-003 ($z_{em} = 3.285$) is separated by 17 arcmin from the bright (V=18.0) quasar Q 0301-005 ($z_{em} = 3.205$). Q 0256-000 is also shown to give an idea of the density of the Lyman α forest at $z \approx 3.2$. The spectra of Q 0256-000 and Q 0302-003 were obtained with the MMT Spectrograph and Big Blue Reticon photon counter, and have resolution $\sim$ 75 km sec^{-1} FWHM. The spectrum of Q 0301-005 was obtained with the Double Spectrograph and TI CCD at the Palomar 5M. Details are given in Bechtold (1990).

There is a gap at z=3.175 with Δz=0.079 in the Lyman α forest of Q 0302-003 (Figure 3). Adopting $\gamma = 2.4$ and $A = 3.0$, $P_> = 0.003$. Thus, this does appear to be a significant void, and the existence of the nearby bright quasar Q 0301-005 with $z_{em} = 3.20$ suggests that it is caused by the proximity effect.

CONCLUSIONS

Lyman-α forest clouds are potentially powerful probes of large scale structure at high redshifts. At present, however, the statistical significance of any observed void is very uncertain, due to the large and uncertain corrections for line-blending which must be made to estimate the cloud number densities. While new data shows that there are absorption lines in the $z = 2.58$ void toward Q 0420-388, they are weak, and below the detection limit of previously published data. However, this gap is probably not statistically significant. Two tantalizing examples suggest that bright foreground QSOs can cause voids in the Lyman-α forest due to the proximity effect.

ACKNOWLEDGEMENTS

Some of the results presented here were obtained in collaboration with S.A. Shectman. This work was supported in part by NSF RII-8800660.

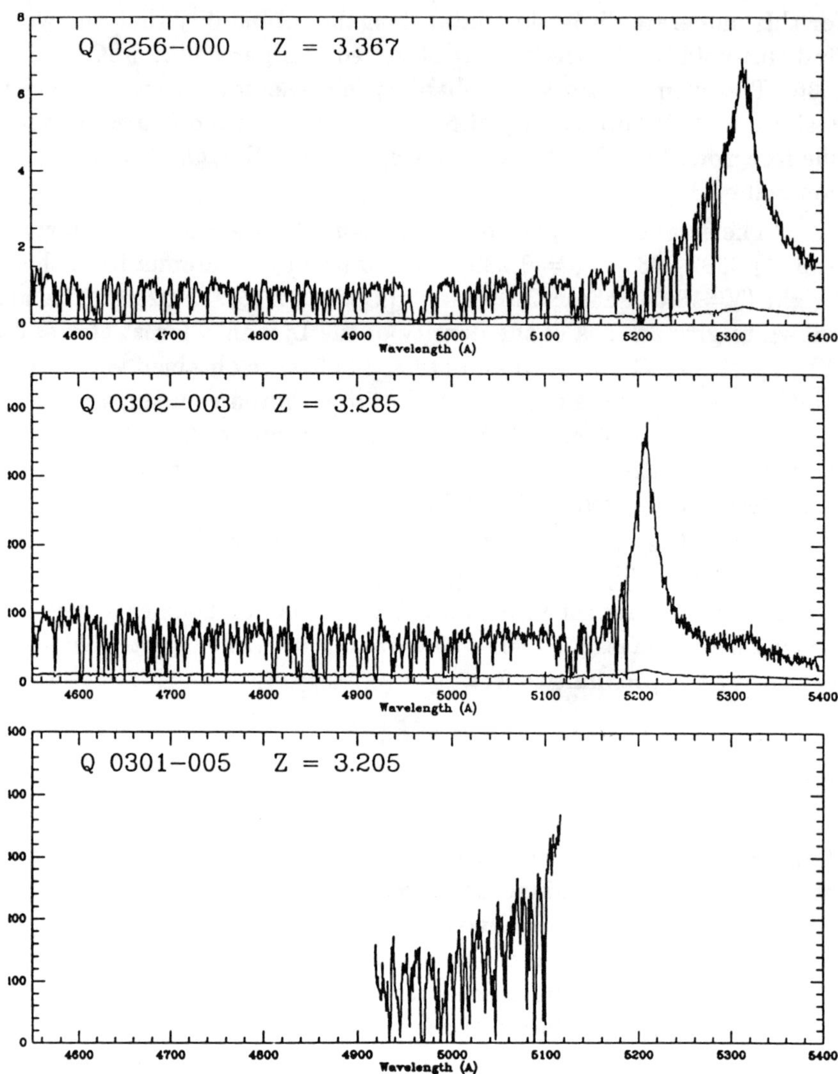

Figure 2. Moderate resolution spectra of Q 0256-000, Q 0302-003 and Q 0301-005.

3a

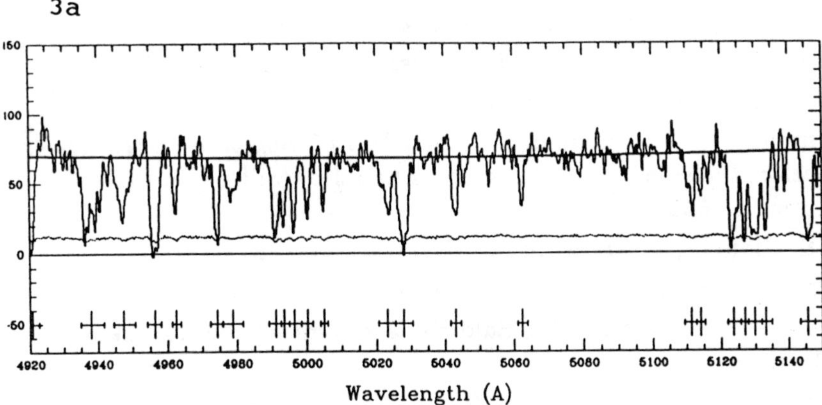

3b

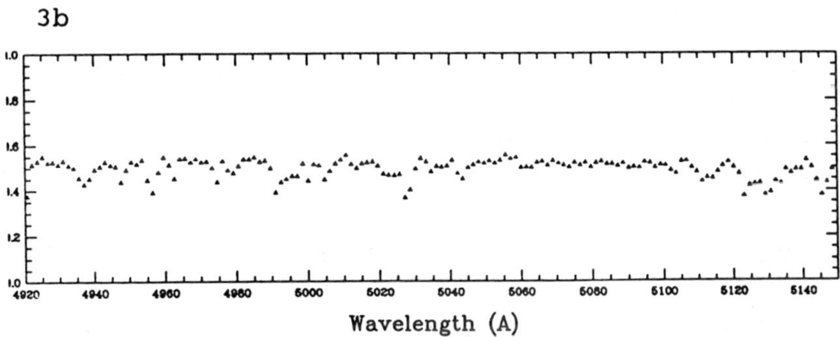

Figure 3. The Void in Q 0302-003. (a) Spectrum from the MMT Blue Spectrograph in the region of the candidate void. (b) The 5-sigma equivalent width limit as a function of wavelength for the spectrum shown in (a), observed.

REFERENCES

Atwood, B., Baldwin, J.A. and Carswell, R.F. 1985 *Ap.J.*, **292**, 58.

Bajtlik, S., Duncan, R.C. and Ostriker, J.P. 1988 *Ap. J.*, **327**, 570.

Bechtold, J. 1990, in preparation.

Bechtold, J. and Shectman, S.A. 1989 in *IAU Symposim 134: Active Galactic Nuclei*, ed. D.E. Osterbrock and J.S. Miller (Dordrect: Kluwer), p. 549.

Bi, H., Bonner, G. and Chu, Y. 1989 *Astr. Ap.*, **218**, 19.

Carswell, R.F. and Rees, M.J. 1987 *M.N.R.A.S.*, **224**, 13P.

Carswell, R.F., Webb, J.K., Baldwin, J.A. and Atwood, B. 1987 *Ap. J.*, **319**, 709.

Cristiani, S. and Shaver, P.A. 1987 in *QSO Absorption Lines: Probing the Universe, A Collection of 37 Poster Papers*, ed. C. Blades, C. Norman, and D. Turnshek. STScI., p. 103.

Crotts, A.P.S. 1987 *M.N.R.A.S.*, **228**, 41P.

Crotts, A.P.S. 1989 *Ap. J.*, **336**, 550.

da Costa, L. N., Pellegrini, P.S., Willmer, C., and Latham, D.W. 1989, *Ap. J.*, **344**, 20.

de Lapparent, V., Geller, M.J., and Huchra, J.P. 1986, Ap.J. (Letters), 302, L1.

Duncan, R.C., Ostriker, J.P. and Bajtlik, S. 1989 *Ap. J.*, **345**, 39.

Kirshner, R.P., Oemler,A., Schechter, P.L. and Shectman, S.A. 1987, *Ap. J.* **314**, 493.

Koo, D.C., Kron, R.G. and Szalay, A.S. 1987 in *13th Texas Symposium on Relativistic Astrophysics*, ed. M.P. Ulmer (Singapore:World Scientific), p. 284.

Kovner,I. and Rees, M.J. 1989 *Ap. J.*, **345**, 52.

Liu, X.D. and Jones, B.J.T. 1988 *M.N.R.A.S.*, **230**, 481.

Ostriker, J.P., Bajtlik, S. and Duncan, R.C. 1988 *Ap. J. (Letters)*, **327**, L35.

Parnell, H.C. and Carswell, R.F. 1988 *M.N.R.A.S.*, **230**, 491.

Pierre, M., Shaver, P.A. and Iovino, A. 1988 *Astr. Ap.*, **197**, L3.

Sargent,W.L.W., Young, P.J., Boksenberg, A. and Tytler, D. 1980, *Ap.J. Suppl.*, **42**, 41.

Sargent,W.L.W., Young, P.J. and Schneider, D. 1982, *Ap.J.*, **256**, 374.

THE HI COLUMN DENSITY SPECTRUM: A SIMULATION PROVIDES EVIDENCE FOR TYTLER'S "SINGLE POPULATION"

CURTIS V. MANNING
Independent researcher
2107 Fifth Street, Berkeley, California 94710 U.S.A.

ABSTRACT The HI column density distribution made by a power law spectrum of gravitationally bound isothermal overdensities can be adequately fit to observations of Lyα systems. Results show that while most clouds are small, most absorption lines are produced by larger systems with diameters >10 kpc; only a few (~1.5%) are made by systems smaller than 10 pc. New data implying clustering on scales in proportion to system size provides the unifying principle for conceiving Lya and metal line systems as a single population.

INTRODUCTION

In his 1987 paper, David Tytler made two startling claims, stimulating a lot of fruitful discussion: (1) the column density spectrum, f(N), fits a power law over seven orders of magnitude in column density, N, and (2) the HI absorbers of both Lyα and metal line systems form a *single population*. While Sargent, Steidel, and Boksenberg (1989, SSB) have confirmed the power law, they dispute the "single population" theory citing their different clustering properties. But it is the lack of a break in f(N) where clouds become optically thick that is the basis for Tytler's argument. That cross-correlations in adjacent QSO spectra have shown a progression in clustering scale with line strength (Crotts 1989) elucidates the nature of the population: that objects are correlated at scales in proportion to their size and that system sizes (i.e., their mass) may thus span a large range. Galaxies, galaxy clusters, and superclusters cluster similarly (Bahcall, Soneira, and Burgett 1986). This concept, plus the suggestion of Rees (1988) that density gradients can cause a power law f(N) are here expanded upon to form a model, and derive f(N).

THE CALCULATION

I have assumed that clouds are spherical, with density varying as r^{-2}. Systems can be scaled to a radius containing a $2\rho_o$ overdensity, $r^*=(GK_s/H^2)^{1/2}$, where K_s is a "system constant" in $M_\odot$/LY, ($r^*=1.215\ K_s^{1/2}h^{-1}$ kpc). The clouds are 10% baryons and 90% cold dark matter. A core radius of $3\times10^{-3}r^*$ is assumed for all clouds; core density is ~0.08 cm^{-3}. The density distribution function of systems of size K_s is asssumed to be $f(K_s)=CK_s^t$. K_s is to range initially from 1 to 10^5 $M_\odot$/LY, the upper bound intended to exclude major galaxies. The calculation of f(N) is complicated by ionization, a function of both baryon density, n_b, and the optical depth in the cloud. For this, the formula of Black (1981) is used (see below). By trial and error, a function describing the attenuation of the intergalactic UV flux, I_o, is developed (see below). The column density spectrum that results is

$$f(N)=\text{Const } N^{2t+2}\int_{Y_{min}}^{Y_{max}} h(Y,N)\, dY, \qquad (1)$$

where the impact parameter $Y=r/r^*$. Y_{min} is the computed distance at which the smallest systems produce the column density N, and Y_{max} is that at which the largest systems produce N. In practice, the Y interval is divided into a number of shells, and the neutral fraction, $f_n=10^{-8}(I/3E\text{-}21)^{-1.22}(n_b/10^{-3})^{1.22}$, where $n_b=0.2\rho_o/3Y^2$, is calculated for each shell. Only within the core, $Y<Y_c$, is the factor I/I_o incorporated. Thus, $h(Y)=Y^{2t+3}/F(Y,N)^{2t+2}$, where $F(Y,N)=f_{n1}(A_1-A_2)+f_{n2}(A_2-A_3)+...+f_{nm}g(Y,N)(Y_c{}^2-Y^2)^{1/2}/(Y_c{}^2/Y)$. If $Y>Y_c$, the last term is then $f_{nm}A_m(Y)$, where $A_m=ATN((Y_m{}^2-Y^2)^{1/2}/Y)$, Y_m being the m-th shell. Within the core, for $\log N\geq 16$, F(Y,N) is enhanced due to diminished I/I_o and is accounted for by the factor $g(Y,Ks)=(1-\exp(-2.5E15(Y/K_s)^2))^{-1.22}$, the average value of the enhancement due to the optical depth encountered with the impact parameter Y in system size K_s. In practice, functions are drawn for parameters Y and K_s, then solved for Y and N so that integration of h(Y,N) is possible.

DISCUSSION

Figure 1 shows the fit to observations for t=-1.8, $I_o=10^{-21.5}$ ergs $cm^{-2}Hz^{-1}sr^{-1}$ at the Lyman limit, and $H_o=75$ km $sec^{-1}Mpc^{-1}$. A better fit might be had with t≈-2.0. The falloff of f(N) below logN=13.75 can be remedied by extending the

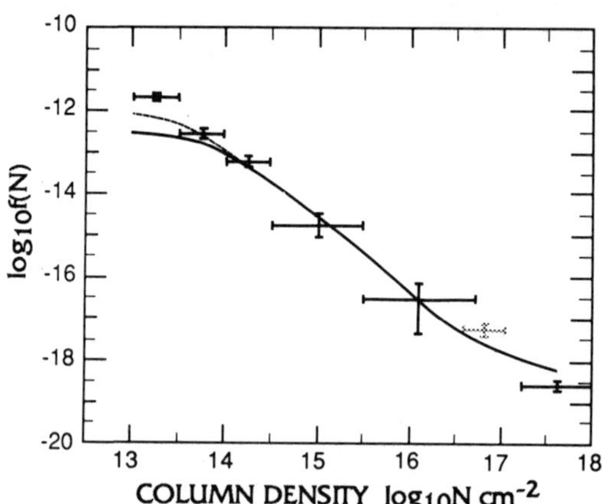

Fig. 1. The modeled HI column density spectrum fit to data from Carswell *et al* (1984), Tytler (1987), and references therein. The screened point has been disputed by SSB. Dashed line uses $(K_s)_{min}=10^{-1}$ $M_{\odot}$/LY.

$(K_s)_{min}$ to $\leq 10^{-1}$ $M_{\odot}$/LY (see Fig. 1), giving minimum diameters of ~2-5 pc. A few clouds smaller than 1 pc in diameter have been found (Carswell 1984), and their scarcity is consistent with this model. The inverse square density profiles make the clouds Jeans stable except at the core, perhaps making such small clouds an enigma. In trials with $I_o=10^{-21.09}$, a dip occurs at logN≈16 which is only partially alleviated by raising $(K_s)_{max}$ to $10^{5.5}$ $M_{\odot}$/LY. This drop is due to the very small Y_{max} for these N under high I_o conditions. Thus this model requires a low I_o to fit observations. At logN≥16.5, a deflection in f(N) caused by increasing optical depth occurs. In the simplified model used here, the slope becomes ~-1, and would be in contrast to the observed -1.35 (Tytler 1987) if it were not that dissipation in LLSs are expected to produce disc-like structures which would steepen the slope toward the expected value. Since Lyα and LLSs evolve at different rates (SSB), a z-dependent bend in f(N) at the meeting point of the two types is to be expected.

Rees' conjecture that profile effects can produce a power law f(N), while partially correct, does not cover more than ~2 orders of magnitude in N for a given cloud size; a range of system sizes is needed. While numerically, most *clouds* are small, most Lyα *lines* are produced by large systems with diameters ≥10 kpc. The work of Foltz *et al.* 1984, on the lensed QSO 2345+007AB provides good direct evidence on system sizes. Widely cited as providing a lower limit of 5-25 kpc, the authors suggest that this range be taken as a characteristic size. Of 32 Lyα lines observed, there are 13 candidate pairs. Of these, five have absorption wavelengths separated by from 0.49-0.89 Å. If these are disallowed as pairs (though they may be "associated"), then there are 8 pairs and 16 individual lines, allowing for the expected ~50% of absorption systems smaller than 10 kpc.

Systems with $W_o \geq 0.4$ Å are found to cluster on scales $<0.7h^{-1}$ Mpc (Crotts 1989). If these systems are produced by objects of size Ks≈10^4 $M_{\odot}$/LY, the clustering scale is ~$5h^{-1}r^*$. Assuming the "single population" hypothesis, this implies a clustering scale of $5h^{-1}$ Mpc for galaxies of substantial size, r*≈1 Mpc; Ks≈$10^{5.8}$ $M_{\odot}$/LY. The possibility of a "universal correlation function" implies objects are somewhat segregated by size in the universe, indicating that the local range of K_s depends upon its environment, but averaged over the whole, it has a very great extent indeed, perhaps $10^{-1} \leq K_s \leq 10^8 M_{\odot}$/LY.

To the extent that this simulation has produced a good fit to observations, it stands as evidence for the underlying assumptions: (1) the density in systems generally varies as r^{-2}, and (2) the number density of systems varies as a power of system size; $f(K_s)=CK_s^t$, and extends at least 6 orders of magnitude in K_s.

REFERENCES

Bahcall, N., Soneira, R. M. , and Burgett, W. S. 1986 *Ap. J.*, **311**, 15.

Black, J. H. , 1981 *M.N.R.A.S.*, **197**, 553.

Crotts, A., 1989 *Ap. J.*, **336**, 550.

Foltz, C. B., Weymann, R. J., Roser, H. J., and Chaffe, F. H., 1984 *Ap. J. (Letters)*, **281**, L1

Tytler, D., 1987 *Ap. J.*, **321**, 49.

Rees, J. J., 1988 in *QSO Absorption Lines: Probing the Universe*, ed. Blades, J.C., Turnshek, D. A., Norman, C. A.

Sargent. W. L. W., Steidel, C. C., and Boksenberg, A., 1988 *Ap. J. Suppl.*, **69**, 703.

SPECTROSCOPY OF LARGE NUMBERS OF FAINT GALAXIES

RICHARD S. ELLIS
Physics Department, Durham University
Durham DH1 3LE, England

ABSTRACT: Multiple object spectroscopy with high throughput spectrographs at last enables us to directly address the origin of the excess faint galaxies seen in the classical number magnitude count studies. Data from two AAT deep spectroscopic surveys show no obvious departures from a no-evolution redshift distribution. Important questions of statistical completeness and clustering are discussed. The results cannot easily be reconciled with models invoking strong luminosity evolution to explain the excess population. Neither are the results compatible with a hitherto undiscovered population of local dwarfs. Evolution must be mostly confined to low luminosity objects undergoing short-term star-formation activity which increases in strength or frequency with look-back time. Together with the existence of early-type objects with present-day colours to much higher redshifts, such activity at recent times provides important evidence for two-tier models of galaxy formation.

1. THE EXCESS POPULATION OF FAINT GALAXIES

Over a decade ago, photographic workers (Kron 1978, Peterson *et al* 1979, Tyson & Jarvis 1979) agreed upon two fundamental results concerning the statistical properties of faint galaxies. Firstly, their number magnitude relation rises more steeply than expected – at a rate $N(m) \propto 10^{0.45m}$ above that required to avoid Olber's paradox. Secondly, the population shows a monotonic trend to bluer colours at fainter limits. These two results could most easily be interpreted in terms of an increase in the past star formation rate for some subset of the galaxy population.

It is important to stress that a diverse set of evolutionary models can be invoked to simultaneously explain the photometric results. Input parameters to such models include the epoch of formation, the star formation history and the initial mass function for each galaxy type. A contrast to possible solutions is illustrative: the model proposed by King & Ellis (1985) provides a good fit to the photometric data by supposing that the Population II stars in spiral bulges and ellipticals evolved strongly with look–back time; this model would predict a significant number of blue high redshift sources to limiting magnitudes now attainable with spectroscopy. Alternatively, Koo (1981,1986) invoked a family of more gradual *mild* evolutionary models involving most of the population; these would produce a less spectacular increase in the mean redshift at a given magnitude. From photometry alone it would be difficult to distinguish between such models or make any quantitative estimate of the past brightening of the galaxy luminosity function.

An entirely different explanation might arise if the local galaxy samples used to construct the *no evolution* predictions were for some reason unrepresentative of the Universe at large. In this Hubble Symposium, we are reminded of the early work on galaxy properties based on volumes that we now know are dominated by the Local Supercluster. Recent determinations of the statistical properties of galaxies (mix of types, luminosity functions - Peterson *et al* 1985, Efstathiou *et al* 1989) have been derived from narrow-cone redshift samples with average depths considerably larger ($\sim 150h^{-1}$Mpc). Yet surprisingly, even now it is not clear whether representative galaxy number densities have been obtained (c.f. Shanks 1989). There remains an apparent inconsistency between estimates based on the southern and northern surveys, which may relate to inhomogeneities in the galaxy distribution on very large scales (c.f. Broadhurst *et al* 1989). The faint end of the luminosity function is particularly prone to difficulties in this regard because of the smaller volumes sampled. The steep slope and blueing trend in the counts could reflect a major underestimate in the number of dwarf galaxies (Kron 1983): a hypothesis that can be readily tested with the acquisition of redshifts for a complete sample of faint galaxies.

Since, as we will see, there is little evidence in the new spectroscopic datasets for any strong evolution, it might be asked whether the *observational interpretation* of the excess counts remains valid in the magnitude range where spectroscopy is now possible ($20< b_J <24$). Figure 1 shows the number magnitude counts of the same calibrated b_J photographic dataset as analysed by 6 independent groups each purporting to measure "total" magnitudes with their own image processing algorithms. This exercise was part of a test carried out by David Koo and I (Ellis & Koo 1990, see Koo *et al* 1989 for a preliminary discussion). Evidently, different groups have different *completeness limits*, but the detection and photometry of objects with b_J <23 are in reasonable agreement. Although the galaxy sample is rather small for accurate estimates of the count slope, $\gamma = dlogN/dm$, it is encouraging to note for the range $20< b_J <23$, the "total" schemes agree very well with a mean value

$$\gamma_{obs} = 0.465 \pm 0.02$$

The isophotal schemes (measuring to 1-2% of sky) typically lose $<0^m.2$ at $b_J(tot) = 23$, which yields a very slight reduction in slope ($\gamma \sim 0.44$).

In the case of field-to-field variations, Jones *et al* (1989) have analysed 6 southern fields to comparable depth in a uniform fashion with one of the algorithms tested. Each area is large enough (~ 1 deg^2) to be statistically reliable. This study reveals a similar mean slope and small scatter:

$$\gamma_{obs} = 0.436 \pm 0.02$$

Interestingly there is little, if any, correlation between the absolute number of galaxies to a given apparent magnitude and slope, confirming γ to be an appropriate observable stable to the presence of clustering. On all accounts the observational slope can be regarded as well-determined.

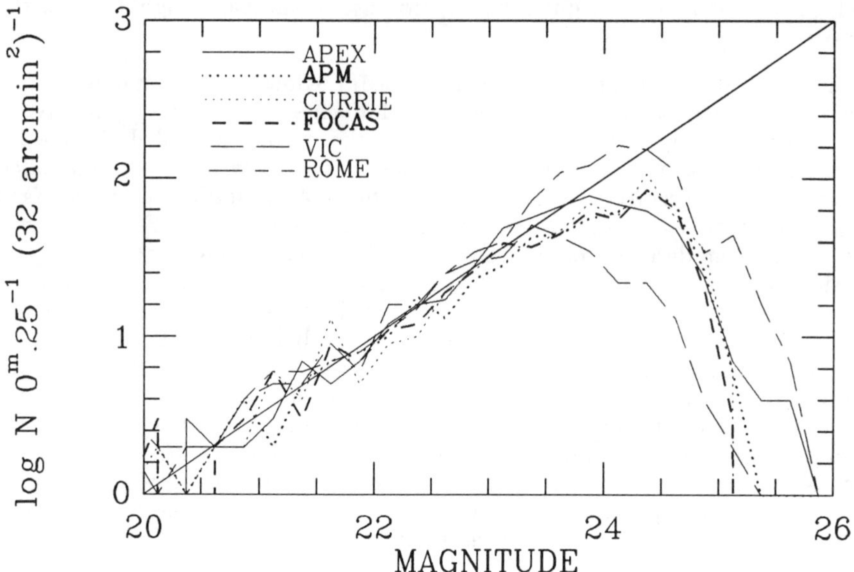

Figure 1: Galaxy number magnitude counts derived by 6 groups independently from the same calibrated test area of a deep IIIa-J 4m plate (after Ellis & Koo (1990).

Uncertainties in the *predicted* no-evolution (NE) slope have been exhaustively discussed in the literature (c.f. Broadhurst *et al* 1988, hereafter BES, and references therein). The most crucial unknowns are the K-correction as a function of galaxy class and the proportion of galaxies associated with each class; both were reviewed by King & Ellis (1985). The K-correction remains uncertain, particularly for very deep blue counts where redshifts z>1 are implied. There are simply no suitable large aperture ultraviolet spectral energy distributions for large samples of spiral galaxies. Broadhurst *et al* examined two independent methods for the no evolution prediction; one based on morphologically-classified luminosity functions derived from the $b_J < 17$ survey of Peterson *et al*; the second used divisions based on b_J–K colours for galaxies from the multicolour survey of Mobasher *et al* (1987). Since there is not a perfectly tight relation between colour and morphology, these alternatives explore the two critical parameters fairly sensitively. In fact, the predicted no evolution slopes for the two procedures both yielded $\gamma_{NE}= 0.32$.

It is important to realise that the above no evolution prediction is entirely empirical; it does not involve any understanding of how galaxies came to have their present day colours. BES simply "replicated" the local Universe out to high z and estimated the N(m) that would result. This contrasts with most other "no evolution" predictions (c.f. Koo 1981, Koo & Szalay 1984, Guiderdoni & Rocca-Volmerange 1987, Yoshii & Takahara 1988, Yoshii & Peterson 1989) where a set of model galaxies with carefully specified parameters (e.g. redshift of formation, cosmological parameters, star formation history) are evolved to match the present day properties of representative galaxies; this final *modelled* population is then the basis of the "no evolution" prediction. Any discrepancy between the two approaches could well reflect on the accuracy with which the models match the present day distribution. More importantly, however, *such no evolution predictions are not unique* and this is a major difficulty in understanding and comparing different conclusions.

Some examples may help reveal the problem. Yoshii & Takahara model 5 classes according to the evolutionary models of Arimoto & Yoshii (1986). The proportions are adjusted to reproduce the present day galaxy population taking into account a wide range of data in the literature (including, but not restricted to, that used by BES discussed above). They find γ_{NE}=0.33 for 20<b_J <23 in close agreement with our empirical approach. In several other respects their NE predictions are similar.

However, Koo & Szalay model 24 classes of galaxies according to the evolutionary models of Bruzual (1983) claiming a complex array is needed to match properly the present day UBV distribution. These models apparently produce galaxies that are consistently redder than the bluest observed today, even if the star formation rate is constant with time. To overcome this problem, an important (~5%) population of blue (B-V<0.46) galaxies with *recent formation* is added. This is bluer than any class in Yoshii & Takahara's inventory and has a detectable effect on the predictions, particularly because these objects are given a much fainter characteristic magnitude ($M^* \sim -17 + 5\log h$* in Schechter's (1976) terminology). Such blue dwarfs provide an Euclidean component to the counts at intermediate magnitudes and thus Koo & Szalay's γ_{NE} is closer to 0.40. Hence only *mild* evolution is required to fit their observations.

What is the evidence for this population of low luminosity blue galaxies? And how should they be treated in a no evolution prediction if they are recently formed as Koo & Szalay specify? Whilst there is clear evidence from local surveys for a lower M^* for blue galaxies (c.f. Ellis 1987) empirical data does not support a γ_{NE} as steep as that proposed by Koo and co-workers. Indeed, their NE model predicts too many low z (<0.15) galaxies in their own deep survey (Koo 1989). Independently, Shanks (1989) has derived Schechter parameters $M^*_B = -20+5\log h$, α=-1.5 for the bluest (B-V<0.6) galaxies in the Peterson *et al* (1985) b_J <17 sample of 350 galaxies, and this yields a γ_{NE}=0.35 – only slightly above that derived from Mobasher *et al*'s b_J-K analysis. Therefore it seems the NE slope is closer to 0.33-0.35 and thus the faint excess population remains significant.

Recently we have seen the emergence of deeper CCD counts (Hall & Mackay 1985, Tyson 1988, Metcalfe *et al* 1987 and Cowie & Lilly 1989) complementing the photographic data in photometric precision and depth. However, establishing statistically reliable counts fainter than b_J=25 is technically demanding.

---

* h is Hubble's constant in units of 100 km sec^{-1} Mpc^{-1}

In particular only 1.5-2^m fainter than the photographic limits, corrections for crowding become necessary. Whilst it is possible to model this problem and derive *corrected counts* somewhat fainter (Tyson 1988), a new approach to photometric analysis based on surface brightness fluctuations (c.f. Hall 1985) might be more appropriate for a proper exploitation of the data. Thus far the CCD counts have primarily shown a continuation of the two effects noted in the photographic data. The slope to b_J=26 is approximately the same as to b_J=24 with no bumps or features (so far as the limited statistics can tell us). Although the bluing trend continues, it is claimed (Cowie & Lilly 1989) that a new flat-spectrum population emerges at b_J ~24. This has aroused much interest and we return to this possibility in §4.

2. DEEP SPECTROSCOPIC SURVEYS

Having established the reality of an excess of galaxies above the no evolution model, we now turn to role of spectroscopy in determining the origin of this excess. Prior to the development of faint object and multiple object spectrographs, recourse was made to multi-band colours. Whilst colours have certainly proved valuable in determining membership in clusters dominated by red galaxies with strong features (c.f. MacLaren *et al*'s 1988 proven success rate in several moderate z clusters), this is clearly because of the *a priori* knowledge of the cluster redshift. For field surveys dominated by spirals and other blue objects, their use as redshift indicators is likely to be quite misleading unless backed up with complete spectroscopy of some sub-set (c.f. Koo 1985).

The major breakthrough in spectroscopic surveys was undoubtedly the implementation of multi-object spectrographs (see the review by Ellis & Parry 1988 for technicalities). Several surveys have been completed or are underway; these are listed in Table 1. I will concentrate here on what has been learnt from the 3 independent surveys conducted at the AAT with my collaborators at Durham.

Table 1

Deep Spectroscopic Surveys

Instrument	N_{sp}	m_{lim}	R Å	Reference
AAT+FOCAP	187g	20< b_J <21.5	4	Broadhurst *et al* 1988
AAT+LDSS	87g+33s	21< b_J <22.5	13	Colless *et al* 1989ab
AAT+LDSS	>150	b_J <24	45	Colless *et al**
KPNO+Cryocam	>400g	R<22	15	Koo & Kron (1988)*
CFHT+FOS	>6g+2s	23<b_J <24	50	Cowie & Lilly (1989)
ESO 3.6m+EFOSC	–	–	–	de Lapparent *et al**

* – in progress

The first survey we completed (BES) was motivated by Kron's suggestion that the faint end of the galaxy luminosity function had been seriously underestimated in the b_J=17 surveys and thus the faint count "excess" might be due to a plethora of low luminosity blue dwarfs. This seemed possible given the abundance of dwarfs in nearby clusters (Binggeli *et al* 1985) although one might have expected many to be classed as stars beyond $b_J \sim 19$. In this context, we note that Morton *et al* (1985), in their stellar survey, failed to find a significant number of compact extragalactic sources to b_J=20.

The BES survey, conducted with the AAT fibre positioner *FOCAP* (Gray 1985), reached b_J=21.5 where the counts are ~30% above the no evolution prediction. However, the survey showed *no* evidence for a departure from no evolution, the mean survey redshift $\bar{z}$=0.216±0.016 (1σ scatter from 5 independent fields) agreeing well with that (0.22) for the NE model. Models with a steep extension of the galaxy luminosity function (Schechter's $\alpha = -2$) or an additional dense population of low-luminosity dwarfs with $M^*_{b_J} = -18.5+5\log h$ such as would be required to steepen the counts, were ruled out since they would lower $\bar{z}$ well below 0.2, in conflict with the survey. Too few low luminosity galaxies were found (none with z<0.03) for local dwarfs to be a major contribution to the faint counts.

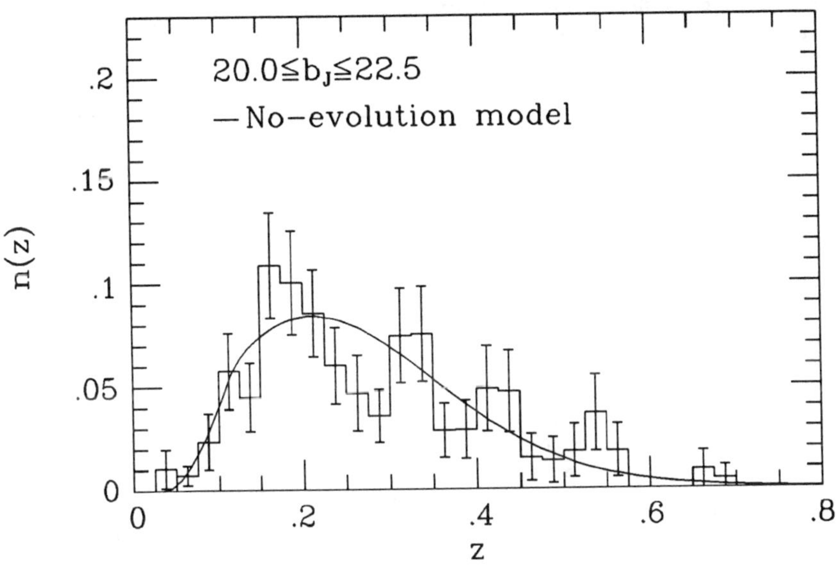

Figure 2: Overall redshift distribution for the combined FOCAP and LDSS AAT galaxy surveys compared with the standard no evolution model.

The absence of luminous high z galaxies to b_J=21.5 might not be surprising given the small count excess at that magnitude. However, a fainter survey completed with a new wide-field multislit spectrograph, the Low Dispersion Survey Spectrograph (LDSS, Wynne & Worswick 1989) reveals a similar result to b_J=22.5 (Colless *et al* 1989ab). Spectroscopic targets were chosen with no regard to image classification – both galaxies *and* stars were surveyed. Here

$\bar{z}$=0.312±0.022 (1σ from 3 independent fields), again close to the no evolution prediction of 0.29, yet the surface density of galaxies at the LDSS limit is almost twice the no evolution prediction. The combined dataset BES+LDSS represents ~300 galaxies with 20< b_J <22.5 and the form of the overall redshift distribution is satisfactorily fit by the NE model (Figure 2). The absence of a significant high or low z extra component to N(z) despite the N(m) excess presents a paradox; the excess galaxies must be hiding in the same redshift range and sharing the same luminosies as galaxies predicted by the no evolution model. What is going on?

At this stage it is important to demonstrate that incompleteness and clustering are not seriously biasing the observed distributions. In both AAT surveys, reliable redshifts could not be determined for about 15% of the sample. In multi-object surveys where each target is given the same exposure time, incompleteness arises from three causes: poor continuum signal/noise, weak features difficult to recognise at low spectral resolutions, or a lack of features in the observing wavelength range because of the redshift or nature of object. BES demonstrated the continuum s/n was significantly worse in vignetted areas of their detector, the failures correlating extremely well with detector position. This accounted almost entirely for the incompleteness observed. For the LDSS survey, incompleteness increases with apparent magnitude as would be expected for s/n reasons, though the lower resolution also makes it harder to detect absorption lines.

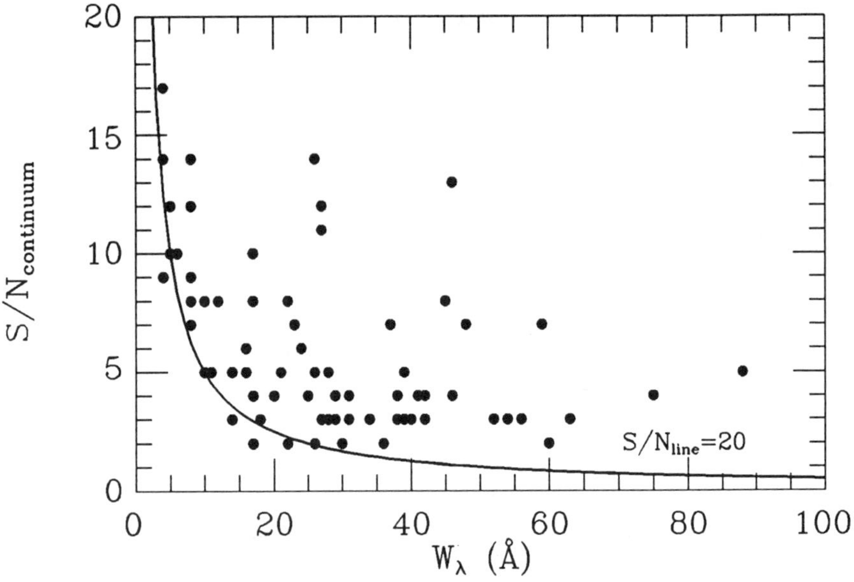

Figure 3: Continuum S/N ratio versus rest-frame equivalent width for detected [O II] 3727 Å emission in the LDSS survey. Points above the smooth curve represent a S/N$_{line}$-limited sample.

100% completeness in apparent magnitude, whilst desirable, is not always practical with multi-object instruments and is not necessary to make progress in understanding the *form* of N(z). The emission line [O II] 3727 Å is present in 85% of the identified galaxy spectra and is an invaluable "beacon" for this purpose. Since the observed N(z) declines well before [O II] shifts out of the observing window (z=0.9 for LDSS), it seems unlikely that the unidentified LDSS objects have 0.6<z<0.9 unless there is a sudden change in the spectral population beyond z~0.6.

This can be turned into a more rigorous statement by imagining a survey statistically complete in terms of a limiting s/n in the emission line, rather than in apparent magnitude (the continuum s/n). Figure 3 shows the measured $s/n_{continuum}$ (at the emission line location) against rest-frame equivalent width W_λ of [O II] for the identified galaxies in the LDSS survey. The spectra are seen to be consistent with a limiting S/N *in the emission line*. This not only confirms the reliability of the line detections but given this threshold and a known $S/N_{continuum}$ over some wavelength range for the *un*-identified galaxies, we can compute the minimum detectable W_λ. If the distribution of W_λ for 0.6<z<0.9 is similar to its observed form for z<0.6, then we find that *over half* of any galaxies in this range should have been detected in [O II]. Such constraints do not, of course, apply outside the range of [O II] visibility and we cannot exclude the possibility that N(z) is bimodal. However, in that case one might expect a rather different class of galaxy to be in the failed category e.g. in terms of colour, but no such evidence is found.

As far as clustering in the deep fields is concerned, the highly structured nature of the redshift distributions observed in the pencil-beam surveys has already been mentioned; in some deep fields over 50% of the galaxies lie in discrete associations (Ellis 1987, Koo & Kron 1988) and new evidence for large scale features in the distribution of these peaks has been presented by Broadhurst *et al* (1989). Clustering will, of course, blur the precision with which small amounts of $\Delta\bar{z}$ evolution might be detected. From the eight independent fields in the BES+LDSS surveys, however, fluctuations of only $\Delta\bar{z}/\bar{z}$ ~0.09 are seen and when examined with K-S tests, the N(z)'s of most fields are consistent both with one another *and* with the no evolution prediction (Figure 2).

3. GALAXY EVOLUTION SINCE Z = 2

Regardless of the redshifts of the missing 15-20% in the LDSS sample, the majority of "extra" galaxies seen in the counts to b_J=22.5 must have z<0.5 and thereby not significantly perturb $\bar{z}$. This provides a very strong constraint on certain forms of luminosity evolution including many originally postulated to explain the steep count slope.

Large amounts of luminosity evolution for some fraction of the population were claimed by several authors in order to raise γ from ~0.32–0.35 to its observed value of 0.44. Shanks *et al* (1987) estimated the evolution implied empirically. By assuming a redshift-dependent evolutionary correction for types earlier than Sbc (representing ~60% of the present population), they derived a fit to the counts implying an enormous $\Delta b_J \sim -3^m$ correction by z = 1. However, attributing the same correction to such a diverse range of types and adopting a polynomial with no astrophysical basis could lead to a significant overestimate of the actual evolution at a given redshift.

King & Ellis (and more recently BES) examined the family of Bruzual's (1983) evolutionary models, adding evolutionary corrections to the empirical no evolution prediction. They found Bruzual's c-models (strong initial bursts with no subsequent star formation at later times) could neither explain the size of the excess nor the gradual onset of the excess faint population. Only models with a strong exponential decline in star formation (so-called μ-models) could raise γ substantially. Such behaviour would have to be attributed to either a substantial portion of the population (c.f. BES) or to the bulge population of each disk galaxy (King & Ellis). Specifically, for the favoured μ=0.5 τ=16Gyr model in a h=0.5 q_0=0.1 cosmology, the evolution correction at $z = 1$ is $\Delta b_J \sim -2^m$. A similar model has been invoked in various forms to explain the colours of high redshift cluster and radio galaxy data (c.f. Djorgovski *et al* 1985).

Such strong evolution of the entire luminosity function (LF) in a form monotonic with redshift can be ruled out by the absence of significant numbers of galaxies to z~0.9 in the AAT surveys. BES proposed a luminosity-dependent evolutionary model whereby *only the low luminosity galaxies get brighter with look-back time*, at least for z<1. This has the desirable effect of increasing γ at the expense of little change in $\bar{z}$. An important feature of BES' hypothesis is the *astrophysical* support for such evolution from a detailed consideration of the *spectra* themselves.

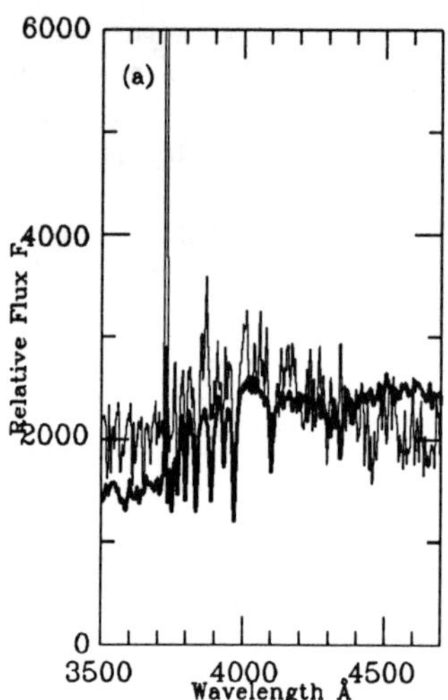

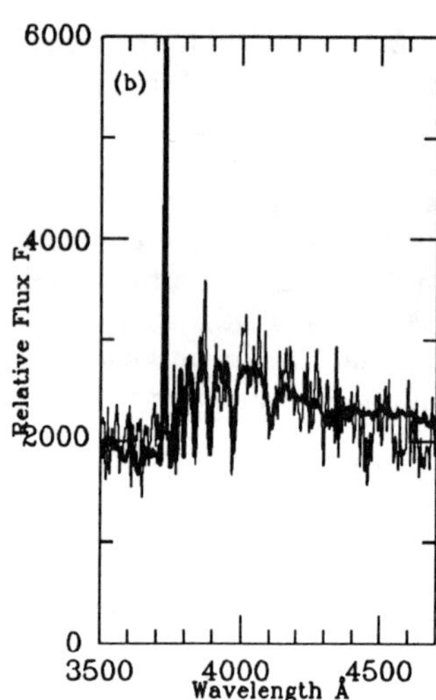

Figure 4: Mean coadded spectrum for Broadhurst et al (1988)'s galaxies with W_λ >35 Å (thin line) compared with models from Bruzual's evolutionary synthesis code (thick line) for (a) a continuous star formation rate viewed 12 Gyr after formation, (b) two-burst burst model with second burst of duration 0.2 Gyr consuming 3% by mass of galactic gas.

An increasing proportion of faint galaxies have unusually strong [O II] emission lines and indeed the excess in the W_λ distribution matches approximately the number required to explain the high counts. Unlike nearby spirals which are accurately modelled by near-constant star-formation rates, the faint galaxies are too blue for their Balmer absorption line strengths and can only be understood in models incorporating short intense bursts of star formation (Figure 4). These galaxies are, in effect, luminous HII regions, examples of which can be found at low redshift but with an abundance far lower than in the faint surveys. It is important to stress here that the strong [O II] effect is much larger than can be accounted for by both aperture effects (the BES aperture is only 30% larger than that used in the local survey, and there is no trend with redshift such as would be expected if such effects were significant) and K-correction bias (which can be calibrated from a well-defined relation between W_λ[O II] and b_J-K for local galaxies).

Just as spiral arms are "highlighted" by HII regions on the passing of a density wave, so the optical and rest-frame uv luminosities of a galaxy increase markedly during an active burst of star formation. Broadhurst *et al* suggested the 30% excess population to b_J=21.5 would be too faint to be detected in their quiescent state and that the steep count slope originates from a gradual increase with look-back time in the occurrence of these short bursts of star formation.

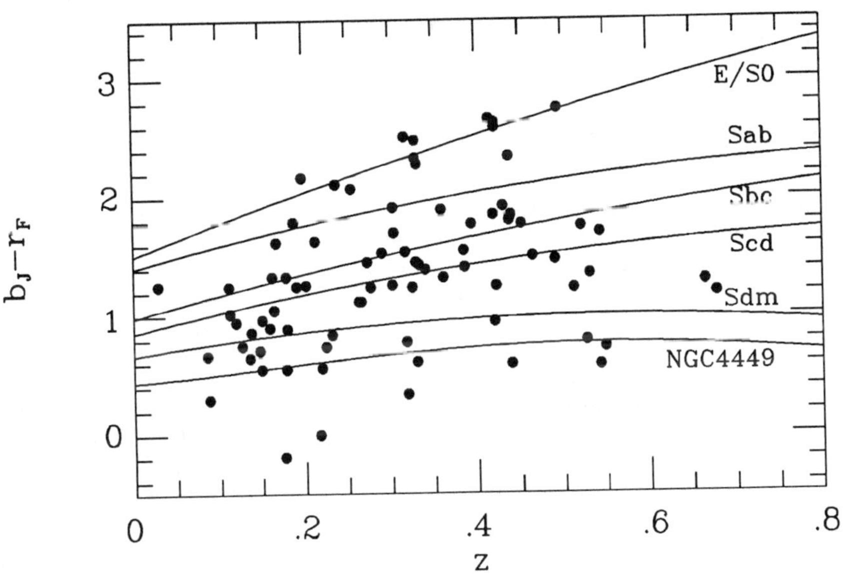

Figure 5: The distribution of b_J-r_F colour with redshift for 81 survey galaxies. The curves represent unevolved colours as a function of redshift for various types including NGC4449, a Sm/Irr exemplary of intense short-term star formation.

The deeper LDSS survey supports the hypothesis via the addition of CCD-calibrated photograph b_J–r_F colours (Figure 5). Two points emerge here. Firstly, there is no blueing at the red end of the to z$\sim$0.5 examples are seen with colours similar to present day E/S0s. Had the counts been dominated by rapidly evolving early-type galaxies some colour evolution would have been predicted – about $0^m.2$ at z=0.5 for the μ=0.5 model discussed earlier. Stricter constraints are now available from similar arguments in higher redshift samples (Hamilton 1985, Persson 1988, Aragon Salamanca *et al* 1990). Secondly, the bluest galaxies are not exclusively at high redshift; indeed the most extreme examples, representative in colour of a significant fraction of the *flat-spectrum* population in the CCD counts (Cowie & Lilly 1989,§4) are *at low z.*

Unpalatable though the burst evolutionary picture may be at first sight, there is strong supporting evidence in at least two independent samples. A recent decline in short-term star formation activity is witnessed from detailed studies in modest redshift rich clusters of galaxies (Butcher & Oemler 1978, Dressler & Gunn 1983, Couch & Sharples 1987, MacLaren *et al* 1988) and from the rising volume density of IRAS galaxies over 0<z<0.2 (Saunders *et al* 1989). Is it possible that the same field galaxy evolution is, in some way, responsible for all three manifestations of galaxy evolution over z<0.5?

Koo (1988) suggested the Butcher-Oemler (B-O) effect might arise from weakly associated blue field galaxies seen in projection on rich clusters at moderate z, but dynamical data (c.f. Couch & Sharples) strongly suggests the blue galaxies are physically associated with the cluster potential. Nonetheless, the similarity between field and cluster evolution is intriguing, particularly when one considers some of the more perplexing aspects of the B-O effect. For example, various stages of bursting covering a lifetime of >2-3 Gyr are simultaneously present within a single cluster (c.f Abell 370 – MacLaren *et al*) and yet the onset in look-back time of the B-O phenomenon is remarkably sharp (< 1 h^{-1}Gyr).

Balmer absorption line strengths and optical/uv colours of z$\sim$0.4 cluster galaxies imply recent bursts and/or truncated star formation which might be the result of gaseous field galaxies falling sporadically into the cluster potential. Since a gaseous system may be susceptible to a burst of star formation in a way that is critically related to its gas content and/or external influences, both of which are likely to be redshift dependent, the rich cluster behaviour may thus merely be mirroring activity in the surrounding field (Bower 1989). Additionally, there might be some *amplification* of the phenomenon by this process which would explain why, in general, bursting galaxies in clusters are slightly more luminous than their field counterparts, producing a larger number of *post-starburst*(E+A)-type galaxies to a fixed absolute magnitude.

The bursting hypothesis is possibly only *one* of several that may account for the combined constraints of counts, redshifts, colours and line strengths. Its strength is that it is astrophysically consistent and has as its baseline the empirical properties of large complete samples at various depths. However, no explanation has been offered for the *physical origin* of the short-term activity. To tackle this, high resolution UV and optical images of representative z$\sim$0.3 strong-lined galaxies are needed (e.g. with HST) to reveal the spatial structure of the star-forming component. The origin of the IRAS phenomenon holds another important clue – a new faint IRAS redshift survey ($f_{60\mu m}$ >0.2Jy reaching z$\sim$0.2) is underway to strengthen the connection between the faint optical and

IRAS samples (Rowan-Robinson *et al* 1990).

Regardless of the curiosities at the faint end of the luminosity function since z~0.5, the absence of *any* galaxies beyond z~0.7 to b_J=22.5 places a interesting constraint on the past appearance of *luminous* (M*) galaxies. Of course, because of survey incompleteness, we cannot eliminate the possibility of a small number of weak-lined z>1 galaxies but N(z) would probably have to be bimodal in view of the S/N arguments of §2. In the context of most evolutionary models, however, high z star forming galaxies should be readily recognisable via strong emission lines and it is therefore illustrative to quantify further how much conventional luminosity evolution at z~1.5-2 is allowed by the non-detection of blue luminous galaxies at these redshifts to b_J=22.5.

Most plausible single-burst (*c*-type) models for luminous galaxies with formation epoch z_f <2.5 would generate uv-strong precursors at z~2 that would be readily visible within the LDSS limit. This constraint had been argued many years ago from the absence of a "bump" in the counts, but the spectroscopic constraint is more rigorous since with the counts one could always appeal to an extended era of formation. The limit is supported further by the "red envelope" argument discussed above; whilst some activity is seen in spheroidal systems since z~1 (Dressler, 1989), the bulk of the star formation formed considerably earlier.

4. FLAT-SPECTRUM GALAXIES

The above spectroscopic studies lie firmly in the photographic domain. Whilst imaging with CCDs has pushed considerably fainter, the grafting of the new CCD data to the well-established photographic data has not been easy. Part of the problem arises from the enormous mismatch in solid angle of the two techniques. In the overlap region at b_J ~24, the properties of ~20,000 photographic galaxies on a single plate are to be compared with <~100 on one of Tyson's CCD frames (and barely 10 on one of Cowie & Lilly's frames). Both Tyson (1988) and Cowie *et al* (1989) claim to have identified the onset of a *new* population of flat-spectrum objects; the fact that this emerges at an apparent magnitude close to the intersection of the two techniques, and that the two teams disagree on the rapidity of onset, leads one to suspect these claims however.

Cowie & Lilly (1989) report the onset is comparable with $\propto 10^{1.5m}$ – a slope 3 times steeper than for the normal population. Yet Tyson (1988) refers to the population's "sub-Euclidean counts (which) suggest they are neither a new class of local objects (z<0.3) nor dwarf galaxies". Evidently the definition of a flat-spectrum object requires some care. Cowie & Lilly use b_J-R<0.6 to illustrate their point and this is conveniently close to the colours used by both Tyson and Colless *et al* in the LDSS spectroscopic survey.

In Figure 6 I compare the onset of a b_J-R<0.6 population for the three samples. For reference I have also plotted the total photographic and Tyson's CCD counts. Whilst the proportion is definitely rising, this would be expected from a gradual bluing. The best fit to Tyson's blue data has γ_{blue}=0.7 – compatible with a Euclidean rise. The absence of any blue objects in Cowie & Lilly's brighter bins can easily be understood in terms of poor statistics. Within the LDSS survey, 7 objects have b_J-r_F <0.5, of which 4 have z<0.35 and 3 are unidentified. If the

3 unidentified LDSS objects to b_J=22.5 are bright examples of Cowie & Lilly's population, then there would be a fairly steep rise ($\gamma_{blue} \sim 0.8$). However, it is reasonable to expect that the 4 identified flat-spectrum objects with low z have fainter counterparts and so this must be an upper limit to the slope. If all 7 LDSS flat-spectrum objects are of this category (as plotted in Figure 6), then a near-Euclidean N(m) slope results. The assumption that all 7 LDSS objects are a common low z class is further supported by limited spectroscopy at $b_J \sim$23.5 reported by Cowie & Lilly (1989); so far *only* z<0.5 galaxies are found.

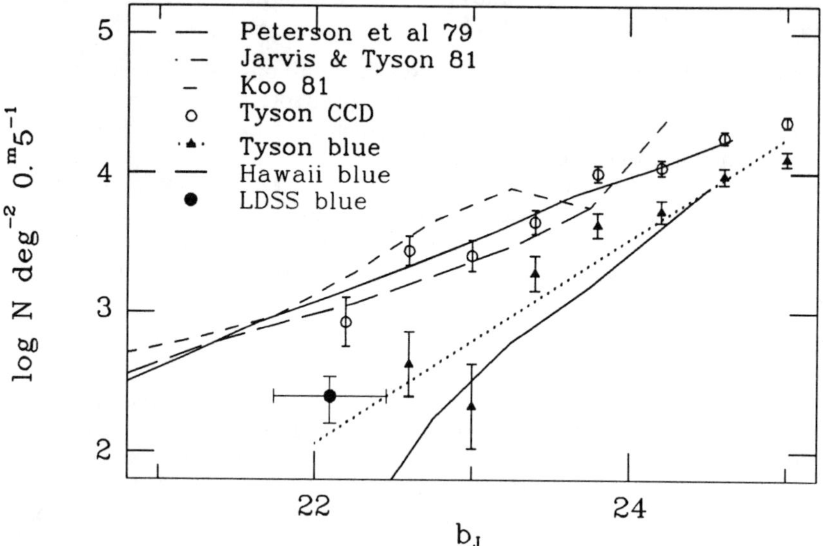

Figure 6: Number-magnitude counts for blue objects with a criterion approximating b_J-R<0.6. The uppermost lines and open circles refer to the photographic and CCD counts for the entire galaxy population. Other points and lines refer to the onset of the blue population. The dotted line is a best fit near-Euclidean slope fitting the Tyson (1988) CCD data and the Colless et al (1989ab) LDSS spectroscopic sample. The line marked 'Hawaii' is the rapid onset suggested by Cowie & Lilly (1989).

There are other conceptual problems with a steep rise in this population. Such a sudden onset is unseen in any other magnitude range for any extragalactic population including QSOs. Cowie & Lilly propose significant discontinuities pass through various filters but this still presupposes a narrow luminosity range. Furthermore why do the total counts not upturn at the point where the flat-spectrum counts meet those of the remainder ($b_J \sim$25)?

It is argued that the low z bursting hypothesis cannot satisfactorily explain the flat-spectrum galaxies on account that the (inferred) low mass of these systems could not sustain star formation at the required rate over the Hubble time.

However, the reservoirs of gas that may surround these little-understood objects could be quite significant. Only a large spectroscopic sample of such objects can resolve this important question.

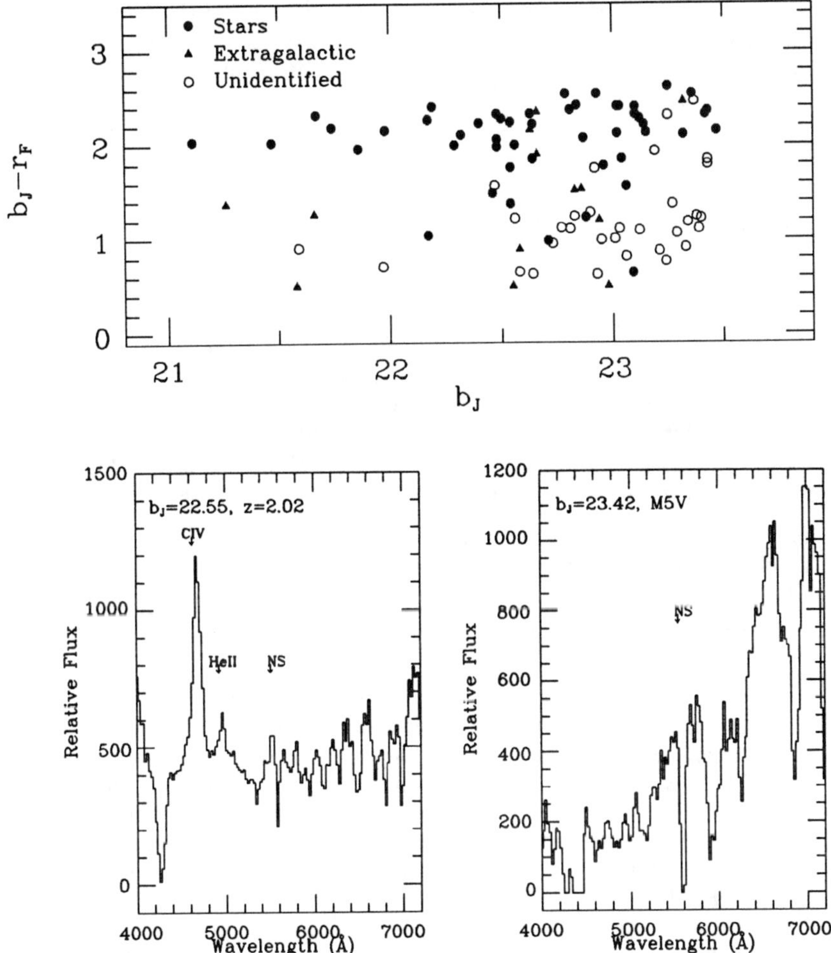

Figure 7: (a) b_J-r_F colours for a low dispersion survey of faint compact objects limited at b_J=23.5. (b) spectra at 45 Å resolution for a z∼2 UVX QSO and a M star close to the survey limit.

Finally, I would like to report on a rather different, serendipitous spectroscopic survey being undertaken with LDSS to investigate the nature of faint b_J <23.5 *compact* objects. This survey, at low dispersion, was really intended to be a faint analogue of an objective prism survey, cataloguing objects that might not be present in brighter surveys. We avoided extended objects on the assumption that all would be galaxies. Out of a sample of over 150 such objects, a complete sample with of 110 with b_J <23.5 can be formed, of which ∼80 have colours so far. The colour-magnitude distribution, success rate and some sample spectra are shown in Figure 7.

Some interesting points emerge. Firstly, there are virtually *no* red extragalactic objects; there is a high success rate for red objects which are almost exclusively M stars. This places a strong constraint on red dwarf galaxies and intergalactic globulars. At $M_B = -10$ the survey probes as far as $\sim$ 50 Mpc and the space density of such red objects must be lower than $\sim$0.5 Mpc^{-3}. Secondly, as in Cowie & Lilly's recent results, of those galaxies which *did* appear in this compact survey, most were further examples of HII galaxies at low z, including several with z<0.05.

5. CONCLUSIONS

Spectroscopic results at faint limits are challenging the dogmas of galaxy evolution established in a decade of photographic and CCD photometry. The samples are always smaller and more incomplete than we would like, but such is the nature of observational cosmology. As Hubble wrote " Eventually, we reach the dim boundary..we measure shadows, and we search among ghostly errors of measurement for landmarks that are scarcely more substantial". That the 'landmarks' at the 'dim boundaries' are mostly [O II] rather than Lyman α has surprised us all; as we probe fainter the cosmological depth is hardly increasing.

We suggest this paradox can be resolved via an astrophysically-consistent picture whereby galaxies undergo bursts of star formation on short timescales, with most of the recent activity being confined to the faint end of the luminosity function. This may *not* be an unique solution, but it is consistent with the data at hand, including accurate co-added spectra of representative faint objects.

It seems that galaxy formation is a two-tier process. To as high a redshift as one can probe with controlled samples, *some* galaxies are old and red – their stellar populations were probably formed before z$\sim$2.5 otherwise their luminous phases would be recognised to the deepest limits now surveyed. Superimposed on this is a more continuous, and possibly erratic, process which may be associated with gaseous star formation truncated at early times in luminous systems and in clusters, but continuing to the present day in low luminosity galaxies. Such gas-rich galaxies will always feed rich clusters, but there may be a rapid decline in their ability to burst strongly which could simultaneously explain the Butcher-Oemler phenomenon, the steep galaxy counts and the rapid rise in IRAS sources with redshift.

Determining the physical origin for the star-formation processes we have witnessed at modest redshifts, and the extent to which this activity can be extrapolated to the truly faint blue CCD populations are the key projects for the 1990's.

Acknowledgements

I thank my colleagues, Tom Broadhurst, Matthew Colless and Richard Bower, for allowing me to quote results determined with their help. It is a pleasure to thank David Koo, Richard Kron, Simon Lilly and Len Cowie for many stimulating discussions and their continuing generosity in communicating results and ideas. Finally, I thank the Astronomical Society of the Pacific for their generous travel funds which enabled me to participate in this historic and enjoyable meeting.

REFERENCES:

Aragon-Salamanca, A, Ellis, R S, Couch, W J & Carter, D 1990, in preparation.

Arimoto, N & Yoshii, Y 1986 *Astr. Ap.*, **164**, 260.

Binggeli, B, Sandage, A & Tammann, G 1985 *Astron. J.*, **90**, 1681.

Bower, R G 1989 Ph D thesis, University of Durham, England.

Broadhurst, T J, Ellis, R S & Shanks, T 1988 *M.N.R.A.S*, **235**, 827.

Broadhurst, T J, Ellis, R S, Koo, D C & Szalay, A, preprint.

Bruzual, G 1983 *Ap. J.*, **273**, 105.

Butcher, H & Oemler, A 1978 *Ap. J.*, **219**, 18.

Colless, M M, Ellis, R S & Taylor, K 1989a in *The Epoch of Galaxy Formation*, eds. Frenk *et al*, p 359, Kluwer.

Colless, M M, Ellis, R S, Taylor, K & Hook, R N 1989b *M.N.R.A.S*, in press.

Couch, W J & Sharples, R M 1988 *M.N.R.A.S*, **299**, 423.

Cowie, L L & Lilly, S J 1989, this volume

Cowie, L L, Lilly, S J, Gardner, J & McLean, I S 1989 *Ap. J.*, **332**, L29.

Djorgovski, S, Spinrad, H & Marr, J 1985 in *New Aspects of Galaxy Photometry*, ed. Nieto, J-L, p193, Springer.

Dressler, A 1989 this volume.

Dressler, A & Gunn, J E 1983 *Ap. J.*, **263**, 533.

Efstathiou, G, Ellis, R S & Peterson B A 1989 *M.N.R.A.S*, **232**, 431.

Ellis, R S 1987 in *Observational Cosmology*, IAU Symposium 124, eds Hewitt, D *et al*, p367, D Reidel.

Ellis, R S & Koo, D C 1990 in preparation.

Ellis, R S & Parry, I R 1988 in *Instrumentation for Ground-Based Astronomy*, ed. Robinson, L, p192, Springer-Verlag.

Hall, P, 1985 Thesis, University of Cambridge, UK

Hall, P & Mackay, C D 1985 *M.N.R.A.S*, **210**, 979.

Hamilton, D 1985 *Ap. J.*, **297**, 31.

Gray, P M 1986 *Proc. S.P.I.E*, **627**, 96.

Guiderdoni, B & Rocca-Volmerange, B 1987 *Astr. Ap.*, **186**, 1.

Jones, L R, Shanks, T, Fong, R, Metcalfe, N M & Ellis, R S 1990, in preparation.

King, C R & Ellis, R S 1985 *Ap. J.*, **288**, 456.

Koo, D C 1981 Ph D thesis, University of California, Berkeley.

Koo, D C 1985 *Astron. J*, **90**, 418.

Koo, D C 1986 *Ap. J.*, **311**, 651.

Koo, D C 1988 in *Towards Understanding Galaxies at High Redshift*, eds Kron, R & Renzini, A, p275, Kluwer.

Koo, D C 1989, this volume.

Koo, D C & Kron, R 1988 in *Towards Understanding Galaxies at High Redshift*, eds Kron, R & Renzini, A, p209, Kluwer.

Koo, D C & Szalay, A S 1984 *Ap. J.*, **282**, 390.

Koo, D C, Ellis, R S & Windhorst, R 1990 in preparation.

Kron, R 1978 Ph D thesis, University of California, Berkeley.

Kron, R 1983 *Vistas*, **26**, 37.

MacLaren, I, Ellis, R S & Couch, W J 1989 *M.N.R.A.S.*, **230**, 249.

Metcalfe, N, Shanks, T & Fong, R 1987 in *High Redshift & Primaeval Galaxies*,

eds. Bergeron, J *et al*, p37, Editions Frontieres.
Mobasher, B, Ellis, R S & Sharples, R M 1988 *M.N.R.A.S.*, **233**, 11.
Morton, D C & Tritton, K P 1985 *M.N.R.A.S.*, **212**, 325.
Persson, S E 1988 in *Towards Understanding Galaxies at High Redshift*, eds. Kron, R & Renzini, A, p251, Kluwer.
Peterson, B A, Ellis, R S, Kibblewhite, E J, Bridgeland, M T, Hooley, T & Horne, D 1979 *Ap. J.*, **233**, L109.
Peterson, B A, Ellis, R S, Efstathiou, G P, Bean, A J, Shanks, T, Fong, R & Zhou, Z-L 1985 *M.N.R.A.S.*, **221**, 233.
Rowan-Robinson, M, Lawrence, A, Broadhurst, T J, Ellis, R S, Parry, I R, Efstathiou, G P, Saunders, W 1990 in preparation.
Saunders, W, Rowan-Robinson, M, Lawrence, A, Efstathiou, G P, Kaiser, N, Ellis, R S & Frenk, C S 1989 *M.N.R.A.S.*, in press.
Shanks, T 1989 in *Extragalactic Background Light*, ed. Mattila, N, Kluwer (in press).
Shanks, T, Stevenson, P R F, Fong, R & MacGillivray, H T 1987 *M.N.R.A.S.*, **206**, 767.
Tyson, J A 1988 *Astron. J.*, **96**, 1.
Tyson, J A 1989 this volume.
Tyson, J A & Jarvis, J F 1979 *Ap. J.*, **230**, L153.
Wynne, C G & Worswick, S 1989 *Observatory*, **108**, 161.
Yoshii, Y & Takahara, F 1988 *Ap. J.*, **326**, 1.
Yoshii, Y & Peterson, B A 1989 preprint.

UBC/LAVAL 2.7M LIQUID MIRROR TELESCOPE (LMT): PROGRESS REPORT

BRAD K. GIBSON and PAUL HICKSON
Dept. of Geophysics & Astronomy, University of British Columbia,
#129-2219 Main Mall, Vancouver, B.C., Canada V6T 1W5

ABSTRACT The status of the 2.7m LMT currently under construction at the University of British Columbia, in collaboration with Laval University, is reviewed. A brief discussion of the engineering design characteristics is given, as well as an overview of th instrument's primary observing program, a deep photometric galaxy redshift survey.

INTRODUCTION

To study the large-scale structure of the universe, and in particular galaxy evolution of a good statistical sample, requires the access to large dedicated survey telescopes. LMTs provide an ideal solution to this restriction in that very large apertures are possible at a fraction of the cost of an equivalent-sized conventional telescope.

The LMT is not a new concept. Its history from the nineteenth century onwards has been reviewed by Olsson-Steel (1986) and will not be repeated here.

INSTRUMENT DESCRIPTION

The principle behind the LMT is very simple. The telescope's primary mirror is the parabolic surface of a uniformly rotating, reflective liquid (in our case, metallic mercury) which develops upon the liquid when it reaches stable equilibrium. Borra, Content, and Boily (1988) have discussed in greater detail,the physical basis behind the operation of an LMT. Borra *et al.* (1989) have also demonstrated that remarkably accurate paraboloidal mirrors can be attained in the laboratory.

Because the symmetry axis of the liquid mirror must remain vertical, the LMT can only be used to observe objects which transit close to the zenith. Lack of mechanical tracking is certainly the instrument's major limitation, but can be minimized by incorporating the drift scan CCD read-out technique (MacKay 1982; McGraw, Cawson, and Keane 1986; Schmidt, Schneider, and Gunn 1986). Drift-scanning offers the additional advantage of improving the photometric accuracy of CCD detectors. Longer integrations can be accumulated by co-adding nightly observations.

Table I lists the principle engineering and detector specifics for the UBC/Laval 2.7m LMT. These data will be elaborated upon in an upcoming paper (Gibson *et al.* 1990).

TABLE I Operational Parameters

Parameter	Value
Latitude	$+49^{\circ}.06$
Seeing (FWHM)	$2''.5$
Sky Brightness	18.0 V^{mag}/arcsec^2
Mirror Diameter	265 cm
Effective Mirror Area	53898 cm^2
Focal Length	500 cm
Focal Ratio	1.89
Mirror Angular Velocity	9.45 rpm
Detector	GEC 385×576 CCD
Read Noise	10 e^-
Quantum Efficiency (5000Å)	40%
Pixel Size	22×22 μm
Pixels Subtend	$0''.908$
Detector Width (NS)	$5'.82$
Detector Height (EW)	$8'.71$
Plate Scale	41.27 arcsec/mm
Integration Time/Galaxy/Night	53.17 sec
Drift Scan Rate of CCD	9.83 arcsec/sec
Survey Area	22.88 deg^2
Data Capture Rate	8.34 kB/sec $\longrightarrow$ 30.01 MB/hr

OBSERVING PROGRAM

The primary observing program for this 2.7m prototype is a year-long narrowband photometric galaxy redshift survey utlizing a series of up to fifty 100Å filters. With a SNR of ten, there is the opportunity to obtain photometric data capable of determining redshifts for approximately 1500 galaxies, for which $V^{mag} < 18.3$, over the course of the year. The goal is to simply explore the viability of the LMT for a much larger redshift survey of this type (eg. $\sim 10^5$ galaxies).

Similar spectrophotometric galaxy surveys have been discussed recently by Koo (1985), and by Loh and Spillar (1986). The application of such a multicolor survey to investigate the evolution of galaxies is also well-documented (eg. Couch *et al.* (1983); Fiala and Rakos (1986)).

Table II briefly lists some of the expected performance estimates related to this prototype's observing program.

TABLE II Narrowband Photometry Performance Estimates

Parameter	Value
Number of Filters	50
Filter Bandwidth	100 Å
Filter Transmission	50%
Typical Galaxy Diameter	5″
Nightly Integrated Limiting Galaxy V^{mag} (SNR=10)	18.3
Number of Galaxies in Survey Area with $V^{mag} < 18.3$	1500

SUMMARY

Upon completion of the UBC/Laval 2.7m prototype LMT in late-1989, the instrument will begin a year-long observing program to accumulate photometric galaxy redshifts and investigate the effects of galaxy evolution on the sample.

While this, and many other projects are available to a telescope of this type, the primary intent of this work is to simply examine the LMT concept, and to determine its viability as a cosmological tool before the development of larger LMTs proceeds.

REFERENCES

Borra, E.F., Content, R, and Boily, E. 1988, Pub. A.S.P., 100, 1399.

Borra, E.F., Content, R., Drinkwater, M.J., and Szapiel, S. 1989, Ap.J. (Letters), in press.

Couch, W.J., Ellis, R.S., Godwin, J., and Carter, D. 1983, M.N.R.A.S., 205, 1287.

Fiala, N., Rakos, K.D., and Stockton, A. 1986, Pub. A.S.P., 98, 70.

Gibson, B.K., Hickson, P., Borra, E.F., and Content, R. 1990, in preparation.

Koo, D.C. 1985, A.J., 90, 418.

Loh, E.D. and Spillar, E.J. 1986, Ap.J., 303, 154.

MacKay, C.D. 1982, SPIE, 331, 146.

McGraw, J.T., Cawson, M.G.M., and Keane, M.J. 1986, SPIE, 627, 60.

Olsson-Steel, D. 1986, J.R.A.S. Canada, 80, 128.

Schmidt, M., Schneider, D.P., and Gunn, J.E. 1986, Ap.J., 310, 518.

THE EVOLUTION OF FIELD GALAXIES: IS $\Omega = 1$?

DAVID C. KOO
University of California Observatories/Lick Observatory,
Board of Studies in Astronomy and Astrophysics,
University of California, Santa Cruz, CA 95064 U.S.A.

ABSTRACT Recent optical observations of the counts, colors, and redshifts of very faint field galaxies have forced a major turning point in the study of the evolution of faint galaxies. In particular, independent counts of faint galaxies are so high that low density (low Ω) universes are strongly favored with reasonable models of the luminosity evolution of distant galaxies. Arguments are made that the bulk of the faint galaxies are not due to extremely low-luminosity dwarf galaxies, despite our relatively poor knowledge of their local number density. The major conclusion is that a high density ($\Omega = 1$) Friedmann universe can be salvaged only if the assumption of the conservation of co-moving volume number density of galaxies is dropped, presumably because of significant merging of luminous galaxies at high redshifts. We explore the effects on number counts and redshift distributions of several simple parametric descriptions of the evolution of number density, luminosity, and shape of the faint end of the luminosity function. Existing redshift surveys are already deep enough to provide powerful constraints on feasible values of these parameters in a high-density universe.

INTRODUCTION

This paper will deal with field galaxies. Since the galaxy formation epoch and process is likely to differ for dissimilar objects, there is danger in assuming that the redshifts and star formation histories of unusual objects are representative of galaxies in general. Traditionally, such peculiar objects as powerful radio galaxies, quasars and N galaxies, central cD cluster galaxies, and perhaps QSO absorption lines, etc., being the only objects observable to redshifts z beyond a few tenths, were used as constraints on the history of ordinary galaxies. With the advent of CCDs and multiaperture spectroscopy, however, more common objects are being surveyed to cosmologically interesting limits.

This presentation will highlight some recent observational CCD results that suggest the need for a major turning point in the study of the evolution of galaxies. The currently fashionable view, especially among physicists, is that the universe consists of roughly 90% non-baryonic matter (many prefer cold dark matter - CDM) and 10% baryonic matter so as to make a flat universe, i.e. one with a deceleration parameter $q_o = 1/2$ or a ratio of mass density to closure density, Ω, exactly equal to 1. Friedmann (pressureless) universes with the cosmological constant $\Lambda = 0$ will be assumed, of course, as well as a Hubble expansion constant , H_o, equal to 50 km/sec/Mpc, that yield universe ages ranging from 13 Gyr to 20 Gyr as Ω is varied from 1 to nearly 0. A value of H_o closer to 100 km/sec/Mpc and $\Omega = 1$ would give a universe age of only 6.5 Gyr, much shorter than the 13 Gyr to 20 Gyr estimates based on fitting the turnoff of the main sequence of observed globular clusters, and thus require, for consistency, acceptance of modifications to stellar evolution codes; of a non-zero Λ for our universe; or perhaps of a new cosmology.

After a brief summary of conclusions from prior research using photographic photometry, we then argue, using assumptions that are largely independent of the details of galaxy evolution models, that faint galaxies are so numerous that an open, low Ω universe is needed to provide enough volume. On the other hand, *if the currently popular* $\Omega = 1$ *cosmology is assumed* , also known as an Einstein-de Sitter model, then we are forced to relinquish three key assumptions of the last generation of galaxy evolution models: 1) that the co-moving volume density of galaxies is conserved with time after the epoch of galaxy formation; 2) that the epoch of galaxy formation occurred essentially at a single brief period of time at high redshifts with no subsequent merging of objects; and 3) that galaxies of similar color or type today have had statistically similar star-formation histories. These three simplifying and interrelated assumptions were necessary to constrain the models by local observations (first and third assumptions) and to constrain the number of free parameters for the second.

To emphasize the robustness of our conclusion, we provide a variety of evidence against the possibility that the excess of faint counts can be explained by a poorly understood population of nearby dwarf galaxies. Then, to sketch the amount of merging needed to salvage an $\Omega = 1$ universe, we will use some parametric descriptions of very simple evolutionary changes to the luminosity function of galaxies to show how faint counts are affected. Finally, we will touch upon the controversy over the redshifts of faint galaxies, some of the results of our redshift surveys, and some suggestions on future approaches to the study of the nature and evolution of faint field galaxies.

COUNTS AND COLORS OF FAINT FIELD GALAXIES: PRIOR WORK

For over a decade, a major arena for the study of galaxy evolution and cosmology has been counts and colors of faint galaxies using 4m class telescopes and fine grain emulsion plates (see summaries by Shanks *et al.* 1984 and Ellis 1988). Based only on counts in the blue to depths of $\approx$ 24th mag, any of the following scenarios could be supported to explain why the observed number of galaxies were in excess of predictions by models which included no evolution, i.e. no changes in the distribution, luminosity, color, size, dynamics, etc. of galaxies in distant volumes of space from those seen locally today:

- The favored view, largely based on the pioneering work of B. Tinsley, was that some galaxies were more luminous in the past, observable to greater redshifts on average, and thus appearing in relatively larger numbers as the depth of observations reached cosmologically interesting limits ($B \gtrsim 20$ mag when typical L* galaxies are beyond z = 0.1). Comparisons of such models and observations generally favored low density (i.e. open, low Ω) cosmologies, but these conclusions were generally accepted as being model dependent to the depth of 24th mag (Bruzual and Kron 1980, Shanks *et al.* 1984, King and Ellis 1985, Koo 1986).

- Galaxies were simply more numerous in the past with no need to invoke luminosity evolution (Koo 1986). Although the required $(1+z)^{1.5}$ increase in comoving volume density, assuming a low density universe, is not totally unreasonable (especially if mergers play a substantial role in the life of most galaxies), this idea had much poorer theoretical underpinnings in the late 1970's and early 1980's than e.g. the luminosity evolution models.

- An alternative explanation was that our local observations of the luminosity function of very blue galaxies were so poor, that a low luminosity, low redshift population of such objects could be dominating over more ordinary galaxies in the faint counts (Kron 1982). This would be a natural result of such galaxies having small or even negative K corrections due to their blue colors; a steep Euclidian rise in counts towards fainter magnitudes since their redshifts are too small to be dominated by cosmological effects; and the need of observations to reach low surface brightness for detection at all.

- Finally, of course, skeptics could raise arguments and problems based on the likeliness that the counts were affected by superclustering; that the galaxy evolution models were inadequate (always true at some level); that the difference of factors of two among counts of various groups were no larger than the expected differences due to evolution; that errors existed in data analysis or calibrations; etc.

Supplementing the blue photographic counts in a single band with additional bands in the ultraviolet (U), red (R), and near-infrared (I) helped somewhat, largely because intrinsic colors and certain ranges of redshifts could be roughly estimated. For example, models with only simple number density increase with lookback time, applied uniformly to all galaxy types, predict far too few blue galaxies; instead, the data suggest an increase in intrinsically very-blue galaxies undergoing active star formation starting at moderate redshifts of only a few tenths (Koo 1986), a view that is gaining more recent support based on faint redshift surveys (Broadhurst *et al.* 1988, Colless *et al.* 1989, Koo and Kron 1988a). The conclusions using more than two passbands were otherwise qualitatively similar to, but strengthened substantially, that from counts alone.

CCD COUNTS AND COLORS OF FAINT FIELD GALAXIES

The advent of CCDs have had a major impact in the field, not only in allowing genuine spectroscopic redshifts to be acquired for galaxies fainter than $B = 20$ mag, but also in vastly improving the reliability of calibrations and accuracy of photometry to faint limits. Such CCD photometry data of faint field galaxies have led to some interesting results. Using multicolor estimates of field-galaxy redshifts to $z \approx 0.8$, Loh and Spillar (1986), for example, conclude that the universe has $\Omega \approx 1$, with little or no luminosity and color evolution to quite faint magnitudes in I of 22nd mag and redshifts $z \approx 0.8$. Tyson (1988a) has pushed the CCD imaging to depths of 27th mag and find colors and counts consistent with substantial luminosity evolution in an open universe, thus strengthening the conclusions based on plates alone. Such CCD imaging also places improved constraints on primeval galaxies, extragalactic optical light, and cosmology in general. Other deep CCD photometry surveys are also underway by Cowie *et al.* (1989) and Metcalfe *et al.* (1987).

Our work in this area has largely been an extension of our multicolor surveys, with concentration in improving the depth of our U and I band data and confronting them against our Bruzual galaxy evolution models. In a project that is a collaboration with R. Kron and S. Majewski, U band photometry has been extended from ≈23rd mag using plates to over 25th mag using CCD's (Majewski 1988, 1989). Such U data provide powerful constraints on the presence of high redshift galaxies, galaxy formation, and cosmology, since $z > 3$ galaxies would be identified by having a very low U flux as the Lyman continuum break at 912Å passes through the band. To $B \approx 24$ mag and within a 10 arcmin^2 field, no galaxies were found that were invisible on the deep U CCD frames, though several were relatively faint. We thus conclude that few galaxies to $B \approx 24$, perhaps a few hundred per deg^2 or less than ≈1% of all galaxies, have redshifts beyond $z \gtrsim 3$. Since the forest of narrow Lyman alpha lines and damped Lyman alpha lines rapidly increase the absorption blueward of rest-frame 1216Å in the continuum of background objects by this redshift, all objects with $z \gtrsim 3$ are expected to appear faint in the blue and ultraviolet, even if an object's intrinsic Lyman continuum break is weak or absent. This conclusion can be contrasted to the claim by Cowie (1989) for about 30% (≈25,000 deg^2) of all galaxies to $I = 24.5$ being at these redshifts, on the basis of 10 out of 33 galaxies with significant drops in U flux in an area of ≈1.3 arcmin^2. Cowie and Lilly (1989) have even measured a spectroscopic redshift of $z = 3.38$ for one such candidate in another field. As seen later in Figure 5, our own Bruzual-type models show that for galaxies brighter than B ≈24th mag, the most probable redshift is less than $z = 0.7$, with surface density about 5 arcmin^2, versus the highly improbable redshifts $z > 3$. Thus the emission line found by Cowie and Lilly (1989) at 5333Å would, on the basis of our models, most likely be [OII] 3727Å at redshift z = 0.43 rather than Lyman alpha 1216Å at z = 3.38.

ARE FAINT BLUE GALAXIES AT Z > 1 ?

Are some of the $B \sim 24$mag galaxies even at high redshift, say z > 1 or are they exclusively at low to moderate redshifts? Taking a slightly different approach from that of Cowie (1988, 1989) or Cowie *et al.* (1988), we show how I–band data is also useful in giving us clues on the redshifts of faint galaxies. We rely heavily on the galaxy evolution models of Bruzual (1983), since empirical color-redshift relations to high redshifts do not yet exist. Our own blue photometry from plates is supplemented with the deep red and near-infrared CCD data from Hall and Mackay (1984).

The key diagnostics are the use of a long baseline color, B-I, and the reliance on the fact that very blue B-I colors associated with active star formation are expected to be visible mainly for redshifts z > 1, when the restframe ultraviolet enters the optical bands. This can best be visualized in Figure 1a, which shows the predicted $J-I$ colors versus redshift of a sample of faint galaxies, where J corresponds to standard B , and I is close to the Kron-Cousins I band. In our bandpass system ($UJFN$ - see Koo 1985), flat spectra (constant in f_ν) have $U-J \approx$ -0.76, $J-F \approx +0.16$, and $F-N \approx +0.30$, so objects with $J-I < +0.46$ are bluer than flat. These are improved values over the estimates given by Koo (1989). Such galaxies are not necessarily very young nor undergoing "starbursts" with ages about 100 Myr (Cowie *et al.* 1988) but can be quite old, up to about 2,000 Myr (see Fig. 2 of Bruzual 1986). Figure 1a shows that *not all*, but the vast majority, of model galaxies that have $J-I \lesssim 1$ (i.e. very blue or close to flat) should have redshifts z > 1. To the current limits of our field galaxy redshift surveys z ≈ 0.8, the observed $J-I$ in Figure 1b appears to match the predicted values well. Note the scarcity of objects with $J-I < 1$ and the excellent agreement of the red envelope colors as a function of redshift.

The expected color-mag distribution of fainter galaxies is shown in Figure 1c, where the model includes mild evolution, namely a high redshift of galaxy formation and a low Ω universe. These models also include photometric errors for the J band (the I band is assumed to have relatively small errors) and incompleteness due to the detection algorithms and the plate limits. If the models are reliable, the presence of substantial numbers of *observed* galaxies in Figure 1d with $J-I \lesssim 1$ strongly support the presence of z > 1 galaxies in our sample for $J > 23$. An independent photometric check that many of these very blue galaxies are indeed at redshifts z > 1 is provided in the two-color plots of Figures 2a and 2b. Here we use the red band (KG3) CCD data from Hall and Mackay to form the equivalent of an (R–I) color; the models use a red band (F) that is close to KG3. The galaxies brighter than 22.5 mag are observed spectroscopically to be mainly at redshifts z < 1 (Colless *et al.* 1989) and appear to occupy the low redshift portion (upper-right) of the distribution of galaxies. Among the galaxies with $J-I < 1$, the model distributions show a small fraction with redshifts less than 1 (starred in Figure 2b), and, as expected, they have $F-I$

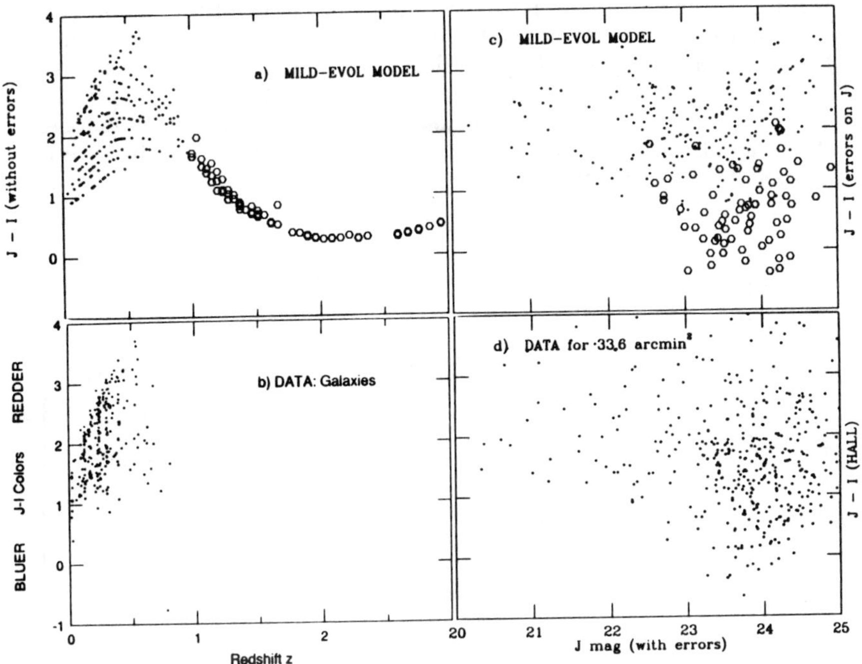

Fig. 1. (a) Plot of $J-I$ colors, without any photometric errors, versus redshift for a sample of model galaxies that should be similar to that detected on sky-limited plates taken with a 4-m telescope over a field of $\sim$34 arcmin^2. The models include evolution of the star formation rate with time and are based on those of Bruzual (1983), with the luminosity function parameters adopted from the work of Koo and Szalay (1984). Circles are those with redshifts greater than one. (b) Plot of observed $J-I$ colors versus redshift for a sample of about 300 faint field galaxies (Koo and Kron 1988a) in Selected Areas 57 and 68. Note the paucity of galaxies with $J-I$ less than 0.8 and the excellent agreement with a) in the red envelope color. c) Plot of model $J-I$ colors versus J mag, where J includes photometric errors, the I-band does not, and the whole sample includes detection incompleteness as measured by Koo (1986). The model includes "mild" evolution, meaning that a low Ω universe is adopted with a very high redshift of formation $z > 5$. (d) Observed $J-I$ colors versus J mag, where the J band was measured from 4-m blue plates and the I-band is from the scanning CCD data of Hall and Mackay (1984). The latter nominally detects objects to $I = 24.4$ at a 4.5 sigma level with standard errors of $\sim$0.2mag at $I = 23$; only objects detected from our plates have been included in the plots.

colors as red as the circled bright galaxies. All the other $J-I < 0.8$ galaxies are predicted to be at higher redshifts. Note that many of these have $F-I$ colors significantly bluer than the lower redshift galaxies. The two-color plot of observed galaxies show such a population of galaxies blue in both $J-I$ and $F-I$, thus lending support to our suggestion that some high-redshift galaxies ($1 < z < 3$) do appear in existing samples of faint field galaxies. Spectroscopy already underway by several groups will hopefully be able to reveal some highly redshifted strong emission lines to confirm this conclusion.

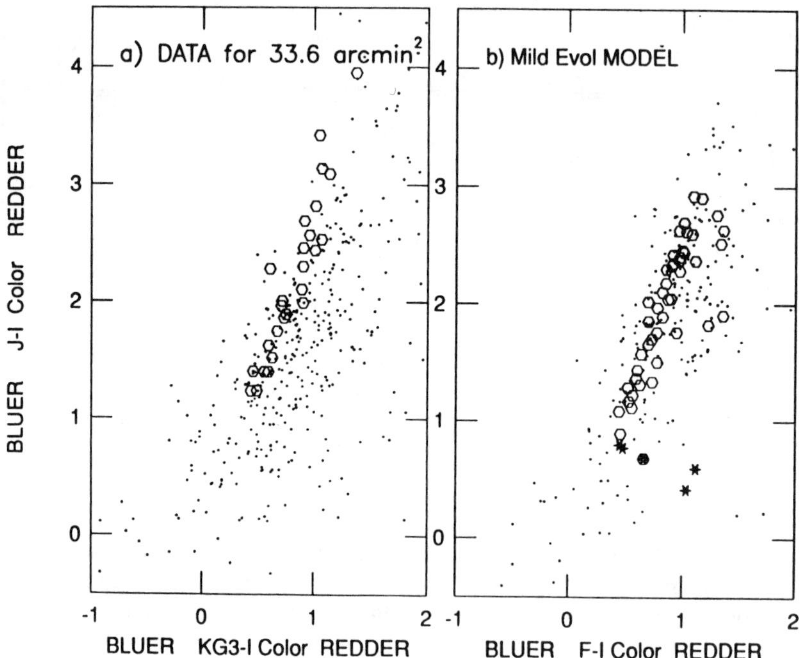

Fig. 2. a) Two-color plot of the galaxy data shown in Figure 1d, where the *KG3* band is the broad red CCD band of Hall and Mackay (1984) and the *J* and *I* are as in Figure 1d. Circled objects are galaxies with *J* between 20 and 22.5 mag. b) Predicted two-plot distribution where *F* is a photographic red band close to KG3. As in Figure 1a and 1c, these models include full photometric errors for the *J* band, one half of the photographic errors for the *F* band to account for the higher precision of the CCD KG3 equivalent, and incompleteness during the original selection of the sample. The starred objects are $J-I < 0.8$ objects with redshifts $z < 1$.

LARGE NUMBER COUNTS AND Ω

The main point of this paper is the strong constraint on the cosmological density parameter Ω now provided by various deep CCD counts.

Such counts serve as a volume test of cosmology, under the assumption that the number of galaxies per comoving volume has largely been conserved. To the extent that luminosity evolution has a relatively small effect on the total number of galaxies after the limiting depth of the survey is fainter than the break of the luminosity function of the most distant galaxies, this test is robust against model assumptions. The key word is limiting depth, for until CCDs exceeded $B \approx 24$th mag limit of plates, both low and high density universes could reproduce the observed counts. Thus conclusions relating the geometry of the universe to counts were indeed model dependent.

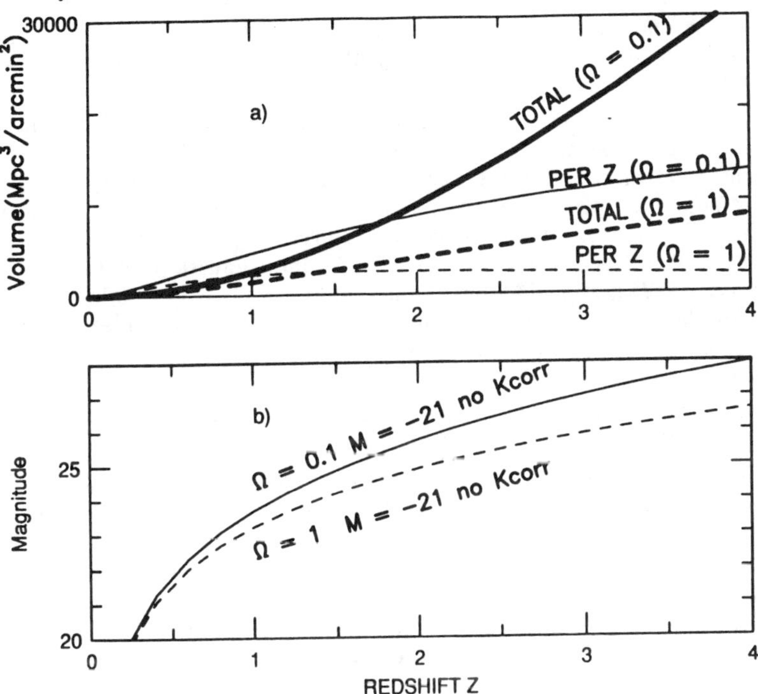

Fig. 3. a) Integral and differential co-moving volume per arcmin^2 in Mpc^3 for two different values of Ω, the density parameter in a Friedmann universe. A value of H_o = 50 km/sec/Mpc has been adopted. b) Apparent magnitude of an object with absolute magnitude M = −21 versus redshift, assuming no K corrections, for two different values of Ω.

What are the expectations and actual observations? Let us start by examining the available volumes shown in Figure 3a. For one square arcmin, the total available volume to z = 1, z = 3 (U Lyman limit), and z = 4 (B Lyman limit) assuming $\Omega = 1$ is 1200 Mpc^3, 6000 Mpc^3, and 8000 Mpc^3, respectively (Ho = 50 km s^{-1} Mpc^{-1}). This should be contrasted with substantially larger volumes of 2000 Mpc^3, 18,600 Mpc^3, and 30,800 Mpc^3, respectively, if the universe is open with $\Omega = 0.1$. As a rough estimate of the number of observable galaxies, we can assume an average observable number density of 0.0015 galaxies per Mpc^3. This is probably an overestimate, considering that the Anglo Australian Redshift Survey at $B \approx 17$ detected half this (Efstathiou *et al.* 1988). Yet, in a universe with $\Omega = 1$, only 9 galaxies per arcmin^2 are expected in the U band and 12 in the B . In contrast, in an open universe with $\Omega = 0.1$, 3x higher densities of 28 to 46 are expected. The raw observed galaxy counts shown in Figure 4a in U and B both peak at 10^5 per mag per deg^2 which corresponds to about 28 galaxies per arcmin^2; additional corrections to the counts due to overlapping objects are only likely to increase this number, perhaps substantially (see Tyson 1988a).

How can we escape the conclusion that the universe does NOT have $\Omega = 1$? Is there some way that the luminosity function of galaxies can evolve which satisfies both the observations and the desire to keep $\Omega = 1$? As a crude estimate of various possible evolutionary scenarios, let us evolve a Schechter-type luminosity function by its three major free parameters, volume density Φ (local value around 0.0015); faint end slope α (local value about -1.25); and M* (local value about -21). K corrections to account for the change in the amount of flux through a given filter as an object's spectrum is redshifted will be assumed to take on the simple form of $-2.5(1 + k)\log_{10}(1+z)$, where k is the power index of the spectrum, where f_ν is proportional to ν^k. Thus a spectrum flat in f_ν has $k = 0$; flat in f_λ has $k = -2$; and QSO-like spectra that give no K corrections have $k = -1$.

First note that luminosity evolution, the traditional physical effect to explain larger number counts, does not help. As seen in Figure 4b or in Fig. 8 of Yoshii and Takahara (1988), the counts in an $\Omega = 1$ universe peaks at about dex 4.7 per deg^2 or 15 galaxies per arcmin^2, regardless of luminosity evolution. (Note that differential counts *per mag* are approximately equal to the integral counts at dex about 0.43, which is coincidentally close to that observed.) Evolution merely pushes the peak to brighter apparent magnitudes rather than higher counts.

Moreover, as argued by Broadhurst *et al.* (1988), the mean redshift of existing redshift surveys to $B \approx 21$ mag already begin to constrain the amount and form of luminosity evolution. For example, compared to the observed mean redshift of about z = 0.2, the best fitting $\Omega = 1$ model (k = 0 in Figure 4b) yields a mean redshift of 0.37 while the $\Omega = 0.1$ model gives a mean redshift of 0.6; both of these are unacceptably high. This does NOT mean that luminosity evolution models are thus excluded as a class. It is true that without major revisions to the luminosity function of galaxies, Tinsley-Bruzual type models that assume $\Omega = 1$ cannot yield enough galaxies to predict a steep count slope that continues from 20th to 25th mag; the amount of luminosity evolution needed to add enough low-luminosity galaxies to the faintest counts would force too many intrinsically brighter galaxies into view at brighter magnitudes. Of course, one could always drop the key simplifying assumption that the low-luminosity galaxies behave in tandem with the high-luminosity galaxies, i.e. one then needs some form of luminosity evolution that depends on mass, a possibility we will address below. Low Ω models,

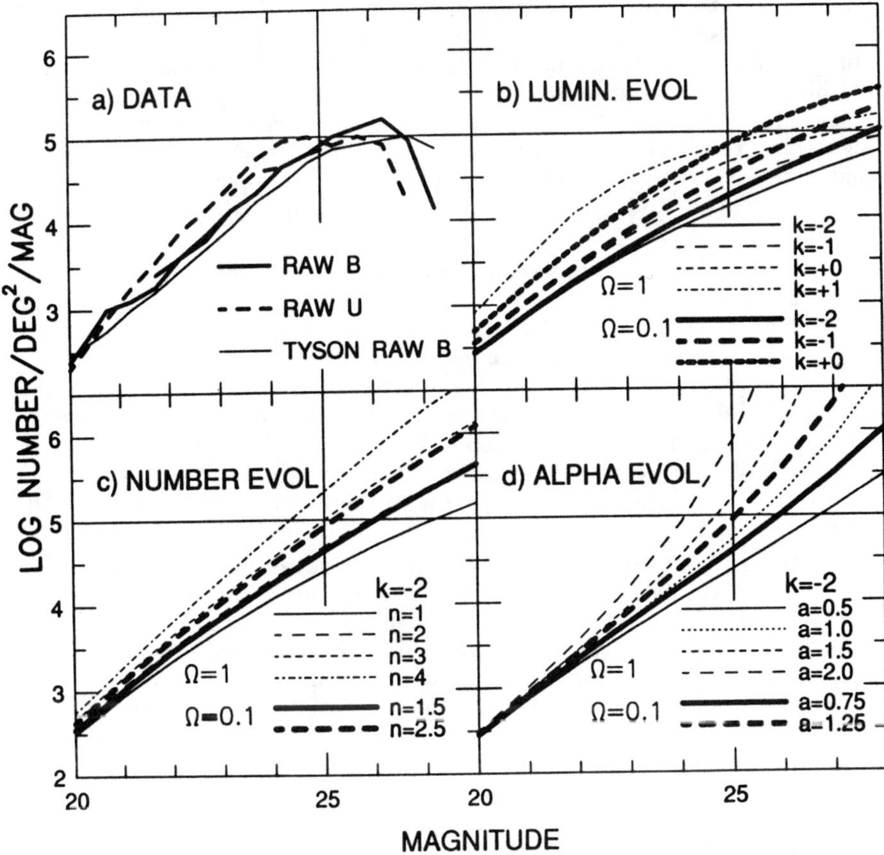

Fig. 4. a) Observed number counts of field galaxies versus apparent magnitude. The raw U and B counts at the bright end are based on 4-m prime focus blue plates taken in SA 57 and SA 68 (Koo 1986). The raw B counts at the faint end were measured by S. Majewski with the FOCAS software package at AT&T using a stack of several such plates in SA 57. The raw U counts at the faint end were measured from 5 hour CCD exposures taken with the 4-m at Kitt Peak National Observatory. The Tyson raw B counts are from Tyson (1988a). The faint raw B counts from plates are competitive with the Tyson CCD counts because the plates had better sampling (0.37 versus 0.67 arcsec). b) Predicted counts assuming a standard Schechter luminosity function with $\Phi^* = 0.0015$, $\alpha = -1.25$, and $M^* = -21$, and various choices for K correction term, k, and Ω, as indicated. See text for meaning of k, which effectively introduces luminosity evolution for values larger than -2 to -3. c) Predicted number counts as in b) except that k is locked to -2 and number density evolution of the form $(1+z)^n$ is added. d) Predicted counts as in b) except that k is locked to -2 and the faint end slope α is multiplied by $(1+z)^a$.

however, can be made to fit the observations within the constraints of all the simplifying assumptions, largely because of the greater volumes accessible. For good fits to the colors and counts, luminosity evolution is slight at low redshifts and the assumed bright phase of early-type (elliptical and large-bulge spirals) galaxies at a single epoch of galaxy formation is introduced at high redshifts $z > 5$. These models are best constrained by redshift surveys that push just beyond the current areas, depth, or completeness.

Number density evolution, i.e. larger values in the past, perhaps as the result of mergers, is another possible explanation. On average, then, each galaxy today would represent the merging of two, and preferably more, galaxies at redshifts z from 1 to 4. To demonstrate the amount of evolution in number density needed to match observations, Figure 4c shows the effects for various values of n, where the density is increased by $(1+z)^n$. With an $\Omega = 1$ universe and $k = -2$, n has be be approx 2 to 3 to yield a count slope close to the observed value of 0.45, without simultaneously predicting too many galaxies at brighter magnitudes. With $\Omega = 0.1$, n needs to be 1.5 to 2.5. In all of these cases, the mean redshift of galaxies at 21st mag is about $z = 0.25$ to 0.27, which is acceptably close to the measured value near $z = 0.23$ by Broadhurst *et al.* (1988).

Recently Broadhurst *et al.* (1988) have used results from their $B = 20$ to 21.5 mag field galaxy redshift survey to argue for evolution of density in the luminosity function of galaxies that depends on luminosity. This form is approximately the same as evolving mainly the faint end slope α. For simplicity, we adopt an evolution in α of the form $\alpha(1+z)^a$, so that positive values of a result in larger numbers of faint galaxies at higher redshifts. As seen in Figure 4d, one needs values of a of 1 to 1.25, almost independent of Ω, to yield good fits to the observed counts. The resultant mean redshift of 21st mag galaxies of $z \approx 0.21$ is close to that observed.

In summary, the above discussion using the number count information in Figure 4 and the mean redshift observed in 21st mag galaxies, shows that unless galaxies were more numerous in the past or the shape of the faint end of the luminosity function was significantly different, an $\Omega = 1$ universe provides too little volume to redshifts z of 3 to 4 to explain the large counts of observed galaxies, even after including large increases in brightness of all distant galaxies. As discussed in some detail by Broadhurst *et al.* (1988), a change in the slope α may be a natural consequence of bursts of star formation. The number density increase, on the other hand, is most easily explained by merging of galaxies.

Today the theoretical basis for believing in substantial mergers is gaining strength, and as these models are more realistically transformed to observables (White 1989), we should be in a better position to evaluate whether mergers remain a viable explanation, without violating other constraints, such as the observed luminosity function of galaxies, especially at the faint end, today or in the recent past. Up to now, galaxy evolution models have always presumed number conservation of galaxies, since the strategy among those working with stellar population models, was to keep free parameters to a minimum. Free parameters were those which cannot be directly constrained by independent methods or observations. If more mundane explanations based on other evidence are not found, including the simplest one of a low Ω universe, new and presumably more complicated models that include dynamics and merging, may be needed to explain the observed counts, colors, and redshifts of faint galaxies.

ARE FAINT BLUE GALAXIES MAINLY LOCAL DWARFS?

An important source of skepticism regarding the measurement of q_o from deep galaxy counts is the concern that the numerous faint objects are merely nearby low-luminosity very-blue dwarfs. In other words, the adopted luminosity functions are not accurate at the faint end. Until redshifts are acquired, proof one way or the other would be difficult, but a variety of arguments favor the view that most of the very blue objects at $B > 24$ are indeed at high redshifts, typically at $z \gtrsim 1$. These include colors, observed luminosity functions of local dwarfs, surface brightness and size, and even some redshift data.

Colors: As we had argued before, galaxies with $J\!-\!I \lesssim 0.6$ are naturally explained by galaxies undergoing active star formation at redshifts $z \gtrsim 1$. Local dwarfs, even those with extremely blue colors, are usually not this blue, as can be seen in Figure 1 for NGC4449 in Koo (1989); this can also be seen in Figure 1b, in which the bluest low-redshift galaxies have $J\!-\!I \gtrsim 0.8$ (corresponds to $B-V \approx 0.1$) and redder such colors up to redshifts of $z \approx 0.5$. Attempts of Tyson and Gullixson (as reported by Tyson 1988b) to measure the colors of local dwarfs indeed yield colors close to $J\!-\!I \approx 1$. The trend of bluer colors with fainter magnitudes is also contrary to expectations if we are viewing dwarfs towards greater redshifts.

Luminosity Function: As reported by Sandage *et al.* (1985) or reviewed by Binggeli *et al.* (1988), the late type dwarfs in the Virgo cluster, which we might imagine are the local counterparts of the presumed dwarfs in the faint counts, actually exhibit a peak and then a drop in their luminosity function fainter than $M_V \approx -17$; in other words, there is no evidence of a continual increase of intrinsically fainter very-blue galaxies (the increase at faint magnitudes is associated with the considerably redder dwarf spheroidals). There is no evidence that this behavior of the luminosity function is unique to Virgo, since Ferguson and Sandage (1988) find similar results in Fornax. Searches by Binggeli *et al.* (1985) or Phillipps *et al.* (1987) for low surface brightness dwarfs outside the high density regions in the Virgo or Fornax clusters yield few such objects (it is worth noting the isotropy of the faint blue counts to within 10% observed by Tyson 1988a, which would be unexpected if these were low redshift dwarfs that were as strongly clustered as those in Virgo or Fornax). Moreover, recent evaluations of the luminosity functions by Efstathiou *et al.* (1988) and Phillipps and Shanks (1987) not only show a flatter ($\alpha \approx -1$) slope for the faint end than the standard value ($\alpha = -1.25$), but also exhibit error bars that are small to 3 mag fainter than the break. Thus any unusual shape to the luminosity function would have to occur fainter than this, corresponding to $M_V > -18$, or $z < 0.1$ in a $B = 21$ magnitude limited survey. As discussed in the next section, a large population of such galaxies is not found in faint redshift surveys. Finally, as noted by Tyson (1988a), even the adoption of a Virgo density worth of $M_J = -16$ galaxies yields a steep (dex = 0.58) count slope from $B \approx 22$ to 27 (a result that supports the possibility of low luminosity galaxies to dominate faint counts) but such a slope is too steep to match the observed slope of dex = 0.45.

Redshifts: As just argued above, very low luminosity objects of very blue colors should yield a Euclidian slope (i.e. dex = 0.6). If indeed the faintest blue objects are mainly such dwarfs, we can lock their contribution to fractions of such galaxies in completed redshift surveys. For example, if most of the faint galaxies at 25th mag are blue low-luminosity dwarfs, we would expect about 25% to 30% of the galaxies in the Durham redshift survey at $B = 20$ to 21.5 to be such objects; the observed results include 6 out of 170 galaxies with $z < 0.1$ and none with $z < 0.03$, which is fully consistent (Broadhurst *et al.* 1988) with extrapolations of the flat faint-end luminosity function as derived by Efstathiou *et al.* (1988). Broadhurst *et al.* (1988) also explored the consequences of adopting much steeper slopes at the faint end of the luminosity function and conclude that results from such modifications are inconsistent with their estimates of the mean redshift. In the models used for my thesis, I actually included a substantial low-luminosity very-blue component (about 3 mag fainter than most galaxies), to force no evolution models to fit the $B \approx 20$ mag counts and colors; such models, however, predict 47 galaxies of our own redshift sample (Koo and Kron 1988a) to have $z \lesssim 0.15$; actually only 7 were seen and most of these were part of the outskirts of the Coma cluster rather than at random redshifts.

Size and Surface Brightness: At sufficient signal to noise to tell the difference, virtually all objects fainter than $B \approx 22$ remain resolved or extended relative to stars; moreover, this extension continues to fainter limits. Since the sizes of the dwarfs of interest should be $\lesssim$1 kpc (Binggeli et al 1985, Phillipps *et al.* 1987), most should rapidly shrink to less than the seeing size as they are observed to redshifts of a few hundredths (at the distance of Coma at $z \approx 0.025$, a 25th mag object would have an intrinsic mag of only −11, and a size of 1.4 arcsec per kpc). Since most of the very-blue stellar-like *galaxies* found in our QSO survey (Koo and Kron 1988b) to $B \approx 22.5$ mag limits have redshifts of a few tenths rather than a few hundredths (and luminosities of typical galaxies, not dwarfs), this redshift survey constrains the possibility that any such dwarfs are hiding in numerous numbers as compact objects.

One remaining possibility is that the dwarf spheroidals today, the only galaxy population with sufficiently high volume density to explain the high counts, were once Im galaxies, and thus brighter and bluer. Pros and cons of this view have recently been mentioned by Ichikawa *et al.* (1988). As was pointed out to me by S. van den Bergh, the presence of both carbon stars indicating recent star formation and older stars such as RR Lyraes indicate that star formation probably did not occur in a single bright event, which is probably what is needed to make any high density of extremely faint spheroidals visible at all at moderately high redshifts. Without dramatic brightening, these galaxies would be viewed at low redshifts, small volumes (see Figure 3), and thus still not be seen in high numbers.

FIELD GALAXY SPECTRA

Although photometric surveys may be very powerful diagnostics of the nature and perhaps distances of faint galaxies, redshifts, perhaps for only a small subsample, are ultimately needed for reliable confirmation. Unfortunately, redshifts are considerably more difficult to measure than broadband colors, unless a large fraction of the flux is being radiated in strong emission lines. To date, several redshift surveys fainter than ~20mag (the depth needed to reach several tenths in redshift and thus a reasonable fraction of the age of the Universe) have been made, including those of the Durham group (see Broadhurst *et al.* 1988; Colless *et al.* 1989 and contribution by Ellis to these proceedings), a group working with the 6m in the USSR (V. Afanasiev, private communication), and ourselves (Koo and Kron 1988a). All three groups are aiming for limits of $B \approx 23$; so far, none of the surveys have achieved high completeness ($\gtrsim$95%) to this limit, but the Colless *et al.* (1989) survey is close to this goal with about 80% completeness for a sample of about 150 objects with B between 21 and 22.5, using their new and very impressive low dispersion survey spectrograph system (LDSS).

Here I would like to mention a few of the consistent results of the Durham (Broadhurst *et. al.* 1988, Colless *et al.* 1989) and our own very deep field galaxy redshift surveys. One pleasant surprise is that both groups agree on the lack of very luminous galaxies at moderately high redshifts of z = 0.5 to 1 or greater in the samples. This result is becoming more of a challenge to some luminosity evolution models as the redshift surveys reach even fainter limits, as discussed in more detail by Colless *et al.* (1989), but as shown in Figure 5, our Bruzual-type evolutionary models remain viable descriptions at current limits. The existing surveys are still not sufficiently extensive, complete, or deep to rule out even these relatively old models dating back nearly a decade (Bruzual and Kron 1980, Bruzual 1981).

We also agree that a substantial fraction of the faint galaxies have [OII] 3727 emission line equivalent widths that are unusually strong when compared to lower redshift samples (Broadhurst *et al.* 1988), but we have not made a quantitative estimate of the statistical significance from our data alone. Further evidence for relatively more extensive star formation in the recent past comes from the large fraction of our faint galaxies with redshifts beyond z = 0.2 which have observed colors corresponding to Sdm or Irr galaxies (see Fig. 2 of Koo and Kron 1988a). One attractive explanation is that many of the blue galaxies are undergoing bursts (Broadhurst *et al.* 1988). Whether this explanation extends to higher redshifts is a crucial problem for observers and theorists alike. As seen in Figure 5, complete samples to B = 23 to 23.5 should resolve the question of whether the Tinsley-Bruzual type of luminosity evolution is, on average, a good description of galaxy properties at high redshift.

Another finding by both groups mentioned above is that few galaxies of low redshift were actually observed. In our survey, the difference between our original models and the data is probably too large to be explained by statistical fluctuations, even if we grant that substantial large scale structure appear in all three of the fields surveyed. Thus our original models (e.g. Koo and Szalay 1984) will require substantial reduction of low-luminosity blue galaxies or that these galaxies must quickly evolve to brighter luminosities by redshifts of only a few tenths. This discrepancy of our models may have arisen because we assumed that the typical galaxies with very blue colors have characteristic luminosities much fainter than redder galaxies, based on the data of Kirshner *et al.* (1979); that galaxies brighter than $B = 21$ have not undergone significant evolution so that the color distributions from our own multicolor data to this magnitude would provide improved normalizations of the relative fractions of galaxies with different types over that provided by Kirshner *et al.* data alone (used as a consistency check); and that the shape of the luminosity function of galaxies with different colors are roughly Schechter-like.

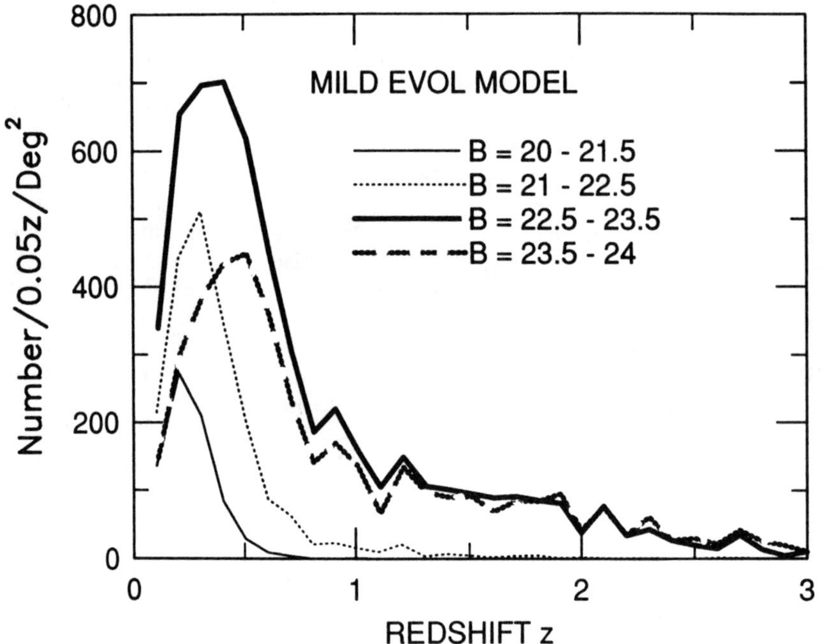

Figure. 5. Redshift distribution for indicated blue (B) magnitude intervals using a mild luminosity evolution model as described in Figure 1. Note that significant numbers of high z > 1 galaxies are expected only at magnitudes fainter than $B = 22.5$.

The first assumption is probably sound, though the actual difference in luminosity remains uncertain until the shape of the luminosity function becomes better defined. The second is being challenged by the aforementioned redshift surveys that all reach at least B $\approx$ 21, which already show a substantial population of very blue galaxies at moderate redshifts of a few tenths rather than very low redshifts much less than $z = 0.1$. The third assumption is probably wrong (see prior discussion of luminosity function of dwarfs), but the effect of inaccurate representations have not been explored in detail. We will be using more powerful techniques in the future to explore viable models (see Bruzual contribution to this conference) and improve our input parameters.

SUMMARY AND FUTURE

We have shown the power of combining counts, colors, and redshifts of very faint galaxies to explore both the geometry of the universe and the birth and evolution of galaxies. A simple model consistent with most of these data is a low density ($\Omega < 1$) universe that initiated galaxy formation at a single epoch at large redshifts $z > 5$, with little or no observable merging with time. Some evidence for more than expected star formation at moderate redshifts ($z \gtrsim 0.3$) is found in the deepest field galaxy redshift surveys. An $\Omega = 1$ Einstein-DeSitter universe can be made consistent with the observations if merging is a common process at redshifts of 1 to 3.

The observational study of galaxy evolution is still very much in its infancy,of course, and we can expect major progress in the near future. The last paragraph of Hubble's *The Realm of the Nebula* remains true: " The search will continue. Not until the empirical resources are exhausted, need we pass on to the dreamy realms of speculation." These resources are far from being exhausted. Besides continuing redshift surveys and searches for primeval galaxies, several technological improvements are bound to have major impacts. Faster computers will allow vastly improved models to compare to data, not only in making more realistic galaxy formation models that allow for the presence of gas, but also in generating much more realistic simulations of two-dimensional images and spectra and in allowing more powerful statistical tests. Larger ground-based optical telescopes and larger CCD's and near-infrared arrays may allow the study of properties of faint galaxies currently too difficult to measure. Ideally, these properties would include mass as measured from velocity dispersions, emission line widths, and even rotation curves or would include metallicity-age estimates using spectral synthesis (Pickles 1989) or Stromgren photometry (Rakos *et al.* 1989). Finally, there is the expected revolution in high resolution imaging using the Hubble Space Telescope and active or adaptive optics, which should provide not only a reliable estimate of the Hubble constant, but also the first clues to evolution in the two other parameters identified with Hubble's name, the structure (Hubble law) and morphology (Hubble sequence) of galaxies. "Look afar and see the end, from the beginning" (from a fortune cookie eaten in 1983).

ACKNOWLEDGEMENTS

Virtually all of the work reported above is based upon data and analysis, some of which is unpublished, shared with my collaborators: R. Kron, G. Bruzual, S. Majewski, J. Munn, and A. Szalay. I would especially like to thank C. Mackay and P. Hall for access to their catalog in SA 57. We are indebted to the staff of Kitt Peak National Observatory and Palomar Observatory for the many years of support needed to undertake these faint surveys. I also had the benefit of helpful discussions with R. Ellis and S. van den Bergh. This work has been partially supported by an NSF Presidential Young Investigator award, AST-8858203.

REFERENCES

Binggeli, B, Sandage, A. , and Tammann, G. A. 1985, *A. J.*, **90**, 1681.

Binggeli, B., Sandage, A., and Tammann, G. A. 1988, *Ann. Rev. Astr. Astrophy.*, **26**, 509.

Broadhurst, T. J., Ellis, R. S., and Shanks, T. 1988, *M.N.R.A.S.*, **232**, 431.

Bruzual, G. A. 1981, Ph. D. thesis, University of California, Berkeley.

———— 1983, *Ap. J.*, **273**, 105.

———— 1986, in proceedings of Erice Workshop on *The Spectral Evolution of Galaxies*, ed. C. Chiosi and A. Renzini, p. 263.

Bruzual, G. A. and Kron, R. G. 1980, *Ap. J.*, **241**, 25.

Colless, M., Ellis, R. S., Taylor, K., and Hook, R. N. 1989, *M.N.R.A.S.*, preprint.

Cowie, L. L. 1988, *The Post-Recombination Universe*, ed. N. Kaiser and A. Lasenby.

———— 1989, *The Epoch of Galaxy Formation*, eds. Frenk, C. S., Ellis, R. S., Shanks, T., Heavens, A. F., and Peacock, J. A. , p. 31.

Cowie, L. L. and Lilly, S. J. 1989, *Ap. J. Letters*, **336**, L41.

Cowie, L. L., Lilly, S. J., Gardner, J., and McLean, I. S. 1988, *Ap. J. Letters*, **332**, L29.

Efstathiou, G., Ellis, R. S., and Peterson, B. A. 1988, *M.N.R.A.S.*, **232**, 431.

Ferguson, H. C. and Sandage, A. 1988, *A. J.*, **96**, 1520.

Hall, P. and Mackay, C. D. 1984, *M.N.R.A.S.*, **210**, 979.

Ichikawa, S., Okamura, S., Kodaira, K., Wakamatsu, K. 1988, *A. J.*, **96**, 62.

King, C. R. and Ellis, R. S. 1985, *Ap. J.*, **288**, 456.

Kirshner, R. P., Oemler, A., and Schechter, P. L. 1978, *A. J.*, **83**, 1549.

Koo, D. C. 1985, *A. J.*, **90**, 418.

———1986, *Ap. J.*, **311**, 651.

———1989, *The Epoch of Galaxy Formation*, eds. Frenk, C. S., Ellis, R. S., Shanks, T., Heavens, A. F., and Peacock, J. A. , p. 71.

Koo, D. C. and Kron, R. G. 1988a, *Towards Understanding Galaxies at Large Redshift*, eds. R. G. Kron and A. Renzini, p. 209.

——————1988b, *Ap. J.*, **325**, 92.

Koo, D. C. and Szalay, A. S. 1984, *Ap. J.*, **282**, 390.

Kron, R. G. 1982, *Vistas Astr.*, **26**, 37.

Loh, E. D. and Spillar, E. J. 1986, *Ap. J. Letters*, **307**, L1.

Majewski, S. R. 1988, *Towards Understanding Galaxies at Large Redshift*, eds. R. G. Kron and A. Renzini, p. 203.

———1989, *The Epoch of Galaxy Formation*, eds. Frenk, C. S., Ellis, R. S., Shanks, T., Heavens, A. F., and Peacock, J. A. , p. 85.

Metcalfe, N., Fong, R., Jones, L. R., and Shanks, T. 1987, *High Redshift and Primeval Galaxies*, eds. J. Bergeron, D. Kunth, B. Rocca-Volmerange, and J. Tran Thanh Van, p. 37.

Phillipps, S., Disney, M. J., Kibblewhite, E. J., and Cawson, M. G. M. 1987, *M.N.R.A.S.*, **229**, 505.

Phillipps, S. and Shanks, T. 1987, *M.N.R.A.S.*, **227**, 115.

Pickles, A. J. 1989, *The Epoch of Galaxy Formation*, eds. Frenk, C. S., Ellis, R. S., Shanks, T., Heavens, A. F., and Peacock, J. A. , p. 191.

Rakos, K. D., Schombert, J. M., and Kreidl, T. J. 1989, preprint.

Sandage, A., Binggeli, B., and Tammann, G. A. 1985, *A. J.*, **90**, 1759.

Shanks, T., Stevenson, P. R. F., Fong, R., and MacGillivray, H. T. 1984, *M.N.R. A.S.*, **206**, 767.

Tyson, J. A. 1988a, *Ap. J.*, **96**, 1.

————-1988b, *Towards Understanding Galaxies at Large Redshift*, eds. R. G. Kron and A. Renzini, p. 187.

White, S. D. M. 1989, *The Epoch of Galaxy Formation*, eds. Frenk, C. S., Ellis, R. S., Shanks, T., Heavens, A. F., and Peacock, J. A., p. 15.

Yoshii, Y. and Takahara, F. 1988, *Ap. J.*, **326**, 1.

K BAND PHOTOMETRY OF A COMPLETE SAMPLE OF FIELD GALAXIES WITH SPECTROSCOPIC REDSHIFTS

MATTHEW BERSHADY, MARK HERELD, AND RICHARD KRON
Department of Astronomy & Astrophysics, University of Chicago, 5640 S. Ellis Avenue, Chicago, IL 60637

DAVID KOO
Lick Observatory, University of California, Natural Science No.2, Santa Cruz, CA 95064

ABSTRACT We present preliminary results from a K band imaging survey of 53 optically selected field galaxies with known redshifts ranging from 0.018 to 0.35. The sample is complete to B < 20.5 and has UBRI photometry. To demonstrate the absolute calibration of our measurements, we report results of a direct photometric comparison of two different near-infrared array detectors. We use the spectroscopic redshifts of our sample to explore the reliability of redshifts measured from broad-band colors. Finally, we illustrate the merits of optical-near-infrared multi-color distributions for studying galaxy evolution. Our results complement deeper near-infrared surveys that do not have spectroscopic redshifts.

DATA

Our sample of field galaxies are a subset of the Koo-Kron spectroscopic redshift survey in three widely separated fields. Near-infrared images were taken with the KPNO IRIM on the Mayall telescope at Kitt Peak and the ARC 2-micron Prototype Camera on the Shane telescope at Lick Observatory. Reduced images have average dark frames subtracted, and are flattened with the median of 5 dark subtracted and normalized, local images. Sky level is determined within annuli outside the photometering aperture in the reduced image. Synthetic apertures are selected to give total magnitudes, and range from 8"-12" in diameter. Other reduction paths, including local sky subtraction and the use of dome flats, yield comparable photometry.

The quoted error in K for each galaxy is the standard deviation in photometry determined from 5 separate 250s integrations with the galaxy moved to widely separated positions on the detector for each exposure. This measure of uncertainty includes sytematics of background subtraction and field flattening, and therefore is fundamentally different than simpy estimating photometric uncertainties from measured pixel-to-pixel variations of sky.

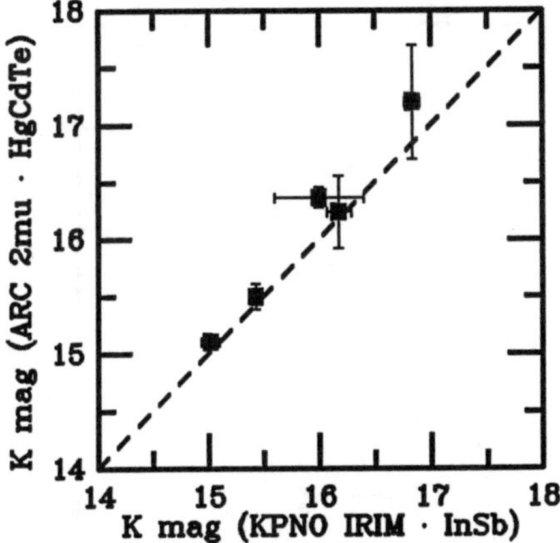

Fig. 1. K band photometry of 5 galaxies imaged on both Rockwell's 64x64 HgCdTe and SBRC's 58x62 InSb near-infrared array detectors demonstrates that these two sytems are photometrically equivalent within the accuracy of the measurements. All 5 galaxies have been reduced and photometered identically for both detectors.

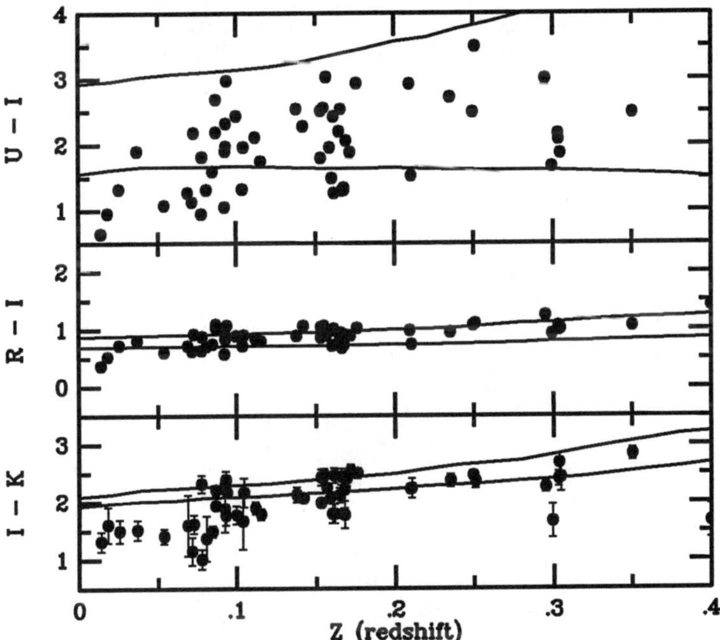

Fig. 2. U-I, R-I, I-K vs redshift are compared to Bruzual models for two different star formation rates with μ = 0.90 (red) and 0.01 (blue) (Ho = 50 km/s, qo = 0, 16 Gyr, Salpeter IMF; Bruzual, G. 1986, in *Spectral Evolution of Galaxies*, ed Chiosi and Renzini, p. 263). Blue objects at low z have optical colors similar to a non-evolving Im galaxy.

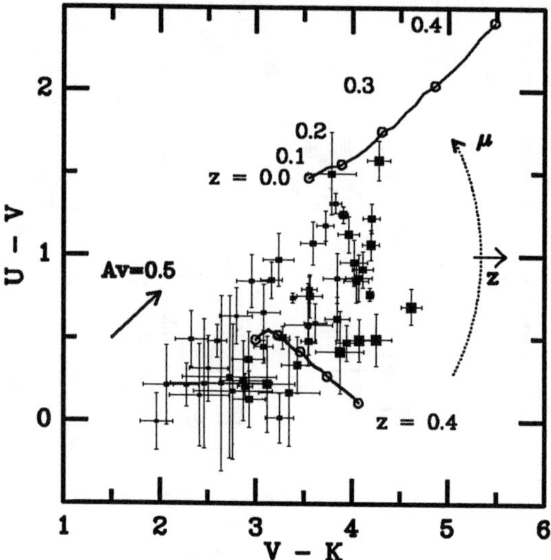

Fig. 3. U-V vs V-K is compared to the same models as in figure 2 (solid curves). Open circles on these curves indicate model redshifts of 0, 0.1, 0.2, 0.3 and 0.4. Box sizes increase with redshift. Galaxies in the UVK plane appear well separated in redshift and SFR.

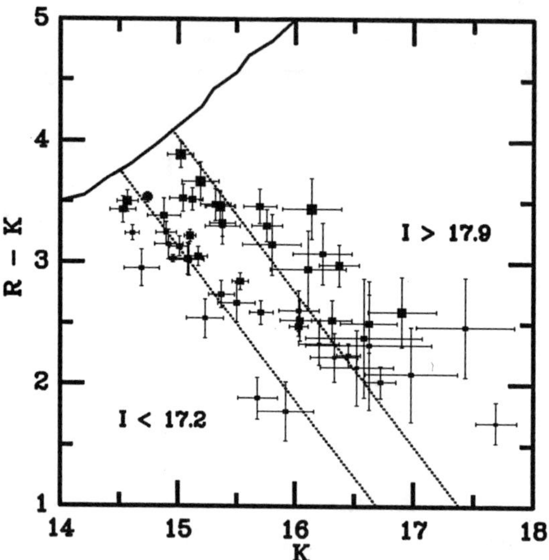

Fig. 4. R-K vs K. Faint galaxies in K with red R-K will only appear in our survey as our sample goes deeper in I, as indicated by the small dispersion in R-I vs I-K. The solid curve represents a non-evolving brightest cluster member (Bruzual, G. 1981, Ph.D. Thesis, University of California, Berkeley).

THE STUDY OF GALAXY EVOLUTION USING A LARGE AREA 2μ SURVEY

Richard Elston
Kitt Peak National Observatory, P.O. Box 26732, Tucson, AZ 85726

George Rieke and Marcia Rieke
Steward Observatory, University of Arizona, Tucson, AZ 85721

ABSTRACT We have surveyed $200 arcmin^2$ of high galactic latitude sky to a 1σ detection limit of $22 mag\ arcsec^{-2}$ at K. This is more than 10 times the area of any previous deep near IR survey. This survey has yielded about 200 galaxies selected by their 2μ emission which will be well suited for studying galaxy evolution at moderate redshift ($z < .8$).

INTRODUCTION

With the recent development of array detectors which operate beyond 1μ, it has become possible for the first time to investigate the sky to very low light levels in the near IR. This has stimulated several deep surveys of the sky at K (Elston 1988, Cowie *et al.* 1988). These surveys are limited since they cover very small areas. To correct this situation, we have surveyed $200 arcmin^2$ of high latitude sky to a 1σ detection limit of $22 mag\ arcsec^{-2}$ at K. Additionally, we have surveyed an area of $30 arcmin^2$ to a 1σ detection limit of about $23 mag\ arcsec^{-2}$ at K. The objectives of this survey are two fold. First, this survey has yielded about 200 galaxies selected by their K band flux. These galaxies should represent a homogeneous sample of galaxies well suited for the study of galaxy evolution. Secondly, the previous surveys found some very unusual objects, but due to their small size, provided only one or 2 examples of each type. Thus, the larger survey should allow us to assess the space density of these objects.

THE SURVEY

This survey was conducted using the Steward Observatory 1.5m telescope at Mt. Bigalow. A Rockwell HgCdTe array with 128 × 128 pixels was used. This array has about 50% quantum efficiency at K, 40 electron read noise and a negligible dark current. Field reduction optics produced a plate scale of $1.8 arcsec\ pix^{-1}$ giving a field of view of $15 arcmin^2$. The telescope was wobbled between two fields during the observations and the two fields were used to flatten each other. For the large area survey 16 fields were observed, each for 100min. For the deeper subset of the survey an integration time of 500min was used for two fields.

To supplement the IR data, we obtained BVRI images of the fields using the Steward Observatory 2.25m telescope. Typical exposure times were 1.5, 1, 1 and 1.5 hours at B,V,R and I respectively under photometric conditions. To classify objects as stars and galaxies we also obtained 15min R band images of the fields using the Techtronics 2048^2 CCD at the Prime Focus of the Mayall 4m telescope with 1″seeing.

GALAXY EVOLUTION FROM THE NEAR IR

The study of faint galaxies in the field is central to understanding the evolution of stellar populations within galaxies and cosmology. Broadhurst *et al.* (1988) and Koo and Kron (Koo 1988) have obtained redshifts for several hundred faint galaxies and find little evidence for luminosity evolution but strong evidence for bluer colors and stronger OII emission at $z > .3$. While selection of the galaxies in the blue (rest frame UV at moderate z) by Broadhurst *et al.* does allow one to identify a strongly evolving population it introduces a strong bias away from finding weakly evolving galaxies. If one wishes to know how the general field population is evolving you must apply highly uncertain corrections to the data to compensate for selecting objects in a band where evolutionary and K corrections are large and differ by several magnitudes for different galaxy types.

In the K band the K corrections are small, uniform over galaxy types and rather immune to evolution. Thus, IR selection insures uniform selection of galaxy types at all redshifts and selection in a band different from the one sensitive to the evolution. This basically relieves the need to apply uncertain corrections to compensate for selection biases. Figure 1 shows the expected redshift distributions for galaxies selected at K. By selecting galaxies to $K = 18$ we should have a fair number of galaxies at z=.5 and beyond. We have begun to use multiple object spectroscopy to obtain redshifts for the 200 galaxies selected by their 2μ emission to study the evolution of their stellar populations.

One potential danger is that evolution of the 2μ emission (the giant branch) could be stronger than is now believed. In our first IR survey we found what appears to be a population of luminous galaxies at high redshift with colors typical of current day ellipticals, luminosities similar to giant ellipticals at the same redshift and a space density close to that of L_* galaxies. Such objects can not be understood by standard galaxy evolution models (Bruzual 1983) which have no evolution of the giant branch. Two test present themselves to test our understanding of IR evolution. First, we can compare IR galaxy counts with those predicted by the models. The counts obtained so far appear consistent with the models. Once redshifts are obtained a comparison with figure 1 would be another test of the standard galaxy evolution models.

REFERENCES

Broadhurst *et al.* 1988, *M. N. .R. A. S.*, , 235, 827.

Bruzual, G. 1983, *Ap. J.*, , 273, 105.

Cowie *et al.* 1988, *Ap. J. (Letters)*, , 322, L29.

Elston, R. 1988, Thesis, The University of Arizona.

Koo, D.C. 1988, in *IAU Sym. No. 130: Large Scale Structure in the Universe* , ed J. Audouz *et al.* , Kluwer Publ.:Dordrecht, 221.

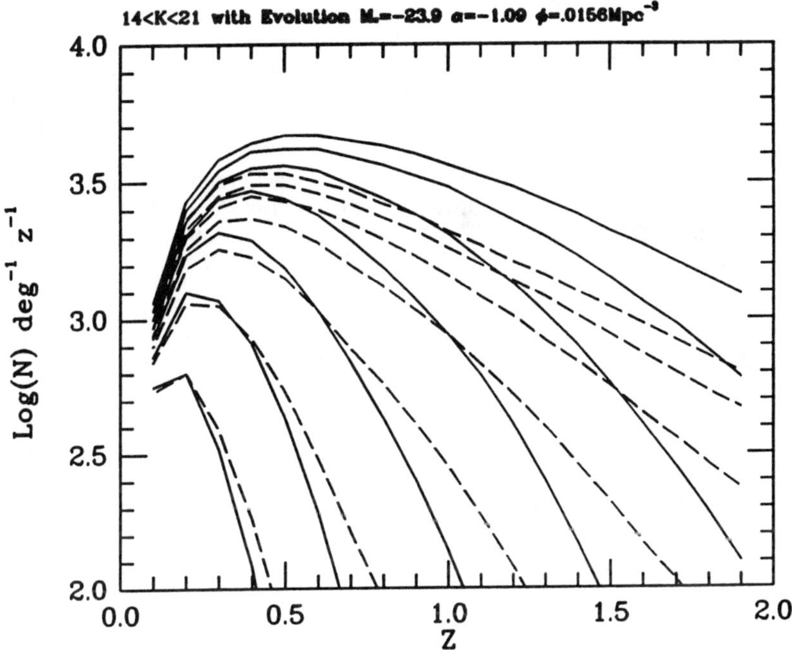

Fig 1. The redshift distributions of galaxies selected at K. The solid lines are for q_o = .1 and the dotted lines are for q_o = .5. From the bottom to the top the lines are for bins centered on K magnitudes of 14.5, 15.5, 16.5, 17.5, 18.5, 19.5 and 20.5. Note that at $K = 18$ the peak redshift is at .5 with a significant tail extending to $z \approx 1$.

IMAGING FAINT GALAXIES

J. A. TYSON
AT&T Bell Laboratories, Murray Hill, NJ 07974

ABSTRACT Deep CCD surveys to 29 mag from .32 to .9 micron wavelength reveal an isotropic population of very blue galaxies. There are over 200,000 of these objects per square degree. The redshift of this population appears to be in the range 1 - 3. New techniques for ultra-deep imaging with CCDs are described. The resulting extragalactic background radiation from the UV to the near-IR due to this population of objects is shown. There apparently was darkness for at least $2x10^9$ years prior to this active phase of stellar burning seen in these galaxies.

1. INTRODUCTION

Deep CCD imaging surveys have revealed a population of faint blue resolved objects [hereafter referred to as galaxies, although they may not be galaxies in the familiar sense] which cover the sky apparently isotropically. There was already evidence of a bluing trend with faintness for field galaxies in the photographic work of Kron (1980) [.2 mag of bluing in B-R from 18 to 23 B magnitude], and by 1979 number counts of these galaxies had been studied (Tyson and Jarvis, 1979). The deep CCD surveys have revealed most of this population now, exhibiting over 2 mag of bluing in B-R to 28 B magnitude. My collaborators in this imaging work have been primarily Raja Guhathakurta and Pat Seitzer. Software for automated photometry was developed in collaboration with John Jarvis, Frank Valdes, and Rick Wenk. Follow-up spectroscopy has been done with Pat Boeshaar.

The apparent brightness of distant galaxies is influenced primarily by galaxy evolution (stellar lifetimes and population evolution), and to a much lesser extent by cosmology (Harrison, 1964). Number-magnitude counts which rise with magnitude like dex [0.4*mag] or faster produce a brighter integrated extragalactic background, and if this slope never falls below 0.4 the surface brightness of the night sky would diverge. These considerations are related to Olbers' paradox: The integrated flux from a given class of objects must not exceed the observed background. The finite luminous lifetime of stellar populations offers a way out of this divergence.

2. CCD SURVEYS

Deep CCD imaging surveys have been carried out in 13 random high galactic latitude fields to 29 B_J, 28 R, and 26 I mag arcsec^{-2} limiting surface brightness (Tyson 1988, Tyson and Seitzer 1988). This B_J, R, I filter system is described by Gullixson, et al (1989). Recently, ultra-deep CCD imaging to 30 mag arcsec^{-2} in

U, B_J, and R has been obtained in three fields (Guhathakurta, Tyson, and Majewski 1989). Limits to the redshift distribution of this faint population of galaxies may be obtained with existing large telescopes, from spectrophotometry and gravitational lens tests. Redshift information for individual galaxies fainter than 24-25 mag will remain impossible until larger telescopes are built.

3. OBSERVING TECHNIQUES

Accurate galaxy surface photometry to 30 mag arcsec^{-2} requires the correction of detector systematic errors to 10^{-4} of the sky background. The dimensional and temporal stability of CCDs make this possible. Guided exposures, long enough to be background limited, are made offset randomly from each other (but overlapping) up to a maximum offset. The maximum offset must be less than about 30% of the image size, for good efficiency. There is also a minimum offset, which must be larger than the angular scale of the largest object in the image. Thus this technique works only for images containing objects (faint galaxies or stars) small compared with the total CCD angular size. The common overlap area is the area of the final cleaned image. This means that we want to use the largest CCD imagers possible. 50-100 sky-limited exposures, with random vector offsets between these limits, are obtained.

This 3-dimensional stack of disregistered exposures contains all the information about the systematic errors in the CCD, telescope and sky, and the true object luminosity distribution -- in separable form. Additive and multiplicative systematics in the detector/ camera are extracted as separate calibration images from this array of raw images (Tyson and Seitzer 1988). A non-local matched filter (Tyson and Wenk 1989) reduces the systematics-corrected stack of disregistered images to a single high resolution image for each color band. The automated detection and photometry software FOCAS (Jarvis and Tyson, 1981; Valdes, 1982) produces a catalog of object multi-band photometric data.

4. NUMBER-MAGNITUDE COUNTS VS WAVELENGTH

The N(m) counts in the B_J band are shown in Figure 1. The dashed line has slope .45 and is the continuation of the counts found at brighter flux levels photographically (see Tyson 1988). Note that this slope of 0.45 is super-critical in the sense that the contribution to the extragalactic background is monotonically increasing with magnitude. The dots show these counts corrected for crowding. This is done with a simple but repetitive Monte-Carlo process in which a faint test galaxy is added to the CCD image. FOCAS is rerun, and the probability of detection and systematic shifts in apparent magnitude can be determined after thousands of such trials. These crowding-corrected counts are accurate to 27 B_J mag. The observed blue number counts depart significantly from a linear relation in log N vs magnitude. Tinsley (1977, 1978, 1980, 1981) was the first to point out that star formation at a given epoch would produce a bump in the number-magnitude counts. The small but highly significant bump seen in our counts around 25 B_J mag is very likely this predicted effect. The slope of the N(m) counts for longer wavelengths is sub-critical, going to a "no-evolution" slope of .3 at .9 micron wavelength. The enhancement of the blue and U band counts is apparently coming from UV-bright objects, in their rest frame. It is likely that we are seeing a major star formation epoch in these galaxies.

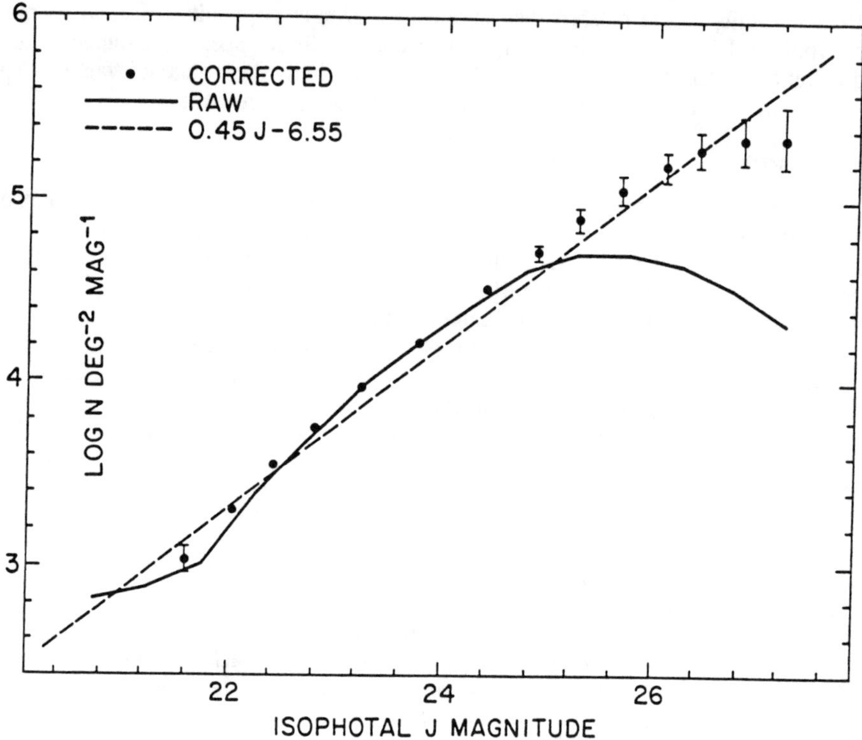

Fig. 1. Number-magnitude counts averaged over 12 deep CCD survey fields. The number of galaxies per square degree per magnitude interval are plotted vs B_J mag, to isophote of 29 B_J mag arcsec^{-2}. The dashed line is the extension of the slope found in the brighter galaxy photographic survey (Tyson 1988). Note the significant bump in counts around 25 mag. This may be due to star formation in primeval galaxies during a 2-3 Gyr long epoch around z=2 or 3. Note also the levelling off of the steep rise in number counts at about 27 B_J mag. The error bars are 3σ, derived from scatter among Monte Carlo crowding simulations on the real data.

The levelling off of the N(m) counts shown in Figure 1 is now known to be real: the recent ultra-deep images of the same fields show few new galaxies. In particular, there are occasional patches relatively devoid of galaxies in the original deep CCD survey which do not fill in with new faint galaxies in the ultra-deep images in the U, B_J, and R bands. The sum of all the imaging data for the south galactic pole survey field is shown in Figure 2 in which the limiting apparent magnitude is about 30th mag. Only a 2.3x2.7 arcmin part of that survey field is shown. Patches of low galaxy count do not appear to fill in with faint galaxies as fainter limiting magnitudes are reached.

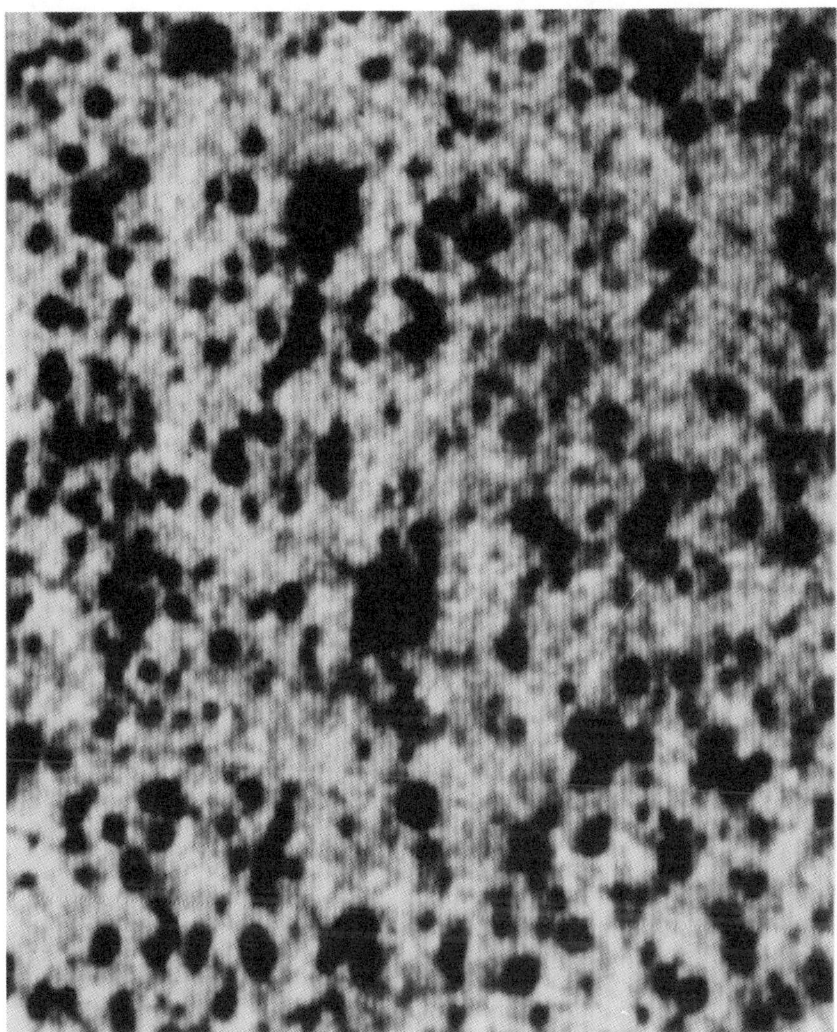

Fig. 2. The sum of all U, B_J, R images of the south galactic pole field. This is one of the three ultra-deep fields produced in the latest CCD survey. A 2.3x2.7 arcmin section is shown here, and the limiting broadband magnitude in this image is about 30 mag. This picture is reversed, with brighter objects shown as black. The "dark" lanes seen on shorter exposures do not appear to fill in with galaxies. If due to dust, it must be very distant dust (otherwise it would appear bright due to reflected disk light [Guhathakurta and Tyson 1989]) with large extinction, implying a massive contribution by dust and gas to Ω. If not due to dust, we may be seeing through tunnels in the galaxy topology out beyond the first shell of stellar burning.

5. THE OPTICAL EXTRAGALACTIC BACKGROUND FROM DISCRETE OBJECTS

Features in the surface brightness distribution of the extragalactic sky detected as discrete objects [size under 1 arcminute] constitute the majority of the extragalactic light. The observed object size distribution is narrowly peaked in the 1-3 arcsec range. Providing the observations go to sufficiently faint surface brightness to catch all the light from these objects, the extragalactic background light (EBL) from all resolved objects can be calculated directly from our galaxy count data. Deep imaging in several wavelength bands can be used to construct a rough spectral distribution for this EBL. Our CCD data, covering the range .3 - 1 micron, can be used to calculate the EBL down to 31 B_J mag arcsec^{-2}.

The flattening of the N(m) counts originally seen in the B_J band (Tyson 1988) have been seen now in the U band, and this represents the outer shell of magnitude contributing significantly to the EBL. At magnitudes where the N(m) slope falls below the critical dLog N/dm < 0.4, relatively little light is contributed to the integrated EBL. Galaxies fainter than 20 B_J mag contribute about 75 percent of the EBL at 4500 Å.

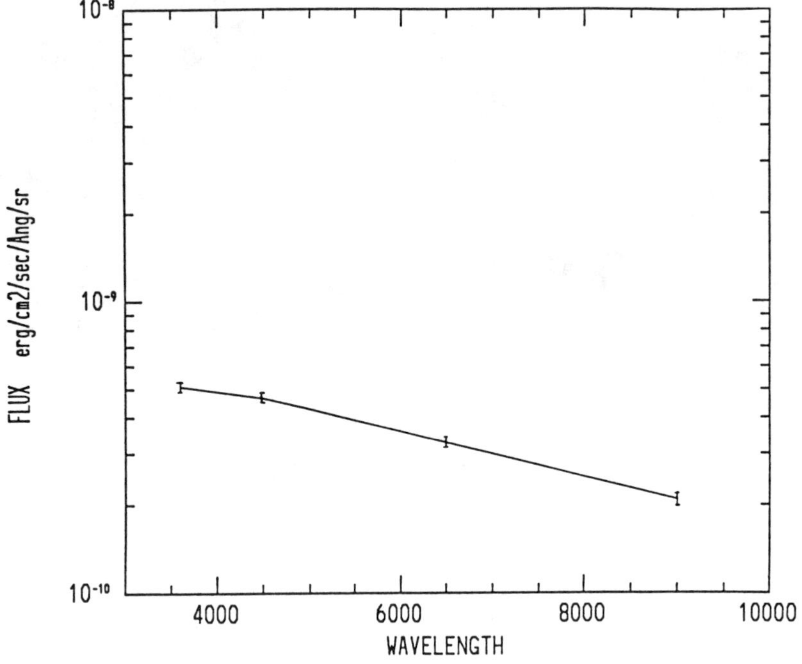

Fig. 3. The sum extragalactic background from the faint galaxy population as a function of wavelength from 3200 Å to 10000 Å. The error bars are 3 σ, arising mostly from field-to-field fluctuations in the brighter galaxy numbers. Note the UV excess of the faint galaxy spectral energy distribution.

5.1 Spectral distribution

Our U, B_J, R, and I surface flux integrals for this EBL are plotted in Figure 3 vs wavelength. Note the blue color of these objects. The error bars on our EBL flux are 3σ, and are mostly due to field-to-field fluctuations in numbers of bright galaxies. Our measured spectral distribution of the EBL is slightly redder than flat in F_ν (see Guhathakurta, Tyson, and Majewski 1989). This SED is more UV-bright than nearby galaxies and is probably due to stellar evolution in this faint galaxy population.

6. ORIGIN OF THE FAINT BLUE GALAXY POPULATION

Speculation on the nature of these faint blue galaxies has ranged from low redshift dwarf galaxies undergoing star formation, to high redshift primeval galaxies forming the majority of their initial stellar population. Although the differential number-magnitude counts will be helpful in constraining theories of the nature of these objects, number-redshift information will be much more effective in narrowing the field of candidates.

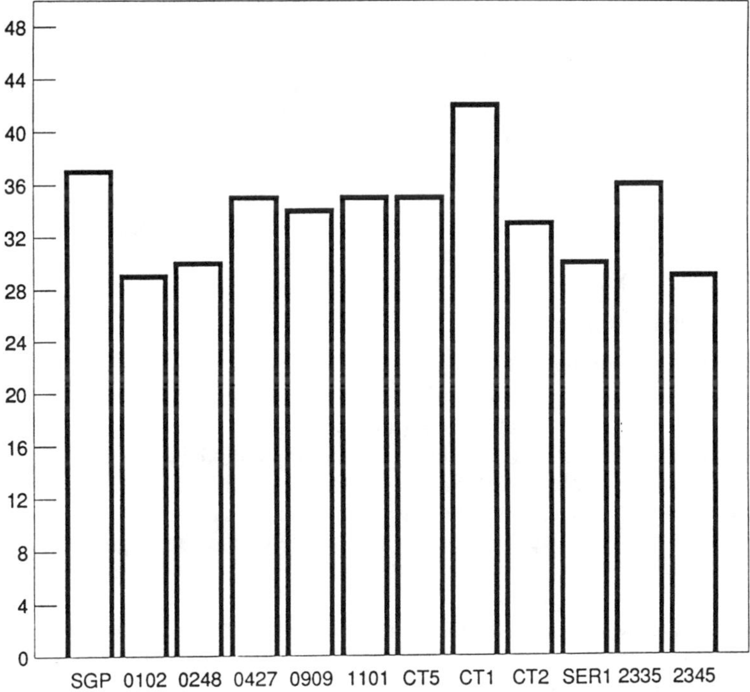

Fig. 4. The isotropy of the number counts of the faint blue galaxies is shown in this plot of the counts in 12 survey fields in the magnitude range 23 < B_J < 26.

6.1 Isotropy

The number counts of these faint blue galaxies appear to be isotropic to within 2σ from field to field. This is consistent with the idea that we are looking through a distribution over a range of luminosity distance. This observed isotropy is shown in Figure 4 for 12 survey fields. It is worth noting that if these galaxies are at redshifts of order 1 or larger the galaxies in any of these survey fields (separated by tens of degrees on the sky) are out of causal communication with those in any other field. However, as mentioned, some cloud-like structure is occasionally seen on less than 30 arcsec scales. The measured two-point angular autocorrelation function for these faint galaxies has the same slope as in brighter more nearby samples, but a smaller amplitude.

6.2 Upper redshift limit

An upper redshift limit for this population of faint blue galaxies is immediately apparent from its spectral energy distribution shown in Figure 3. If a significant fraction of these galaxies were at redshift greater than 2.8, the Lyman break would be shifted through the U passband, causing these galaxies to either drop out in the U band or to have a large drop in flux between B_J and U. Lyman breaks from most stars in these primeval galaxies are probably at least a factor of two. In addition, these galaxies would have even more hydrogen than present galaxies, which would absorb all the Lyman continuum photons, causing these galaxies to be black in the U-band for redshifts larger than 3 (Guhathakurta, Tyson, and Majewski 1989). We find an upper limit to the redshift distribution of $z = 3$ for 95% of these galaxies. Our technique relies on two key elements: (1) uniform faint completeness limits in each wavelength band from .33 to .72 micron, and (2) automated detection and multicolor photometry of over 2000 galaxies. Only a few percent of the galaxies observed in U/B_J/R drop out in the U band. The galaxies appear to be undergoing relatively recent evolution with rest frame spectra that are approximately flat down to the Lyman limit.

6.3 Lower redshift limit

A lower redshift limit for this faint population can be derived from its response to a gravitational lens placed in front of it. A gravitational lens distorts background objects, stretching their images along circles centered on the lens. A first test of the lower redshift limit of this blue galaxy population is provided by the dark matter lens in the cluster 1409+52 at a redshift of 0.46. The same number of faint blue galaxies per square arcminute are seen behind this red cluster, but most of their images are stretched by the gravitational lens and are aligned orthogonal to the vector to the cluster center (Tyson, Valdes, and Wenk 1989). The distorted images of background galaxies appear as mini-arcs, small parts of circles centered on the foreground cluster. The distribution of alignments of these blue background galaxies is contrasted with that of the red cluster galaxies in Figure 5. A significant excess of faint blue background galaxies is seen to be aligned, by the distortion of the gravitational lens, tangent to circles centered on the foreground cluster lens. The excess in the tangent alignment bin can only occur if most of these galaxies have redshifts greater than 0.9. Figure 6 shows the critical surface mass density for lensing vs the redshift of the background population of galaxies, for cosmological density parameter .2 and 1.

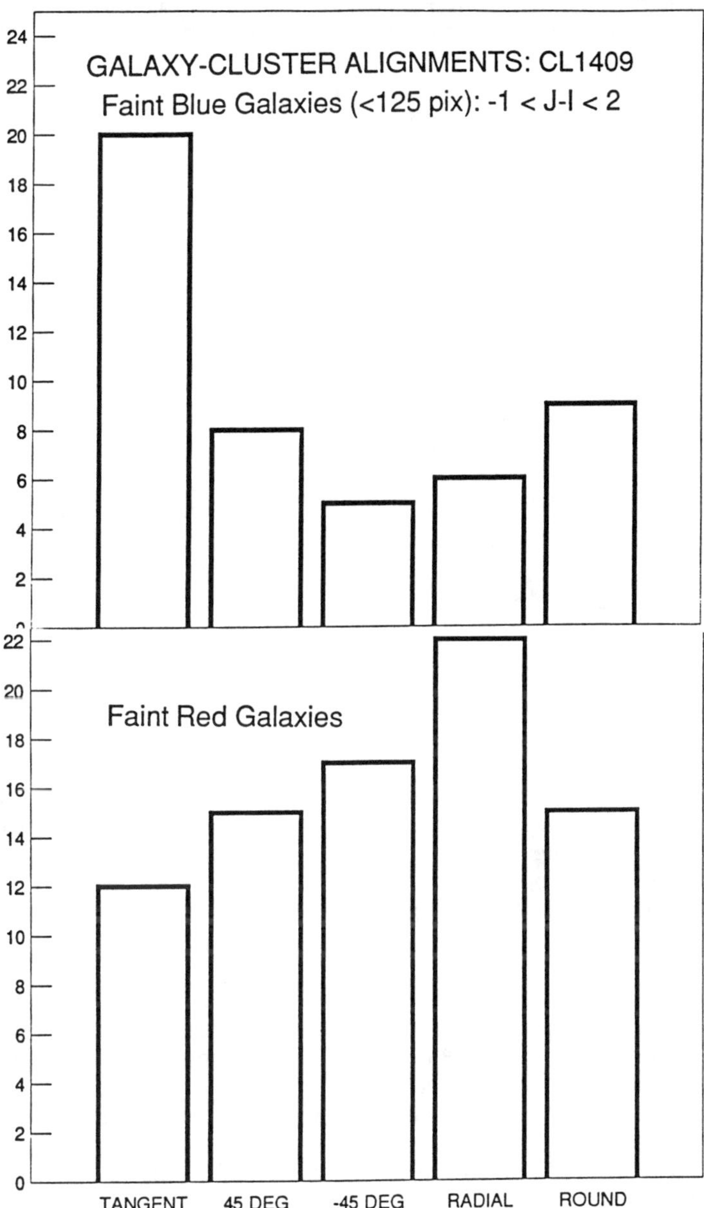

Fig. 5. A test of the lower redshift limit of the majority of the galaxies producing the extragalactic background light: Gravitational lens distortions from a lens at redshift 0.46. The distribution of alignments of the blue galaxies is contrasted with that of the red cluster galaxies. The excess blue galaxy count in the tangent alignment bin can only occur if the majority of these galaxies have redshifts greater than about 0.9.

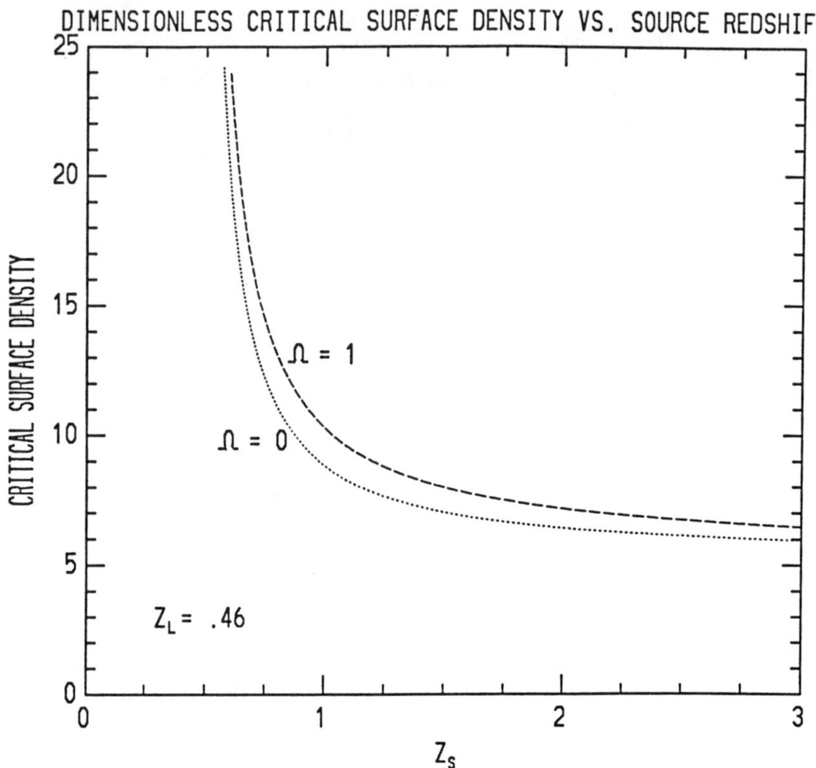

Fig. 6. The critical surface mass density for a gravitational lens at redshift 0.46 to focus a background galaxy as a function of redshift is shown. The asymptotic surface density at high redshift, only slightly dependent on Ω, is about 0.8 g cm^{-2}. For background redshifts much lower than 1, unphysically large mass densities would be required in the cluster lens for this case.

6.4 Protogalaxies

Primeval galaxies undergoing star formation at $z < 7$ should be observable in this deep CCD survey, and their low-level surface brightness morphology is within reach of present CCD/telescope technology. The signal-to-noise ratio of these CCD observations is more than sufficient to detect star formation in galaxies forming solar metallicity for all redshifts less than 7. It is possible that we are seeing the first epoch of star formation in these galaxies, and in that sense these blue galaxies would be primevals.

We do not find many extended red objects with R-I > 2 at faint I magnitudes. Either there was negligible star formation in field proto-ellipticals and proto-bulges for z>5, or there was adequate dust from a very much earlier star formation epoch to dim that luminous phase below detection. Ostriker and Heisler (1984) show that dust obscuration could become severe at redshifts 3-4. If star formation began at very high redshift, it may be unreasonable to hope to detect either protogalactic

Lyman-α clouds or very early primevals in the optical. Recently, Baron and White (1987) have made numerical simulations of inhomogeneous galaxy formation in the CDM model. In their hierarchical clustering model they find that 60 percent of the stars in a galaxy form before collapse of the initial galaxy-scale perturbation. If, as our data suggest, early star formation in galaxies is continuous and spread out over at least 10^9 years, earliest galaxy formation redshifts can be high (say 5-10) but much star formation may be ongoing at a redshift of 2 or lower.

It is interesting to speculate on the relation of this population of faint blue galaxies in the redshift range 1-3 to the population of objects giving rise to quasar damped absorption line features in that same wavelength range (Wolfe, et al. 1986). The area density on the sky, about 15-20% of the sky, is nearly the same for both sets of observations. However, the density in redshift of the damped absorption line systems is probably higher than the blue galaxy population.

All current theories of galaxy formation (e.g., Frenk, et al. 1985) predict a high degree of clumping for primevals at moderate to low redshift. Unfortunately, such clumps of low surface brightness galaxies could appear as a single diffuse "protogalaxy". Nevertheless, our observed galaxy surface brightness distribution appears to be more nucleated than current cold dark matter simulations predict.

6.5 Morphology of faint blue galaxies

There are enough bright stars in each deep CCD survey field to accurately define the stellar point spread function. The centroided sum of all galaxies in a magnitude bin has sufficient signal/noise ratio to permit extraction of average metric scale length information. Model radial profiles for these "statistical" faint galaxies are fit to the observed average profiles over the range .1 to 10 arcsec. After seeing deconvolution (Tyson and Wenk 1989) galaxies in our sample fainter than 26 B_J mag are found to have average metric exponential scale length 0.2-0.3 arcsec. Over the magnitude range of 17 to 27 B_J mag of our CCD sample there is no evidence for a significant tail past 3 arcsec radius in the galaxies, and all but two of the faint galaxies with central surface brightnesses fainter than 24 B_J mag $arcsec^{-2}$ have scale lengths less than 1 arcsec. Galaxies of 26th B_J mag appear to be no more extended in comoving metric size than brighter galaxies at low redshift. To 30 B_J mag $arcsec^{-2}$, there is no evidence of bimodality in the angular size distribution of galaxies.

It is possible to extract one higher moment for the shapes of these galaxies. The blue galaxies in our deep sample have considerably higher ellipticities than those in clusters. Only about 10% of the faint galaxies have observed (uncorrected for seeing) ellipticities less than 0.2. When corrected for seeing, the a/b ratios will increase even further. Comparison of this observed flattening with a nearby population of galaxies is in progress. We may be seeing spiral galaxy disks in formation.

6.6 Models

For evolving galaxies with formation redshifts ranging from 3 to 5 (and nearly independent of deceleration parameter) Yoshii and Takahara (1988) estimated an integrated extragalactic background flux of 6×10^{-6} erg cm^{-2} sec^{-1} $ster^{-1}$ μ^{-1} at .45 μ wavelength. We have run various models, using different evolution scenarios (including various stellar burning histories and merging), galaxy luminosity functions, stellar initial mass functions, and cosmological parameters. We find that, since the luminosity function is not known, little if any cosmological information can be uniquely determined from the number-magnitude counts. Redshift information is needed. Various evolutionary scenarios for the stellar populations in galaxies have been proposed (Tinsley 1980, 1981, Bruzual 1981, 1983a,b, Scalo 1986, Wyse 1985, Gunn 1982, 1989). All of these model galaxy evolution scenarios have many adjustable parameters: the stellar initial mass function, the star formation rate for each population, the stellar burning onset scenario, dust content as a function of redshift, the galaxy luminosity function (shape and characteristic luminosity for the various galaxy types as a function of redshift), bulge to disk ratio, dissipation and collapse histories in the primeval galaxies, merging, etc. Under very restricted model scenarios in which the shape of the galaxy luminosity function is invariant, model stellar formation occurs at some universal epoch (different for spirals and ellipticals), and with the arbitrary assumptions of no merging and no episodic star formation (Bruzual 1981), various authors have remarked that it is difficult to fit the observed galaxy counts with a high value of the cosmological parameter q_o (Yoshii and Takahara 1988, Tyson 1988). This result is very model dependent.

Changes in stellar burning scenario affect the apparent brightness (hence N(m)) of galaxies more than changes in cosmology. Changes in cosmology affect the numbers of galaxies seen at some redshift N(z). In general, galaxy number-magnitude counts N(m) are relatively more sensitive to evolution, and number-redshift counts are more sensitive to cosmology. We will probably never be able to conclude much of interest cosmologically from galaxy number-magnitude counts without also having good number-redshift information. Even then, we will have to know something about the shape of the galaxy luminosity function vs redshift (Caditz and Petrosian 1989). We have experimented with Bruzual-type evolution models, relaxing the assumptions of invariant luminosity function, non-episodic SFR, or merging. We find that while it is possible to fit the observed $N(m,\lambda)$ with low q_o and the traditional models (Tyson 1988), it is also possible to get a good fit with q_o=.5 by varying the faint tail of the luminosity function vs look-back time and/or introducing galaxy number non-conservation (Guhathakurta and Tyson 1989).

This sensitivity to evolution is particularly strong at wavelengths where much of the redshifted UV from early star formation appears. Differential galaxy counts in the blue part of the spectrum are relatively more sensitive to evolution. The sum total EBL, being an integral over the flux times number counts at each wave band, is even less sensitive to cosmology. Simply put, without good N(z) restrictions for redshifts covering the peak evolution epochs, galaxy evolution models have more adjustable parameters than necessary for a good fit to the $N(m,\lambda)$ data. This situation improves when we have high redshift constraints on the N(z) distribution, and it will perhaps be possible in coming years to obtain corresponding rough constraints on q_o using these $N(z, m, \lambda)$ constraints.

7. SUMMARY

Multi-band ultra-deep CCD images to 30 mag arcsec^{-2} have been obtained in sparse high galactic latitude fields and in fields with known galaxy clusters in the foreground. We find an isotropically distributed population of UV-excess galaxies in a redshift range 1 - 3. Their number density is high: over 200,000 per square degree. There appears to be a leveling off of the N(m) counts at 26 mag in the blue and U wavelength bands. Models based on number vs magnitude counts are insufficiently constrained to yield useful information on cosmology separate from evolution. There appears to have been a "dark ages" for at least $2x10^9$ years prior to this active phase of stellar burning seen in these galaxies. The possiblility exists that these faint blue galaxies are medieval rather than primeval. If there was an earlier phase of star formation, and if that population of stars was clumped on galactic scales, then such a primeval epoch will be detectable in ultra-deep 2 micron surveys.

REFERENCES

Baron, E. and White, S. D. M. 1987, *Ap. J.*, **322**, 585.

Bruzual, G. 1981, Ph.D. thesis, U.C. Berkeley.

Bruzual, G. 1983a, *Ap. J. Supp.*, **53**, 497.

Bruzual, G. 1983b, *Ap. J.*, **273**, 105.

Caditz, D. and Petrosian, V. 1989, *Ap. J.*, **337**, L65.

Frenk, C. S., White, S. D. M., Davis, M., and Efstathiou, C. 1988, *Ap. J.*, **327**, 507.

Guhathakurta, P. and Tyson, J. A. 1989, *Ap. J.*, in press.

Guhathakurta, P., Tyson, J. A., and Majewski, S. 1989, *Ap. J. Lett.*, submitted.

Gullixson, C. A., Boeshaar, P. C., Seitzer, P., and Tyson, J. A. 1989, submitted.

Gunn, J. E. 1982, in "Astrophysical Cosmology", eds. H. A. Bruk, G. V. Coyne, M. S. Longair (Pont. Acad. Sc., Vatican).

Gunn, J. E. 1989, in *Epoch of Galaxy Formation*, Eds. Frenk, C. S., Ellis, R. S., Shanks, T., Heavens, A. F., and Peacock, J. A. (Kluwer, Norwell, MA) p. 167.

Harrison, E. R. 1964, *Nature*, **204**, 271.

Jarvis, J. F., and Tyson, J. A. 1981, *Astron. J.*, **86**, 476.

Kron, R. G. 1982, *Vistas in Astronomy*, **26**, 37.

Ostriker, J. P. and Heisler, J. 1984, *Ap. J.*, **278**, 1.

Scalo, J. 1986, *Fund. Cosm. Phys.*, **11**, 1.

Tinsley, B. M. 1977, *Ap. J.*, **216**, 349.

Tinsley, B. M. 1978, *Ap. J.*, **220**, 816.

Tinsley, B. M. 1980, *Ap. J.*, **241**, 41.

Tinsley, B. M. 1981, *Ap. J.*, **250**, 758.

Tyson, J. A. and Jarvis, J. F. 1979, *Ap. J.*, **230**, L153.

Tyson, J. A. 1988, *Ap. J.*, **96**, 1.

Tyson, J. A. and Seitzer, P. 1988, *Ap. J.* , **335**, 552.

Tyson, J. A., Valdes, F., and Wenk, R. A. 1989, *Ap. J.*, in press.

Tyson, J. A. and Wenk, R. A. 1989, submitted.

Valdes, F. 1982, Proc. *SPIE*, **331**, 465.

Wolfe, A. M., Turnshek, D. A., Smith, H. E., and Cohen, R. D. 1986, *Ap. J. Supp.*, **61**, 249.

Yoshii, Y. and Takahara, F. 1988, *Ap. J.*, **326**, 1.

Wyse, R. F. G. 1985, *Ap. J.*, **299**, 593.

DEEP CCD IMAGING OF FIELD GALAXIES IN U, B_J, AND R: CONSTRAINTS ON GALAXY EVOLUTION

P. GUHATHAKURTA
Institute for Advanced Study, Olden Lane, Princeton, NJ 08540

J. A. TYSON
A. T. & T. Bell Laboratories, 1D335, Murray Hill, NJ 07974

S. R. MAJEWSKI
Yerkes Observatory, University of Chicago, Williams Bay, WI 53191

Deep optical CCD imaging of three isolated high latitude fields reveals a high surface density of faint blue galaxies on the sky approaching the confusion limit. Using the CTIO 4-m telescope and an 800×800 TI CCD we have obtained images in U, B_J, and R bands, reaching a limiting surface brightness of 29 mag arcsec^{-2}, with reliable detections at 27 mag in each of the three bands. A total of nearly 2000 galaxies are detected in at least two of the three bands in the 27 arcmin^2 covered by the three fields.

The $U - B_J$ and $B_J - R$ colors of a typical galaxy are only slightly redder than constant F_ν. The colors get bluer as one goes to fainter magnitudes (Figs. 1a and 1b; Guhathakurta, Tyson, and Majewski 1989, hereafter GTM). For all but the faintest bins, the spread in color is much greater than the uncertainties in photometry. There are very few objects with blue $B_J - R$ colors *and* very red $U - B_J$ ($\leq 5\%$) as shown in the color-color diagram (Fig. 2; GTM; Guhathakurta 1989). In fact, most galaxies show no evidence for any breaks between U, B_J, and R. For a break of a reasonable size at the Lyman limit, this rules out any significant fraction of galaxies in our sample beyond $z \sim 3$ for which the Lyman break would have redshifted past the U band. If instead, most of the galaxies were at $z < 3$, their Lyman breaks would have to be very small ($\leq 40\%$). This seems unlikely given that: (a) synthetic spectra over a wide range of IMFs have a factor of 2–5 drop in flux at 912 Å, and (b) a small quantity of neutral gas would absorb Lyman continuum photons (GTM).

The U band counts have a slope of over 0.5 over the range U = 21.5–24.5 (Fig. 3), which is a little steeper than the B_J count slope of 0.45. While the number-magnitude counts in the R and I bands are consistent with no-evolution models (slope $\sim$ 0.39 and 0.34, respectively — Tyson 1988), the relatively steep slope at bluer wavelengths suggests fairly recent ($z \lesssim 3$) evolution in these field galaxies with the ultraviolet flux from massive young stars being redshifted into the U and B_J bands. This is consistent (qualitatively, at least) with the observed lack of U band Lyman break candidates.

The distribution of galaxies on the sky is not uniform and the galaxy-galaxy angular correlation function has a slope similar to that found for more nearby samples. The "holes" in the projected galaxy distribution do not fill in with fainter galaxies as one goes to deeper limiting magnitudes. Further, there is some evidence, though marginal, that we may be seeing an *actual* turnover in the B_J number counts, even after a correction is made for the effects of crowding (Tyson 1988). This is a hint that we may be seeing the far end of the optical galaxy distribution. It has been suggested that, if the light from these blue galaxies is dominated by star formation, there is enough of them to explain all the heavy elements seen in the present Universe.

REFERENCES

Guhathakurta, P. 1989, *Ap. J.*, in preparation.
Guhathakurta, P., Tyson, J. A., and Majewski, S. R. 1989, *Ap. J.* (*Letters*), submitted.
Tyson, J. A. 1988, *A. J.*, **96**, 1.

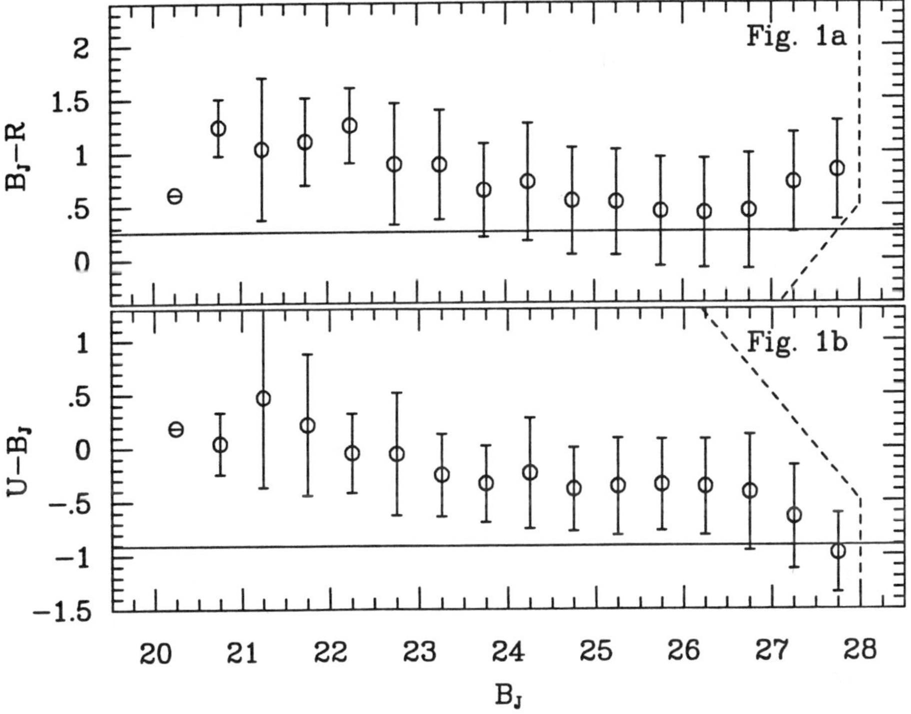

Fig. 1. (a) $B_J - R$ versus B_J for the $\sim$ 1400 galaxies in our sample that are brighter than the detection thresholds of 27.5 U, 28.0 B_J, and 27.5 R (dotted lines). The bars indicate the spread in color in each bin, and is typically much larger than the measurement error. The solid horizontal line at $B_J - R = 0.26$ marks flat F_ν. The redward trend beyond $B_J = 27$ is due to color selection bias. (b) Same as (a) showing $U - B_J$ versus B_J. A flat spectrum source corresponds to $U - B_J = -0.91$ (solid line).

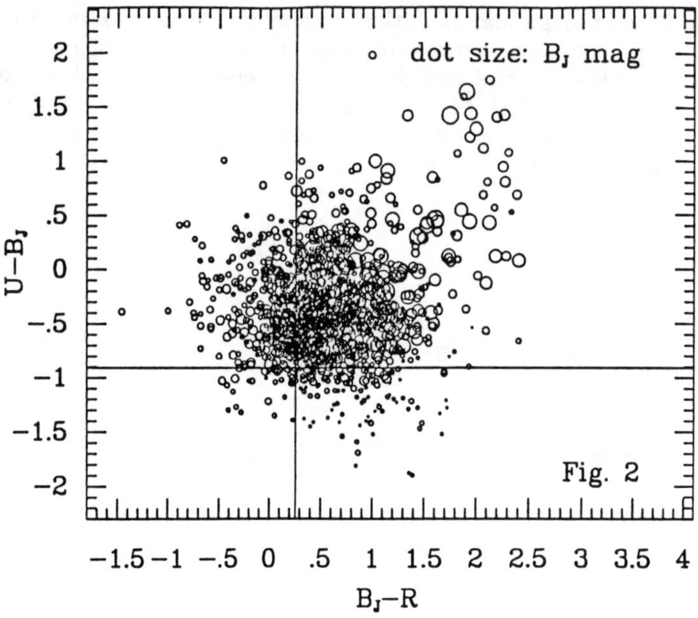

Fig. 2. $U - B_J$ versus $B_J - R$ for the same set of galaxies shown in Fig. 1. The cross marks flat F_ν, and the dot size is proportional to the B_J mag (bigger dot ⇒ brighter galaxy). Very few galaxies have a blue $B_J - R$ *and* a red $U - B_J$ — these would be U band Lyman break candidates.

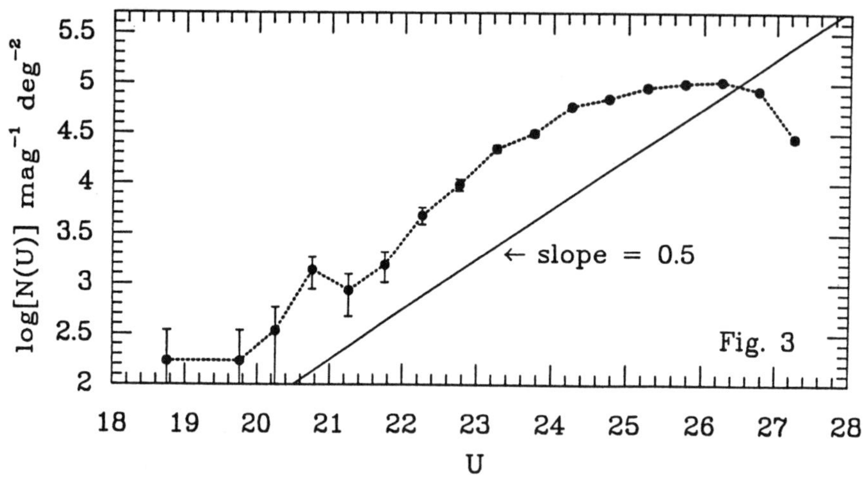

Fig. 3. The U band number-versus-magnitude counts. The error bars indicate $\pm 1/\sqrt{N}$ variation in the counts. The solid line has a slope $d\log N/dm$ of 0.5 (with an arbitrary intercept), and the count slope exceeds this for U $\lesssim$ 25. The counts have not been corrected for undercount caused by crowding, which produces a roll over beyond U $\sim$ 25.

LYMAN α EMISSION FROM YOUNG GALAXY DISKS

ROSS D. COHEN AND HARDING E. SMITH
Center for Astrophysics and Space Sciences, C–011, University of California, San Diego, San Diego, CA 92093

ABSTRACT We have conducted an imaging survey of Lyα–disk *absorption* systems for Lyα *emission* from the absorbing objects, which are candidate young disk galaxies. We set limits to the flux in Lyα from these objects of approximately 10^{-16} ergs cm^{-2} s^{-1} (3σ), which corresponds to a limit on the star formation rate of 2–7 $M_{\odot}$ yr^{-1}, in the absence of dust. This suggests that although disk structures form early, rapid star formation does not occur immediately, or else occurs in episodic fashion.

INTRODUCTION

Ultimately, we wish to understand the formation and evolution both of large scale structure in the universe and the formation and constitution of galaxies. As steps toward this, we must determine when formation of most disk galaxies occurred, as well as the subsequent rate of star formation and consequent metal enrichment. However, until recently, most of the objects known at redshifts of interest have been active galactic nuclei of some sort. While interesting in their own right, there is no *a priori* reason to believe these are typical objects, or indeed, that they form in typical regions of the universe. Unfortunately, non–active, high–redshift objects are extremely difficult to detect.

In a previous work, (Wolfe *et al.* 1986, Paper I) we conducted a search for disk galaxies at high redshift by looking for strong self–damped Lyα absorption systems seen against bright background QSOs. These absorption features are an unambiguous signature of galaxy disks. As described below, we have identified a sample of galaxy disks at z =2–3 which can be identified with all of the luminous baryonic matter that we see locally. The absorption properties are those we expect from young galaxy disks. In this work, we present an attempt to detect the absorbing material directly in Lyα emission. Actively star forming objects should have strong emission at Lyα; successful detection would yield both the star formation rate and the size of the region over which star formation is occurring.

Among the myriad absorption lines seen in the spectra of QSOs, members of one class stand out from the others; they possess narrow metal lines but broad hydrogen lines. These lines are broadened not by the motion of the gas, but by natural broadening, and thus indicate a very high column density, such as we find in disks of spiral galaxies today. Based on the work of Bosma (1981), we consider a Lyα absorption system to be indicative of a galaxy disk if it has a neutral hydrogen column density $N_{H\ I} \geq 2\times10^{20}$ cm^{-2}, comparable to the H I column density along a path perpendicular to the plane of the Galaxy or other nearby late–type spiral galaxies; we refer to these as damped–Lyα systems and Lyα–disk systems. The prototype system is at $z = 2.3094$ in PHL 957 (Black, Chaffee, and Foltz 1987). This object shows an H I column density of 2.5×10^{21}, a carbon abundance $[C/H]_{\odot} \sim -2$, and a molecular hydrogen fraction $N(H_2)/N(H\ I) \lesssim 10^{-5}$ that of the Galaxy, suggesting a galaxy in which only a small amount of star formation has gone on and which possesses very little dust.

The Lick Survey

Our survey for damped Lyα lines (Paper I) was conducted using the IDS on the Lick 3 m telescope. The final sample contained spectra of 68 quasars between a minimum redshift defined by the atmospheric cutoff, absorption at the Lyman limit, or poor signal–to–noise ratio, and a maximum redshift defined as 3000 kms^{-1} short of Lyα at the emission redshift. The spectra were analyzed over a total redshift path $\Delta z = 56$ between $z = 1.65$ and $z = 3.26$ to generate candidate line lists and yield candidate systems with $W_{rest} \geq 5$Å. Assuming a one–to–one correspondence with spiral galaxies at the present epoch, extrapolating the local density of spiral galaxies using a Schechter luminosity function (Kirshner *et al.* 1983), and assuming that $R(H\ I) = 1.5 \times R_{Holmberg}$ we expect a maximum of 3.5 systems with $N(HI) \geq 1.8 \times 10^{20}$cm^{-2}. We detected 47 candidate systems for follow up observation at higher resolution and signal–to–noise ratio.

High Resolution Observations

To date 18 candidates are confirmed damped–Lyα systems (Turnshek *et al.* 1989, Paper III; Wolfe, Turnshek, and Lanzetta 1989, Paper IV; Smith, Cohen, and Bradley 1986, SCB). The criteria used by Turnshek *et al.* were as follows:

1. The Lyα line should be well fit by a damping profile;
2. The difference between the redshift determined from the metal lines and Lyα should be smaller than the velocity range covered by the metal lines, which in turn should be much smaller than the width of Lyα;
3. The equivalent widths and profiles of the higher Lyman lines should be consistent with the derived properties of the system.

Although there is no direct evidence identifying the absorbing objects as disks of spiral galaxies, there is considerable indirect evidence supporting this identification.

1. H I column densities are similar to column densities through late type nearby galaxies perpendicular to the plane: $2 \times 10^{20} \leq N_{HI} \leq 8 \times 10^{21}$ cm^{-2} (SCB, Paper III).

2. Some metal enrichment has occurred in all systems: $-2 \lesssim [X/H]_{\odot} \lesssim 0$, and all systems have low or mixed ionization state.

The low–ionization states dominate the high–ionization states (e.g. C II $\gg$ C IV), and approximately one half of the systems would have been missed through searches for C IV alone (Paper III).

3. The gas shows two components, a quiescent (disk) component, $\sigma_v \lesssim 10$ km s^{-1}, containing the bulk of the low ionization gas and giving rise to the damped Lyα and 21 cm absorption, and a turbulent (halo) component, $\sigma_v \gtrsim 20$ km s^{-1}, the properties of which resemble the optically thick Mg II absorption systems. In some cases only the low velocity dispersion component is present (Paper III).

4. At least one damped–Lyα system intervening toward a radio QSO (Pks 0458–020) shows 21 cm absorption. This material shows multiple cloud structure with $\sigma_v \approx 6$ km s^{-1}, $T_s \approx 100$ K, and structure extended over at least several kpc on the sky (Briggs *et al.* 1989).

5. In one case where a direct limit on the density can be made from observations of the C II fine structure line, there is evidence that the gas may be self–gravitating with scale height of the order of 300 pc (Smith, Turnshek, and Wolfe 1983).

6. Examination of the Lyα equivalent width distribution shows that the absorbers represent a unique population, distinct from the 'Lyman α forest' and 'Lyman α–metal systems'. These systems cover approximately 20% of the sky between z ≈ 2–3 and contribute a mean density of the order of 0.003 the critical density, comparable to the total baryonic density observed at the current epoch (Wolfe *et al.* 1990, Paper II).

IMAGING OBSERVATIONS OF LYMAN α–DISK SYSTEMS

The discovery of objects with such a strong identification with galactic disks or disk progenitors, at a look back time of 10^{10} yr, leads directly to our search for evidence of star formation. If young galaxies ($\tau \approx 2$ Gyr) are rapidly forming stars, there will be hot stars, H II regions and Lyα emission. Although any continuum emission will be lost in the glare of the background quasar, they should be visible in Lyα emission; the quasar spectrum itself is dark there, because of the absorption from H I in the galaxy disk itself. Thus we have undertaken a program of narrow–band imaging to detect Lyα emission from the Lyα absorption systems.

Observations

We used narrow–band ($\Delta\lambda \approx 20$Å) filters and the Lick CCD imaging spectrometer to isolate the regions of Lyα absorption in each quasar's spectrum to look for Lyα emission from the disk absorption systems. Figure 1 shows the transmission curve of one of our filters overlaid on a spectrum of PHL 957 (kindly provided by Dr. Craig Foltz). It demonstrates how the region of the quasar spectrum is effectively excluded from the narrow–band image, although some quasar light does leak through in the wings of the filters. A typical observation requires about 60 m until sky noise dominates read–out noise. Table I lists our substantially completed observations.

Other studies (see below) have used spectroscopic techniques to perform similar searches; results of these searches have been largely confined to a region centered on the quasar. Although these studies *may* be more sensitive than ours, we have no *a priori* knowledge of the location and extent of any star formation. In fact, since the absorption cross sections are higher in the outer parts of galaxies, we might expect any star formation to be offset from the quasar.

TABLE I Observations of Lyman α in Emission

QSO	z_{em}	z_{abs}	N_{HI} (cm^{-2})	λ_c (Å)	Filter FWHM (Å)	Int. Time (m)
PHL 957	2.69	2.309	2.5×10^{21}	4024	24	300
0458-020	2.29	2.039	5×10^{21}	3689	30	330
0528-250	2.77	2.812	2×10^{21}	4632	35	210
0836+113	2.70	2.466	4×10^{20}	4217	13	630
1215+333	2.61	2.001	1×10^{21}	3653	23	360
1337+113	2.92	2.796	8×10^{20}	4616	20	140
1347+112	2.70	2.471	2×10^{20}	4217	13	225
UM 184	3.03	2.616	2×10^{21}	4378	21	300

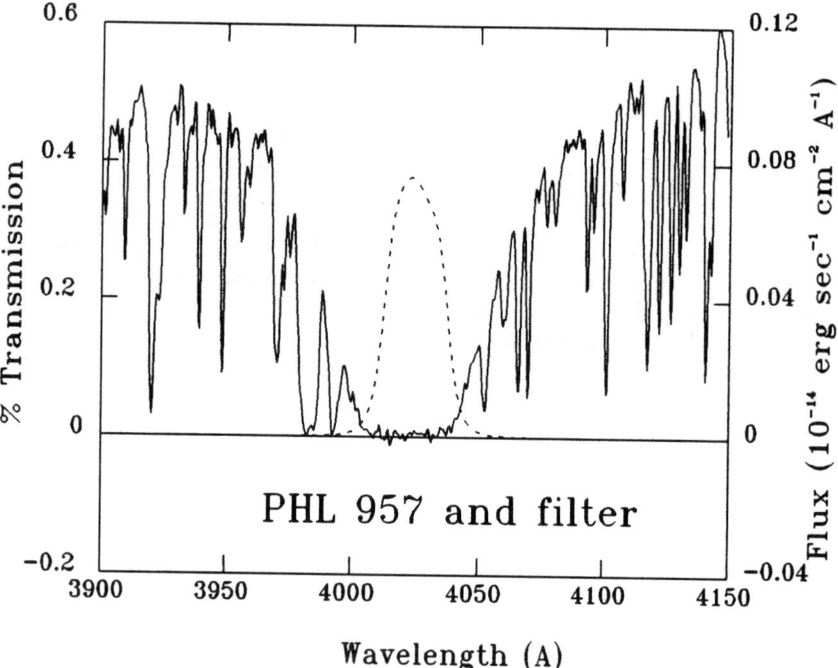

Fig. 1 1Å resolution MMT spectrum (Black, Chaffee and Foltz 1987) of PHL 957 (solid line) with our measured narrow–band filter response (dashed line) superposed. The filter width is 24Å (FWHM) centered at 4024.5Å with a peak transmission of 38%. The filters are blocked from the UV through 1.2μm, although the PHL 957 filter was discovered to show a small but significant leak in the green.

Our broad- and narrow-band images are shown in Figures 2-4, and our results are listed in Table II. These results were derived by measuring the flux in a 5 × 5 pixel (3."5 × 3."5) square box centered on the QSO, after subtracting sky (measured by fitting a second order surface nearby).

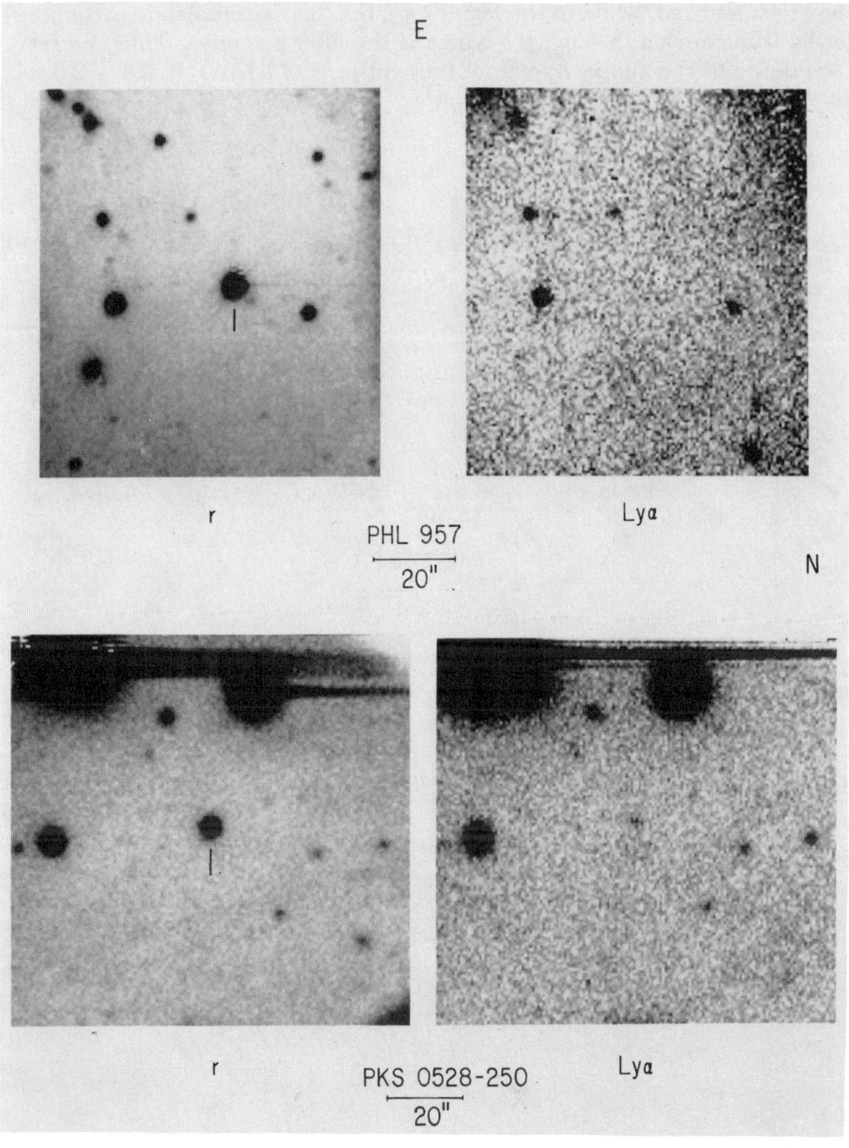

Fig. 2 Images of the fields of the QSOs PHL 957 and Pks 0528–250. On the left are broad-band "r" images; on the right are the narrow-band Lyα images. North is at the right and east is up. The image scale is $\sim$ 0.7 arcsec per pixel. Leakage through the PHL 957 filter has been subtracted from the image as described in the text, producing the artificial looking profile.

The (1σ) error is the standard deviation of the 8 surrounding 5×5 pixel boxes. In only 3 cases, PHL 957, Pks 0528–250 (Smith *et al.* 1989) and 0836+113 (Cohen *et al.* 1989) was any positive flux detected. In the former, this is accounted for by the flux through three small (< .03%) leaks in the green and red, while in the latter two, the flux is completely accounted for by transmission through the wings of the filter response. Thus, we set (3σ) limits to the flux in Lyα from these objects of l(Lyα) $= 0.3 - 2.0 \times 10^{-16}$ ergs cm^{-2} s^{-1}.

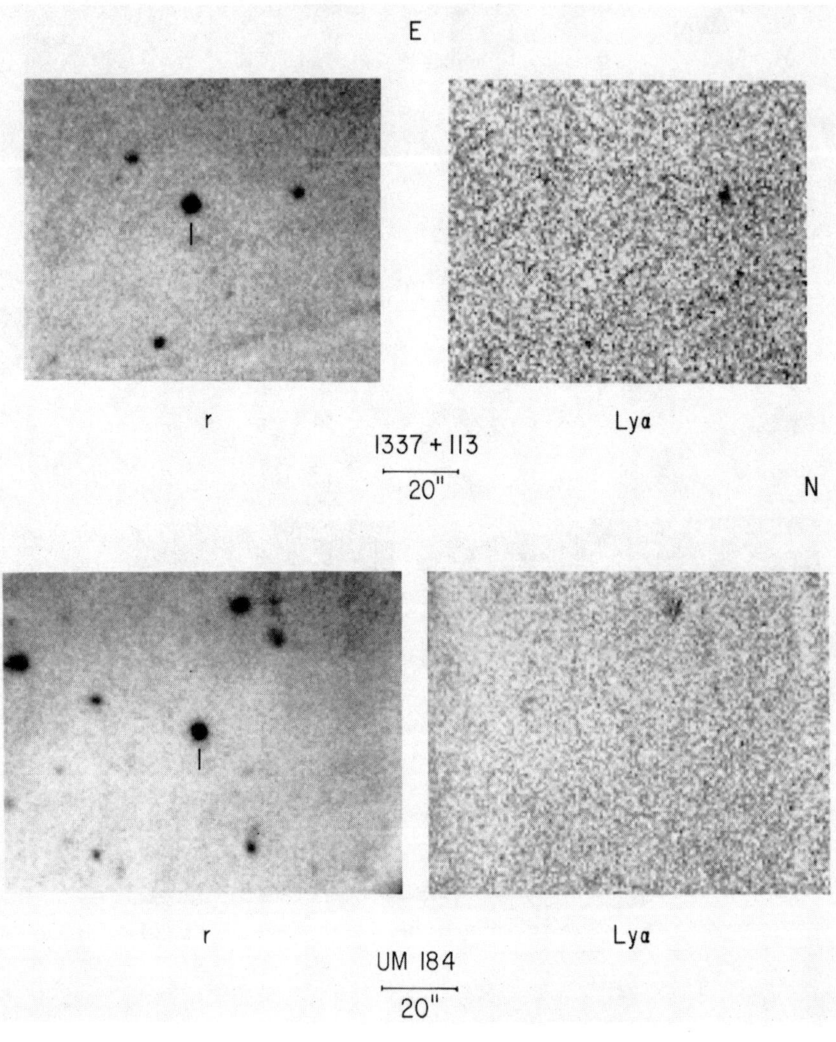

Fig. 3 Images of the fields of the QSOs 1337+113 and UM 184. On the left are broad–band "r" images; on the right are the narrow–band Lyα images. North is at the right and east up. The image scale is ~ 0.7 arcsec per pixel.

For a reasonable range of cosmological parameters ($H_0 = 50$ km s^{-1} Mpc^{-1} and $q_0 = 0$; $H_0 = 100$ km s^{-1} Mpc^{-1} and $q_0 = 0.5$), this limits the Lyα luminosities of these systems to: $L(Ly\alpha) \lesssim 10^{42-43}$ erg s^{-1}. For the purposes of detailed comparisons we will adopt $H_0 = 50$ km s^{-1} Mpc^{-1} and $q_0 = 0.5$ elsewhere in this paper; our limits then represent Lyα luminosities:

$$L(Ly\alpha) \approx 1 - 7 \times 10^{42} \text{erg s}^{-1}.$$

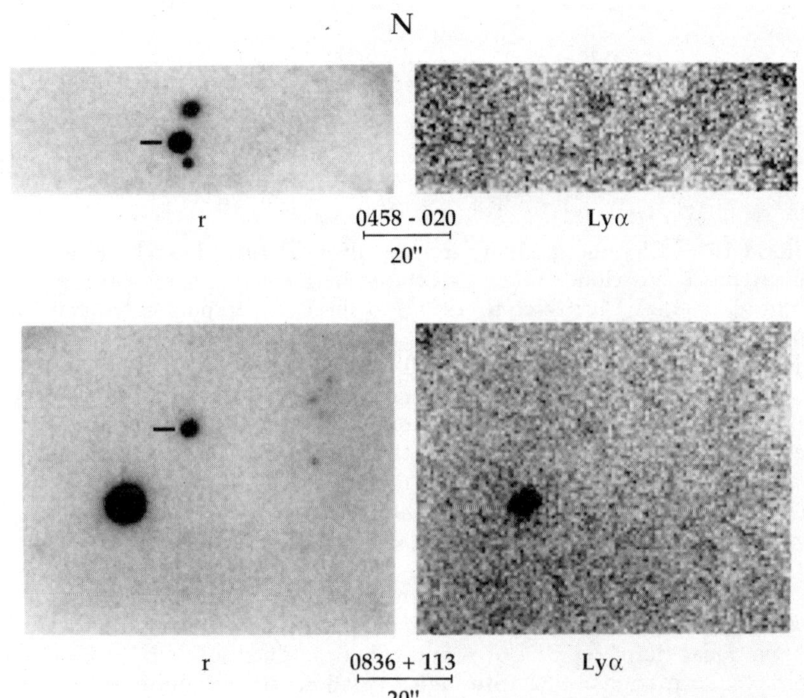

Fig. 4 Images of the fields of the QSOs Pks 0458–020 and 0836+113 On the left are broad–band "r" images; on the right are the narrow–band Lyα images. North is up and east is at the left. The image scale is ~ 0.7 arcsec per pixel.

Other results from imaging are presented in Wolfe (1989) who presents similar results for Pks 0458–020. Spectroscopic upper limits through a 2.″5 aperture centered on the QSOs PHL 957 and MC 3 1331+170 (Foltz, Chaffee, and Weymann 1986) and spectroscopic detections through a narrow slit centered on the QSOs PHL 957 and 0836+113 (Hunstead and Pettini 1989; Pettini, Boksenberg and Hunstead 1989) are consistent with our limits.

Comparison with Other Objects and Models

In Table III, we compare our Lyα limits and those of the other authors (above) to Lyα fluxes measured for 3C radio galaxies (Spinrad *et al.* 1985;

TABLE II Observed Lyman α Fluxes

QSO	Measured Flux Limit (3σ) (ergs cm^{-2}s^{-1})	Luminosity (ergs s^{-1})
PHL 957	$<0.8 \times 10^{-16}$	$<3 \times 10^{42}$
0458-020	$<2.0 \times 10^{-16}$	$<7 \times 10^{42}$
0528-250	$<1.2 \times 10^{-16}$	$< 7 \times 10^{42}$
0836+113	$<0.3 \times 10^{-16}$	$< 1.3 \times 10^{42}$
1337+113	$<0.6 \times 10^{-16}$	$< 4 \times 10^{42}$
UM 184	$<0.4 \times 10^{-16}$	$<2 \times 10^{42}$

Spinrad 1988; Chambers, Miley, and van Breugal 1988; Lilly 1988) and the extended Lyα clouds found associated with some of these galaxies (Spinrad 1988, McCarthy *et al.* 1987), to the Lyα companions to two QSOs (Spinrad 1988, Djorgovski *et al.* 1985, Schneider *et al* 1986), to a high–redshift normal galaxy (Cowie and Lilly 1989), and to nearby star forming galaxies (Hartmann *et al.* 1988). In most cases, the high redshift objects emit considerably more Lyα than do our Lyα–disk absorption objects. Only the nearby star forming galaxies and the high–redshift field galaxy emit Lyα in amounts comparable to our upper limits!

Alternatively, we can compare our results to the Lyα emission predicted by various models of young galaxies. We compare our results to three somewhat different models.

Meier's (1976) models follow Larson's (1974) hydrodynamical models of galaxy collapse employing stellar population synthesis to estimate the observational characteristics of young galaxies. The models have epochs of rapid star formation $z > 2$ with high Lyman continuum luminosities. Such systems, if optically thick to Lyman continuum radiation, will exhibit Lyα emission with L(Lyα) $\approx 10^{44}$ erg s^{-1}.

Shull and Silk (1979) consider radiative shocks from supernovae during bursts of star formation in young galaxies. They find that, depending on the initial mass function, the Lyα emission from SN shocks can exceed that from H II regions around hot young stars. There will be a steady state with a network of overlapping supernova remnants, and if there is comparable emission from H II regions, there will be strong Lyα emission: L(Lyα) $\gtrsim$ 10^{44} erg s^{-1} in these models.

Cox (1985) calculates the behavior of the ISM during the formation of a *disk galaxy.* Disk formation begins at approximately t $\approx 10^9$ yr after the Big Bang, corresponding to redshifts of order z = 5–10 for the cosmological parameters considered here. At this point the disk would have an extent of the order of R $\approx$ 20 kpc, a scale height h $\approx$ 200 pc, and a column density of the order of $N_{HI} \approx 10^{22}$ cm^{-2}. For such a system the Lyα emission is: L(Lyα) $= 2 \times 10^{45}$ erg s^{-1}.

In all cases the galaxy formation models predict much higher Lyα fluxes for young galaxies than we observe.

TABLE III Lyman α Galaxies

Object	z	l(Lα) erg cm^{-2}s^{-1}	L(Lα) erg s^{-1}
3C Radio Galaxies	2–3	10^{-15}–10^{-14}	10^{43}–10^{45}
3C Lα Clouds	~1.8	10^{-15}–10^{-14}	$\sim 2 \times 10^{44}$
QSO Lα Companions	3.2	$3\text{–}4 \times 10^{-16}$	$\sim 3 \times 10^{43}$
Nearby Star-Forming Galaxies	0.3–0.5	$\sim 10^{-13}$	10^{41}–10^{42}
High z "Normal" Galaxy	3.38	$\sim 10^{-16}$	$\sim 10^{43}$
Lα Disks-Lick Imaging	2.0–2.8	$\leq 2 \times 10^{-16}$	$<1\text{–}7 \times 10^{42}$
Lα Disks-Other Works	2.0–2.5	$\leq 4 \times 10^{-16}$	$\leq 10^{42}$–$\leq 10^{43}$
Model Calculations	1–3		10^{44}–10^{45}

Star Formation

In order to convert our Lyα upper limits to star formation rates, we must use some prescription. We assume:

1. All $\lambda < 912$Å UV photons produced by the stellar population are absorbed by the gas and result in the emission of a Lyα photon;
2. That the Lyα photons are not absorbed by dust or destroyed by other mechanisms;

and we follow the prescription developed by Kennicutt (1983) for estimating star–formation rates from Hα emission from nearby late–type galaxies, employing a modified Miller–Scalo (1979) initial mass function which has a slope of 2.5 for stars more massive than one solar mass and mass limits $0.1 \leq M \leq 100\ M_\odot$. We convert to Ly$\alpha$ fluxes assuming a Case B ratio of Lα/Hα. This prescription yields a "steady state" star–formation rate:

$$\mathrm{SFR}(M_\odot\ \mathrm{yr}^{-1}) \approx L(\mathrm{Ly}\alpha)/10^{42}\ \mathrm{erg\ s}^{-1}.$$

Our Lyman α limits translate directly into star formation rates of approximately 1–7 $M_\odot$ yr^{-1}.

Effects of Dust

The greatest uncertainty in the above star formation rates lies in the effects of dust. The presence of a "normal" amount of dust would have severe effects on the emission of Lyα from the disk galaxy and would also redden and dim the QSO continuum. Following the relation between H I column density and extinction, we would expect values of $E_{B-V} \approx 0.1\text{–}0.5$. Arguments against the presence of this much extinction are qualitative:

1. The low metal abundance and lack of H_2 argue against the presence of grains;
2. The abundance pattern argues against the selective depletion of refractory elements, again suggesting a lack of grain formation (Pettini, Boksenberg, and Hunstead 1989; Meyer, Welty, and York 1989);
3. PHL 957 has one of the bluest spectra in a sample of QSOs (Soifer *et al.* 1983), despite that fact that a column density of $N_{H\ I} = 2.5 \times 10^{21}$ would produce an amount of reddening of more than several tenths between λ1216 and visual wavelengths in the Lyα–disk system;

4. Normal amounts of reddening would select against the inclusion in magnitude limited samples of QSOs with Lyα–disks along the line of sight, and there are already too many of such systems.

However, the presence of an otherwise undetectable amount of dust can still have large effects on the Lyα emission. Fall, Pei and McMahon (1989) have analyzed the continuous energy distributions of the QSO sample of Sargent, Steidel and Boksenberg (1989), to search for evidence for reddening in the spectra of QSOs with damped–Lyα absorption systems. They find evidence for a small amount of reddening in the QSOs with damped–Lyα lines. At the 3σ significance level, they detect reddening corresponding to a dust–to–gas ratio between 0.05 to 0.25 that of the Galaxy, for extinction laws appropriate to metal–poor or normal metallicity systems.

For $N_{H\ I} = 1 \times 10^{21}$ and a reddening law appropriate to the SMC, this produces a reddening of $E_{B-V} = 0.011$. Following Bonilha *et al.* (1979), we may calculate the escape of Lyα from the midplane of an optically thick, plane parallel slab with dust. If we take a value $E_{B-V} = 0.01$, only 13% of the Lyα photons escape.

The star–formation rate consistent with our Lyα flux density limits will then be correspondingly larger, of the order of 8–50 $M_{\odot}\ yr^{-1}$, if there is a small, but significant amount of dust. This amount of star formation is roughly consistent with the models of star formation discussed above.

SUMMARY

The work on the damped–Lyα systems supports a number of conclusions and inferences about the formation and early history of galactic disks:

1. Gaseous disk–like structures form early, at $z \gtrsim 2$–3 and are probably a factor of 2–3 (or more) larger in diameter than current disks;
2. Star formation in these structures is not actively occurring, is well hidden, occurring in very clumpy, dusty regions such that Lyα does not escape, or is episodic with a duty cycle smaller than about 10%;
3. Some enrichment has occurred by this epoch, but abundances may be low. Enrichment in this scenario might be produced by the formation and evolution of the spheroid;
4. The typical dust–to–gas ratio is apparently much lower than that of the solar neighborhood.

Some Evolutionary Considerations

Conventional closed–box disk galaxy evolution models in which the star–formation rate is proportional to the available gas mass have star–formation rates which are considerably in excess of the limits set here (Rocca–Volmerange and Guiderdoni 1987), which are similar to those in the Galaxy today. Other well known problems include the paucity of low–metallicity late–type dwarfs in the solar neighborhood. There are, however, relatively conventional global models which *are* roughly consistent with the scenario developed here. Pagel (1989) has described simple analytical models which he has used to model the chemical evolution of the solar neighborhood. These models enrich the early disk material with nucleosynthesis products from the evolution of the halo and allow for infall to solve the problem of the dwarf abundance distribution. Finally, depending on the prescription for infall, the gas mass, and hence the star–formation rate, reaches a maximum a few Gyr after the onset of disk star–formation. We have calculated models (Cohen *et al.* 1989) with the following properties:

1. The halo evolves over the first 2 Gyr enriching the disk to an abundance about 0.1 solar;
2. The initial disk star–formation rate, taken to begin at the "end" of halo formation, is approximately equal to the current global SFR, reaching a maximum rate about a factor of three higher after 3–4 Gyr;
3. The model disk reaches solar abundance after about 8 Gyr and a gas fraction of about 10% at about 12 Gyr and "looks" much like a current disk galaxy. Due to infall the final disk mass is about five times the mass of gas in the original proto–disk.

Alternately, if we consider models in which the gas *surface density* determines the SFR (Dopita 1985) then the inference that these early disks have large surface area would mean lower surface density and a correspondingly lower early SFR. Finally we note the increasing body of evidence that suggests that the SFR is not directly coupled to the gas density, but is determined by stochastic processes in the disk (*c.f.* Scalo 1989).

ACKNOWLEDGEMENTS

We wish especially to thank Barbara Uchida, David Moore, and Joseph Burns for their considerable help with this work. We also wish to thank the staff of Lick Observatory in developing the CCD system and assisting with these observations. Special thanks are due to Dr. Craig Foltz for providing us with high resolution MMT observations of PHL 957 and Pks 0528–250. This work was partially supported by the California Space Institute, by NSF grant AST84–14658, by NASA contract NAS 5 29293 for the HST Faint Object Spectrograph, and by NSF grant AST86–14510 to Lick Observatory in support of CCD development.

REFERENCES

Black, J. H., Chaffee, F. H., and Foltz, C. B. 1987, *Ap. J.*, **317**, 442.

Bonilha, J. R. M., Ferch, R., Salpeter, E. E., Slater, G., and Noerdlinger, P. D. 1979, *Ap. J.* **233**, 649.

Bosma, A. 1981, *Astron. J.*, **86**, 1825.

Briggs, F. H., Wolfe, A. M., Liszt, H. S., Davis, M. M., and Turner, K. L. 1989, *Ap. J.*, **341**, 650.

Chambers, K. C., Miley, G. K., and van Breugel, W. J. M. 1988, *Ap. J. (Letters)*, **327**, L47.

Cohen, R. D., Smith, H. E., Uchida, B., and Moore, D. 1989, in preparation.

Cowie, L. L. and Lilly, S. J. 1989, *Ap. J. (Letters)*, **336**, L41.

Cox, D. P. 1985, *Ap. J.*, **288**, 465.

Dopita, M. A. 1985, *Ap. J. (Letters)*, **295**, L5.

Djorgovski, S., Spinrad, H. McCarthy, P., and Strauss, M. A. 1985, *Ap. J. (Letters)*, **299**, L1.

Fall, S. M., Pei, Y. C., and McMahon, R. G. 1989, *Ap. J. (Letters)*, **341**, L5.
Foltz, C. B., Chaffee, F. H., and Weymann, R. J. 1986, *Astron. J.*, **92**, 247.
Hartmann, L. W., Huchra, J. P., Geller, M. J., O'Brien, P., and Wilson, R. 1988, *Ap. J.*, **326**, 101.
Hunstead, R. W. and Pettini, M. 1989 in *The Epoch of Galaxy Formation*, ed. C. S. Frenk, R. S. Ellis, T. Shanks, A. F. Heavans, and J. A. Peacock (Dordrecht: Klewer), p. 115.
Kennicutt, R. C. 1983, *Ap. J.*, **272**, 54.
Kirshner, R. P., Oemler, A., Schechter, P. L., and Shectman, S. A. 1983, *Astron. J.*, **88**, 1285.
Larson, R. B. 1974, *M.N.R.A.S.*, **166**, 585.
Lilly, S. J. 1988, *Ap. J.*, **333**, 161.
McCarthy, P. J., Spinrad, H., Djorgovski, S., Strauss, M. A., van Breugel, W., and Liebert, J. 1987, *Ap. J. (Letters)*, **319**, L39.
Meier, D. L. 1976, *Ap.J.*, **207**, 343.
Meyer, D., Welty, D., and York, D. 1989, *Ap. J. (Letters)*, **343**, L37.
Miller, G. E. and Scalo, J. M. 1979, *Ap.J. Suppl.*, **41**, 513.
Pagel, B. E. J. 1989, *Rev. Mex. Astron. y Astrofis.*, in press.
Pettini, M., Boksenberg, A., and Hunstead, R. W. 1989 in *The Epoch of Galaxy Formation*, ed. C. S. Frenk, R. S. Ellis, T. Shanks, A. F. Heavans, and J. A. Peacock (Dordrecht: Klewer), p. 107.
Rocca-Volmerange, B. and Guiderdoni, B. 1987, *Astr. Ap.*, **175**, 15.
Sargent, W., Steidel, C., and Boksenberg, A. 1989, *Ap. J. Suppl.*, **69**, 703.
Scalo, J. M. 1989, preprint.
Schneider, D. P., Gunn, J. L., Turner, E. L., Lawrence, C. R., Hewitt, J. N., Schmidt, M. and Burke, B. F. 1986, *Astron. J.*, **91**, 991.
Shull, J. M. and Silk, J. 1979, *Ap. J.*, **234**, 427.
Smith, H. E., Cohen, R. D., Burns, J. E., Moore, D. J., and Uchida, B. 1989, *Ap. J.*, **347**, in press.
Smith, H. E., Cohen, R. D., and Bradley, S. E. 1986, *Ap. J.*, **310**, 583.
Smith, H. E., Turnshek, D. A., and Wolfe, A. M. 1983, in *Quasars and Gravitational Lenses*, ed. J. P. Swings (Liege: U. of Liege), p. 567.
Soifer, B. T., Neugebauer, G., Oke, J. B., Mathews, K., and Lacy, J. H. 1983, *Ap. J.*, **265**, 18.
Spinrad, H. 1989 in *The Epoch of Galaxy Formation*, ed. C. S. Frenk, R. S. Ellis, T. Shanks, A. F. Heavans, and J. A. Peacock (Dordrecht: Klewer), p. 39.
Spinrad, H., Filippenko, A.V., Wyckoff, S., Stocke, J. T., Wagner, M., and Laurie, D. G. 1985, *Ap. J. (Letters)*, **299**, L7.
Turnshek, D. A., Wolfe, A. M., Lanzetta, K. M., Briggs, F. H., Cohen, R. D., Foltz, C. B., Smith, H. E., and Wilkes, B. J., 1989, *Ap. J.*, **344**, 567 (Paper III).
Wolfe, A. M. 1989 in *The Epoch of Galaxy Formation*, ed. C. S. Frenk, R. S. Ellis, T. Shanks, A. F. Heavans, and J. A. Peacock (Dordrecht: Klewer), p. 101.
Wolfe, A. M., Turnshek, D. A., and Lanzetta, K. M. 1989, in preparation (Paper IV).
Wolfe, A. M., Turnshek, D. A., Smith, H. E., and Cohen, R. D. 1986, *Ap. J. Suppl.*, **61**, 249 (Paper I).
Wolfe, A. M. *et al.* 1990, in preparation (Paper II).

A SEARCH FOR PRIMEVAL GALAXIES AT Z=2

C.J. PRITCHET and F.D.A. HARTWICK
Department of Physics and Astronomy, University of Victoria, P.O. Box 1700, Victoria, BC V8W 2Y2, CANADA

ABSTRACT We report on the first search for primeval galaxies at redshifts lower than 2. An area of 0.1 deg^2 has been searched for emission line objects at 3500Å; none have been found down to a threshhold of 2.7×10^{-17} erg cm^{-2} s^{-1} arcsec^{-2} (1σ). From this limit constraints have been placed on the formation of galaxies at redshifts 1.8 or greater. The resultant constraints are marginally in conflict with models for the formation of galaxies in Universes dominated by Cold Dark Matter.

1. INTRODUCTION

Based on the seminal work of Eggen, Lynden-Bell, and Sandage (1962) and Larson (1974), galaxy formation was viewed in the 1970s as a rapid collapse of large monolithic structures, with a relatively early formation epoch. Most of the effort in searching for primeval galaxies (PGs) over the past decade or so has therefore concentrated on large redshifts (z $\gtrsim$4), and very tight observational constraints (*e.g.* Cowie 1988, Pritchet and Hartwick 1987) have now been placed on models that invoke rapid collapse at early epochs (Larson 1974, Meier 1976). The reader is referred to a review by Koo (1986) for more details on this work.

Meanwhile, a wide variety of observational and theoretical evidence now seems to favour the view that galaxy formation is a relatively low redshift phenomenon. *(i)* Theoretical models of galaxy formation in Universes dominated by cold dark matter (CDM) predict that the formation of luminous galaxies is delayed until relatively recent epochs (*e.g.* Baron and White 1987, Silk and Szalay 1987). *(ii)* Several massive star forming galaxies have been discovered near $z \approx 1.8$ (McCarthy *et al.* 1987, Djorgovski 1987). The properties of these objects (Ly α flux, star formation rate, lumpy structure, luminosity) is in fact in reasonable accord with theoretical models (Baron and White 1987). These objects are certainly very young, and may be close to the PG epoch. *(iii)* The QSO redshift distribution peaks near $z = 1.5 - 2$ (Crampton *et al.* 1987), possibly indicating a maximum in the merger and accretion rate of small lumps at that epoch. *(iv)* The cutoff in the luminosity function of faint white dwarfs (Winget *et al.* 1987) suggests that the Galactic disk is at most 10 Gyr old, but the globular cluster system is closer to 15 Gyr. This argues that the bulk of star formation that built up the Milky Way must have taken place more recently than z=2.

None of the above arguments is compelling by itself, and there is some evidence that at least *some* galaxy formation must have taken place at earlier epochs (*e.g.* Lilly 1988). Nevertheless, the combined evidence is suggestive, and the hypothesis that galaxy formation is primarily a low ($z < 2$) redshift phenomenon must be seriously considered.

In this paper we report on the first attempt to search for Ly α emission from PGs at redshifts below 2. The details of this work will appear elsewhere (Pritchet and Hartwick 1990).

2. OBSERVATIONS

Our observations were made with a TI 800×800 CCD on the 4m telescope at CTIO. We imaged a total of 0.10 deg^2 through broadband U and a 3500Å interference filter ($\Delta\lambda = 100$Å). Exposures were 3×10 min (shifted) through the 3500Å filter, and 10 min in U. Initial flat fielding was done with dome flats. Final flat fielding was done with a median night sky flat (calculated separately for each night of observation). Other preprocessing techniques were standard.

To search for emission line PGs, we blinked the 3500Å and U frames (after appropriate intensity normalization). The $35 - U$ color of stars is less than 0.1 mag for all spectral types, so that objects that "blink" are almost certainly emission line objects.

3. RESULTS

We found *no* objects in our survey with emission line surface brightness that was stronger than 2.7×10^{-17} erg cm^{-2} s^{-1} arcsec^{-2} (1σ). For objects with structure similar to that of 3C326.1, the limiting flux is 7×10^{-16} erg cm^{-2} s^{-1}. Fainter than these limits there *may* be emission line objects and even diffuse structure, but it is difficult to decide on the reality of this structure without deeper observations; such observations are planned for Sep 1989.

4. DISCUSSION

Here we compare our result to the theoretical predictions of Baron and White (1987). Although we surveyed a fairly narrow redshift range (our bandwidth corresponded to $1.84 \leq z \leq 1.92$), we are actually sensitive to galaxy formation over a relatively wide redshift interval. In the Baron and White simulations galaxies remain luminous (at about the same emission line luminosity) from $t_{coll}/5$ to t_coll. Hence objects that started to coalesce as early as $z = 7.5$ ($q_o = 0.5$) would still be visible in their bright phase at z=1.8.

Following the Baron and White procedure, we calculate the expected surface density of emission line PGs in our survey. The result is somewhat dependent on the assumed cosmology and collapse time t_{coll}. For H_o in the range 50 – 100 km s^{-1} Mpc^{-1}, $q_o = 0.1 - 1$, and collapse times between 1 and 2 Gyr, we find that our limits for 3C326.1-like objects just intersect the model predictions. If the objects are stellar then our limits fall *below* model predictions by a factor 30 or more.

There are sufficient uncertainties in the models that discrepancies less than 1 order of magnitude probably should not be taken too seriously. In addition there are the well known caveats regarding the observability of Ly α in starburst galaxies (*e.g.* Hartmann *et al.* 1984), although at some point early enough in the history of a PG the dust to gas ratio must be sufficiently low that Lyα will emerge. (The very existence of 3C sources with strong Lyα provides confirmation of this argument.) We conclude that it is too early to say whether our observations really violate the predictions of CDM simulations; clearly we need to observe a larger area with greater sensitivity to be sure.

REFERENCES

Baron, E., and White, S.D.M. 1987, *Ap. J.*, **322**, 585.
Cowie, L. 1988, in *The Post-Recombination Universe*, ed. N. Kaiser and A.N. Lasenby (Dordrecht: Kluwer), p. 1.
Crampton, D., Cowley, A., and Hartwick, F.D.A. 1987, *Ap. J.*, **314**, 129.
Djorgovski, S. 1987, in *Towards an Understanding of Galaxies at High Redshifts*, ed. R. Kron and A. Renzini (Dordrecht: Reidel), p. 259.
Eggen, O., Lynden-Bell, D., and Sandage, A. 1962, *Ap. J.*, **136**, 748.
Hartmann, L.W., Huchra, J.P., and Geller, M.J. 1984, *Ap. J.*, **287**, 487.
Koo, D. 1987, in *The Spectral Evolution of Galaxies*, ed. C. Chiosi and A. Renzini (Dordrecht: Reidel), p. 419.
Larson, R. 1974, *M.N.R.A.S.*, **166**, 585.
Lilly, S. 1988, *Ap. J.*, **333**, 161.
McCarthy, P.J., Spinrad, H., Djorgovski, S., Strauss, M.A., van Bruegel, W., and Liebert, J. 1987, *Ap. J.*, **319**, L39.
Meier, D.L. 1976, *Ap. J.*, **207**, 343.
Pritchet, C.J., and Hartwick, F.D.A. 1987, *Ap. J.*, **320**, 464.
Pritchet, C.J., and Hartwick, F.D.A. 1990, in preparation.
Silk, J., and Szalay, A. 1987, *Ap. J.*, **323**, L107.
Winget, D.E. *et al.* 1987, *Ap. J.*, **315**, L77.

THE EVOLUTION OF GALAXIES AND GALAXY CLUSTERS ASSOCIATED WITH QUASARS

H. K. C. Yee
Department of Astronomy, University of Toronto,
Toronto, Ont., M5S 1A7, Canada

ABSTRACT Results from a number of systematic imaging surveys of fields around quasars are described and summarized. From these surveys, not only can correlations between the environments of quasars and quasar properties be derived, properties of associated galaxies and galaxy clusters at moderately high redshifts can also be measured. Quasars in general are found to be situated in regions of enhanced galaxy density. Specifically, radio-loud quasars are found in environments significantly richer than those of radio-quiet quasars. However, there is some preliminary evidence that some radio-quiet quasars are located at the edges of rich clusters. A significant fraction of radio-loud quasars are located at or near the centers of rich clusters of galaxies. Furthermore, compared to quasars found in poorer environments, quasars situated at the centers of rich clusters have undergone a much more rapid luminosity evolution in the redshift range between 0.6 and 0.4. This may be a direct consequence of the evolution of the physical conditions in the centers of rich clusters. Preliminary results from the study of clusters associated with quasars show that they may systematically have properties which are believed to be favorable to the existence of quasar activity. The luminosity function (LF) of the associated galaxies at $z{\sim}0.6$ is also measured, and is found to be brighter than the local LF by $\sim 0.9 \pm 0.3$ mag, which can be interpreted as luminosity evolution in early-type galaxies in moderately rich environments.

1. INTRODUCTION

Quasars, by virtue of their high redshift and hence implied large cosmological distance, provide a unique opportunity for us to sample the universe at an earlier evolutionary time. Because galaxies have a tendency to cluster, quasars, being the nuclei of galaxies, can also serve as excellent markers for locating less remarkable objects at similar redshifts, allowing us to study the evolution of more ordinary galaxies up to redshifts as high as 1. In the past few years, a number of collaborators and I have been systematically surveying the environment of a large number of quasars via direct imaging and multi-object spectroscopy. The motivations for such observations included confirming the cosmological interpretation of quasar redshift, searching for relationships between environment and quasar activity, and studying galaxies and galaxy clusters at high redshifts.

The initial survey of fields around quasars (Green and Yee 1984, and Yee and Green 1984, hereafter, paper II) consisted of two samples of optically-bright quasars with no selection restriction on the redshift – a radio-loud sample with quasars brighter than ~17.5 mag, and the Palomar-Green sample of optically-selected quasars brighter than ~16.5 mag. These fields were observed with the Palomar 1.5 m telescope with a SIT camera. The average completeness limit for the detection of galaxies is ≈ 21.5 mag, allowing one to study galaxies associated with quasars up to a redshift of ~0.4. Subsequently, the quasars in the redshift range of 0.30 – 0.65 were re-observed with the Steward 2.3 m telescope using CCD detectors (Yee, Green, and Stockman 1987, and Yee and Green 1987, hereafter, paper IV) obtaining an average completeness magnitude of about 22.5 mag, enabling us to determine the environment of quasars up to a redshift of ~0.6. In this paper, a summary of published results that we have obtained from these imaging surveys will be presented.

We have since expanded the survey to include deeper observations at the CFHT of the quasars with redshifts between 0.5 and 0.9 from the paper II samples. In addition, an imaging survey of Parkes radio-loud quasars has been carried out at CTIO. More recently, to complement the samples of bright quasars, a large sample of optically-faint quasars, with an apparent magnitude limit of 19.5, has been surveyed using the Steward 2.3 m (Ellingson 1989; also see Ellingson, in these proceedings). Furthermore a sample of faint optically-selected quasars from the AAT-fibre quasar survey has also been imaged at La Palma using the INT 2.5 m (Boyle, Shanks, and Yee 1988). Preliminary results from these new surveys will also be reported.

The new observations strongly support the evidence from the previous surveys of luminous quasars that quasars situated in rich clusters have evolved substantially between the redshift of 0.6 and 0.4, implying a strong evolution in the properties of galaxy clusters hosting quasars. We will also present recent results from the investigation of clusters of galaxies associated with quasars in an effort to identify conditions favorable for the formation of quasars in rich clusters. Finally, luminosity functions (LF) of the associated galaxies are derived, with results indicating a significant brightening of the LF at z~0.6.

In this paper, unless stated otherwise, $H_o = 50$ km sec^{-1}Mpc^{-1} and $q_o = 0$ are used throughout.

2. QUASAR ENVIRONMENTS

2.1 Method of Analysis

Due to the lack of information on the actual spatial distribution of galaxies in the quasar fields, the quantitative analysis of the richness of the environments of quasars from direct imaging data is a complex process involving many assumptions. For data, one requires catalogs of magnitudes and positions of the galaxies detected, well established completeness limits, and well determined background galaxy counts as a function of magnitude. To derive the absolute richness of the associated groups or clusters of galaxies, we need to correct for two sampling effects in counting galaxies: different sampling depths in the galaxy luminosity function and different metric areas imaged around the quasars. These effects are due to both the fact that the quasars have different redshifts, and that the observations were made under varying conditions and with different instruments. Thus, one needs to know or assume an averge LF and a spatial distribution of the associated galaxies in order to correct the measures of excess galaxy counts from different fields to the same standard depth and metric area. Furthermore, at $z \gtrsim 0.3$, the choice of cosmological models and galaxy evolution becomes significant, and it is thus important to choose self-consistent models for the analysis.

The method we adopted for obtaining quantitative measures of the absolute richness of quasar environment is the quasar-galaxy covariance function. The details of the method, essentially that used by Longair and Seldner (1978) in their study of the environment of radio galaxies, are outlined in paper IV. Specifically, we derive for each quasar the amplitude B_{gq} of the spatial covariance function $\zeta(r)$ as defined by:

$$\zeta(r) = B_{gq} r^{\gamma}$$

where γ, the power-law index of the spatial covariance function, is taken to be the canonical -1.77 found for the galaxy-galaxy covariance function. (*e.g.*, Seldner and Peebles 1978). The average surface density profile of the excess galaxies around the quasars is found to be consistent with such a spatial distribution. The B_{gq} amplitude for each quasar is computed by normalizing the angular covariance function amplitude (which is simply a measure of the fractional excess galaxy counts within a circular region around the quasar corrected for the sampling metric area) with a normalized galaxy LF. The galaxy LF is derived by averaging the actual magnitude distributions of excess galaxies in the quasar fields (see Section 3). The measured relative LF is normalized by fitting galaxy count models to the background counts, providing self-consistent sets of cosmological models and galaxy LFs. It is found that as long as one uses such self-consistent parameters, the covariance amplitudes derived are not a strong function of the cosmological model adopted.

2.2 Summary of Previous Results

The analysis of the Palomar SIT survey of fields around quasars showed conclusively that there is a significant enhancement in galaxy density around quasars (paper II). Furthermore, the distribution of the two-point angular covariance amplitudes is a function of redshift, in that there is no significant enhancement of galaxy counts for fields around quasars with $z \gtrsim 0.5$. Since the completeness limit of the survey is $\sim$21.5 mag, equivalent to first-ranking galaxies at $z \sim 0.5$, this dependence of the angular covariance amplitudes on z is interpreted as being consistent with the cosmological interpretation for the redshift of quasars.

The data from the first survey was of sufficient depth to obtain reasonable estismates of B_{gq}'s for low-redshift quasars. Applying the method of analysis outlined in Section 2.1 to the quasars with $z < 0.4$, it is found that, on average, quasars tend to be situated in regions with higher than average galaxy density. However, with the exception of one or two out of a sample of 56, none is found in regions comparable to the centers of Abell class 1 clusters.

The images obtained from the Steward 2.3 m CCD allowed us to derive B_{gq} up to a redshift of $\sim$0.6 (paper IV). For the $0.3 < z < 0.5$ subsample, both the radio-loud and radio-quiet quasars have environments similar to those of low-redshift quasars from the SIT survey. Using an average galaxy-galaxy covariance amplitude, B_{gg}, of 67.5 $\mathrm{Mpc}^{1.77}$ (Davis and Peebles 1983) as the fiducial reference, we found that for radio-loud quasars with $z \lesssim 0.5$, the average $< B_{gq}/B_{gg} >$ is about 2.8, and for the radio-quiet quasars, 2.0. For the $0.5 < z < 0.65$ small subsample of radio-loud quasars, however, we found a surprising and significant increase in the average richness of the environment, with $< B_{gq}/B_{gg} > \approx 8.0$. As many as half of the quasars are found in regions equivalent to the centers of clusters of richness of Abell class 1 or higher.

This apparent change in the preferred environment of radio-loud quasars can be interpreted as an indication that we are observing two different populations of optically bright radio-loud quasars. The population that is situated in the centers of rich clusters has undergone a drastic evolution in the 2 billion years or so between the redshifts of 0.6 and 0.4. The average luminosity of quasars in rich clusters has dimmed by several magnitudes causing them to be excluded in an apparent magnitude limited sample; whereas radio-loud quasars situated in poorer environments have not evolved in such a significant fashion, and hence dominate the sample at $z \lesssim 0.5$.

This result has two important implications. One is that quasar evolution may be intimately tied to the physical conditions of their environment. The other is that the physical conditions in rich clusters harboring bright quasars may have undergone significant changes between the epoch of $z \sim 0.6$ and $z \sim 0.4$. These conclusions will be elaborated upon and strengthened in the following sections.

2.3 Further Evidence for the Evolution of Quasar Environment

Motivated by the results of the Steward survey, we have expanded our sample to increase the number of quasars, and to cover both a larger redshift and luminosity range. The CTIO imaging survey of Parkes quasars is intended to cover a wider range of properties of radio-loud quasars to search for possible correlations between radio emission properties and the environment. Preliminary results from a subset of quasars with $0.7 \gtrsim z \gtrsim 0.5$ are available at this time. The B_{gq}'s for 8 radio-loud quasars in this range have been derived, almost doubling the sample in this redshift range from that of the Steward survey. The additional data support the conclusion of paper IV. The average $< B_{gq}/B_{gg} >$ ratio is 7.8 compared to 8.0 for the Steward sample. The distributions of the B_{gq}'s of the radio-loud quasars in the two redshift samples are plotted as histograms in Figure 1 to illustrate the dependence of the environment on redshift. Combining the two samples and dividing them at $z = 0.5$, the higher z sample is found to have a higher average B_{gq} at the 99% confidence level.

An equally important step in the further investigation of this evolutionary effect is expanding the sample to fainter magnitudes. If the interpretation of the drastic dimming of quasars in rich clusters is correct, at lower redshift, we should discover fainter radio-loud quasars located in rich environments. Furthermore, such observations will allow us to estimate the approximate time scale in the disappearance of these quasars. Thus, a survey of faint quasars (down to ~19.5 mag) was carried out by Ellingson as part of her thesis work. I will report briefly on the main results which are given in more detail in Ellingson (1990).

The faint Steward sample are found to have B_{gq}'s similar to those from the brighter quasars in the same redshift range. However, a small number of radio-loud quasars at $z < 0.5$ with absolute magnitudes fainter than −25 were found in regions with $B_{gq} > 500$, representing environments as rich as Abell class 1 clusters. This is consistent with the scenario that quasars situated in the centers of rich clusters have dimmed between redshift of 0.6 and 0.4. A powerful way to illustrate the difference in the evolution of radio-loud quasars in clusters and poorer environments is shown in Figure 1 in the paper by Ellingson (1990) in these proceedings. The quasars are divided into two samples according to the environment: those with $B_{gq} > 500$ which are assumed to be located in clusters as rich as Abell class 1, and the ones with $B_{gq} < 500$. The figure plots the absolute magnitudes of the quasars, M_{qso}, versus redshift for the two samples. Here, we see that the rich environment sample has a very different distribution from that of the small B_{gq} sample, in that the former shows a drastic dimming in M_{qso} as a function of redshift. This difference in the distribution demonstrates conclusively that the change in the average B_{gq} for radio-loud quasars is not due to luminosity selection effects in the quasar sample. If we assume that the points of the large B_{gq} subsample represent the envelope of the highest luminosity quasars in rich clusters, the drop in the ensemble luminosity is approximately 1 magnitude per 0.5 billion years.

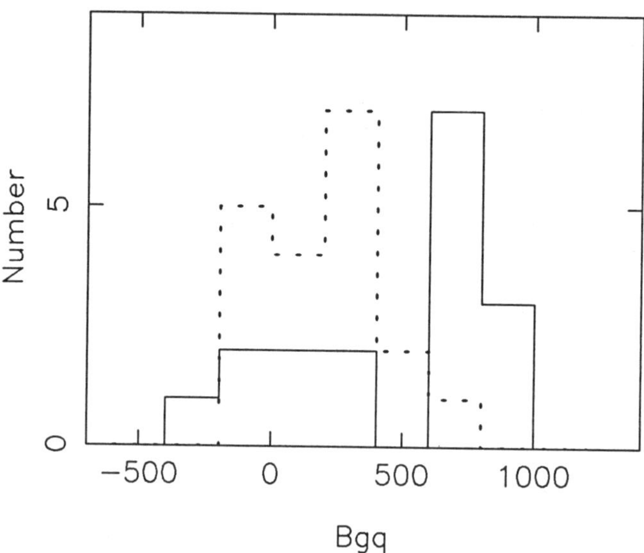

Figure 1. Histograms of the distributions of B_{gq} for radio-loud quasars for $z < 0.5$ (dotted lines), and $z > 0.5$ (solid lines).

We note that similar investigations by Hill and Lilly (Hill 1989) and Yates, Peacock, and Miller (1989) on the environment of Fanaroff-Riley (FR) class II radio galaxies – which have similar morphologies to steep-spectrum radio-loud quasars – have also found results that are consistent with ours, lending support to the conclusion that there may have been significant evolution in the environment of rich clusters over this redshift range.

This scenario of rapid fading of quasars in clusters at $z \sim 0.5$ leads to an interesting question: Where are the remnants of these once bright quasars that used to be the central objects in rich clusters? One possible candidate is the relatively weak FR class I radio galaxies which are often the brightest members of nearby rich clusters and are normally considered optically quiescent. To study these objects, high signal-to-noise ratio spectra of the nuclei of a small sample of 6 3CR radio galaxies in cluster environment have been obtained at CFHT. We found that contrary to the original classification of them as "absorption line" galaxies they all have weak Seyfert-like emisssion-line spectra; including broad Hα (Yee and De Robertis 1989). Hence, these radio galaxies may represent the faintest example of active nuclei occurring in the centers of rich clusters. Detailed comparisons between the properties of these clusters with those of their counterparts at higher redshifts and hosting more luminous active objects may provide important clues concerning the relationship between quasars and their environment.

2.4 Environments of Radio-Quiet Quasars

In papers II and IV, we found a marginally significant difference between the environments of low-redshift radio-loud quasars and radio-quiet quasars in that the former are situated in slightly richer environment than the latter. With the addition of the new data from the considerably larger sample of the faint low-redshift quasar survey, it is found that we can conclude at a 99% confidence level that the average environment of radio-loud quasars is richer than that of radio-quiet quasars (Ellingson 1990). This implies that radio emission from quasars is a strong function of the environment. The result is consistent with the suggestion from investigations of the fuzz component of quasars that the host galaxies of radio-loud quasars are more similar to elliptical galaxies, whereas those of radio-quiet quasars are more like spiral galaxies. (Malkan 1984, Smith *et al.* 1987).

A further difference between radio-loud and radio-quiet quasars is that, from either the bright or faint Steward sample, there is no evidence that the preferred sites of radio-quiet quasars have undergone the same change at $z \gtrsim 0.5$ as found for radio-loud quasars. However, this conclusion is based on a small number of quasars as there is a shortage of available bright optically-selected quasars in the redshift range between 0.5 and 0.8 due to selection effects in quasar surveys. To remedy this situation, an imaging survey of faint optically-selected quasars with $z < 0.9$ from the AAT fibre survey (Boyle *et al.* 1987) has been carried out using the La Palma Issac Newton 2.5 m Telescope. Preliminary results have been reported in Boyle, Shanks, and Yee (1988); and a more complete analysis will appear in Yee, Boyle, and Shanks (1990). With a small sample of 8 relatively faint optically-selected quasars (absolute magnitudes ~ -24), we found the average B_{gq} computed by counting galaxies within a radius of 500 kpc to be similar to that obtained from the optically-selected quasars in the two Steward samples, confirming that at $z \sim 0.6$, radio-quiet quasars, unlike some radio-loud ones, are not found in centers of rich clusters. However, the CCD camera at INT has a total field of about $6' \times 4'$, allowing us to examine the environment of these quasars out to a radius as large as 1.5 Mpc at $z \sim 0.6$. We found the surprising result that the average B_{gq} computed by counting galaxies in a radius of 1 Mpc around the quasars increases by a factor of almost 3. This finding is also manifested by the average radial distribution of galaxies derived with the quasars as centers. This distribution shows a significant bump in excess galaxies around 1 Mpc away from the quasars. The simplest interpretation of this excess is that some of these faint radio-quiet quasars are preferentially situated at the edges of rich clusters, whereas a significant fraction of radio-loud quasars at $z \gtrsim 0.5$ are found at or near the centers of rich clusters. Unfortunately, at this point, even if this tantalizing result based on a small number of quasars proves to be true in general, we cannot say whether there is an evolutionary effect similar to that found for radio-loud quasars. The reason is that due to both a considerably smaller field and lower average redshift for the Steward low-redshift sample, we can sample the environment out to a limit of only 250 to 500 kpc radius. It would be of great interest if the environment of the low-redshift sample can be re-examined with a CCD camera with a much wider field.

3. PROPERTIES OF GALAXY CLUSTERS ASSOCIATED WITH QUASARS

From our extensive imaging data, we have shown first that quasars do occur in the centers of rich clusters. Moreover, with the additional data we have recently obtained, we also confirmed that the occurrence of bright quasars in rich environments has been drastically curtailed since the epoch corresponding to $z \sim 0.5$. This evolutionary effect and its dependence on environment provides us with great opportunities to isolate conditions that are conducive to quasar activities by comparing the physical conditions in the clusters containing quasars that have different levels of activity with those of more normal clusters.

There are several theoretical ideas on the dependence of quasar evolution on environment. Stocke and Perrenod (1981) specifically predicted that quasars exist in the centers of rich clusters at an earlier epoch when intracluster gas density is low and gas stripping of the host galaxies of the quasars is not efficient. Other models use the general idea that interaction and merger are the main mechanisms for inducing quasar activity in a host galaxy that already habors a supermassive blackhole. (e.g., Roos 1981, De Robertis 1985, and more recently, Carlberg 1989). The idea is that the merger/interaction rate declines with the age of the universe and this change produces a general evolution of the quasar luminosity function. More specifically, in the case of clusters, Roos (1981) pointed out that the merger rate at the center of a rich cluster is expected to decrease very rapidly as the cluster virializes, implying that quasars in rich clusters evolve differently from those in the field. If these scenarios are correct, we may expect that clusters hosting bright central quasars represent clusters at an early stage of their evolutionary history. Thus, these clusters are not only important for understanding the relationship between quasar activity and environment, but also for the study of the evolution of galaxy clusters.

With these goals in mind, we are studying the clusters associated with quasars using multi-object spectroscopy and multi-color imaging. Here, I report on some of our preliminary findings concerning the correlation between cluster properties and quasar activity. Our data so far have lent support to these scenarios.

There are several tantalizing pieces of preliminary evidence suggesting that a cluster hosting a quasar in its center may be systematically different from the more "normal" clusters. First, although we do not have rigorous statistics at present because the CTIO Parkes quasar survey has not been fully analysed, we find that most quasars located in rich clusters have classical double or triple morphology (FR II). Second, from a small number of four clusters associated with quasars having redshifts from 0.2 to 0.65 (Yee *et al.* 1989), we found that they have blue galaxy fractions similar to or slightly higher than those found by Butcher and Omeler (1985) for ordinary clusters up to redshifts of 0.5. However, unlike the other clusters where almost all blue galaxies are found outside the centers (Gunn and Dressler 1988), these appear to have a substantial population of blue galaxies near the centers. Results from our faint quasar imaging survey also support this conclusion. Third, these clusters all have a large density spike of galaxies centered around the quasars, with empirical King-model core radii of radial galaxy distributions between 50 and 200 kpc, much smaller than those expected for Abell class 1 clusters. Finally, in a detailed study of the cluster around 3C206, a relatively bright low redshift ($z = 0.2$) radio-loud quasar situated in a rich cluster, we have obtained spectra of 10 associated galaxies (Ellingson *et al.* 1989). The velocity dispersion of the cluster is around 500 km s^{-1}, somewhat lower than the 800 km s^{-1} found for low redshift rich clusters (*e.g.*, Struble and Rood 1985). Further evidence that a low velocity dispersion exists between quasars and their associated galaxies has also been found by combining the velocities from a large sample of galaxies associated with different quasars (Ellingson 1989).

Although most of these results are very preliminary and are based on a small number of objects, nevertheless, they are consistent with the scenarios described above. High galaxy density and low velocity dispersion are conditions that are conducive to high interaction and merger rate. The existence of blue galaxies near the centers of these clusters may be indicative of inefficient stripping due to a relatively low density intracluster medium. This condition is also favorable for the formation of FR II morphology of the radio emission. All these properties are ones expected in a relatively young cluster. As a cluster evolves, we generally believe that it virializes and increases its intracluster gas density. Although it may be difficult to ascertain which process is more important, both can effectively starve a bright quasar in a cluster center, leading to the change in the average environment of bright radio-loud quasars that we observe.

4. EVIDENCE FOR LUMINOSITY EVOLUTION OF GALAXIES ASSOCIATED WITH QUASARS

The determination of the LF of the galaxies associated with quasars provides us with one of the most efficient methods currently available for estimating the luminosity of a substantial number of galaxies at moderately high redshifts. In addition, it is also an essential step in obtaining absolute measures of the richness of the environment of quasars. The details of the method used are presented in paper IV, and summarized here.

The quasar sample is nominally divided into subsets grouped by redshift. To form the average LF of the excess galaxies for each group, all galaxies in each quasar field brighter than the completeness limit are binned according to the absolute magnitudes computed using the redshift of the quasar and K-corrected to the *observed* r band of a fiducial redshift which is the mean of the quasar redshifts in the group. By computing the LFs in the observed r band, we can circumvent the problem of uncertain K-corrections for individual galaxies which arises from the fact that we neither know the morphology, nor whether a specific galaxy is a background or an associated galaxy. The corresponding background counts are subtracted from the magnitude-binned data, leaving us with a net distribution in luminosity for the excess galaxies. The data are then summed over the subsample to form an average LF for the redshift bin. Because of the inherent poorer nature of the environments of radio-quiet quasars, reasonable LFs can be determined only for the galaxies associated with radio-loud quasars.

For the purpose of inter-comparing LFs derived from different redshift bins, averaged relative K-corrections weighted according to approximate estimates of morphological types are then applied. For comparing with LFs of local galaxies from the literature which are normally better determined, we compare the derived LFs with model LFs created from summing K-corrected zero-redshift LFs of different morphological types. Using the associated galaxies from the Steward radio-loud sample, we found that at $z{\sim}0.6$ there is a brightening of M^*, the characteristic magnitude obtained by fitting a Schechter function to the LF, of 0.9 ± 0.5 mag when compared with the local LF as determined by King and Ellis (1985) from the AAT redshift survey.

The additional data from the $z{\sim}0.6$ sample of Parkes quasars observed at CTIO produce a similar result. The best fitting M^* for the excess galaxies in the 8 Parkes quasars fields corrected to an observed r band at $z = 0.61$ is -21.73 ± 0.32 mag; identical to the -21.74 ± 0.47 mag obtained for the Steward sample. Combining the two samples, we derived an M^* in the observed r band of -21.62 ± 0.23 mag for the associated galaxies at $z{\sim}0.6$. The result is plotted in Figure 2. This is compared with an M^* of -20.86 mag obtained by summing the K-corrected LFs of different morphological types from King and Ellis (1985) and fitting one Schechter function down to an absolute magnitude of -19.0.

Can we generalize this observed luminosity change to a general evolution of galaxies? Because of the strong correlation between galaxy density and morphological types (Dressler 1980), it is likely that this sample is dominated by early-type galaxies, since the main contribution to the excess disproportionately comes from the rich associations. Thus, a more valid and conservative comparison can be made with the LFs of cluster galaxies. Since clusters are dominated by E and S0 galaxies and their LFs are similar to those of the field (*e.g.*, Bingelli, Sandage, and Tamman 1988), we can adopt simply the LF of ellipticals and S0's as an upper limit comparison. K-correcting the E + S0 LF obtained by King and Ellis to $z = 0.61$, we obtain an M^* of -21.0 mag, not greatly different from that of the LF summed over morphological types. This is because we are observing in a red band and to a depth of only approximately 2 magnitudes past M^*, so that even at a redshift of 0.6, galaxies of Sa or earlier morphologies still dominate the counts. Furthermore, we can also make the internally consistent comparison with the LF of galaxies associated with low-redshift quasars (paper IV). With the additional data for the $z \approx 0.6$ bin, we find that the change in M^* corresponds to a brightening of 0.6 ± 0.4 mag between $z = 0.24$ and $z = 0.6$, a 1.5σ effect.

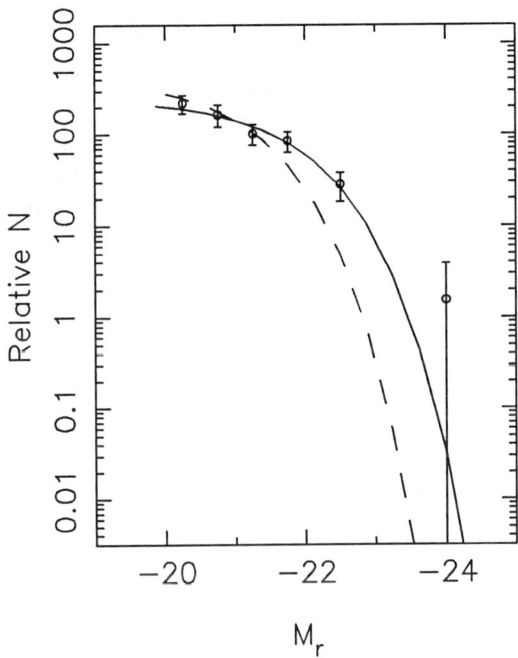

Figure 2. Luminosity function of galaxies associated with $z \sim 0.6$ radio-loud quasars in the observed r frame. The solid line represents the best fitting Schechter function with a fixed $\alpha = 1.0$. The dashed line shows the best fit using the K-corrected LF from local galaxies.

Comparing with other observations, the amount of excess luminosity in M^* that we obtained is very similar to that derived for first rank-cluster galaxies by Kristian, Sandage, and Westphal (1978) in an open universe, giving support that we are observing a general brightening of at least early-type galaxies in the earlier epoch. We note that this amount of evolution of $\sim$0.6 to 0.8 mag is also consistent with reasonable passive evolution models of Tinsely (1980) and Bruzual (1983). However, it should be noted that the deep Durham redshift survey reported by Ellis in these proceedings has found that the redshift distribution of galaxies with $B_J < 21.5$ mag is consistent with no significant evolution.

5. SUMMARY

Using principally direct imaging data with proper background galaxy count corrections and reasonable assumptions, we are able to derive quantitative information about the environment of quasars and their associated galaxies over an interesting redshift range. We learn that quasars in general are located in regions of higher than average galaxy density; and in some cases, in the centers of very rich clusters, and perhaps also in the outskirts of clusters. There is strong evidence that quasars in rich environments have evolved much more rapidly compared to their counterparts in poorer environments. These clues support the notion that interaction and merger are important mechanisms in triggering quasar acitivty. Furthermore, the observed change in the average environment of luminous radio-loud quasars is direct evidence that quasar evolution can be intimately connected with environmental conditions, and hence it is unlikely that there is a single universal luminosity evolutionary law applicable to *all* quasars.

By studying galaxy clusters associated with both bright and faint (and perhaps dying) quasars, we may be able to isolate specific conditions that are favored by quasar activity and hence gain insights into the mechanisms of quasar formation and sustenance in more detail. Thus far, the sparse amount of data suggest that both high rate of interaction or merger (as indicated by high core galaxy density and low velocity dispersion), and low intracluster gas (hinted by double-lobed radio morphology and blue galaxies in the cluster cores) may be important. Considerably more data are required before we can differentiate definitively between different scenarios.

In both the associated galaxies and clusters, there is some evidence for evolution. The LF of the associated galaxies shows a significant brightening of luminosity at $z{\sim}0.6$ which is consistent with models of passive evolution of early type galaxies. The large change in the average luminosity of quasars associated with clusters over a period of a billion years can be taken as indirect evidence for the evolution of rich clusters. More detailed study of these clusters may allow us to find direct evidence for evolution.

Acknowledgement The various results presented in this paper were obtained in conjunction with a number of collaborators, including B. Boyle, M. De Robertis, E. Ellingson, R. Green, C. Pritchet, T. Shanks, and P. Stockman.

References

Bingelli, B., Sandage, A., and Tamman, G. A. 1988, *Ann. Rev. Astron. and Astropys.*, **26**, 509.

Boyle, B. J., Fong, R., Shanks, T., and Peterson, B. A. 1987, *M. N. R. A. S.*, **227**, 717.

Boyle, B. J, Shanks, T., and Yee, H. K. C. 1989, in *The Proceedings of the IAU Symposium 130: Large Scale Structure and Motions of the Universe*, eds. J. Adouze *et al.* , (Dordrecht: Reidel), page 577.

Bruzual, G. 1983, *Ap. J.*, **273**, 105.

Butcher, H., and Oemler, A. 1978, *Ap. J.*, **226**, 559.

Carlberg, R. G. 1989, preprint submitted to *Ap. J.*.

Davis, M., and Peebles, P. J. E. 1983, *Ap. J.*, **267**, 465.

De Robertis, M. (1985), *A. J.*, **235**, 351.

Dressler, A. 1980, *Ap. J.*, **236**, 351.

Ellingson, E. 1989, Ph. D. thesis, University of Arizona.

Ellingson, E. 1990, in these proceedings.

Ellingson, E., Yee, H. K. C., Green, R. F., and Kinman, T. D. 1989, *A. J.*, **97**, 1539.

Green. R. F. and Yee, H. K. C. (1984), *Ap. J. Suppl.*, **54**, 495.

Gunn, J. E., and Dressler, A. 1988, in the *Proceedings of the* 5th *Erice Workshop on Towards Understanding Galaxies at Large Redshift*, ed. R. G. Kron and A. Renzini, (Dordrecht: Kluwer), page 227.

Heckman, T., Bothun, G., Balick, B., and Smith, E. P. (1984), *A. J.*, **89**, 958.

Hill, G. 1989, Ph. D. thesis, University of Hawaii.

King, C .R. and Ellis, R. S. 1985, *Ap. J.*, **288**, 456.

Kritstian, J., Sandage, A., and Westphal, J. A. 1978, *Ap. J.*, **221**, 383.

Longair, M. S. and Seldner, M. 1979, *M. N. R. A. S.*, **189**, 433.

Malkan, M. 1984, *Ap. J.*, **287**, 555.

Roos, N. 1985, *A. A.*, **104**, 218.

Seldner, M. and Peebles, P. J. E. 1978, *Ap. J.*, **225**, 7.

Smith, E. P., Heckman, T. M., Bothun, G. D., Romanishin, W., and Balick, B. 1987, *Ap. J.*, **306**, 64.

Stocke, J. T. and Perrenod, S. C. 1985, *Ap. J.*, **245**, 375.

Struble, M. F. and Rood, H. J. 1985, *Ap. J. Suppl.*, **63**, 681.

Tinsley, B. M. 1980, *Ap. J.*, **241**, 41.

Yates, M., Peacock, J. A., and Miller, L. 1989, *M. N. R. A. S.*, in press.

Yee, H. K. C., Bolye, B. J., and Shanks, S. 1990, in preparation.

Yee, H. K. C. and De Robertis, M. M. 1989, in *The Proceedings of the IAU Symposium 134: Active Galactic Nuclei*, eds. D. Osterbrock and J. Miller, (Dordrecht:Reidel), page 457.

Yee, H. K. C., Ellingson, E., Green, R. F., Pritchet, C. J. 1989, in *Proceedings of the NATO Advance Research Workshop on The Epoch of Galaxy Formation*, eds. C. Frenks *et al.* , (Dordrecht: Kluwer), page 185.

Yee, H. K. C. and Green, R. F. 1984, *Ap. J.*, **280**, 79 (Paper II).

Yee, H. K. C. and Green, R. F. 1987, *Ap. J.*, **319**, 28 (Paper IV).

Yee, H. K. C., Green, R. F., and Stockman, H. S. 1987, *Ap. J. Suppl.*, **62**, 681.

THE EVOLUTION OF QUASARS IN GALAXY CLUSTER ENVIRONMENTS

E. Ellingson
Steward Observatory, University of Arizona
Tucson, AZ 85721

ABSTRACT In previous studies of bright, radio-loud quasars, it has been found (Yee and Green 1987) that these objects at redshifts greater than ~ 0.6 are often found situated in galaxy cluster environments as rich as Abell richness class 1 or greater but not at lower redshifts. At low redshifts, however, optically bright radio-loud quasars are only found in poorer environments. This result may be interpreted as evidence that quasars in these environments have faded since that epoch. Imaging of the environments of an optically faint sample of radio-loud quasars at $0.3 < z < 0.6$ has identified several faint quasars in rich galaxy clusters, suggesting a statistical fading timescale of 900 million years. This result indicates that the population of radio-loud quasars in rich environments evolve much more rapidly than those in poor environents and that environment plays a crucial role in the triggering and sustenance of quasar activity. The environments of radio-loud and radio-quiet quasars are also compared, providing strong evidence that radio-quiet quasars are not found in these rich environments, and that there is a difference in the underlying physical properties of these two types of objects.

Quasars have long been known to be associated with groups and clusters of normal galaxies, and there have been many indications that the quasar environment may provide many clues towards the understanding of how quasar activity is triggered and sustained (see Yee 1990 for a review of current work). Here we examine the optical evolution of quasars found in rich galaxy environments. Yee and Green (1987, hereafter YG87) found that the preferred site of bright radio-loud quasars has evolved in recent epochs. Quasars with $z \sim 0.6$ are often found in regions of galaxy density comparable to Abell richness class 1 clusters, while bright quasars at lower redshift are not. This observed lack of quasar-cluster associations implies that the quasars have faded appreciably since these epochs.

To test this interpretation, the environments of a sample of 66 faint quasars with apparent magnitudes between 17.5 and 19.5 and redshifts between 0.3 and 0.6 were investigated. Direct CCD images of the quasar environments were obtained using the Steward Observatory 2.3-m and KPNO 2.1-m telescopes. Each field was observed in Gunn r to an average limiting magnitude of 23.5.

Object finding, classification and photometry were performed using exactly the same methods as were used for the YG87 sample.

The quasar-galaxy spatial covariance amplitude, B_{gq} , was calculated for each field. This quantity is a measure of the overdensity of galaxies in space near the quasar, and is normalized for both the expected spatial and luminosity distributions of galaxies at the redshift of the quasar. In the sample of 32 radio-loud quasars, four were found to have B_{gq}'s greater than 500, indicating environments as rich as Abell class 1 clusters. The detection at $z \sim 0.4$ of these rich environments associated with faint quasars but not with bright quasars suggests that rich environments at this epoch, unlike at $z \sim 0.6$ are not favorable for optically bright quasar activity. Evolution of the cluster properties during this interval may therefore be responsible for the observed evolution of radio-loud quasars.

In a flux-limited sample, objects at higher redshift will have preferentially higher luminosities than lower-redshift objects. It is therefore important to test whether the observed evolution of quasar environment for objects of a given luminosity may be due to a correlation between environment and luminosity, and not true evolution. For the combined sample of high and low luminosity quasars, no such correlation was found. Likewise, no correlation was found between B_{gq} and radio power at 20 cm. The change in observed quasar environments therefore seems unlikely to be caused by selection effects in the quasar samples.

The fading time for radio-loud quasars in rich environments is illustrated in Figure 1. The axes plotted are redshift and optical absolute magnitude of the quasar. Solid dots are radio-loud quasars found to have B_{gq}'s greater than 500, and open dots are radio-loud quasars in poorer environments. Both the YG87 and the faint quasar samples are plotted, and the curves describing the YG87 limiting apparent magnitude of 17.5 and the faint quasar sample limit of 19.5 are marked. A Kolmogorov- Smirnov test for two-dimensional data (Press and Teukolsky 1989) indicates that the distribution in redshift-luminosity space for quasars in rich environments differs from that of quasars in general at a 95% confidence level. If the quasars observed in these samples represent objects at luminosities characteristic of the population of quasars to be found in rich environments at a given redshift, the fading of the quasar luminosity function for quasars in rich environments can be determined from these data. The dashed line represents a best fit to the quasar fading relation, indicating an e-folding time of approximately 900 million years. This timescale is approximately a factor of 6 times shorter than is found for samples of optically selected quasars (see Green 1989, and references therein). The evolution of the population of radio-loud quasars in rich environments, therefore, appears to be much more rapid than for quasars in general, indicating strongly that environment is crucial in the formation or sustenance of radio-loud quasars.

The quasar-galaxy spatial covariance amplitude was also calculated for the 34 radio-quiet objects in the faint quasar sample. These objects were found to have different environments from the radio-loud sample in that there is no evidence that radio-quiet quasars are ever found in very rich environments at any redshift or luminosity. The radio loud and radio quiet samples were compared using a Kolmogorov-Smirnov test; the samples were found to be drawn from different distributions of environment at the 99% confidence level. This result indicates that the two types of objects may be physically different, and in particular is consistent with the suggestion that radio-loud quasars are situated in elliptical host galaxies and radio-quiet quasars in spiral host galaxies.

REFERENCES

Green, R. F. 1989, in *The Epoch of Galaxy Formation*, ed C. S. Frenk, Kluwer Academic Press, p. 121.

Press, W. H. and Teukolsky, S. A., 1989, preprint.

Yee, H. K. C 1990, this volume.

Yee, H. K. C. and Green, R. F. 1987, *Ap. J.*, **319**, 28.

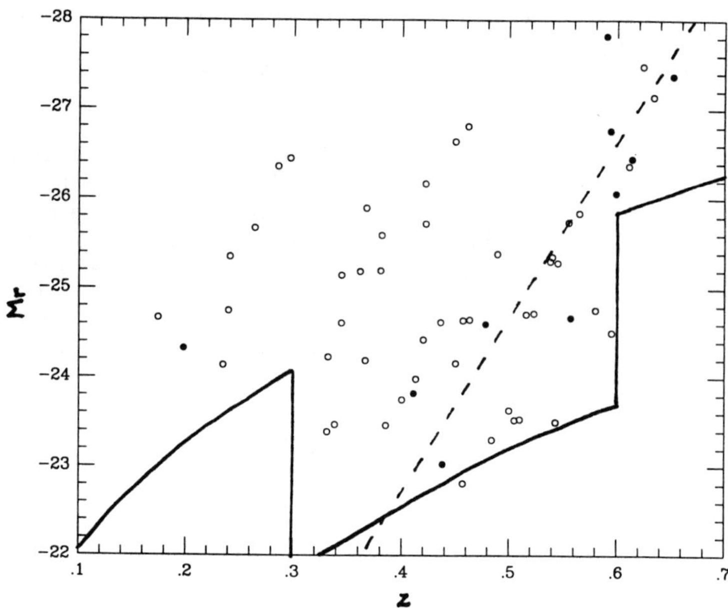

Figure 1. Quasar absolute magnitude versus redshift for radio -loud quasars in the YG87 bright quasar survey, and the faint quasar survey. Solid dots represent quasars in environments as rich or richer than Abell richness class 1 clusters,and open dots represent objects in poorer environments. Sample limits for the two surveys are indicated by solid lines. The dashed line represents the fading time for radio-loud quasars in rich environments.

TESTING THE BINARY BLACK HOLE MODEL OF OJ 287

GEORGE J. CORSO, RONALD W. HARRIS, RICHARD FOX
Loyola University of Chicago, Dept. of Physics
Chicago, Illinois 60626

JOSEPH SCHULTZ
Cernan Earth and Space Center, Triton College
Rivergrove, Illinois 60171

ABSTRACT We have tested one prediction of the binary black hole model of the BL Lacertid object OJ 287 proposed by Sillanpaa et al (1988). The composite B light curve shows no clear evidence for the predicted double minimum of 1987-88.

INTRODUCTION

Based on their analysis of the V light curve of OJ 287 Sillanpaa et al (1988) proposed a model which involves a precessing binary system of massive black holes with accretion disks. They find major outbursts at intervals of 11.65 years and a periodic double minimum every 11.05 years with an interval of about 1 year between the minima. The next predicted major outburst does not occur until 1994, however a double minimum should have occured between March of 1987 and June of 1988. Here we combine data obtained in blue light with the 1 meter reflector of Northwestern's Lindheimer Astronomical Research Center with observations made by other investigators to test this prediction.

OBSERVATIONS

Table 1 contains the list of our observed B magnitudes and their Julian Date. As with our previous observations of OJ 287 (see Corso, Ringwald, and Harris 1988) each was obtained useing a 103aO plate and a GG13 filter. Reduction was accomplished useing a Cuffey Iris Astrophotometer and a comparison sequence provided by Penston and Wing (1973) which was supplemented by a faint comparison star provided by Mc Gimsey et al (1976). Fig. 1 is the composite B light curve for the

1987-88 observing season obtained by combining our data from Table 1 with all other observations known to us. Fig. 2 is the composite B light curve for all observations known to us that were made between November of 1986 and May of 1987. The filled circles in Figs. 1 and 2 are LARC observations by us.

TABLE I Observed Blue Magnitudes

LARC Plate No.	UT DATE	JULIAN DATE (2440000+)	m_b
8802	Jan. 27	7187.6188	16.47±.10
8805	Feb. 14	7205.5906	16.58±.10
8807	Feb. 17	7208.6021	16.32±.06
8809	Feb. 28	7219.5993	16.25±.11
8810	Mar. 6	7226.5969	16.17±.05
8812	Mar. 20	7240.5701	16.78±.09
8815	Mar. 22	7242.6035	16.71±.08
8816	Apr. 9	7260.5833	16.46±.12
8818	Apr. 12	7263.6000	16.65±.10
8820	Apr. 26	7277.5819	fainter than 15.9 (plate limit)
8821	Apr. 30	7281.6007	16.79±.11
8823	May 11	7292.6149	16.80±.09
8825	May 14	7295.6194	fainter than 16.8 (plate limit)
8827	May 26	7307.6188	16.73±.15
8828	June 4	7316.6340	fainter than 16.6 (plate limit)
8830	June 7	7319.6354	16.50±.20 (eye estimate of poor plate)

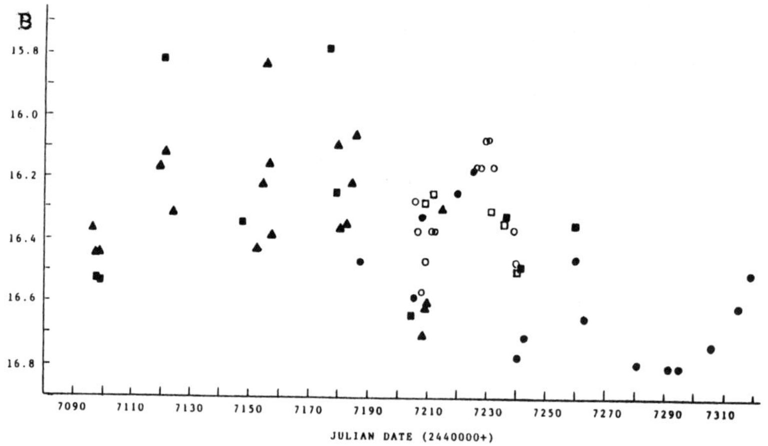

Fig. 1. Composite B Light Curve For 1987-88

We note the presence of several gaps in the coverage (on the order of about two weeks duration each) and the overall spikey appearance of Figs. 1 and 2.

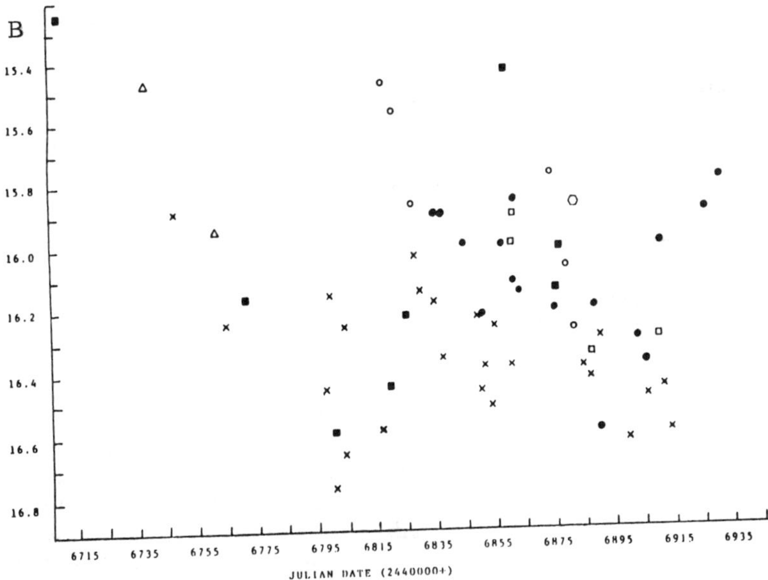

Fig. 2. Composite B Light Curve For 1986-87

DISCUSSION AND CONCLUSIONS

There appears to be no clear evidence in Figs. 1 or 2 for anything other than random variations superimposed on a slowly varying underlying continuum. A key feature of the model proposed by Sillanpaa et al is the unique periodicity of the eclipses. Upon constructing the composite light curve for the 1985-86 observing season we noted the presence of an event which occured during early 1986 when OJ 287 became as faint as B=17.5 and which looks more eclipse like than anything seen in either 1987 or 1988. Whatever their cause we can no longer regard the deep flux minima as strictly periodic phenomenon.

REFERENCES

Corso, G.J., Ringwald, F., and Harris, R.W. 1988, Astron. Astrophys., 195, 25

Mc Gimsey, B.Q., Miller, H.R., and Williamson, R.M. 1976, A.J. 81, 750

Penston, M.V. and Wing, R.F. 1973, Observatory, 93, 149

Sillanpaa, A., Haarala, S., Valtonen, M.J., Sundelius, B., and Byrd, G.G. 1988, Ap.J., 325, 628

REDSHIFTS AND THE SMF MODEL FOR QUASARS AND AGNS

HOWARD D. GREYBER
10123 Falls Road, Potomac, MD 20854

ABSTRACT: The extended strong magnetic field model(ESMF) explainsArp's deduction of a partly nonvelocity redshift in quasars. It predicts a possible few percent variation in the redshift of blob-emitting quasars over one cycle.

The Strong Magnetic Field model (SMF) is a distinct physical model for the "engine" that powers the enormous energy output from quasars, BL Lac objects and active galactic nuclei(AGN), (Greyber, 1961-1990). Detailed evidence for SMF is listed briefly in Table I. The explanation for Kellerman's (1985) comment, "Typical morphology shows a compact core with one or more blobs or components moving away in a single direction" is illustrated in Fig. 1. Notice the topological magnetic mirrors.

Most models for the central core assume a maximum magnetic corresponding to equipartition, yielding typically 10^{-7} to 10^{-1} gauss. Also mechanisms for particle acceleration must be found in spite of the Inverse Compton problem. In direct contrast SMF does not assume equipartition, and does not need exotic acceleration mechanisms since ultrarelativistic electrons are stored in a gravitationally bound current loop (GBCL) and scattered out causing various phenomena. The general morphology and dynamics is determined by the ratio of magnetic field energy to rotational energy after gravitational collapse. When the intense dipole magnetic field anchored at the nucleus by its GBCL is extremely large, quasars with jets and radio galaxies with double lobes are seen. When the field is smaller, comparable to rotational energy, spiral arms are formed.

THE REDSHIFT CONTROVERSY

Halton Arp (1987) gives in elaborate detail his arguments that quasars are associated in space with nearby galaxies and that their large redshifts are partly intrinsic, not wholly cosmological. But the establishment opinion is that all the unusual alignments found by Arp are merely statistical fluctuations, and that quasar redshifts are cosmological.

However the small minority on Arp's side of the controversy includes two distinguished astrophysicists, G. Burbidge and Fred Hoyle. Therefore it occurred to me to study an ESMF model to explain Arp's deductions. (It should be clear that the original SMF model is separate and fits perfectly with the cosmological explanation for redshifts.)

AN EXTENDED STRONG MAGNETIC FIELD MODEL (ESMF)

One must modify an SMF assumption to read "galaxies and groups of galaxies formed by gravitational collapse of some pregalactic gas cloud". This implies that in many cases a cloud of 10^{13} to 10^{14} solar masses collapsed. Presumably several plasma concentrations, most with their own GBCL, would form during collapse close to the most massive galactic concentration with the largest, most intense dipole magnetic field and GBCL.

Those compact objects near the dipole field axis of the most massive galaxy would then tend to be expelled due to the "toothpaste tube" plasma acceleration effect from an increasing dipole magnetic field on a diamagnetic plasma. Klein and Brueckner (1960) analyzed this process for laboratory plasmas. While not efficient for laboratory plasmas, it is very efficient for the extremely high conductivity of astrophysical plasmas. This would explain the tendency of quasars to be found somewhat aligned along the minor axis of a massive galaxy, deduced by Arp.

Now ESMF must explain extragalactic partly nonvelocity redshifts. The gravitational effect of electromagnetic radiation was illustrated by a great physicist, John A. Wheeler, in a study of "geons". For ESMF we notice that the most intense magnetic field (see Fig.1) is on the major axis, compressed between the thin toroid of plasma frozen to the GBCL and the bulk plasma around the black hole, in a narrow belt-like region.

ESMF requires we make two assumptions: (a) the dominant radiation mechanism in the optical region of the spectrum is a strong positive function of the magnetic field intensity, and (b)the magnetic field in the narrow belt is so strong, i.e. has such a large equivalent mass, as to have a strong gravitational effect on the radiation produced in the transition zone (which is the order of a Debye length wide) between the plasma and the magnetic field. A simple calculation with semi-reasonable assumptionsand a magnetic field intensity of 10^{16} gauss in the narrow belt gives a redshift, Z, of 2.4. Varying parameters can yield a wide range of redshift values due to gravitation. Due to assumption (a), the optical spectral lines will not be washed out, but simply displaced by the gravitational redshift.

CONCLUSIONS

The ESMF model can explain Arp's deductions by use of a novel concept, the gravitational effect on spectral lines of a narrow region of very intense magnetic field in the central core of quasars. When, as in 3C120, a radio blob is ejected every 18 months or so, the GBCL must readjust slightly for the temporary small drop in plasma pressure of the bulk plasma (accretion soon refills the bulk plasma around the central object).

If the gravitational magnetic field redshift is a significant factor, ESMF predicts one might observe a small fluctuation cyclicly in the redshift of 3C120 and similar objects over the cycle between successive emission of radio blobs. Such careful repeated redshift measurements on blob-emitting quasars within one cycle should be done. It should be emphasized that the original SMF model is distinct from ESMF and is wholly compatible with the quasar redshifts being cosmological.

ACKNOWLEDGEMENTS

It is a pleasure to acknowledge the late Donald H. Menzel and and John A. Wheeler for stimulating advice, and Gart Westerhout for permission to use the U.S. Naval Observatory library. Early parts of this research were sponsored by the U.S.A.F.O.S.R. and N.A.S.A..

REFERENCES

Arp, H. 1987, Quasars, Redshifts and Controversies, Int. Media

Greyber, H.D. 1961, Trans. Int.Astron. Union 11B, 332

Greyber, H.D. 1962, U.S.A.F. Off. of Scient. Research Report No. 2958

Greyber, H.D. 1964 in Quasi-stellar Sources and Gravitational Collapse, Univ. of Chicago Press, Chap. 31, 389

Greyber, H.D. 1967 in Instabilite Gravitationelle- -, Memoirs of the Royal Society of Sciences of Liege, XV, 189

Greyber, H.D. 1967, Ibid., 197

Greyber, H.D. 1967, Publ. Astron. Soc. Pacific, 79, 341

Greyber, H.D. 1984, Ann. New York Acad. Sciences 422, 353

Greyber, H.D. 1988, in Supermassive Black Holes, Cambridge Univ. Press, 360

Greyber, H.D. 1989, "The Importance of Strong Magnetic Fields in the Universe" in Comments on Astrophysics, 13,#4, 201

Greyber, H.D. 1989 in The Center of the Galaxy, ed. M. Morris, Proc. I.A.U. Symp. No. 136, 335

Greyber, H.D. 1990 in Proc. 14th Texas Symp. on Relativistic Astrophysics, to be published

Kellerman, K. 1985, Comments on Astrophysics, 11, no.2, 69

Klein, M.M. & Brueckner, K. 1960, Jour. Appl.Physics, 31, 1437

Wheeler, J.A. 1955, Phys. Review 97, 511

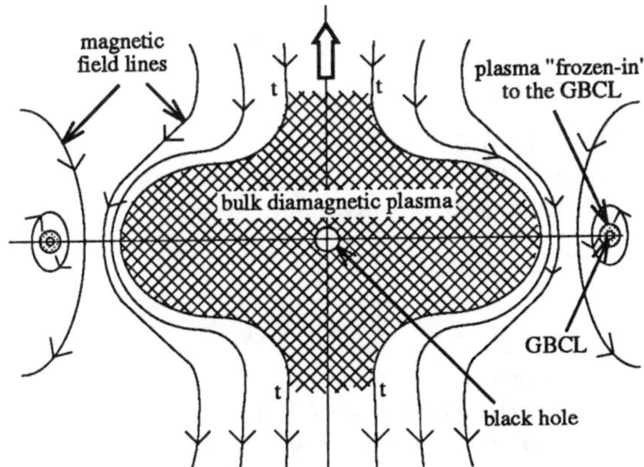

Figure 1. SMF model for a quasar, BL Lac object or AGN when a very large accretion of matter around the black hole has occurred. t-t is the throat of the magnetic mirrors. Arrow shows the direction of the motion of the explosively ejected hot plasma blob after upper throat suddenly opens. As accretion continues, process repeats every few years.

TABLE 1 ; COMPARISON OF THE STRONG MAGNETIC FIELD MODEL (SMF) WITH THE ROTATING ACCRETION DISK MODEL

	Strong Magnetic Field Model (SMF)	Rotating Accretion Disk Model
ENERGY SOURCE	Gravitational Collapse + Accretion	Accretion
ENERGY STORAGE	Gravitational Bound Current Loop	Velocity of Accreting Material
EXPLAINS:		
Very frequent injection of high energy relativistic electrons	Yes	No
Millions of points where relat. electrons inject into strong mag. fields	Yes	No
Production of successive radio blobs	Yes	No
Shape of giant double lobe radio sources	Yes	No
One-sided jets in strong sources	Yes	No
Strong magnetic fields in very young (i.e. high Z) galaxies	Yes	No
Largest polarization near the innermost radio blob	Yes	No
Core-halo polarization difference in core-halo radio galaxies	Yes	No
Morphology of objects of galactic dimension	Yes	No
Extremely rapid time variations observed (50-100 seconds)	Yes	No
Remarkable straightness of jets	Yes	?
Remarkable confinement of jet material over huge distances	Yes	No
Lack of evidence for rotation in the spectra of quasars	Yes	No
Production of jet	Yes	Yes-- with assumed funnel
Radio-quiet quasars	Yes	?
Stability of configuration	Yes	No
Generation mechanism	Yes	Yes
Existence for times of interest	Yes	?
Very few spirals in rich clusters	Yes	No

HIGH REDSHIFT RADIO GALAXIES: EVIDENCE FOR EARLY GALAXY FORMATION

SIMON J. LILLY
Institute for Astronomy, University of Hawaii,
2680 Woodlawn Drive, Honolulu, Hawaii 96822.

ABSTRACT. The evidence that the radio galaxies recently identified at high redshift are massive galaxies (10^{12} $M_{\odot}$ of stellar material) in which the bulk of the stars are of order 10^9 years or more old is discussed. Emphasis is placed on the M/L ratio of the stellar population rather than its age, since it is the masses of the galaxies which provide the most direct confrontation with galaxy formation theory. A key observation is the small dispersion in the $K-z$ Hubble relation since it is hard to imagine how the very low M/L ratios required to reduce the masses of the galaxies would not produce a much larger dispersion in the absolute magnitudes of the population than is observed. New data on a potential counter-example, the proto-galaxy candidate 3C326.1 at $z \sim 1.8$, suggest that this object is not a galaxy seen in the process of formation. Finally, the implications of this idea for theories for the formation of massive galaxies in general are discussed.

1. Introduction - Old radio galaxies at high redshift

The idea that the radio galaxies that have recently been found at $z > 2$ are massive galaxies (of order 10^{12} $M_{\odot}$ of stellar material) in which the bulk of the stars are of order 1 Gyr or more old (Lilly 1988, 1989) is of considerable interest for theories of galaxy formation. In this paper, I will review the evidence in favor of this idea, discuss some of the issues that have recently been raised concerning it, and then describe some of its implications for theories of galaxy formation.

In this context, clear distinction should be made between the ages of the *stellar populations*, which may in principle be constrained by the readily determined colors of the galaxy, and the *dynamical* ages of the galaxies, which will reflect the time since the constituent mass was assembled in place. This can only be inferred from much more difficult morphological or kinematic observations. Unfortunately, galaxy formation theories generally make more specific predictions for the latter than for the former, since most hierarchical scenarios can accommodate star-formation in small units at very high redshifts, if required. Correspondingly, the interest in the radio galaxies for galaxy formation centers as much on the high *masses* that are implied for these objects by the high M/L ratios of 1 Gyr stellar populations, rather than on the precise ages themselves.

1.1. The spectral energy distributions at high z

The optical-infrared colors of radio galaxies at $z \sim 1$ are (a) generally rather red and (b) show a large dispersion, with $4 \leq (r - K) \leq 6$ (Lilly and Longair 1984). This is illustrated in Figure 1a, which shows the rest-frame spectral energy distributions (SEDs) of 3C65 and 3C368, which both have $z \simeq 1.15$, compared to the SED of a present-day gE galaxy. The *uniformity* of the infrared colors and luminosities, and the *variation* in optical-infrared colors that are illustrated by this diagram were ascribed to varying amounts of ultraviolet activity in an otherwise 'normal' gE-like galaxy (Lilly and Longair 1984). In the case of 3C65, this must be at least 2 Gyr old.

The spatially-extended flat-spectrum ultraviolet continuum has generally been ascribed to a burst of star-formation, probably associated with the radio source. This ultraviolet emission is usually aligned with the radio source axis (McCarthy *et al.* 1987b, Chambers *et al.* 1987) and its strength relative to the red component is correlated with both the radio spectral-index and with the strength of the radio-optical alignment effect and the emission line spectrum (Lilly 1989). A variety of theoretical models have been constructed for jet-induced star-formation (Rees 1989, De Young 1989, Daly 1989). This 'burst' interpretation has generally been preferred over the alternative scenario wherein the range of colors reflects a range of ages of the overall galaxy because of the lack of correlation between $(r - K)$ color and K absolute magnitude, over a range corresponding to almost an order of magnitude variation in age.

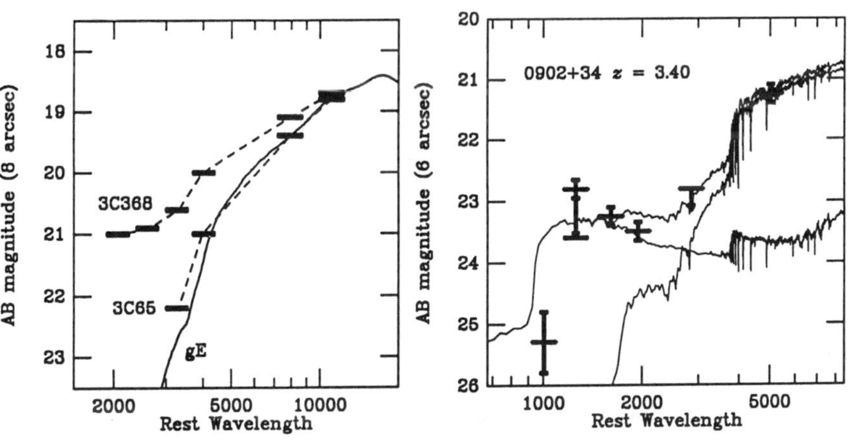

Fig 1: The rest-frame SEDs of (a) 3C65 and 3C368 at $z \sim 1.15$ and (b) 0902+34 at $z \sim 3.4$. The model consists of 99% of a 1 Gyr population plus 1% of a 0.1 Gyr population.

The recent observation of high wavelength-dependent optical polarization in 3C368 by di Serego Alighieri *et al.* (1989) suggests that dust scattering of an obscured nucleus may play an important role in the rest-frame ultraviolet in at least some objects. However, the nature of the ultraviolet continuum is at some level irrelevant to the arguments concerning the identification of the 'red bump' with a mature stellar population.

The new galaxies found at $z \geq 2$ have similar SEDs. The large aperture SED of the '1 Jansky' radio galaxy 0902+34 at $z \sim 3.4$ (Lilly 1988) is shown in Figure 1b. Note that this has a reduced amplitude of the 'red bump' relative to the SED given by Lilly (1988) because of an incorrect estimate of the small-aperture K flux density in that work. The SED of 0902+34 is intermediate between the extremes of 3C65 and 3C368 in Figure 1a. The SED of 4C41.17 at $z \sim 3.8$ (Chambers and Miley, this volume) is remarkably similar to that of 0902+34. Using the stellar population models of Bruzual (1983), the overall SED can be modeled by a small population of very young stars, contributing as little as 1% to the mass of the system and required to extend the continuum out to Lyman α, and a much larger population of older stars of minimum age of around 1 Gyr (Figure 1b).

The minimum age of the 'old' red component is determined by the steepness of the rise between I and K, or 1900 A and 5000 A in the rest-frame. The steepening of the model SEDs with age in this wavelength interval is produced primarily by the burning off of the main sequence. The precise I/K ratio adopted for the old component (and hence its age) depends on the fraction of the I-band light that is contributed by the 'hot' component that is responsible for the flat part of the SED that extends out to Lyman α at 1216 Å. If the 'hot' component has a reasonably flat SED, then around 50% of the I-band light must be from this component and the resulting I/K ratio for the cool component of around 18 produces the minimum age of 1 Gyr mentioned above. Nevertheless, halving this ratio, i.e. permitting all the I-band light to come from the cool component, would allow rather younger ages of around 0.5 Gyr for the stars in 0902+34.

For a Hubble's constant of 50 km s^{-1} Mpc^{-1} , an age of 1 Gyr at $z \sim 3.4$ implies a formation redshift for the bulk of the stars in this galaxy of around 5 for $\Omega_0 \simeq 0$ and around 10 for $\Omega_0 \simeq 1$. However, at these high redshifts, a small uncertainty in the age can produce large uncertainties in the implied redshift of formation (e.g. an age of 0.5 Gyr implies $z \sim 5$ for $\Omega_0 \simeq 1$).

The question of whether the K-band light is really stellar at all is best addressed by the form of the $K - z$ relation for these galaxies.

1.2. The $K - z$ magnitude-redshift 'Hubble' diagram

The remarkable feature of the $K - z$ diagram for these galaxies is the very small scatter in absolute magnitude seen at a given redshift. This implies that radio galaxies at high redshift are a remarkably uniform population when viewed between 5000 A and 1 μm in the rest-frame.

Figure 2 represents the $K - z$ relation for all the radio galaxies with spectroscopically determined redshifts from the 3C and 1 Jansky samples that were observed in the infrared by Lilly and Longair (1984), Lilly *et al.* (1985) and Lilly (1989). For interest, I have also included 4C41.17 at $z \sim 3.8$ at $K = 18.5$ (Chambers and Miley, private communication). The data have been plotted in the form of an 'equivalent visual Reduced Absolute Magnitude' (see Schneider *et al.* 1983) by applying K-corrections and other distance related corrections. A perfect standard candle would have constant RAM in a $q_0 = 0.5$ cosmology. Evolution of the stellar population and/or dynamical evolution of the masses of the galaxies through accretion, changes in the selection function as a $f(z)$, and different geometries will all produce deviations from a horizontal relation. For the present purposes, however, the validity of the corrections and assumptions involved in constructing Figure 2 are *irrelevant* since it is the *scatter* in absolute magnitude at a given redshift which is of primary concern.

The $K - z$ relation clearly stays remarkably tight to the highest redshifts sampled, and the dispersion about the mean relation stays approximately constant at around 0.4 magnitudes (1σ) (see Lilly 1989) over the interval $0 \leq z \leq 2$. The few systems known at $z \geq 3$ also maintain this impression of homogeneity.

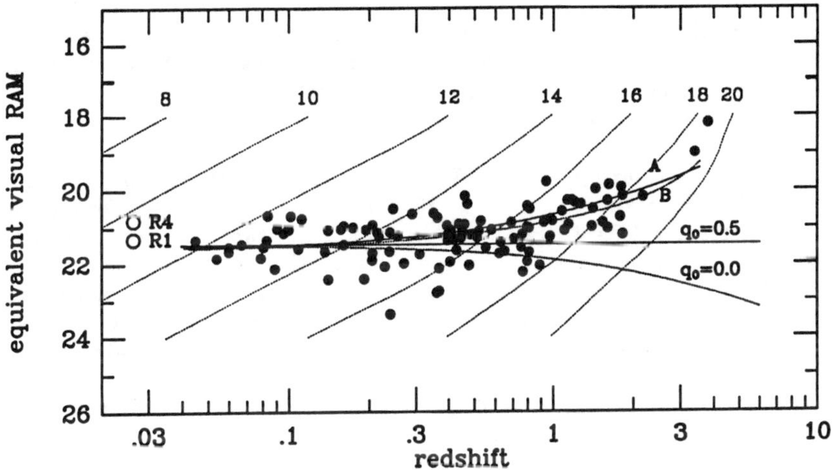

Fig. 2. The magnitude-redshift relation for 3C and 1 Jansky radio galaxies (squares and circles). The K data has been plotted in the form of an equivalent visual Reduced Absolute Magnitude, Lines of constant observed K magnitude are shown. The apparent brightening trend at $z \geq 1$ can be adequately accounted for by a range of passively evolving models for the stellar population (A: $q_0 = 0.5, z_F = 10, x = 1.3$ and B: $q_0 = 0, z_F = 5, x = 0$). R1 and R4 indicate the mean RAM of the 1st ranked galaxies in Abell clusters of Richness Class 1 and 4 (Schneider *et al.* 1983).

It should be stressed that the 3C sample used in Figure 2 is 'complete' in the sense that all the statistically complete Laing *et al.* (1983) LRL sample is included, and we can be sure that essentially no objects are missing. The less powerful 1 Jansky sample that is also plotted on the diagram is less complete in the sense that, although essentially all sources are identified (Lilly 1989), not all have had redshifts measured so far. Nevertheless, the exclusion of these last sources is unlikely to seriously bias the diagram (see Lilly 1989). The small dispersion in the Hubble relation is compelling since the infrared photometry for the $z \geq 1$ objects was generally published in advance of the redshifts being measured, and is surprising on account of the range of phenomena seen in these objects. The following effects, which might be expected to add scatter to the $K - z$ relation were discussed at the Berkeley Workshop on these galaxies that preceded this conference:

(a) A factor of six difference in radio luminosity at a given redshift between the 3C (10 Jansky at 178 MHz) and '1 Jansky' (at 408 MHz) samples.

(b) A range of 2 magnitudes in $(r - K)$ color, $4 \leq (r - K) \leq 6$, present in both the 3C and 1 Jansky samples (Lilly and Longair 1984, Lilly 1989), and a similar range of emission line strengths.

(c) A bewildering variety of optical morphologies (McCarthy *et al.* 1987b, Chambers *et al.* 1987) and (probably) infrared morphologies (Chambers *et al.* 1988) plus the usual range of radio morphologies, sizes and spectral indices.

(d) Possible gravitational lensing and amplification (Le Fevre *et al.* 1987).

(e) Possible observational errors in these difficult observations, e.g. contaminating objects in the infrared beams, usually 8-12 arcsec in size, or errors in the positioning of the aperture.

(f) Possible non-cosmological redshifts.

The small dispersion in the $K - z$ relation at high redshift $z \sim 2$ indicates that the rest-frame light at 8000 A is dominated by uniform stellar populations from galaxy to galaxy. It is remarkable that the basic dispersion in the (one-parameter) absolute magnitudes of the radio galaxies of 0.4 magnitudes is comparable to that found in (two-parameter) relations such as the Tully-Fisher and Faber-Jackson relations. It is hard to see how non-stellar alternatives could produce such a small scatter in view of the wide range of radio and other properties present within the sample. Given then that the light is stellar in origin, the small dispersion in absolute magnitude presumably reflects a tight mass-selection effect (operating directly or indirectly) coupled with a small dispersion in M/L ratio.

Since it is easier to envisage a tight mass-selection effect operating in a stable dynamical environment, the small dispersion in mass in the high redshift radio galaxies suggests to me that these are probably not dynamically young systems. Whatever is producing the radio emission (central engine, environment or whatever) knows the total mass of stars within 30 kpc, and furthermore requires that mass to be constant to within 40% (independent of the radio luminosity).

The *curvature* of the $K - z$ relation provides a rather weaker constraint because of the uncertainties involved in the many corrections applied to construct the RAM. Nevertheless, Figure 2 illustrates the fact that the radio galaxies at $z \geq 1$ are systematically brighter than the low redshift systems, despite the fact that the latter are already close to the top of the galaxy luminosity function. This brightening could be caused by a number of possible effects. As Lilly and Longair (1984) pointed out, it can be readily accounted for in terms of the passive evolution of a simple (i.e. coeval) stellar population, and a number of models with plausible parameters reproduce the observed trend. This is illustrated on Figure 2 by two curves that have been constructed using Tinsley and Gunn's (1976) formula, $\Delta M_V = (1.3 - 0.3x)\ \ln\ \tau$. Indeed, the range of acceptable models and uncertainties therein are so large that they dominate all the other potential effects (such as q_0). Nevertheless, Figure 2 provides a powerful impression of continuity in radio galaxies from the present out to very high redshifts. It is tempting to deduce from this diagram that the radio selection criterion is operating in a broadly similar way at all epochs and is selecting out galaxies lying in a relatively narrow mass range about 2 magnitudes above L*.

1.3. Radio galaxy masses, ages and evolving M/L ratios

A radio galaxy such as 0902+34 at $z \sim 3.4$ and $K = 18.8$ has a stellar mass of between $1 - 4 \times \Upsilon \times 10^{12}$ $M_\odot$ (for $0.5 \geq q_0 \geq 0$), where Υ is the visual M/L ratio in solar units ($\Upsilon \sim 6$ at the present epoch for the stellar cores of gE galaxies). As is clear from the model curves in Figure 2, stellar populations of age 1 Gyr can easily have an order of magnitude reduction in Υ (depending on the IMF and the logarithmic age), and indeed the radio galaxies at the highest redshifts *require* such a change if they are not to have very high mass ($\gg 10^{12}$ $M_\odot$). However, an evolving M/L of this type would be expected to produce a scatter in the $K - z$ relation unless there was only a small range of ages present at a given redshift. Lilly (1989) estimated $\Delta\tau/\tau \leq 0.25$ at $z \sim 1.5$ for the models that account for the overall brightening observed, the upper limit implying that there is no variation in mass of the host galaxies. If the M/L ratio was evolving even more rapidly, as would be required to reduce the masses of the high redshift galaxies down to L* levels (10^{11} $M_\odot$) then an even smaller dispersion in age is required.

Consequently the small dispersion in the $K - z$ relation requires either that the M/L ratio Υ be slowly evolving with age (which then implies relatively high Υ and high masses for the galaxies) or that the range of ages is exceptionally finely tuned. *Hence, the high masses inferred for the radio galaxies (i.e. of order* 10^{12} $M_\odot$ *of stellar material) can therefore be viewed as a direct consequence of the small dispersion in the* $K - z$ *relation.*

Finally, it should be noted that these direct arguments concerning the masses of the radio galaxies at high redshift are actually quite consistent with the fundamentally different estimates for the masses of the host galaxies of high z luminous quasars made by Efstathiou and Rees (1988).

2. Discussion

If it is correct, then the foregoing has potentially important implications for our understanding of galaxy formation. Ascribing the 'red bump' to stellar populations of age around 1 Gyr implies that there existed at these redshifts massive galaxies (10^{12} $M_\odot$ of stellar material) that were composed of stars formed at significantly higher redshifts. In the last year, attention has been focused on three issues.

2.1. Alternative mechanisms for the 'red bump'

In terms of the SED alone, it has always been appreciated that there exist alternative mechanisms that will reproduce red optical-infrared colors aside from the conventional interpretation of 'old' stars. These are discussed below in three generic forms which cover the range of possibilities. All of them might be expected to produce much more scatter in the $K-z$ relation than is observed.

Firstly, we may be dealing with unusual stellar populations. The Bruzual models used in the previous discussion are conventional in having a continuous IMF extending from very high to very low masses. Within these models the conclusions are not sensitive to the precise slope of the IMF because for most plausible IMFs, the colors are dominated by the most massive stars that are at, or are just past, the main sequence turn-off point. Nevertheless, radically altered IMFs could *in principle* reproduce the observed SED at very much younger ages than 1 Gyr in one of two ways. Firstly, if there were only low mass stars (e.g. cool G dwarfs) present, then the SED could be fixed at an acceptable shape for the first several billion years. However, the M/L ratio should be even higher in this case than with a continuous IMF, and so the reduction in age of the stellar population is purchased at the price of an undesirable increase in the mass of the system. A second alternative which could drastically reduce the required masses and which has recently been advocated by Bithell and Rees (1989, preprint) is to have the red light coming from a population of very massive stars going through a high luminosity red supergiant phase at an age of order 10^7 years. This could reduce the mass requirements to very small levels on account of the very low M/L values of the stars involved. This may seem attractive since 10^7 years is a natural timescale for radio sources and odd IMFs could well be produced in the high pressure regimes of a radio source. The problem here is that individual massive stars spend much longer as blue supergiants (more than 80% of their lives) than as red ones (e.g. Chiosi and Maeder 1986, and references therein). Hence, given the strong mass-lifetime relation for massive stars, one has to have (a) a very narrow range of mass, less than about 10% $\delta M/M$ *and* (b) observe this restricted population in the final 20% of its life, in order to have a RSG dominated population. Apart from the contrived nature of this scenario, it also leads to a very rapidly varying M/L ratio which would be expected to produce a large scatter in the $K-z$ relation.

The second and more plausible class of scenario would arise if an intrinsically hotter component is reddened by dust (the thermally reprocessed radiation would itself be emitted at longer wavelengths, $\lambda_{rest} > 1\mu$m). One possibility is to slightly redden an already cool component, thereby increasing the derived age of the stellar population as measured by the SED by some amount. The potential depression in M/L thus produced would however be offset by the apparent reduction in rest-frame visual luminosity. Alternatively, one could have substantial reddening of either a quasar-like central nucleus or of a conventional flat-spectrum star-burst. In the latter case could be extended well away from the nucleus. For either quasar or star-burst it is hard to see how such a component could have such a small dispersion in reddened luminosity. One might expect to see at least the dispersion in radio luminosity (a factor of six between the 3C and 1 Jansky samples at a given redshift) within the sample being considered. In addition, the strong Lyman α emission and very blue ultraviolet continua seen in these objects does not support the idea of a dusty object and at the very least would require careful spatial segregation of the different components.

Finally, there are a number of 'exotic' explanations involving synchrotron radiation from the radio hotspots, Comptonization of the CMB, electron scattering of an anisotropically radiating central quasar. Many of these have been discussed in the context of 3C368 by Chambers *et al.* (1988). Most of these fail to plausibly account for the SED shape or luminosity and in any case predict a larger dispersion in rest-frame optical luminosity than is observed. It is less easy to rule out some of these possibilities for the hot component, where the dispersion in properties is much larger.

2.2. The morphologies at infrared wavelengths

The association of the 'red bump' with an 'old' stellar population places some constraints on the spatial morphologies that would be expected on deep K-band images. Clearly one would like to see nice symmetric galaxies. However, there is now clear evidence in at least one case, 3C368 (Chambers *et al.* al 1988) and preliminary evidence in several others (Eisenhardt *et al.* 1989, private communication) that the infrared images of many these galaxies show morphological peculiarities that are at least qualitatively similar to the aligned structures seen in the rest-frame far-ultraviolet (McCarthy *et al.* 1987b, Chambers *et al.* 1987). This is potentially the largest embarrassment for the 'old galaxy' interpretation of the high redshift radio galaxies. However, two points should be borne in mind, and to my mind this issue is by no means settled.

Firstly, even in the basic two-component model, some significant fraction of the K-band light is contributed by the 'hot' component that dominates at shorter wavelengths and produces the strong alignments seen at optical wavelengths. Lilly (1989) estimated that this was around 20% in typical 3C radio galaxies and as high as 40% in extreme objects such as 3C368. Although these additional components would not be sufficient to perturb the small scatter in the $K - z$ relation, they could produce noticeably unusual morphologies and some residual 'alignment' would be produced even if the red component was perfectly symmetrical. Many of the objects observed so far in the infrared (Chambers *et al.* 1988, Eisenhardt *et al.* 1989, private communication) are amongst the most spectacularly aligned in the optical waveband and would be expected (see Lilly 1989) to have higher than average contamination from the 'hot' component at K. Accurate color maps are required to determine the colors of the aligned component to see if they are generally as red as the overall SEDs of Figure 1. The fact that the 'aligned' galaxies in the McCarthy et al (1987b) sample are bluer in $(r - K)$ than the unaligned galaxies (Lilly 1989) suggests that the aligned component is usually bluer than the rest of the galaxy. Occasional red components, such as the northern lobe in 3C368 (Chambers et 1988), could possibly be ascribed to an 'old' companion moving into the path of the jet.

Secondly, it is clear from the previous section that there are ways to generate red SEDs from alternative mechanisms. Hence, even if the aligned components (which are presumably not 'old stars') are shown to be as red as the main bodies of the galaxies, this does not *per se* invalidate the conclusion that these galaxies are old, since that is strongly based on the small scatter in $K-z$ as well as on the form of the individual SEDs. Consequently, the clear and testable prediction of the 'old galaxy hypothesis' is that *within the population as a whole*, not more than 20-30% of the K-band light should come from structures which are demonstrably not old. It is the uniform component in the $K-z$ diagram which should be unaligned.

2.3. The proto-galaxy candidate 3C326.1 at $z = 1.82$

The most obvious potential counter-example to the idea that essentially all radio galaxies are mature systems composed of stars of order 1 Gyr or more old and assembled well prior to the epoch of observation is 3C326.1 at $z = 1.82$, which was proposed by McCarthy *et al.* (1987a) as a promising proto-galaxy candidate. This object is outside the 'statistically complete' LRL 3C sample discussed above but clearly nevertheless merits considerable attention.

In the radio, this source has an unremarkable classical double structure, albeit with large brightness assymmetry between the two lobes. In the rest-frame far ultraviolet, several faint objects were seen embedded in and around a large and apparently diffuse Lyman α cloud. McCarthy *et al.* (1987a) argued that the absence of a centrally condensed object similar to that seen in other radio galaxies at similar redshifts suggested that this was a galaxy seen in the process of formation. New observations of this object however argue against this idea.

Firstly, a deep K-band image of 3C326.1 shows that one of the faint continuum objects, 'B', dominates the system at rest-frame optical wavelengths (Lilly and McLean 1989, Figure 3a). This object has $K = 19$ and while it is thus fainter than the mean K magnitude at $z = 1.8$, the shortfall is only 0.9 magnitudes, about twice the standard deviation within the population. Given that there are now 20 or so radio galaxies known at $1 \leq z \leq 2$, it is therefore not anomalous. The $(R-K)$ color of this object is also typical of radio galaxies at $z \geq 1$. The location of 'B' between the radio lobes supports 'B' as the site of the active nucleus and its similarity in SED and in absolute magnitude to the other radio galaxies at $z \geq 1$ removes much of the reason for believing this object to be a proto-galaxy. If 'B' is the host galaxy, then 'C', which is colocated with the western hotspot, is an extreme (and 'detached') example of the radio-optical alignment effect. It is considerably bluer in $(R-K)$ than the main galaxy.

Secondly, a new deep exposure in Lyman α on the 2.2m UH telescope (Figure 3c) suggests that the line emission is also rather more centrally concentrated around the lobe component 'C' than had been previously thought. Although there may be some modest continuum contamination through the filter, $\Delta\lambda \sim 300$ Å, this is unlikely to be significant except for the brighter 'A' component, which may be entirely continuum emission. While smoothing reveals a faint 'bridge' back towards 'B', there is no detectable Lyman α emission exterior to the hotspot.This suggests that the Lyman α gas is much more intimately associated with the radio source than had previously been thought, and that the high velocities observed in the gas (McCarthy *et al.* 1987a) may tell us more about the energetics of the radio source than the collapse of a proto-galactic gas cloud.

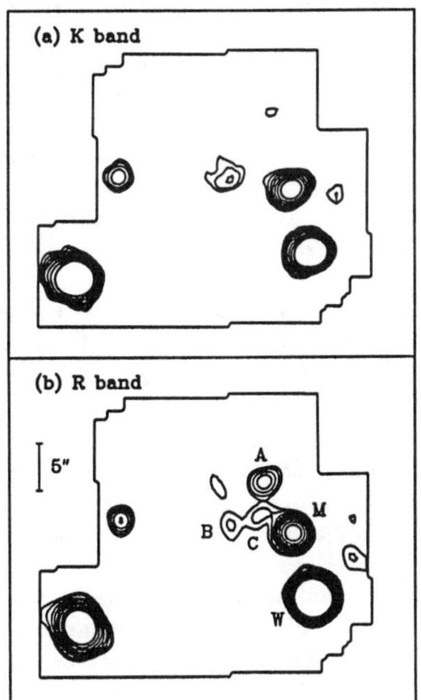

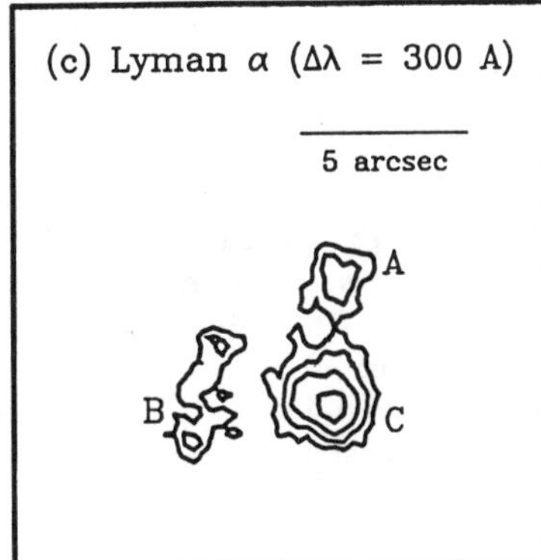

Fig 3. New images of the protogalaxy candidate 3C326.1: (a) K-band and (b) coregistered and smoothed R-band, showing component 'B' dominates at restframe 8000 Å. This component has a K magnitude and $(R-K)$ color that are similar to other radio galaxies at this redshift. (b) A $\Delta\lambda = 300$ A filter centered on Lyman α. Contour intervals are logarithmic with an interval of $\sqrt{2}$. This image suggests that the gaseous emission is more concentrated on the lobe component 'C' and hence more closely associated with the radio source, than has previously been thought. Emission from 'A' is probably continuum.

3. The relevance of radio galaxies to 'normal' galaxy formation

3.1. The comoving number density of radio galaxies

The radio luminosity function for powerful sources is reasonably well-defined for $1 \leq z \leq 2$ (e.g. Peacock 1985). At $z \sim 2$, the comoving density of the extremely powerful sources ($P_{408} = 10^{27-28}$ WHz^{-1}sr^{-1}) that are represented in the 3C and 1 Jansky samples is still very low, around 10^{-9} Mpc^{-3}. Since the duty-cycle of a radio source is unlikely to be unity, the radio galaxies are probably sampling a somewhat larger population of galaxies. Even so, the powerful sources that have been studied so far therefore represent only a tiny fraction of the present day galaxy population. Even massive gE galaxies (roughly 5-10 L*), which are the hosts of local radio sources and which are plausibly (from the continuity of Figure 2) the evolutionary endpoint of the high z radio galaxies, have a present density a couple of orders of magnitude higher at $\rho \sim 10^{-7}$ Mpc^{-3} (King and Ellis 1985).

However, the properties of the host galaxies of sources do not show strong correlations with radio power above the 'break' in the RLF at around $P_{408} = 10^{25}$ WHz^{-1}sr^{1} at either $z \sim 0$, or at $z \sim 0.5$ (Hill and Lilly, in preparation), and if the RLF at $z \sim 2$ is integrated down to this point, then the comoving density of all 'powerful' radio sources rises to of order 10^{-7} Mpc^{-3} (e.g. Peacock 1988), comparable to the density of present day massive galaxies. These radio sources will be found in 'milli-Jansky' radio samples which are only now being pushed to high redshifts, $z \geq 1$ (Windhorst, this conference). It remains to be seen whether these fainter radio galaxies appear to be as old and as massive as the more powerful 3C and '1 Jansky' sources discussed above, noting only that most 'milli-Jansky' radio galaxies at $z \leq 1$ do indeed appear to be broadly similar to the 3C and '1 Jansky' galaxies at the same redshift (Windhorst *et al.* 1985 and references therein, Hill and Lilly, in preparation).

If the 'milli-Jansky' sources are indeed found to be similar to the rarer and more luminous 3C and 1 Jansky sources discussed above, and if it is accepted that these objects are indeed the pregenitors of present day massive galaxies, then the densities are high enough that we would be forced to accept that the radio galaxy results must be telling us about the formation of a large fraction of all massive galaxies.

3.2. The significance of the peak in AGNs at $z \sim 2$

There has been considerable speculation concerning the possible link between the peak in comoving density of quasars and radio sources at $z \sim 2$ and the epoch of galaxy formation.

The observation that all known radio galaxies at this redshift (including now 3C326.1 at $z \sim 1.8$) contain dominant stellar populations that appear to be at least 1 Gyr old and are amazingly uniform from galaxy to galaxy, argues against this idea. These do not appear to be 'forming' galaxies. Given that no radio galaxies at $z > 1$ are found with ages demonstrably greater than about 2 Gyr, two possibilities present themselves: Either the progenitors of all radio galaxies formed at high redshift $z \geq 5$, and the peak in radio activity at $z \sim 2$ then represents a significant phase of the post-formation evolution of galaxies, or else the distribution of redshifts of active galaxies around the peak at $z \sim 2$ does indeed reflect the formation of massive galaxies, but with a time-lag of order 1 Gyr or more. Objects like 0902+34 and 4C41.17 that lie well beyond the peak could be systems that were triggered earlier than most, or could be galaxies that actually formed earlier than most.

Distinguishing between these possibilities will not be easy, which is unfortunate because it is important for confronting theories of galaxy formation. A population of 10^9 year old $10^{12.5-13}$ $M_\odot$ galaxies (including some Dark Matter) at $\rho \sim 10^{-7}$ Mpc^{-3} at $z \sim 2$ (i.e. assuming that the milli-Jansky sources are indeed the same as the 3C and 1 Jansky sources) is just about reconcilable with CDM (see Efstathiou and Rees, 1988, Peacock

1988), if the bias parameter $b \sim 1.5$ and all $10^{12.5-13}$ $M_\odot$ objects are associated with galaxies rather than groups. However, if this population is pushed to $z > 4$, this becomes much more difficult to accommodate. Despite the conflict with this popular theory, I suspect that the *continuity* of the Hubble relation from $0 \leq z \leq 3$, suggests an early formation epoch for *all* the galaxies that are subsequently seen as radio sources, and that this may then be telling us about the formation of essentially all massive elliptical galaxies.

4. Summary

It has been argued that the properties of well-defined samples of radio galaxies at $z \geq 1$, and in particular the remarkable uniformity of their infrared (rest-frame optical) properties, indicates that they are massive galaxies that formed well prior to the epoch of observation.

References

Bithell, M., and Rees, M.J.; 1989, *preprint.*

Bruzual, G.A.; 1983, *Ap.J.*, 322, 585

Chambers, K., Miley, G.K. and van Breugel, W.; 1987, *Nature*, 329, 609

Chambers, K., Miley, G.K., and Joyce, R.R.; 1988, *Ap. J. (Letters)*, 329, L79.

Chiosi, C. and Maeder, A.; 1986, *Ann. Rev. Astron. Astrophys.*, 24, 329.

Daly, R.; 1989, *preprint*

De Young, D.S.; 1989, *Ap. J. (Letters)*, 342, L59.

di Serego Alighieri, S., Fosbury, R.A.E., Tadhunter, C.N. and Quinn, P.J.; 1989,*In "Extranuclear activity in galaxies"* (eds. R.A.E. Fosbury and E.J.A. Meurs), *in press.*

Efstathiou, G., and Rees, M.J.; 1988, *M.N.R.A.S.*, 230, 5P.

King, C.R., and Ellis, R.S.; 1985, *Ap.J.*, 288, 456.

Laing, R.A., Riley, J.M., Longair, M.S.; 1983, *M.N.R.A.S.*, 204, 151.

Le Fevre, O., Hammer, F., Nottale, N., Mathez, G.; 1987, *Nature*, 326, 268.

Lilly, S.J., and Longair, M.S.; 1984, *M.N.R.A.S.*, 211, 833.

Lilly, S.J.; 1988, *Ap.J.*, 333, 161.

Lilly, S.J.; 1989, *Ap.J.*, 340, 77.

Lilly, S.J., and McLean, I.S.; 1989, *Ap. J. (Letters), in press.*

McCarthy, P.J., Spinrad, H., Djorgovski, S., Strauss, M.A., van Breugel, W. and Liebert, J.; 1987a, *Ap. J. (Letters)*, 319, L9.

McCarthy, P.J., van Breugel, W., Spinrad, H., Djorgovski, S.; 1987b, *Ap. J. (Letters)*, 321, L29.

Peacock, J.A.; 1988, In *"The epoch of galaxy formation", (eds. C.S. Frenk, R.S. Ellis, T. Shanks, A.F. Heavens, J.A. Peacock)*, Kluwer NATO ASI Series C, Vol. 264, p.391.

Rees, M.J.; 1989, *M.N.R.A.S.*, 239, 1P.

Schneider, D.P., Gunn, J.E., and Hoessel, J.G.; 1983, *Ap.J.*, 268, 476.

Tinsley, B.M and Gunn, J.E.; 1976, *Ap.J.*, 203 52.

Windhorst, R.A., Miley, G.K., Owen, F.N., Kron, R.G., Koo, D.C.; 1985, *Ap.J.*, 289, 494.

EVOLUTIONARY MODEL UNCERTAINTIES IN THE K-BAND HUBBLE DIAGRAM

W. N. WEIR and S. DJORGOVSKI
Division of Physics, Mathematics and Astronomy, California Institute of Technology, Pasadena, CA 91125

GUSTAVO BRUZUAL A.
Centro de Investigaciones de Astronomia (CIDA), Apdo. Postal 264, Merida 5101-A, Venezuela

ABSTRACT We explore the effects of variations in evolutionary model parameters on the K-band Hubble diagram, and compare the predictions of the older and the newer population synthesis models. The predictions of the old and the new Bruzual models differ significantly; this is tractable to a problem in combining the input stellar libraries. The new models are sensitive to a number of evolutionary parameters, and there is some coupling with the cosmological parameters. Therefore, the K-band Hubble diagram still cannot be used as a reliable cosmological tool with the state-of-the-art evolutionary population synthesis models.

The Hubble diagram is a classical cosmological test (Sandage 1988), and the last refuge of scoundrels (Lilly 1989, private communication). There is now a revival of interest in it, spurned by the recent discoveries of galaxies at redshifts where the cosmological effects should become noticeable, and by the advances in the infra-red observing technology. Samples of radio galaxies are now complete out to $z \sim 2.5$ and reach up to $z \simeq 3.8$, and even though these objects are highly unusual, they give us some glimpses about the behaviour of galaxies at large look-back times. Optically selected BCM's are known up to $z = 0.92$. The differences between $q_0 = 0$ and $q_0 = 1/2$ models in that redshift regime may be $0.7 - 1$ mag, not including the K-corrections.

Studies of radio galaxies have shown that the evolutionary effects in the visible bands dominate over the cosmological differences by several magnitudes at such large look-back times. The uncertainties of evolutionary models and their interpretation make it impossible to use Hubble diagrams in the visible regime as a viable cosmological tool. These difficulties arise mainly because the data in the observed visible light probe the rest-frame UV, which is dominated by young stars whose short lifetimes reflect and amplify any changes in the star formation rate (SFR); fluctuations in the upper end of the IMF and intrinsic extinction differences in these galaxies only worsen the situation. In addition, non-thermal light contributions in the rest-frame UV may be important. For low mass stars, which are the progenitors of red giants, and which dominate in numbers and are thus less sensitive to the fluctuations of the IMF, the relevant time scales are long (a few Gyr), and any flicker or even a long-term change in the SFR is smoothed/convolved with the stellar evolutionary time scales.

IR imaging/photometry should sample these relatively old stellar populations in distant galaxies, which change slowly. The IR magnitudes are thus potentially a robust estimator of galaxy luminosity at large look-back times. This has been tentatively confirmed by the IR data available so far: the K-band Hubble diagrams appeared to be remarkably insensitive to all the parameters of evolutionary models by Bruzual (1983), but retained their sensitivity to cosmology (Spinrad and Djorgovski 1987). Thus it was hoped that it may be possible to use distant galaxies in the near-IR as "standardizable candles". It is still too early to do this: the cosmology will have to wait for well-understood samples of high-redshift galaxies, and for more reliable population synthesis models which will account better for post-giant branch evolution and metallicity effects. However, we can make at least preliminary studies of the IR Hubble diagram.

Our first goal is to understand the near-IR Hubble diagram from the operational/procedural and theoretical points of view, without actually attempting serious measurements of cosmological parameters. We explore the sensitivity of the K-band Hubble diagram on the evolutionary model parameters, using the available K-band magnitudes of BCM's and radio galaxies from the literature as a representative data set. The purpose of this investigation is to locate the most critical evolutionary input parameters, which need to be better determined before the diagram can be used as a cosmological tool. We are *not* trying to determine the evolutionary and cosmological parameters yet – the data are not sufficiently well understood. We plan to explore the effects of dynamical evolution, selection effects and possible correlations of IR and radio luminosities at large redshifts in future papers.

A new version of Bruzual galaxy spectral evolutionary models was used to derive predictions for the K magnitudes vs. redshift. The stellar library used in Bruzual's (1983) original models was modified in the optical and IR ranges in order to build an improved set of evolutionary models. In the optical range (3600 - 10000 Å) the stellar SED's were taken from the Gunn and Stryker (1983) library. The KAO spectral library of Strecker and collaborators (1986, private communication) was used to complete the stellar SED's from 1 to 2.56 μm with 0.02 μm resolution. The matching of the IR and optical SED's was performed by means of standard relationships (Bruzual and Persson 1989, in preparation). In the UV range we used the same stellar SED's as in Bruzual (1983). The overall properties of the new set of models are similar to that of the original models. However, in the IR, the new models represent a considerable improvement over the previous ones, which only included broad band fluxes in this range. The predictions for the IR properties of galaxies derived from the new models should thus be more reliable.

As a working sample of objects, we use the available data on 3CR and 1-Jy radio galaxies, and optically selected brightest cluster members. The measurements are from Lilly and Longair (1984) and Eisenhardt and Lebofsky (1987). It is reasuring that these different samples show the same behavior in the K-band Hubble diagram, with a reasonably small cosmic scatter. However, we do not propose that this sample is homogeneous, complete, or even remotely adequate for cosmological studies. The origin of the IR light in all these objects, and powerful radio galaxies in particular, is still not well understood.

We explore the effects of changes in some of the more important parameters of evolutionary models, i.e., the star formation history, the power-law slope of the IMF, and the mass cutoffs of the IMF. The star formation rate is assumed to decline exponentially (Bruzual μ-models). A simple Friedman cosmology is assumed. The model K magnitude curves were normalized by requiring them to bisect the distribution of data points in the redshift range $0.1 < z < 0.5$ (the aperture corrections are uncertain at lower z's, and the evolutionary and cosmological effects become important for the higher z's).

The most interesting result so far is that whereas the old models fitted the IR data rather well, as indicated in the preliminary study by Spinrad and Djorgovski (1987), the new models do not fit the data as well for the same range of parameters, and show a systematically different behavior. The new models are systematically too bright at large redshifts with respect to the old ones, for the same parameters. Alternatively, one can say that the data are too faint at large redshifts. Such an effect may be produced, e.g., by dynamical evolution (mergers and accretion), for which we do not account.

The $\mu = 0.5$ models can fit the data only in long-time-scale cosmologies, e.g., $H_0 = 50$, $z_{gf} > 10$, and $q_0 \simeq 0$. Steepening of the IMF, and/or biasing towards the *low* mass stars helps, but only slightly. For an "intermediate" set of cosmological parameters, $\mu = 0.1$ models provide a better fit, even though they fail in the optical-UV. Thus, there is still a substantial coupling between the cosmological and evolutionary parameters. We are at this point unable to recommend a good set of evolutionary model parameters.

This illustrates the inherent uncertainty in the state-of-the-art population synthesis models. The problem is tractable to a fundamental calibration difficulty in matching of the input data: the optical-UV spectra are difficult to combine with the near-IR spectra, which are obtained using quite different technologies. Work is now in progress to understand better this process, and produce a more reliable evolution code.

These preliminary results indicate that the potential of the IR Hubble diagram as a cosmological test is less certain than it appeared only a couple of years ago. This conclusion is based upon the variations and ambiguities inherent in the evolutionary models, independent of any problems which may exist with the data. Selection effects, light from nonthermal sources, and other complications can only make the situation worse.

We acknowledge a partial support from California Institute of Technology and Alfred P. Sloan Foundation (SD), the NSF (WNW), and CIDA (GBA).

REFERENCES

Bruzual, G. 1983, Ap. J., 273, 105.
Eisenhardt, P., and Lebofsky, M. 1987, Ap. J., 316, 70.
Gunn, J.E., and Stryker, L. 1983, Ap. J. Suppl., 52, 121.
Lilly, S., and Longair, M. 1984, M.N.R.A.S., 211, 833.
Sandage, A. 1988, Ann. Rev. Astr. Ap ., 26, 561.
Spinrad, H., and Djorgovski, S. 1987, in Observational Cosmology, IAU Symposium 124, eds. A. Hewitt *et al.* , p. 129. (Dordrecht: Reidel.)

RELATIONS BETWEEN RADIO AND OPTICAL PROPERTIES IN DISTANT RADIO GALAXIES

WIL J. M. VAN BREUGEL*
Lawrence Livermore National Laboratory,
Institute for Geophysics and Planetary Physics,
P.O. Box 808, L-413, Livermore, CA 94450.

PATRICK J. MCCARTHY
Observatories of the Carnegie Institution of Washington,
813 Santa Barbara St., Pasadena, CA 91101.

ABSTRACT We review recent discoveries of close relationships between the radio and optical properties of powerful, high redshift 3CR radio galaxies. These include the alignments of distant galaxies and their associated radio sources, and the extremely good correlation between radio lobe and emission-line gas asymmetries. The possible source of ionization of this gas and the radio spectral and polarization properties are discussed briefly. These results show that the origin and evolution of radio galaxies, their environment, and the interaction between them are inextricably connected.

1. INTRODUCTION

Until recently much of the motivation for studying distant radio galaxies has come from their potential use as cosmological probes (e.g. Spinrad and Djorgovski 1987). Multi-wavelength studies over the past few years have shown, however, that many of their optical and radio properties are closely related. This suggests that (radio) galaxy activity plays an important role in shaping the physical conditions in these young galaxies. It is important to better understand the nature of these various interrelations if one is to use the optical data of radio galaxies to constrain galaxy evolution and cosmological models. Such investigations might also provide further insight in the origin of galaxy activity itself, and could be used to explore possible relationships to quasars, the most prominent class of active high redshift objects.

Here we discuss recent progress in the studies of these radio galaxy relationships. Where needed we will assume $H_0 = 50$ km s^{-1} Mpc^{-1}, $q_0 = 0.0$.

---

* Formerly at Radio Astronomy Laboratory, University of California, Berkeley.

2. SAMPLE

Our primary sample for studying the radio/optical relationships in distant radio galaxies is taken from the 3CR catalog (Bennett 1962). During the past years all of the radio galaxies in this sample have been identified and all but 4 of these have had their redshifts determined (Spinrad *et al.* 1984; Spinrad 1987; Djorgovski *et al.* 1988). The object with the largest redshift is now a radio galaxy (3C 257, $z = 2.48$) rather than a quasar (3C 9, $z = 2.012$).

Optical continuum and emission-line ([OII] λ3727, Ly α) images have been obtained at Lick Observatory and Kitt Peak National Observatory of most of the radio galaxies with $z \gtrsim 0.5$ (McCarthy 1988). Supplemented with similar data obtained by Baum *et al.* (1988) at lower redshifts ($z \lesssim 0.2$), emission line images now are available for $\sim$ 50% of the 3CR sample. Modern radio maps of many of the sources are available in the literature and most 3CR radio galaxies with $z \gtrsim 1$ have recently been observed with the VLA (van Breugel *et al.* 1990). Thus it is now possible to examine the optical and radio morphologies of 3CR radio galaxies over a large range in redshift.

3. RADIO/OPTICAL ALIGNMENTS: CONTINUUM

It is known from broad-band images that at large redshifts many 3CR galaxies are elongated and exhibit multimodal structure (Le Fèvre *et al.* 1988, and references therein). By comparing these optical images with our VLA observations and radio data published in the literature we have found that the long axes of distant galaxies are generally aligned with those of their associated radio sources (McCarthy *et al.* 1987*b*). This was independently discovered also by Chambers *et al.* (1987) using a sample of 4C sources which were suspected, and subsequently confirmed, to have a large fraction of distant radio galaxies on the basis of their very steep radio spectra. The alignments also occur at infrared wavelengths (Chambers *et al.* 1988; Eisenhardt and Chokshi 1989), and at the highest redshifts known (Chambers *et al.* 1989).

Since our earlier paper we have obtained considerably more information on the optical continuum and radio morphologies for radio galaxies with $z \gtrsim 0.5$ (McCarthy 1988). These data confirm and improve on our previous results. An example is shown in Figure 1*a*. It is also clear that the alignment between galaxies and their associated radio sources becomes better with increasing redshift, as shown in Figure 1*b*.

This alignment effect at high redshifts should be contrasted with that of nearby ($z \lesssim 0.2$), low luminosity radio galaxies for which the opposite relationship seems to be true, especially when the sample is restricted to very large sources (Palimaka *et al.* 1979; Baum and Heckman 1989) or to sources with relatively bright radio cores (Kapahi and Saikia 1982).

Most of the high redshift radio galaxies have blue optical and IR colors, suggesting that they are undergoing large scale bursts of star formation and a straightforward interpretation of the radio/optical alignments is then that these star bursts have been triggered by the bowshocks and backflows associated with the radio hotspots (McCarthy *et al.* 1987*b*).

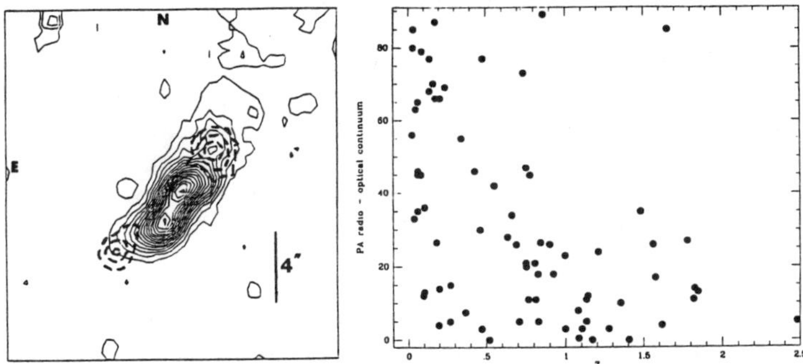

Figure 1a. A superposition of our 6 cm VLA map of 3C 256 (0.4″, dashed contours) on top of a R band CFHT image (0.7″, solid contours) from Le Fèvre *et al.* 1988). emission. *Figure 1b.* A plot of the radio/optical position angle differences vs. redshift for 3CR radio galaxies.

Other explanations such as gravitational amplification (Le Fèvre *et al.* 1988), fuelling along the radio axes (Djorgovski 1988), inverse Compton scattering (Chambers *et al.* 1988), Thompson scattering (Fabian 1989) and dust scattering (di Serego Alighieri *et al.* 1989) appear less attractive in light of the current data. An interesting new explanation, which merits futher study, was proposed by Quinn (1989) who suggested that the elongated ("bar") galaxies might have formed in asymmetric, dark halos and that they might contain acrction disks with the their spin axes (= radio jet directions) aligned with the galaxy major axes. A complication in the interpretation of the continuum light from distant radio galaxies might arise if some of this continuum light would be polarized, as has been reported to be the case for the radio galaxy 3C 368 (di Serego Alighieri *et al.* 1989).

At present we believe that the most plausible model to explain the alignments is that the radio sources enhance star formation in their surrounding (proto-?) galactic environments, perhaps somewhat similar to the jet-induced star formation observed in the nearby, *low* luminosity source PKS 0123-016 and its associated "Minkowski's Object" (van Breugel *et al.* 1985*b*). It is conceivable that the backflow and bowshocks associated powerful double radio galaxies may trigger the collapse of dense regions in a clumpy gaseous medium which is likely present in a young, forming galaxy. Several such models have recently been proposed (De Young 1989; Daly 1989; Begelman and Cioffi 1989; Rees 1989). The key problem for these models lies in the interpretation of the IR (K-band) data, which also show the alignment effect. The question is whether or not there is an extended old (1–2 Gyr) stellar population present along the radio source axes, which would be difficult to understand on dynamical grounds in jet-induced starformation models. A further discussion the stellar populations in distant radio galaxies and consequences for the interpretation of the alignment effect is given by Chambers (these Proceedings).

4. RADIO/OPTICAL ALIGNMENTS: LINE-EMISSION

Most of the distant, powerful radio galaxies have extended emission line regions, nearly all of which are parallel to the radio axes (McCarthy *et al.* 1987*b*; McCarthy 1988). In a number of radio galaxies extended optical line emission is also found *beyond* the radio hotspots and a good example is shown in Figure 2 (3C 435*A*, z = 0.461). The line emission overlaps in part with a large, diffuse continuum object. This is presumably a galaxy at the same distance as 3C 435*A* which happens to be located along the direction of the radio source.

The presence of emission-line gas along the radio source axis well beyond radio hot spots is interpreted as evidence that radio galaxy nuclei may have an anisotropic source of (photo-) ionization along the radio source direction. Diagnostic emission-line ratios are consistent with such an interpretation (Section 7).

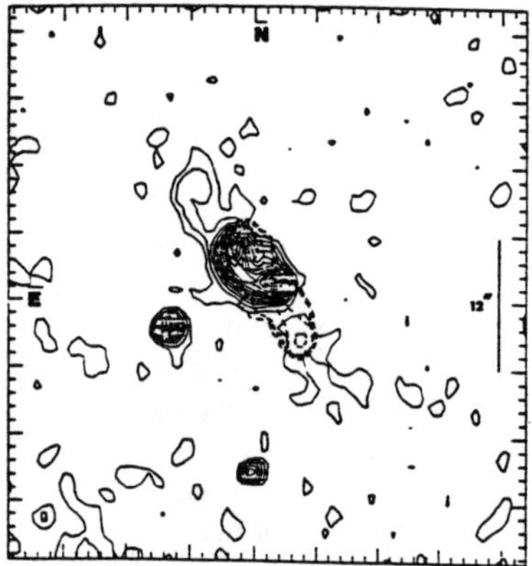

Figure 2. Contour map of the extended optical line emission (O[II]) associated with 3C 435*A*. Overlaid (dashed lines) is a 21 cm VLA map with 1.2″ resolution. The line emission extends ~ 50 kpc beyond the radio hot spots.

5. *MIS-* ALIGNMENTS

While the great majority of the distant, powerful 3CR radio galaxies show the above described alignments, there exist some where the position angles between the radio and optical major axes (continuum and/or line emission) differ significantly. An example is shown in Figure 3 (3C 294). In 3C 294 no extended optical continuum was detected but the Ly α line emission is very elongated and directed along the inner radio axis. Another mis-aligned source is 3C 124, where we have detected faint, diffuse radio lobe well outside its bright inner double. The hotspots of this small inner double are slightly resolved and together with the diffuse emission the overall source structure appears S-shaped. The optical continuum and line emission of this radio galaxy are both directed along the inner radio source axis.

The S-shaped radio morphologies of 3C 294, 3C 124 suggest that the ejection axes of their radio nuclei have changed direction, for example through precession. Precession is often inferred to occur also in other radio galaxies and quasars. It may be associated with the formation and evolution of binary black holes during the late stages of galaxy merging and may be accompanied by an increase of galaxy activity (Roos 1988).

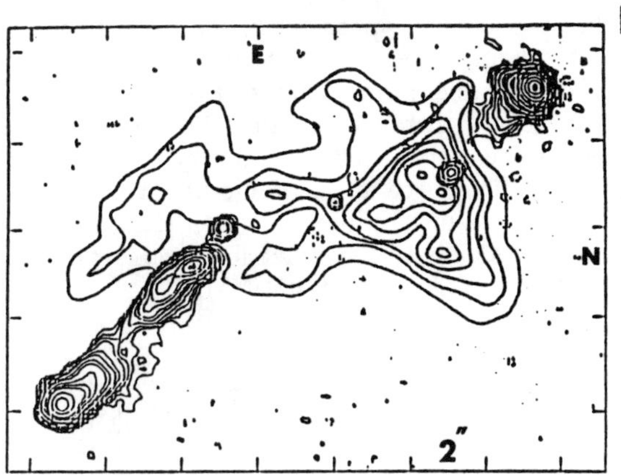

Figure 3. Radio map of 3C 294 (0.4″, 6 cm) with the Ly α emission (thick contours; McCarthy *et al.* 1989) overlaid.

6. RADIO-LOBE AND EMISSION-LINE GAS ASYMMETRIES

From our radio and emission-line images we find that powerful 3CR radio galaxies are very asymmetric in their radio lobe and emission-line gas properties (McCarthy and van Breugel 1989). These asymmetries increase with redshift, and are correlated.

The lobe distance ($R_{ldist} \equiv D_{max}/D_{min}$), flux density ($R_{fluxd} \equiv F_{max}/F_{min}$; at 6 cm observed) and emission-line brightness ($R_{elbr} \equiv B_{[OII],max}/B_{[OII],min}$) ratios are all larger for $z \gtrsim 0.7$ than is observed locally. The radio asymmetries R_{ldist} and R_{fluxd} are (weakly) statistically correlated, as found before (e.g. Macklin 1981).

More importantly, however, radio galaxies which have extended optical line emission are generally the most asymmetric in their radio lobe properties. In particular we find that radio galaxies with relatively large emission-line regions have the brightest line emission essentially always on the side of the *closer* of the two radio lobes (see Figure 4).

This asymmetry relationship is extremely robust seems to be largely independent of the linear sizes of the radio sources, their lobe flux densities and redshift or radio power. Additionally, a preliminary investigation of data available in the literature shows that similar results hold for double lobed 3CR quasars. The close-lobe/bright-line-emission correlation even holds for very small lobe distance asymmetries, and very large radio sources. A few typical examples for various redshifts and R_{ldist} values are:

3C 294 ($z = 1.779$, $R_{ldist} = 1.40$; lin. size = 183 kpc)
3C 280 ($z = 0.998$, $R_{ldist} = 1.27$; lin. size = 139 kpc)
3C 330 ($z = 0.550$, $R_{ldist} = 1.08$; lin. size = 526 kpc)
3C 445 ($z = 0.056$, $R_{ldist} = 1.05$; lin. size = 855 kpc)

In the past attempts have been made to interpret radio lobe asymmetries in terms of light travel time (projection) effects, which could then be used to constrain radio lobe advance speeds (e.g. Ryle and Longair 1967). More recently it has become increasingly clear that this can not be the dominant effect, however. For example, if one assumes that the sources are intrinsically symmetric, then large lobe distance asymmetries should be correlated with small projected linear sizes. This has not been found (Fokker 1986). Alternatively, if the radio source nuclei would be intrinsically *a*-symmetric, exhibiting alternating-side or "flip-flop" ejection (Rudnick and Edgar 1984), then one would expect a deficit of symmetric sources. This was not confirmed by later analyses (Ensman and Ulvestad 1985).

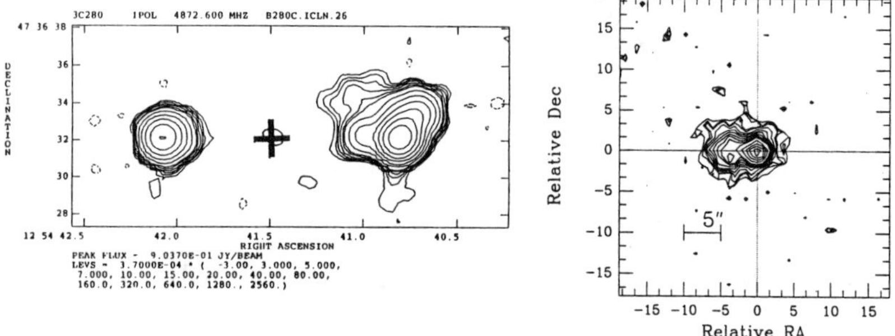

Figure 4. A 6 cm radio map (1.2″, top) and emission-line image ([OII]λ3727, bottom) of 3C 280 ($z = 0.998$) The galaxy nucleus is indicated by a cross. Note the large lobe distance and fluxdensity asymmetries ($R_{ldist} = 1.27$; $R_{fluxd} = 3.70$), and the one-sided optical line emission which is coincident with the lobe closest to the galaxy nucleus. 3C 280 is also an example where lobe distance and fluxdensity asymmetries are *anti*-correlated.

The very good correlation between lobe distance and emission-line gas asymmetries therefore can best be explained by a non-uniform gaseous environment in these radio galaxies. This would also be consistent with evidence that powerful radio sources might be triggered by galaxy interactions (Heckman *et al.* 1986), which disturb the interstellar media of the radio galaxies and produce tidal tails, shells etc. One would then assume that the side with the extended optical line emission is also the side with the densest gas, which would slow down an expanding radio lobe the most. A potential problem with this is that the correlation is extremely good and even holds for very large, relatively old sources with rather small lobe distance asymmetries. It is not immediately clear whether the lifetimes of the various constituents (radio source, emission-line gas and non-uniform environment) are compatible, and what the role is of AGN/ionization source asymmetries (if any).

7. EMISSION-LINE RATIOS AND IONIZATION

Using close pairs of emission-lines to minimize the effects of extinction, phenomenological diagrams can be constructed to help distinguish between possible sources of ionization such as photoionization by hot stars or powerlaw UV emission from AGN's and collisional ionization by shocks (e.g. Baldwin *et al.* 1981). Such line ratio diagnostics for low redshift radio galaxies have shown that they constitute a fairly homogeneous class of objects for which photoionization by a non-stellar source best reproduces the spectra and, in particular, that the off-nuclear emission-line regions follow the same trend as the narrow lined galaxy nuclei (e.g. Robinson *et al.* 1987).

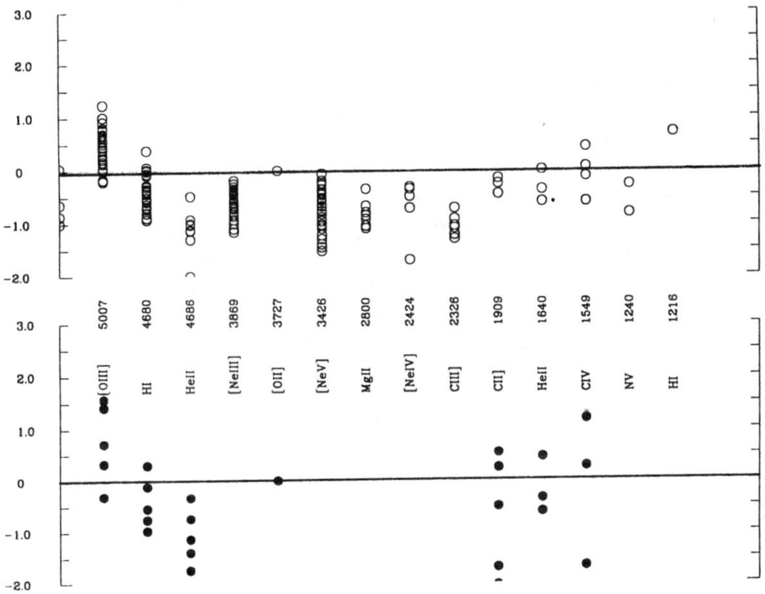

Figure 5. The upper panel shows the logarithm of the strength of various emission lines relative to [OII]λ3727 for the high redshift 3CR radio galaxies. Each point corresponds to a measurement for an individual galaxy, the ensemble displays the range of line ratios with the sample as a whole. The lower panel shows the same quantities for the dilute photoionization models of Ferland and Netzer (1983). The individual points in the model correspond to a range of ionization parameter from Log U = −1.5 to −3.5. The spectral index of the ionizing continuum is $\alpha = -1.5$. Emission lines with no data were not tabulated in Ferland and Netzer (1983).

For our high redshift objects most of the standard spectral diagnostics are redshifted out of the groundbased observing window and different emission-line diagrams had to be used. Using line ratio diagrams such as [NeV]/[OII] vs. [NeIII]/[OII] we arrive at similar conclusions as for the low redshift objects (McCarthy *et al.* 1988; McCarthy *et al.* 1990). At the highest redshifts, z ~2, we find the classic signature of power-law photoionization, namely a wide range of ionization states with roughly equal strength in a single species (e.g. CII], CIII], CIV). Models involving stars, radiative shocks, non-equilibrium shocks, and cosmic ray heating all fail to reproduce the observed line spectra. Thus the emission line gas is photoionized, and a key point is that the off-nuclear material is spectroscopically indistinguishable from the nuclear emission line material.

Such an interpretation of the spectroscopic data is consistent with the morphological information which shows that the line emission is nearly always in the same direction as the radio sources. Sometimes the line emission extends beyond the radio hotspots, but in most cases it is smaller than the radio source and is centrally concentrated. Thus also from this standpoint the active galaxy nuclei seem to be natural sources of ionizing photons, which then appear to be collimated or beamed along the same direction as the radio jets powering the lobes. The correlation between radio and emission-line luminosity (Baum and Heckman 1989; McCarthy 1988) can be interpreted as further indirect evidence for such a model.

Some problems still exist with such an interpretation. In particular it is difficult to account for the more than 10^{55} ionizing photons that are needed. Unlike in nearby radio galaxies where often (but certainly not always) one can reasonably argue that there might be enough UV continuum emission in the galaxy nuclei to supply the required photons (Baum and Heckman 1989), the AGN photon deficit in very luminous, high redshift objects is substantial. The obvious solution for this problem is to assume that not all the nuclear optical continuum can be observed, because of central obscuration or beaming effects. Such a scenario would be neither new nor unique since many of the properties of "classic" active galaxies such as Seyferts and quasars can be understood this way (e.g. Fosbury 1989) and the so-called "unifying schemes" of radio galaxies and quasars would naturally lead to such a picture (e.g. Browne 1989; Barthel 1989).

Additional complications exist however because in many cases the strong radial gradients in the ionization state, predicted by AGN photoionization models, have not been observed (i.e. for radio galaxies embedded in cooling X-ray halos, Heckman *et al.* 1989), and in other objects *local* sources of ionization are known to exist (3C 277.3, van Breugel *et al.* 1985*a*; PKS 2152-69, Tadhunter *et al.* 1988).

8. RADIO SPECTRAL PROPERTIES

Since early radio surveys it has been recognized that there seems to be a correlation between the radio spectral index (defined as $S_\nu \sim \nu^{-\alpha}$) and power and/or redshift, and that very steep spectrum sources ($\alpha \gtrsim 1$) tend to be associated with blank fields on Palomar Sky Survey plates (e.g. Blumenthal and Miley 1979). Taken together these correlations indicate that steep spectrum sources may be identified with very distant galaxies, dramatic proof for which has recently been found (Chambers *et al.* 1989; Chambers, these Proceedings).

With the virtual completion of the identifications of 3CR radio sources and the determination of their (large) redshifts by Spinrad and collaborators, and the succesful usage of the steep spectrum selection criterion to find high redshift radio galaxies, it is of interest to re-examine the $\alpha - z$ / P correlations. In particular, it is now possible to use the many new redshifts of 3CR sources to determine spectral indices in the galaxy *restframes*.

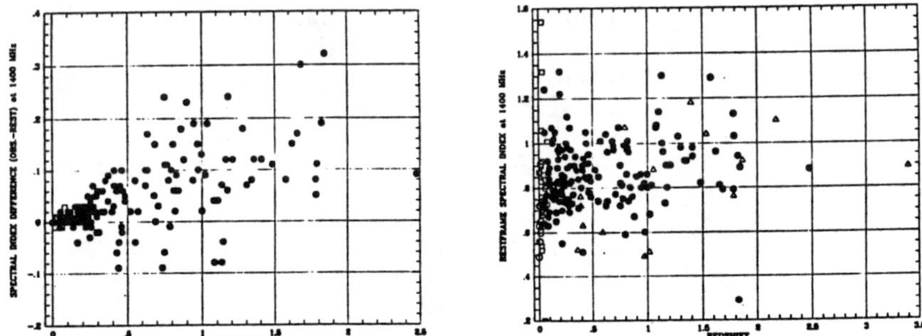

Figure 6a. The spectral index difference $\alpha_{obs} - \alpha_{rest}$ at 1400 MHz as function of z for all 3CR radio galaxies. *Figure 6b.* α_{rest} at 1400 MHz as function of z for 3CR radio galaxies (circles and squares), including sources from the B2/1Jy sample with known redshifts larger than ~ 0.4 (e.g. Allington-Smith *et al.* 1988; open triangles).

For example, if distant radio galaxies would resemble the nearby and powerful source Cygnus A (z = 0.0565), then one would expect their radio spectra to be steep. As is well known, Cygnus A exhibits a rapidly steepening spectrum for frequencies above $\sim$ 750 MHz changing from $\alpha \sim 0.7$ to $\alpha \sim 1.2$ and this seems to be a rather general property of powerful radio sources (e.g. Laing and Peacock 1980). This suggests that an $\alpha_{obs} - z$ correlation might be largely due to a "K-correction effect". Also, the $\alpha_{obs} - P$ correlation might then follow naturally because of selection effects (Kapahi, *priv. comm.*) since distant radio galaxies are usually faint with flux densities near radio-catalog confusion or sensitivity limits.

We have determined α_{obs} and α_{rest} at 1400 MHz using single dish data from Kellermann *et al.* (1969) for the entire 3CR sample. To avoid introducing any bias in our assumptions of the radio spectra we used a "free-form" (spline) interpolation scheme to derive α_{rest}. As shown in Figure 6*a* the differences between α_{obs} and α_{rest} become larger as function of redshift, indicating that spectral curvature or K-correction effects are increasingly important at high redshifts.

Figure 6*b* shows α_{rest} as function of redshift. We find that for z $\lesssim$ 1 there seems to be a correlation between α and z, as known before, but for z $\gtrsim$ 1 this correlation seems to disappear. A similar conclusion, using different and smaller radio source samples, was also reached by Gopal-Krishna (1988). These results emphasize the importance of using radio source samples for which complete redshift information exists, and which are not biased in their radio spectral properties, if one wants to study spectral index related properties.

In this respect it is very interesting that the most distant source in the combined 3CR and B2/1Jy samples, 0902+343 (z = 3.395, Lilly 1988), has only a moderately steep spectrum ($\alpha_{rest} \sim 0.9$) and that its radio structure is different from less distant radio galaxies (relatively bright core and one-sided jet, plus an overall S-shaped morphology; Figure 7). Indeed, in many ways its radio morphology resembles that of quasars and one could speculate, assuming the radio galaxy unification model of Barthel (1989), that 0902+343 is perhaps a source which is precessing towards the observer. These results also suggest that the steep spectrum selected sample studied by Chambers *et al.* may consist of a rather extreme class of radio sources.

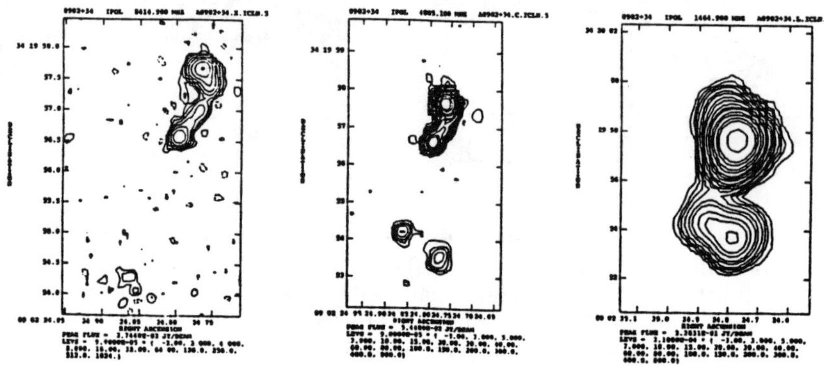

Figure 7. VLA maps of 0902+343 at 8.4 GHz, 4.9 GHz and 1.5 GHz with a resolution of 0.25″, 0.4″ and 1.2″ respectively (panels left-to-right).

9. RADIO POLARIZATION PROPERTIES

In at least two cases, detailed multifrequency radio observations of high redshift 3CR radio galaxies (3C 326.1, z = 1.825; and 3C 256, z = 1.819) have shown that their hotspots have a low degree of polarization (a few %) and large *restframe* rotation measures. *Minimum* restframe rotation measures for the W and NW hotspots in 3C 326.1 and 3C 256, are $\sim$ 2500 rad m^{-2}. In both cases these hotspots are embedded in regions with extended optical line (Lyα) emission. In 3C 326.1 the E hotspot does not exhibit significant Faraday rotation. Interestingly, this is also the side of the source where the optical line emission is very faint. Preliminary analysis of multifrequency data for 3C 294 shows that also in this object no significant (restframe) Faraday rotation occurs in radio components which do not overlap with optical emission-line regions. Regions which do overlap are unpolarized.

Together these data strongly suggest that the Faraday rotation (and depolarization) occurs in magnetoionic gas associated with these emission line regions. By analogy to the nearby radio galaxies this gas may form an irregular Faraday screen surrounding the radio hotspots.

Gas densities in the interstellar media of young galaxies such 3C 326.1 and 3C 256 may be relatively large. Also large magnetic fields might exist, generated through dynamo action (Pudritz and Silk 1989). The entrainment and compression of such a medium by the bowshock of a powerful source like

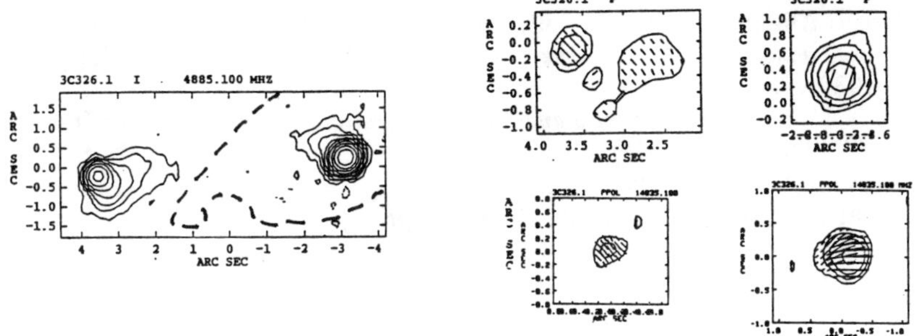

Figure 8. Contour maps of 3C 326.1 at 6 cm and 2 cm, all with 0.4″ resolution. The left panel shows a 6 cm total intensity map with the approximate extent of the extended Lyα emission superimposed (dashed line). The right panels show blow-ups of the E and W hotspots. The top panel shows the 6 cm polarized intensity + position angles, while the bottom panel shows the same fields at 2 cm. Note that the Lyα emission is associated with the brightest, western of the two hotspots (McCarthy *et al.* 1987*a*), which also shows the largest Faraday rotation (see text).

3C 326.1 might result in a dense magnetoionic screen around the hotspot and in the back flowing radio lobes.

REFERENCES

Allington-Smith J. R. *et al.* 1988, *M. N. R. A. S.*, **234**, 1091.

Baldwin, J. A. *et al.* 1981, *Publ. A. S. P.*, **95**, 5.

Barthel, P.D. 1989, *Ap. J.*, **336**, 606.

Baum, S. A. *et al.* 1988, *Ap. J. Suppl.*, **68**, 833.

Baum, S. A., and Heckman, T. M. 1989*a*, *b*, *Ap. J.*, **336**, 681, 702.

Begelman, M. C., and Cioffi, D. F. 1989, (preprint).

Bennett, A. S. 1962, *Mem. R. A. S.*, **68**, 163.

Blumenthal and Miley 1979, *Astron. Astroph.*, **80**, 13.

Browne, I. W. A. 1989, *Proc. ESO workshop on "Extranuclear Activity in Galaxies"; R. A. E. Fosbury and E. J. A. Meurs, Eds.*, (Garching, May 1989).

Chambers, K. C., Miley, G. K., and van Breugel, W. J. M. 1987, *Nature* **329**, 604.

Chambers, K. C., Miley, G. K., and Joyce, R. R. 1988, *Ap. J. Lett.*,**329**, L79.

Chambers, K. C., Miley, G. K., and van Breugel, W. J. M. 1989, (in preparation).

Daly, R. A. 1989, (preprint).

De Young, D. S. 1989, (preprint).

di Serego Alighieri, S. *et al.* 1989, *Proc. ESO workshop on "Extranuclear Activity in Galaxies"; R. A. E. Fosbury and E. J. A. Meurs, Eds.*, (Garching, May 1989).

Djorgovski 1988, *in Starbursts and Galaxy Evolution*, (Editions Frontières Paris, Ed. Th. Montmerle).

Djorgovski, S., *et al.* 1988, *A. J.* **96**, 836.

Eisenhardt, P. and Chokshi, A. 1989, (preprint).

Ensman, L. M., and Ulvestad, J. S. 1984, *A. J.*, **89**, 1275.

Fabian, A. C. 1989, *M. N. R. A. S.*, **238**, 41P.

Ferland, G. J., and Netzer, H. 1983, *Ap. J.*, **264**, 105.

Fokker, A. D. 1986, *Astron. Astroph.*, **156**, 315.

Fosbury, R. A. E. 1989, *Proc. ESO workshop on "Extranuclear Activity in Galaxies"; R. A. E. Fosbury and E. J. A. Meurs, Eds.*, (Garching, May 1989).

Gopal-Krishna 1988, *Astron. Astroph.*,**192**, 37.

Heckman, T. M. *et al.* 1986, *Ap. J.*, **311**, 526.

Heckman, T. M. *et al.* 1989, *Ap. J.*, **338**, 48.

Kapahi, V.K., and Saikia, D.J. 1982, *J. Astron. Astroph.*, **3**, 161.

Kellermann, K. I. *et al.* 1969, *Ap. J.*, **157**, 1.

Laing, R. A., and Peacock, J. A. 1980, *M.N.R.A.S.*, **190**, 903.

Le Fèvre, O. *et al.* 1988, *Ap. J. Lett.*, **324**, L1.

Lilly, S. 1988, *Ap. J.*, **333**, 161.

Macklin, J. T. 1981, *M. N. R. A. S.*, **196**, 967.

McCarthy, P. J. *et al.* 1987*a*, *Ap. J. Lett.*, **319**,L39.

McCarthy, P. J. *et al.* 1987*b*, *Ap. J. Lett.*, **321**,L29.

McCarthy, P. J. 1988, *Ph.D. Thesis, UC Berkeley.*

McCarthy, P. J., *et al.* 1989, *Ap. J.*, (submitted).

McCarthy, P. J. and van Breugel, W. J. M. 1989, *Ap. J.*, (submitted).

McCarthy, P. J. *et al.* 1990, in preparation).

Palimaka, J. J. *et al.* 1979, *Ap. J.*, **231**, L7.

Pudritz, R. E., and Silk, J. 1989, *Ap. J.*, **342**, 650.

Quinn, P. 1989, *Proc. ESO workshop on "Extranuclear Activity in Galaxies"; R. A. E. Fosbury and E. J. A. Meurs, Eds.*, (Garching, May 1989).

Rees, M. J. 1989, (preprint).

Robinson, A. *et al.* 1987, *M. N. R. A. S.*, **227**, 97.

Roos, N. 1988, *Ap. J.*, **334**, 95.

Rudnick, L., and Edgar, B. K. 1984, *Ap. J.*, **279**, 74.

Ryle, M., and Longair, M. S. 1984, *M. N. R. A. S.*, **136**, 123.

Spinrad, H., Djorgovski, S., Marr, J., and Aguilar, L. A. 1984 1984, *Publ. A. S. P.*, **97**, 932.

Spinrad, H., and Djorgovski, S. 1987, in *Observational Cosmology*, Proceedings of IAU Symposium 124. eds. A. Hewitt, G. Burbidge and Li Zhi Fang (Dordrecht: Reidel).

Spinrad, H. 1987, in *High Redshift and Primeval Galaxies*, J. Bergeron, D. Kunth, B. Rocca-Volmerange and J. Tran Thanh Van (eds.), Paris: Editions Frontières.

Tadhunter, C. N. *et al.* 1988, *M. N. R. A. S.*, **235**, 403.

van Breugel, W. J. M., *et al.* 1985*a*, *Ap. J.*, **290**, 496.

van Breugel, W. J. M., *et al.* 1985*b*, *Ap. J.*, **293**, 83.

van Breugel *et al.* 1990, (in preparation).

INFRARED IMAGES OF DISTANT 3C RADIO GALAXIES

P. EISENHARDT
MS 245-6, NASA/Ames Research Center, Moffett Field, CA 94035

A. CHOKSHI
Department of Astrophysical Science, Princeton University, Princeton, NJ 08544

M. DICKINSON
Department of Astronomy, University of California, Berkeley, CA 94720

S. DJORGOVSKI
California Institute of Technology, 105-24C Pasadena, CA 91125

P. MCCARTHY
The Observatories of the Carnegie Institution of Washington, 813 Santa Barbara St., Pasadena, CA 91023

H. SPINRAD
Department of Astronomy, Univeristy of California, Berkeley, CA 94720

ABSTRACT We have obtained J and K images of radio galaxies with redshifts of up to 3.4,including 3C 41, 54, 124, 194, 256, 257, 265, 294, 326.1, 356, 437, 441, & 454.1, and B2 0902+34. The observations were made using the Infrared Imager (IRIM) on the KPNO 4m telescope. Stellar image sizes (FWHM) are from 0.85 to 1.2 arcseconds, roughly 10 kpc for H_0=50. The limiting sensitivity (3σ in 1 square arcsecond) of the deepest images is approximately K=21 (2.7μJy).

The galaxies were selected from those with the largest known redshift, primarily 3C sources whose R band and redshifted [OII] 3727 Å narrow band images showed extended asymmetric emission with multiple components, frequently aligned with the radio lobe axis (McCarthy *et al*. 1987 Ap.J. Lett. **321**, L29). Such an alignment could be explained by star formation induced by the interaction of a radio jet with the ambient medium. The R band measures emission at 0.2 to 0.4μm in the rest frame for our sample. Infrared imaging allows us to examine emission at longer wavelengths which presumably arises from red giant stars and is more representative of the bulk of the stellar population.

In general, we find that the infrared morphologies of these galaxies are just as peculiar as their optical morphologies. For most of the galaxies, when asymmetric structure is present in the optical, structure with the same orientation is seen in the infrared, although each object has its own quirks. For example, 3C356 (z=1.08) has a K light distribution much like that seen in redshifted [OII] 3727 Å, while emission at J is more similar to that in broad band R. It may be difficult to account for the infrared emission in terms of young stars whose formation is triggered by radio jets. We find B2 0902+34 to be substantially fainter at K than reported by Lilly (Ap.J. Lett. **333**, L161, 1988), which reduces the minimum required age for this object.

HIGH REDSHIFT RADIO GALAXIES AND THE ALIGNMENT EFFECT

K.C. Chambers
The Space Telescope Science Institute and The Johns Hopkins University, 3700 San Martin Dr., Baltimore, MD, 21218

G. K. Miley
Sterrewacht, Postbus 9513, 2300RA, Leiden, The Netherlands

ABSTRACT We have identified a portion of a sample of 4C radio sources selected for their ultra-steep radio spectrum. Most are at high redshift, including the most distant galaxy known, 4C41.17 at $z = 3.8$. The extended optical/infrared continuum of high redshift radio galaxies is generally aligned along the radio axis. The spectral energy distributions can be nearly flat (f_ν) in the rest frame ultraviolet, but rise significantly into the red. While the nature of the optical/infrared continuum is controversial, it is likely that aligned radio galaxies are dominated by starlight, and that at early epochs powerful radio sources were capable of stimulating vast amounts of star formation.

INTRODUCTION

To our eyes, the few hundred brightest objects in the night sky are all nearby stars. At radio frequencies, the few hundred brightest objects in the sky include objects so distant that we are observing them when the universe was a fraction of its present age. In a classic conspiracy of nature, the luminosity function of radio sources falls off with a slope nearly equal to the rise in volume sampled in a Euclidian universe. So, contrary to intuition, the observed radio flux of an object tells us almost nothing about its distance. A source of 1 or 2 Jy's at 178 MHz can be anywhere in the universe. This is a measure of both the power and the handicap of radio surveys for finding distant galaxies. Therefore it is desirable to develop techniques for preselecting the best candidates from the large number of radio sources which become available as we push to fainter and fainter flux limits.

We have developed such a method based on the observed correlation between the radio spectra and luminosities of radio sources. Many authors have remarked on such a possible correlation (e.g. Heeschen, 1960, Veron *et al.* 1972, Macleod and Doherty, 1972, Bridle *et al.* 1972), but the most striking manifestation of this correlation was found as the byproduct of an investigation of a sample of 4C radio sources having ultra-steep spectra (Tielens *et al.* 1978). It was found that the fraction of radio sources which had counterparts on the Palomar Sky Survey was a strong function of the radio spectral index, with almost no identifications for the steepest spectrum sources (Blumenthal and Miley 1979). See Figure 1. On the reasonable assumption that the unidentified ultra-steep spectrum radio sources were distant galaxies, fainter than the Survey limits, we began a multi-spectral investigation (Chambers, Miley, and van Breugel, 1987, Chambers, 1989) with the most sensitive instrumentation now available, of the ultra-steep spectrum radio source sample of Tielens et al. This comprised 4C sources known to have spectral indices of $\alpha < -1$, between 178 and 5000 MHz, where $S_\nu = k\nu^\alpha$.

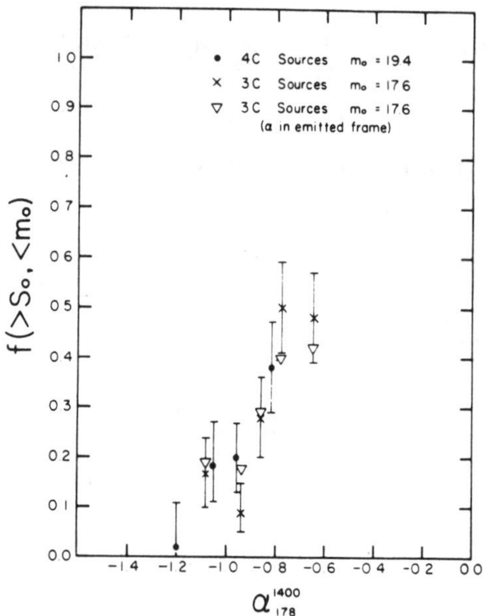

Figure 1: *The fraction of radio sources optically identified on the Palomar Sky Survey plotted against the observed metre-wave radio spectral index (from Blumenthal and Miley, 1979).*

BACKGROUND

The study of high redshift radio galaxies is currently undergoing a dramatic growth, due in part to new technologies and new techniques which have led to surprising discoveries. With the advent of CCDs in the optical, InSb arrays in the infrared, the high sensitivity and dynamic range of the VLA in the radio region of the spectrum, as well as the software to obtain the optimum performance of these instruments, our ability to identify and study these distant stellar systems has increased dramatically. Previous work on distant radio galaxies has centered on various complete samples from flux-limited surveys, e.g. the 3CR sample of sources with $S_{178} > 10 Jy$ (Laing, Riley and Longair, 1983) the "1 Jy" sample (Allington-Smith 1982, Lilly 1988a, 1989, this conference), and the LBDS sub-millijansky sources (Windhorst *et al.* 1985, this conference). Reviews on identifications and optical/infrared properties of high redshift radio galaxies prior to use of the ultra-steep spectrum technique and before the discovery of the alignment effect are given by Spinrad (1986), and Longair and Lilly (1985). Begelman, Blandford, and Rees (1984) reviewed the theory of extragalactic radio sources, Miley (1980) has reviewed the radio observations, and Bridle and Perley(1984) reviewed the situation for radio jets.

The 3CR sample is unique in that it is now nearly completely identified, and the years of investigation of these objects by Spinrad and collaborators has resulted in the discovery of ~ 30 radio galaxies with redshifts larger than one (Spinrad *et al.* 1985; Spinrad 1987; Djorgovski *et al.* 1988; McCarthy 1989; Spinrad 1989, van Breugel, this conference). The most distant 3CR galaxy is the recently identified source 3C257 at $z = 2.474$ (Spinrad, private communication), the most distant "1 Jy" galaxy is 0902+34 at $z = 3.395$ (Lilly, 1988b), and the LBDS has discovered at one galaxy Herc 202 with $z = 2.39$ (Windhorst, this conference).

With the discovery of 4C41.17 and 0902+34 the infrared Hubble diagram has now been extended out to very high redshifts (Lilly, this conference). However, it must be emphasized that one should not discuss high redshift radio galaxies (or plot their infrared K magnitudes) without mentioning their most remarkable property: the alignment between their optical/infrared continua and their radio axes (Chambers *et al.* 1987, 1988b; McCarthy *et al.* 1987b). (See Figures 3, 4, and 5.) This phenomena has recently stimulated a number of observers and theorists to explain it, and we will review some of that work here. The alignment effect is important not only because of its intrinsic interest as new astrophysical phenomena, but also because understanding the alignment effect is crucial to any attempt to use radio galaxies for measuring cosmological constants. Previously, it could be assumed that the host galaxies were otherwise normal giant ellipticals whose properties were divorced from the powerful radio sources by which they were selected. And while there is surprisingly little evidence for a correlation between radio power and infrared luminosity (Lilly 1988a), an intimate relationship between the optical/infrared properties and the radio properties such as the alignment effect is an important clue to some sort of physical interaction.

ULTRA-STEEP SPECTRUM SURVEY

The current status of our 4C ultra-steep spectrum survey is encouraging. Out of 33 4C ultra-steep spectrum sources, we have obtained 31 optical identifications,

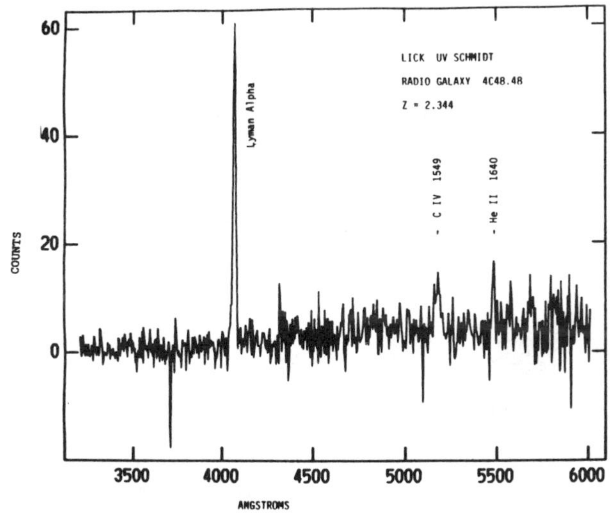

Figure 2. Spectrum of 4C48.48 at z=2.344

attempted optical spectroscopy of 32, have found 16 with strong emission lines and with $z > 0.5$, 8 of these with $z > 2$ including 4C40.36 at $z = 2.267$ (Chambers *et al.* 1988a) and 4C41.17, currently the most distant galaxy known, at $z = 3.800$ (Chambers *et al.* , 1989). In Figure 2 we present a spectrum of 4C48.48 at a redshift of 2.371. It has a strong Lyman α emission extended along the radio axis, CIV 1549, and HeII 1640, typical of the high redshift radio galaxies.

We are also currently extending our ultra-steep spectrum survey to fainter flux levels using sources from the Parkes, Texas, Molonglo, 6C, and 8C surveys (Miley *et al.* 1989).

THE ALIGNMENT EFFECT

There are several observational aspects of the alignment effect (Chambers *et al.* 1987, 1988b; McCarthy *et al.* 1987b) which need to be emphasized. First, we prefer to call the phenomena the "alignment effect" rather than a morphological correlation because the optical/infrared emission in general is *not* coincident with the radio lobes. Often the radio source is larger, sometimes much larger, than the optical/infrared continuum, but there are examples, e.g. 4C41.17, where the aligned continuum and/or emission line gas extends beyond the radio lobes. The failure to make this distinction has resulted in untenable explanations which require spatial coincidence between the radio emission and the optical/infrared emission. While the radio emitting regions and the optical/infrared emitting regions occassionally overlap, the crucial characteristic is that they have a common axis. (See Figure 3.) Second, although a

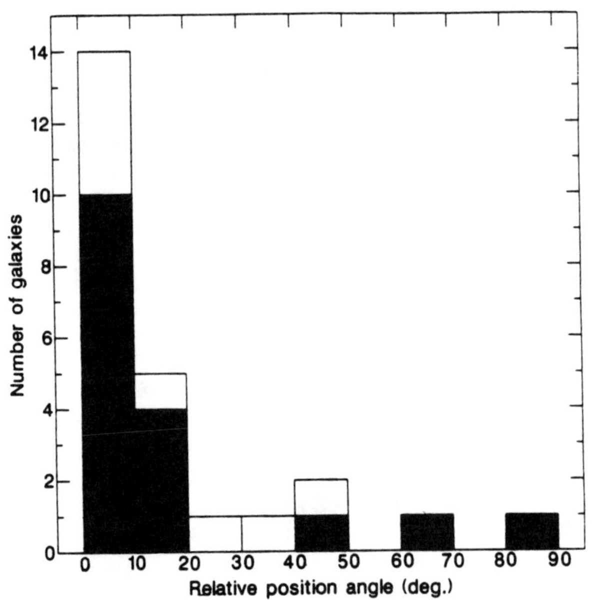

Figure 3. Histogram of the angle between the optical and radio axes in high redshift radio galaxies. From Chambers etal. 1987.

statistically significant phenomena, there are exceptions to the alignment effect. Some systems, even though they are reasonably elongated, are not well aligned with the radio axis. This is worth keeping in mind, since some explanations would seemingly require no exceptions. Third, the alignment is seen both in the continuum and in strong extended emission lines, but the detailed morphology of the emission lines and the continuum often differ. Fourth, there appear to be small color gradients in the continuum of some objects, although contamination of the continuum by emission lines and different ways of displaying low signal-to-noise images can exaggerate these. But it is important to note that the infrared alignments (Chambers *et al.* 1988b, Eisenhardt and Chokshi 1989) have now been detected in enough objects that it is reasonable to conclude that in general the infrared morphologies resemble the optical morphologies. There are *no* observed counter-examples which support a popular conjecture that these objects are really nice round ellipticals in the red, and only aligned in the ultraviolet. Fifth, while there are exceptions, (e.g. 0902+34 van Breugel, this conference) the radio sources generally do not have peculiar morphology or polarization structure. In general they have the Fanaroff-Riley type II edge-brightened double morphology of normal powerful radio sources.

ALIGNED RADIO GALAXIES 3C368 AND 4C41.17

There is danger in assigning a prototype to any particular class of object, particularly if it is really an archetype, having certain characteristics to an extreme degree. Nonetheless, 3C368 has become the classic aligned radio galaxy (Djorgovski *et al.* 1987, Chambers *et al.* 1988b, LeFevre *et al.* 1988). It is worth noting however that because of its extreme elongation and relatively low redshift ($z = 1.132$) that it is not typical of aligned radio galaxies. In Figure 4 we show a VLA radio image overlayed with a 2.2 micron image to demonstrate the infrared alignment. This infrared image was one of the first deep images obtained with an InSb array and was not deep enough to detect the southern component that is visible in optical images. Techniques have improved dramatically in the last year, and deeper images are now possible.

The most distant stellar system known today is 4C41.17 at a redshift of $z = 3.800$ (Chambers, Miley, van Breugel, 1989). The radio source was mapped with the VLA as part of our ultra-steep spectrum survey, and identified with a faint extended optical object, whose spectrum showed two emission lines corresponding to Lyman α and CIV 1549 redshifted by $z = 3.800$. The Lyman alpha emission is elongated along the radio axis and includes an elliptical halo that extends more than 100 kpc, considerably beyond the radio source. It is clumpy and has a velocity dispersion of ~ 500 km/s, but with some gas in regions well correlated with the radio source having velocities at least as great as ~ 2000 km/s. The Lyman α has a rest frame equivalent width of (~ 270 Å). The optical and infrared continua have similar properties to those of other high-redshift radio galaxies, i.e. extended along the same axis as the radio source. In 4C41.17 however the continuum extends beyond the eastern radio lobe, although subsequent deep radio imaging has detected a diffuse very steep spectrum component to the northeast. In Figure 5 we show a radio image overlayed with the 2.2 micron image. In the rest frame this is rougly equivalent to a V band image, i.e., sampling the continuum longward of the 4000 Å break. The rest frame spectral energy distribution is rougly flat in the ultraviolet and exhibits a "red bump" longwards of 4000 Å (see Figure 6.) For more images of other aligned radio galaxies see McCarthy (1989), Lefevre *et al.* (1988a, 1988b, 1988c), and for infrared images, Eisenhardt and Chokshi (1989).

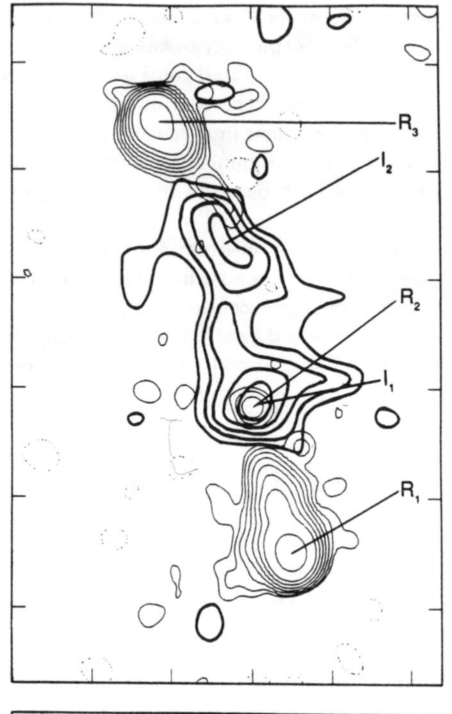

Figure 4. 3C368 K band image (dark contour) and VLA radio image (light contour). Adapted from Chambers et al. 1988b.

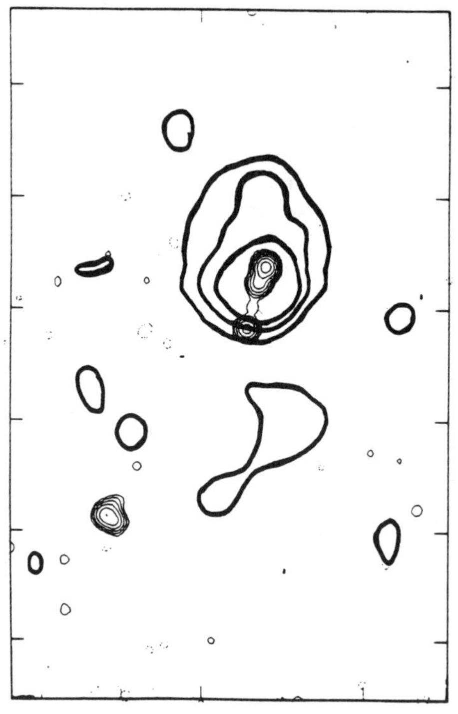

Figure 5. 4C41.17 K band image (dark contour) and VLA radio image (light contour). Adapted from Chambers et al. 1989.

THE NATURE OF OPTICAL/IR CONTINUUM

Is the optical and infrared emission from aligned radio galaxies dominated by starlight? The best evidence that the extended optical/IR continuum emission in such objects is produced by stars is the existence of stellar absorption features recently found in the summed ultraviolet spectra of high redshift radio galaxies by Chambers and McCarthy (1989). However over the past two years other mechanisms for producing the optical/ir continuum have been discussed in the literature (e.g. Chambers *et al.* 1987, 1988b, McCarthy 1987, LeFevre 1988, Fabian 1989, Rawlings and Eales 1989, di Serego Alighieri *et al.* 1989, Tadhunter 1989), and currently there is no consensus. In light of the recent work advocating various alternative explanations for the optical/ir continuum from distant radio galaxies we will discuss the various ideas and arguments. The ability of each senario to explain various observations is summarized in Table 1. This table is meant as guide to where both the observations and the simple models need additional work.

(a) Gravitational Lensing.

In a series of papers Le Fevre, Hammer, and collaborators (1988a, 1988b, 1988c) have discussed the role of gravitational lensing (both multiple imaging and amplification) in contributing to the luminosity and morphology of high redshift radio galaxies. It is unlikely that the observed morphology is due to multiple gravitational images. Some of these sources have radio cores, yet no multiple images of the radio cores have been observed. Furthermore the lens hypothesis cannot account for the alignment effect since the often very large and extended radio lobes cannot be multiply imaged, and there is no reason to prefer multiple images along the background radio axis. Since these sources are dominated in the radio by such large extended radio structure, it is implausible that gravitational amplification has significantly biased the radio selection of these objects. In addition some optical spectral lines have been shown to have spatially distributed velocity gradients, and different gradients in different lines (e.g. Djorgovski *et al.* 1987). This is difficult to account for since a gravitational lens is achromatic. There are however aligned radio galaxies with obvious foreground galaxies near the line of sight. These might provide some optical luminosity amplification, as opposed to multiple imaging. And while such an explanation is neither necessary nor sufficient to account for the morphology and high luminosity of the aligned radio galaxies, some contribution to their luminosity from gravitational lenses can not be ruled out.

(b) Nonthermal Mechanisms.

The high energy tail of the radio synchrotron spectrum in some nearby radio sources has been observed in the infrared, optical, and X-rays (e.g. M87, 3C273, 3C66, see Miley 1980) However such extrapolations are too faint by several orders of magnitude to account for the optical/ir emission from the high redshift aligned radio galaxies which have the luminosities surface brightnesses greater than or equal to that of the largest giant ellipticals. Moreover, the rest frame spectral energy distributions (SEDs) show large departures from a nonthermal power law.

Inverse Compton radiation from scattering of the microwave background by the radio-emitting relativistic electrons would require extreme departures from equipartition.

Finally explanations involving mechanisms related directly to the radio synchrotron photons suffer from the inability to account for the extreme differences in the radio and optical morphologies.

(c) Electron Scattering.

Fabian (1989) has proposed electron scattering of an obscured QSO nucleus as the mechanism for the alignment effect and Rawlings and Eales (1989) have favored this explanation for their data. Polarized emission has been detected from 3C368 (di Serego Alighieri *et al.* 1989), but their data favor dust scattering and will be discussed below. Electron scattering would not modify the overall SED of the nucleus, but the observed SEDs of aligned radio galaxies (e.g. Lilly 1988a, 1988b, Chambers and Charlot 1989) are very different from the power-law continuum of QSO's. Furthermore, while such a senario has been adjusted to fit a particular geometry (i.e. 3C356) as a general explanation of the alignment effect it would require either an extreme mass in ionized hydrogen, or QSO (blazar) luminosities far above what is observed directly. For example, for an extended region 10 kpc across, Thompson scattering requires a mass of ionized hydrogen $M \sim \tau A/\sigma \sim 10^{16} M_{\odot} \tau (R/10\mathrm{kpc})^2$, where $A = R^2$ is the apparent area of the emiting region, τ is the optical depth, and $\sigma = 0.4\mathrm{cm}^2/\mathrm{gram}$ is the Thompson cross section for fully ionized hydrogen. The optical depth τ must be greater than than 0.01 to produce absolute magnitudes of $M_v \sim -25$ for an extended region or the central (obscured) quasar would be much brighter than the brightest known qso's. This requires a minimum mass of $10^{14} M_{\odot}$ in ionized hydrogen, which is much too extreme for the extended gas kinematics. The alternative is that there is a central blazar in all these objects which is several magnitudes brighter than any object observed. Two additional arguments against electron scattering come from the high densities required. Rawlings and Eales put a lower limit of $n_e = 5cm^{-3}$ for the thermal plasma in their electron scattering model, but for a typical equipartition magnetic field ($\sim 10^{-4}$ gauss) this would preclude the detection of any polarized radio emission from a lobe spatially coincident with aligned optical flux (e.g. the southern lobe in 3C368 or the eastern lobe in 4C41.17). Furthermore, cool emission-line clouds in pressure balance with such a hot dense phase would require densities far greater than what is derived from emission line ratios of the extended narrow line emission (McCarthy 1989).

(d) Dust Scattering.

Because of their very high Lyman α luminosities, aligned radio galaxies are not expected to be dusty objects. However, if the the emission line gas is clumped in a substantially different way from the dust, then dust scattering of light from a hidden nucleus is another mechanism for producing aligned blue flux. Some evidence that dust scattering may play a role is provided by the recent observation of polarized blue light from 3C368 (di Serego Alighieri *et al.* 1989). However, because of its inefficiency in scattering red light, dust scattering cannot explain the infrared alignments (Chambers *et al.* 1988b and Eisenhardt and Chokshi 1989) and thus one must propose two carefully tuned alignment processes acting in concert: one for the blue component and one for red. An alternative single explanation is preferable.

There are additional arguments against dust scattering dominating the extended emission. For example, typical qso broad line lyman alpha equivalent widths are $\sim 100\mathring{A}$ with ~ 0.8 in the broad component (Kinney, 1987,1989). Therefore if the

TABLE 1

Are proposed senarios consistent with the observed properties of Aligned Radio Galaxies ?

	Young Stellar Population			Old Stars	Obscured AGN (consquence of orient.)		Nonthermal		Grav. Lens
Observed Properties	Young	Reddened Burst	Massive Stars	Old + Burst	Electron Scattering	Dust Scattering	Synchro-tron	Inverse Compton	Grav.
Optical Alignment Effect	Yes	Yes	Yes	Yes	Yes	Yes	No	No	No
Infrared Alignment Effect	Yes	Yes	Yes	No	Yes	No	No	No	No
SEDs	Yes	Maybe	Maybe	Yes	No	No	No	No	Yes
Stellar Absorption Features	Yes	Yes	Yes	Yes	No	No	No	No	Yes
High Surface Brightness	Yes	Yes	Yes	Yes	No	Maybe	No	No	Yes
Lyα Eq. Width	Yes	Maybe	Yes	Yes	Yes	Maybe	Yes	Yes	Maybe
Continuity of Hubble diagram	Yes	Yes	No	Yes	No	No	No	No	No
Low Dispersion of Hubble dia.	Yes	Yes	No	Yes	No	No	No	No	No
Kinematics	Yes	Yes	Yes	Yes	Yes	Yes	Yes	Yes	No
Extended Polarization	Maybe	Maybe	Maybe	Maybe	λ ind.	λ dep.	Yes	Yes	No
Nuclear Polarization	Yes	Yes	Yes	Yes	Yes	Yes	Yes	Yes	No
Low Limit on Broad Lyα	Yes	Yes	Yes	Yes	Yes	No	Yes	Yes	Yes
Faint Radio Cores	Yes	Yes	Yes	Yes	Maybe	Maybe	Yes	Yes	Yes
Inferred Properties									
z_F	1 - 5	1 - 5		7 - 30					
IMF	Normal	Normal	$M_s > 5M_\odot$	Normal					
Mass									
Radio Source Induced Star Form.	Yes	Yes	Yes	Yes	No	No	No	No	No
AGN anisotropic abs. Mag.					$M_v < -30$	$M_v < -30$			

continuum is dominated by dust scattering of a typical QSO nucleus, broad Lyman alpha components with an equivalent width of $\sim 80\AA$ ($\sim (1+z)80\AA$ in observed frame) should have been detected. Such a feature should be even more apparent when one considers that the broad component should be enhanced in a dusty environment due to the resonant destruction of Lyman α in the core of the line.

This last argument can only be avoided if the central object has no broad lines (a high redshift BL Lac or featureless blazar) and effectively stays that way for $\sim 10^5$ years. Under optimistic conditions (e.g. Tadhunter, 1989) the dust scattering of a blazar can produce enough blue (but not red) surface brightness if the blazar has an absolute magnitude of $M_v \sim -30$, i.e. marginally above the brightest blazar known.

Finally, all senarios which endeavor to explain the alignment effect as a manifestation of the orientation of radio loud quasars (e.g. Barthel 1989), including all scattering senarios (e.g. Fabian, 1989; Rawlings and Eales, 1989; di Serego Alighieri *et al.* , 1989; Tadhunter, 1989) suffer from the inability to explain the observed continuity and low dispersion of the infrared Hubble diagram for radio galaxies (Lilly, 1988a and this conference). As emphasized by Lilly (1988b), the continuity argument is quite powerful; it is difficult to imagine why alternative senarios would happen to give just the right luminosities at just the right redshifts to join smoothly on to the Hubble diagram (both in amplitude and in slope) with intermediate redshift giant ellipticals. A corallary to this argument is that the high redshift counterparts to intermediate redshift radio galaxies must exist somewhere, if they are not the aligned radio galaxies, where are they? Although the low dispersion of the Hubble diagram for radio galaxies is curious, it is certainly clear that intrinsic quasar luminosities show far greater dispersion (e.g. Hewitt and Burbidge, 1987). Presumably any scattering mechanism could only add to the quasar dispersion; hence in the context of the scattering senarios it is hard to understand how the radio galaxies could be drawn from the same population as the quasars. On the other hand, if the alignment effect is not due to scattering effects, then there is at present no inconsistency with the Barthel (1988) hypothesis.

(e) Massive Stars.

One early explanation for the alignment effect was that the radio source was some how triggering star formation along its axis (Chambers *et al.* 1987, McCarthy *et al.* 1987). Mechanisms of this kind will be discussed below; here we consider whether consistent stellar population models can account for the observed SED's of the aligned continuum. The models have been constructed using an updated version of Bruzual's code (1983). (See Chambers, Miley, van Breugel 1989, and Chambers and Charlot 1989 for details of the models.) Figure 5 shows how four of these models are fit to the observations of 4C41.17. These models are not unique, and are meant only to investigate various possibilities.

First we examine an IMF restricted to massive stars as an efficient way to produce a large luminosity with relatively few stars. This works well in the ultraviolet, but the challenge is to explain the alignment of the observed red and near infrared continuum (the "red bump" $\sim$ 4000 to 10000 Å in the rest frame). Chambers *et al.* (1988b, 1989) and Bithell and Rees (1989) investigated the possibility that the infrared continuum in aligned radio galaxies is due to a large population of red supergiants. The idea acquired new motivation with the discovery of objects like 0902+34 and 4C41.17 at very high redshift as a mechanism to avoid the large cosmic age suggested by Lilly (1988b).

However, our present modelling (Chambers, Miley, van Breugel 1989) indicates that it is difficult to produce a plausible massive star senario that fits the spectral energy distributions for the simple reason that massive stars spend more time being blue than red. The situation is a bit complicated, since the proportion of time that a massive star spends being blue to the time spent being red depends crucially on the details of evolution models for massive stars (e.g. mass loss and convective overshooting) and even the current models are not well matched by the galactic star counts. There is a large and rich literature on the topic which is presently advancing rapidly, not the least because of SN 1987a (e.g. Woosley 1988.) However, with any currently envisioned blue to red supergiant ratio, the simplest model, constant (steady state) massive star formation, is too blue to fit the red bump in the SED of aligned radio galaxies.

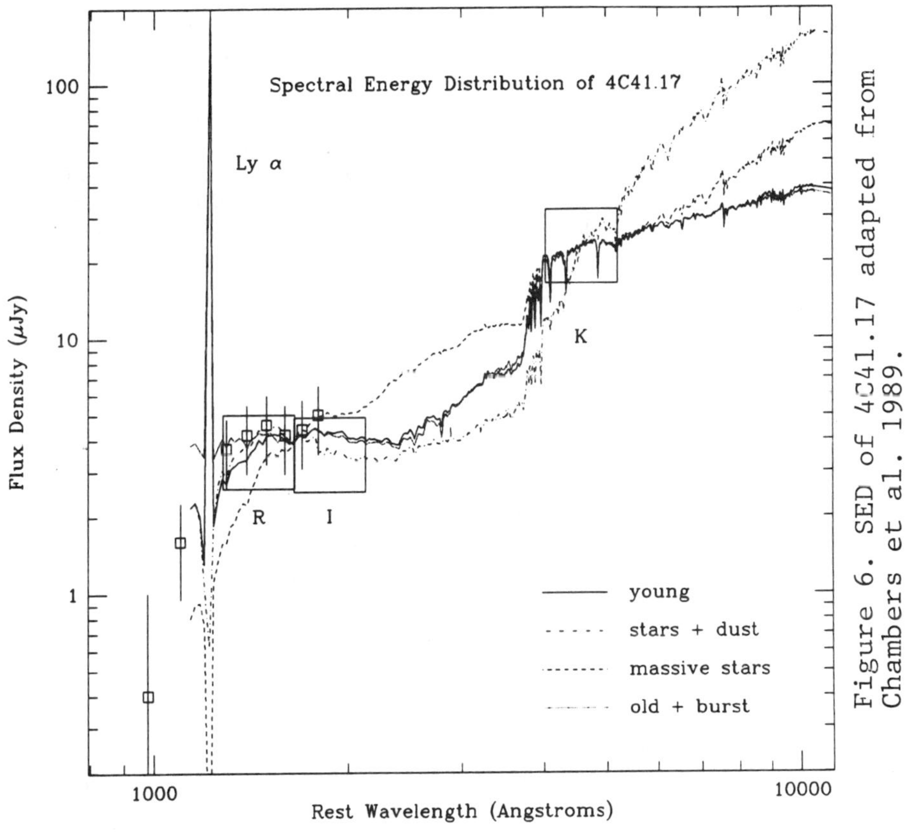

Figure 6. SED of 4C41.17 adapted from Chambers et al. 1989.

A plausible attempt to get around this problem is to use an evolving model, since a burst of massive star formation is eventually dominated by red supergiants with a very red SED. Various evolving massive star models with different star formation rates have been studied (Chambers *et al.* 1989, Bithell and Rees 1989), with similar results. Since the smallest of the high mass stars lives the longest the evolution of the SED at late times is controlled totally by the assumed low mass cutoff. This causes two fundamental difficulties for massive star models. First, the model SEDs fit the data only for a very small percentage of its luminous phase, spending a much greater fraction of its lifetime ($\sim$ 90%) with a very blue SED, which has never been observed, or being extremely red ($\sim$ 5% of its luminous lifetime). Second, while it has an extremly low M/L, and is thus very efficient, the luminosity has dropped by a factor of $>$ 100 from its burst luminosity by the time it fits the observed SED's. Since aligned radio galaxies are already very luminous, such a model would predict extreme luminosities for their burst phase.

Finally, as with the scattering models, a massive star senario is incapable of explaining the continuity and low dispersion of the infrared Hubble diagram.

(f) Old Population plus Small Burst.

Lilly (1988b, this conference) has argued that the red bump in the SED is due to an old stellar component. Figure 5 shows a similar model where it is assumed that the bulk of the stars are formed in a continuous burst lasting 1 Gyr. Then the star formation halts and the population evolves. At some later epoch (the epoch of observation) a second star burst is assumed to occur, perhaps associated with the radio activity. This second burst is responsible for the flat ultraviolet component, and its high star formation rate need only last for a short period of time. The young starburst is taken to last 0.1 Gyr and a mass fraction of 0.04 is great enough to produce the observed ultraviolet if the burst happens this fast. Using the stellar population models we find that the older generation had to be at least 1.3 Gyrs old (or 0.3 Gyrs after the end of the 1 Gyr constant burst) in order to match the red bump. Clearly an even older population would fit as well, but 1.3 Gyr was taken as the youngest possible age in this senario which will give an acceptable fit to the SED of 4C41.17.

There are two major difficulties with the old stellar population model. First, the large ages asserted for the highest redshift objects become a powerful constraint on the epoch of formation of these galaxies, and on cosmological models as well. If Lilly is correct, and 4C41.17 is older than $\sim 1.3 \times 10^9$ years, then this excludes an $H_0 = 50, q_0 = 0.5$ cosmology where the universe is only 1.25×10^9 years old. While the problem is not as extreme for low q_0 cosmologies, large ages for the highest redshift radio galaxies imply very high redshifts of formation $z_F \sim 10 - 30$. Second, the most crucial objection, is that old stellar population model can not explain the infrared alignments. The ages of the galaxies in this model are far larger than than is reasonable for the radio source. Thus one must assert that the radio jet (which must be moving much faster than escape velocity) is sensitive to the mass distribution, a difficult proposition to believe. One might argue that the infrared alignments are due to selection effects, i.e. we only find the radio objects whose luminosity has significantly brightened because the radio jet happened to hit a nearby companion galaxy. However, if the radio luminosity were so dramatically effected by such an interaction, then it is difficult to understand why an enormous asymmetry in the radio luminosity (or spectral index) is not observed along with the size asymmetry which is observed and is due to a density asymmetry (McCarthy 1989).

(g) Young Stellar Population.

Chambers and Charlot (1989) have shown that the observed SED's of high redshift radio galaxies (0902+34 and 4C41.17 in particular) can be produced with a normal (Scalo, 1986) IMF in less than $\sim 3 \times 10^8$ yr if the galaxies formed the bulk of their stars on slightly shorter timescales ($\lesssim 1 \times 10^8$ yr). In this connection, it is interesting to note that the star formation timescales are comparable to the probable free-fall times of the galaxies. This suggests that these objects may be relatively young and that we are observing them (at least the ones with the flat UV) near the time in which they formed most of their stars. The "epoch" of this type of galaxy formation would have extended over a fairly wide range of redshifts, from $z \sim 1$ to $z \sim 5$. Therefore they do not require the large ages suggested by Lilly (1988b). This is primarily due to the short timescale of formation, although the addition of an observationally derived AGB contribution contributes $\gtrsim 20\%$ of the red light on the timescales considered.

The ability to obtain "flat UV plus red bump" SED's on short timescales and maintain a roughly constant red luminosity during the first Gyr makes the Chambers and Charlot (1989) senario the only model so far capable of explaining both the infrared alignments while at the same time maintaining the observed small dispersion and continuity of the Hubble diagram. Such a senario is also consistent with both the observed ultraviolet stellar absorption features recently discovered in the summed spectra of aligned radio galaxies by Chambers and McCarthy (1989), and their suggestion that the overall shape of the ultraviolet continuum indicates that the star formation was substantially greater in the immediate past.

The only difficulty with the young stellar population senario seems to be relativly minor. The coherent extended polarization in 3C368 reported by di Serego Alighieri *et al.* is large; however it does presuppose substantial dust, and therefore star formation extended $\sim$ 50 kpc along radio source anyway. Hence a small contribution to extended emission by dust scattering would be consistent with the young stellar population model, whereas scattering alone cannot explain the infrared alignments.

(h) Stars plus Dust.

Aligned radio galaxies are not expected to contain much dust because of their large Lyman α surface brightnesses. Even a small amount of dust can extinguish most resonance line photons such as Lyman α. However, the inferred star formation rates are large ($\gtrsim 100 M_{\odot}$/yr) and the dust could be clumped differently from the emission line gas. Figure 5 shows a burst SED (rougly flat) reddened by an extinction curve derived from the Small Magellanic Clouds. A similar result could be obtained using the Galactic extinction curve, except that it would exhibit the pronounced feature at 2000 Å . Dust would mainly affect the rest-frame "flat UV" part of the spectrum, but hardly affect the red or near infrared emission. Thus it would not change the conclusions of the young galaxy picture discussed above. The greatest affect of ignoring reddening would be to *over-estimate* the ages derived from the observed SED's.

RADIO SOURCE INDUCED STAR FORMATION

Can a radio source trigger vast amounts of star formation at early epochs? At low redshift slight amounts of star formation associated with the jet of Cen A were suggested by Osmer (1978) and DeYoung (1981). The ability of radio jets to interact with, entrain, and accelerate ambient media was demonstrated by Miley and collaborators in a series of papers (e.g. Miley *et al.* 1981, Heckman *et al.* 1982, 1984 van Breugel *et al.* 1984, 1985a, 1986). Then Minkowski's Object (van Breugel 1985b, Brodie 1985) was investigated, and it appears to be the best nearby example of a starburst triggered by a radio source. With the discovery of the alignment effect at high redshift, Chambers *et al.* 1987 and McCarthy *et al.* 1987 suggested that the effect might be due to star formation stimulated by the associated radio sources.

Recently, Rees (1989) and Begelman and Cioffi (1989) have independently proposed a specific mechanism for efficient radio source induced star formation at early epochs. (See also DeYoung 1989 and Daly 1989). While galaxies are in the process of forming, the development of a two phase structure is a nearly inevitable consequence of inhomogenities (Fall and Rees, 1985). Infalling material is heated to the virial temperature, $10^6 - 10^7$ K; embedded in it are clouds or filaments at $\lesssim 10^4$ K. If then the cocoons of shocked gas that are expected to surround powerful radio sources (Scheuer, 1974) engulf and compress the circumgalactic clouds, this can drive them over the Jeans limit and trigger star formation. Rees (1989) estimates that the star formation rate can be enhanced by the radio source over what one would otherwise expect from normal gas dynamical considerations (e.g. Fall and Rees, 1985) by an amount comparable to the Mach number of the radio jet, i.e. by a factor of $10 \sim 100$. This suggests a dramatic star formation rate during the expansion of the radio source through the galactic environment. Begelman and Cioffi (1989) give a prescription for the evolution of the radio cocoon and derive a comparable star formation rate.

Can the timescale of the passage of the radio source be associated with a reasonable prescription for the star formation rate? First, in the Scheuer "Dentist's Drill" model the radio hot spots track the instanteous point of impact of the jet, and not the time averaged bow shock of the entire cocoon. The observed speed of a hot spot, or jet material (e.g. Cygnus A) may not reflect the true speed of the advance of the cocoon. Hence it may be difficult to estimate in practice the crossing time of the cocoon. Spectral index age estimates are unreliable, particularly in the ultra-steep spectrum sources where other mechanisms for steepening the radio spectrum are probably at work (Chambers *et al.* 1989). Secondary star formation processes may also be at work. For instance the passage of the radio source could be identified with the fast rise-time of the star formation rate of Chambers and Charlot (1989) and the tail of star formation could then be due to effects steming from the starburst, e.g. supernovae driven shocks. The results of Chambers and McCarthy (1989) suggest that the high radio luminosity sources (3C and 4C) do not achieve their maximum radio luminosity until somewhat after the bulk of the star formation has occured.

SUMMARY

Selecting radio sources for study on the basis of their ultra-steep radio spectrum has proven to be an extremely successful technique for locating very high redshift galaxies. This in turn has helped discover a remarkable property of high redshift radio galaxies, that their optical/infrared continua are generally aligned with their radio axes. The extended optical and infrared continuum of high redshift radio galaxies is probably dominated by starlight, and radio source induced star formation is the most plausible explanation for the alignment effect. The young galaxy model discussed is the only idea so far that explains all the observations, especially the infrared alignments. Powerful radio sources appear to be stimulating substantial star formation at early epochs.

Much work remains to be done, including addressing the question of how much of a role radio sources may have played in the formation of ordinary galaxies.

ACKNOWLEDGEMENTS

We would like to thank Hy Spinrad and Wil van Bruegel for organizing the Berkeley conference "Radio Galaxies at High Redshift, a Multi-Spectral Synthesis" which provided a great deal of stimulating discussion amongst all the participants.

REFERENCES

Allington-Smith, J. R. 1982, *M.N.R.A.S.*, 199, 611
Barthel, P.D., 1989, *Ap. J.*, 336, 606
Begelman, M.C., and Cioffi, D.F., 1989 *Ap. J. (Letters)*, 345, L21
Begelman, M. C., Blandford, R.D., Rees, M.J., 1984, *Reviews of Modern Physics*, 56, 255
Bithell, M., and Rees, M., 1989 *preprint*
Blumenthal, G., Miley, G.K., 1979, *Astr. Ap.*, 80, 13
Bridle, A.H, Perley, R.A., 1984, *Annual Reviews of Astronomy and Astrophysics*, 22, 319
Bridle, A.H., Kesteven, M.J.L. Guindon, B., 1972, *Ap. J. (Letters)*, 11,27
Brodie, J.P., Bowyer, S., McCarthy, P.J., 1985, *Ap. J. (Letters)*, 293, L59
Bruzual, G. 1983 *Ap. J.*, 273, 105.
Chambers, K.C., 1989, Ph.D. Thesis
Chambers, K.C., and Charlot, S., 1989, *Ap. J. Letters, in press*
Chambers, K.C., and McCarthy, P.J., 1989, *Ap. J. Letters, in press*
Chambers, K.C., Miley, G.K., and Joyce, R. R. 1988b, *Ap. J.*, 329, L75
Chambers, K.C., Miley, G.K., and van Breugel, W. 1987 *Nature* , 329, 604.
Chambers, K.C., Miley, G.K., and van Breugel, W. 1988a, *Ap. J.*, 327, L47
Chambers, K.C., Miley, G.K., and van Breugel, W. 1989, *Ap. J.* submitted
Daly, R., 1989 ,*Ap. J. in press*
DeYoung, D., 1981 *Nature* , 293, 43
DeYoung, D., 1989 *Ap. J. (Letters)*, 342, L59
Djorgovski, S., 1986, in *Starbursts and Galaxy Evolution*, ed. T. Montmerle, p. 401.
Djorgovski, S., Spinrad, H., Pedelty, J., Rudnick, L., Stockton, A., 1987, *A. J.*, 93, 1307
Djorgovski, S., Spinrad, H., McCarthy, P., Dickinson, M., van Breugel, W., Strom, R., 1988, *A. J.*, 96, 836
Eisenhardt, P., Chokshi, A., 1989, *Ap.J. Letters*, submitted
Fabian, A.C., 1989, *M.N.R.A.S.*, 238, 41p
Fall, S.M., and Rees, M.J. 1985, *Ap. J.*, 298, 18.
Heckman, T.M., Miley, G.K., Balick, B., van Breugel, W.J.M., and Butcher, H.R., 1982, *Ap. J.*, 262, 529
Heckman, T.M., van Breugel, W.J.M., and Miley, G.K., 1984, *Ap. J.*, 286, 509
Heeschen, D., 1960, *Pub. A.S.P.*, 72, 368
Hewitt, A., Burbidge, G., 1987, *Ap. J. Suppl.*, 63,1
Kinney, A., Huggins, P.J., Glassgold, A.E., Bregman, J.N., 1987, *Ap. J.*, 314,145
Kinney, A., 1989 *private communication*
Laing,R.A., Riley,J.M., Longair, M.S., 1983, *M.N.R.A.S.*, 204, 151
Le Fevre, O., Hammer,F., Nottale, L., Mazure, A, Christian, C., 1988a, *Ap. J. (Letters)*, 324, L1
Le Fevre, O., Hammer,F., Jones, J., 1988b, *Ap. J. (Letters)*, 331, L73

Le Fevre, O., Hammer,F., 1988c, *Ap. J. (Letters)*, 333, L37

Lilly, S. J. 1988a, *Ap. J.*, 340, 77.

Lilly, S. J. 1988b, *Ap. J.*, 333, 161.

Lilly, S. J., Longair, M. S. 1984, *M.N.R.A.S.*, 211, 833

Longair, M. S., Lilly, S. J., 1984, *Journal of Astrophysics and Astronomy, 5, 349*

McCarthy, P. J., Van Breugel, W., Spinrad, H., and Djorgovski, S. 1987, *Ap. J. (Letters)*, 321, L29

McCarthy, P. J., Spinrad, H., Djorgovski, S., Strauss, M. A., van Breugel, W. J. M., and Liebert, J. 1987a *Ap. J. (Letters)*, 319, L39

McCarthy, P.J., 1989 , Ph.D. Thesis

Macleod, J.M., Doherty, L.H., 1972, *Nature* , 238, 88

Miley, G.K., 1980, *Annual Reviews of Astronomy and Astrophysics*, 18, 165

Miley, G.K., Heckman, T.M., Butcher, H.R., and van Breugel, W.J.M., 1981, *Ap. J.*, 247, L5

Miley, G.K., Chambers, K.C., Hunstead, R., Macchetto, F., Roland, J., Rottgering, H., Schilizzi, R., 1989, *ESO Messenger*, 56, 16.

Osmer, P.S. 1978, *Ap. J. (Letters)*, 226, L79

Rawlings, S. and Eales, S., 1989a, Second Wyoming Conference, *The Interstellar Medium in External Galaxies*, in press.

Rees, M.J. 1989, *M.N.R.A.S.*, 239, 1p.

Scalo, J.M, 1986, *Fundamentals of Cosmic Physics*, **11**, 1.

Scheuer, P.A.G., 1974, *M.N.R.A.S.*, 166, 513

di Serego Alighieri, S., Fosbury, R, Tadhunter, C., 1989, *Nature*, in press

Spinrad, H. and Djorgovski, S. 1984, *Ap. J. (Letters)*, 285, L49

Spinrad, H. Djorgovski, S., Marr, J., Aguilar, L. 1985a, *Pub. A.S.P.*, 97, 932

Spinrad, H., *et al.* 1985b, *Ap. J. (Letters)*, 299, L7

Spinrad, H. 1986 *Pub. A.S.P.*, 98, 269

Spinrad, H., Djorgovski, S. 1987 IAU Symposium 123, *Observational Cosmology*, ed. G. Burbidge, (Dordrecht:Reidel), P. 129

Tadhunter, C.N., Fosbury, R.A.E, di Serego Alighieri, *Proceedings of the Como Conference on BL Lac Objects: 10 Years After*, ed. L. Maraschi, in press

Tielens, S., Miley, G., Willis, A., 1979, *Astr. Ap. Suppl.*, 35,153

van Breugel, W.J.M., Heckman, T.M., Butcher, H.R., and Miley, G.K., 1984, *Ap. J.*, 276, 79

van Breugel, W.J.M., Miley, G.K., Heckman, T.M., Butcher, H.R., and Bridle, A., 1985a, *Ap. J.*, 290, 496

van Breugel, W.J.M., Filippenko, A.V., Heckman, T.M., Miley, G.K., 1985b, *Ap. J.*, 293, 83

van Breugel, W.J.M., Heckman, T.M., Miley, G.K., Filippenko, A.V. 1986, *Ap. J.*, 311, 58

Veron, M.P., Beron, P., Witzel, A., 1972, *Astr. Ap.*, 18, 82

Windhorst, R.A. Miley, G.K., Owen, F.N., Kron,R.G., Koo, D.C., 1985, *Ap. J.*, 289, 494

Woosley, J., 1988, *Ap. J.*, 330, 218

THE EVOLUTION OF WEAK RADIO GALAXIES AT RADIO AND OPTICAL WAVELENGTHS

ROGIER WINDHORST, DOUG MATHIS and LYMAN NEUSCHAEFER
Department of Physics and Astronomy, Arizona State University,
Tempe, AZ 85287-1504

ABSTRACT We review the statistical properties of radio sources. The linear sizes of gE radio galaxies increase mildly with radio power, and decrease drastically with redshift. Combined with the redshift distribution, this causes angular sizes (Ψ_{med} –S) to decrease continuously towards fainter fluxes. Nonetheless, the fraction of large sources (Ψ >15") is not negligible at *any* flux level, requiring a revision of the resolution correction to the source counts. The 1.4 GHz counts are now well established (within ~5%) from 10,575 sources in 24 different surveys that cover 100 μJy < $S_{1.4}$ <10 Jy. This allows to study field-to-field anisotropies. None are found in the wide area surveys for $S_{1.4}$ >10 mJy, and only marginal anisotropies in some small-area surveys. These are compared to the redshift distribution and the angular correlation function of radio sources.
The low frequency spectral index – flux density relationship is reviewed and is consistent with the observed 0.6 and 1.4 GHz counts. Combined with the available redshift information, the effects of radio power and redshift on α_{med} can now be disentangled. The *observed* median spectral index depends more strongly on redshift than on radio power, but the former can be largely attributed to the radio K-correction. This explains why very high redshift galaxies have been found so successfully amongst very steep spectrum strong radio sources. Finally, an example is given of a very high redshift radio galaxy, that recently showed up in the Leiden Berkeley mJy sample, and resembles M87 in several respects.

1. INTRODUCTION

Radio galaxies evolve in their radio and optical properties on timescales much longer than the human, but smaller than the Hubble time. Hence, we can only observe the evolution of an entire population with cosmic epoch, not that of individual objects. When faint objects are selected from deep radio or optical images, they generally occupy a handful of independent pixels, so that the studied parameters are usually limited to: 1) their morphology or characteristic size; 2) their luminosities; 3) colors or spectra; and 4) their space density (or luminosity function). Each of these parameters may appear to evolve with cosmic epoch, either because of poorly understood systematic errors and selection effects, or because of good physical reasons, or both.

Very little is known about the morphological evolution of galaxies at *optical* wavelengths. However, the evolution of *radio* sizes is not only well established observationally, but also the only property of the radio source population whose cosmic evolution has a good physical explanation. The evolution of the radio luminosity function (RLF(z), or cosmological evolution) has been rather well established over the last two decades, but is still poorly understood physically. While there are very good theoretical reasons why the optical luminosity of galaxies should evolve with cosmic time (because stars form and evolve), their luminosity evolution (or OLF(z)) is not yet well determined from the galaxy counts and redshift distributions. Nonetheless, observational programs in the last decade have indicated modest evolution in the integrated optical spectra of field and cluster galaxies, in many cases the consequence of brief episodes of star formation. The spectral evolution of sources at radio wavelengths has neither been measured unequivocally, nor been understood theoretically. This paper will present new data that address this problem.

The evolution of the radio source population in its various properties has been reviewed recently by, e.g., Condon (1984b, 1989b), Danese et al. (1987), Kellermann and Wall (1987b), Oort (1987a), Peacock and Miller (1988), Wall et al. (1986), and Windhorst (1985, 1986). It is now generally agreed upon that the most powerful radio sources, giant ellipticals and quasars, have undergone strong cosmological evolution in their RLF. They were more luminous and/or more frequently radio sources in the past. This RLF evolution is somewhat less for radio galaxies of lower powers (but exceeding the characteristic break in the RLF, $P > P^*$). There are several indications that radio galaxies with powers below P^* have undergone modest cosmological evolution as well, although this is still not well established. We believe that the cosmological evolution of radio galaxies is a *direct consequence* of the process of galaxy formation itself. Galaxies evolve, often through episodic starbursts that are triggered by dynamic disturbances (such as mergers), which also indirectly trigger the central engine to produce a radio source (see also Windhorst 1984). Therefore, in order to understand the physics of the cosmological evolution of radio sources, we must first comprehend the process of galaxy formation and evolution itself.

2. ANGULAR SIZES AND LINEAR SIZE EVOLUTION

With the advent of large synthesis radio telescopes, especially the VLA, the measurement of radio source structure and angular sizes has become feasible at resolutions often even inaccessible to optical telescopes. The main result of these programs is summarized in Figure 1. The median radio source angular size decreases monotonically from ~20" around 1 Jy to 2" below 1 mJy. The data points below $S_{1.4} = 3$ Jy in Figure 1 are all consistent with:

$$\Psi_{med} = 2.0'' \, S_{1.4}^{0.30}, \quad (S_{1.4} \; in \; mJy) \tag{1}$$

indicated by the dotted line. Although radio sources become steadily smaller towards fainter fluxes, a comparison of 8 deep surveys at different flux levels demonstrates that the fraction of sources with large angular sizes ($\Psi \geq 15'' >$ FWHM in all surveys) is *not negligible* at any flux level. This fraction increases from $\gtrsim 3\%$ at $S_{1.4} = 200$ μJy to 18% at 5 mJy (see Windhorst 1989b for a review). The Leiden Berkeley Deep Survey points of Oort (1988b) had to be slightly revised

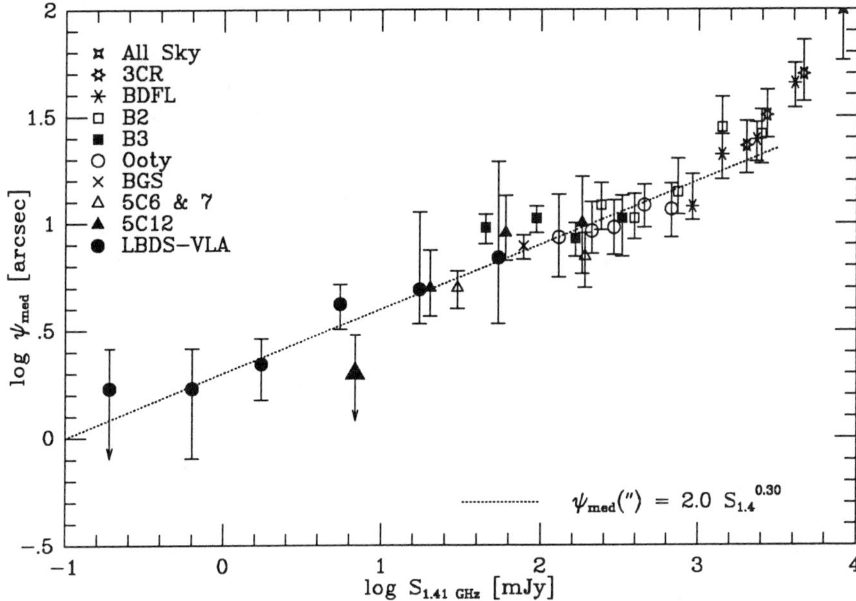

Fig. 1. Median angular size – flux density relation. Surveys not at 1.4 GHz were transformed to 1.4 GHz with the relation of Fig. 7.

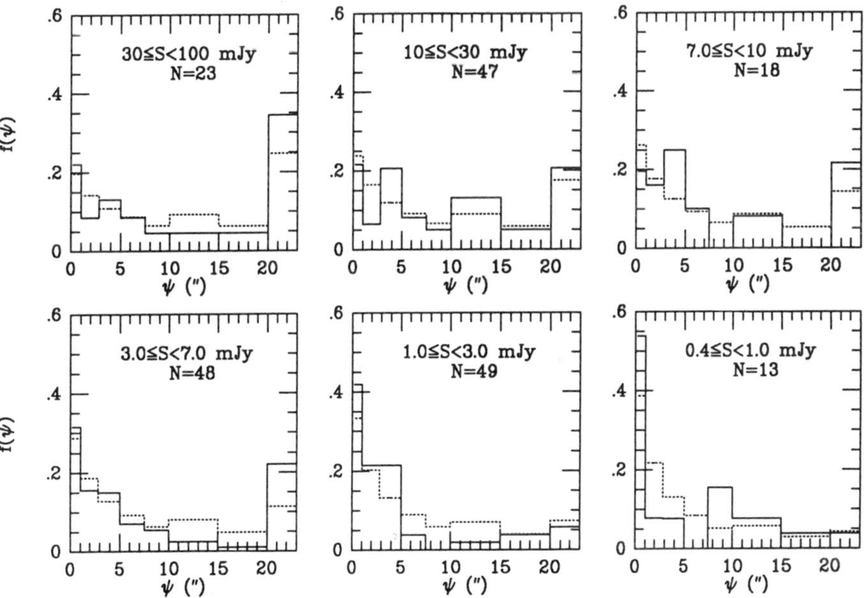

Fig. 2. Full drawn: Oort's (1988b) differential angular size distribution. Dotted: predictions from eq. (2) used for the resolution correction.

upward, because faint low surface brightness (SB) radio sources were missing ab initio in the LBDS. Figure 2 shows that the *shape* of the angular size distribution is rather flux dependent. This has long been uncertain, and hampered the faint radio source counts, since it determines the resolution correction (for missing low SB sources) to the counts. The following flux dependency represents the (integral) angular distribution very well (dotted line in Fig. 2):

$$h(\Psi) \;=\; exp\,[\,-ln\;2\;(\Psi/\Psi_{med})^{0.62}\,], \tag{2}$$

with Ψ_{med} from (1). The steady decline in angular size at faint radio fluxes cannot be due to selection effects alone. It occurs because almost all blue radio galaxies that dominate the counts at (sub-)mJy levels have subarcsecond sizes at 1" resolution. Their radio spectra are in general steep (α_{med} ~0.7, Donnelly et al. 1987), ruling out radio emission of thermal origin. Several are variable on timescales of month to years (Oort and Windhorst 1985), show low frequency turnovers in their radio spectra (below 0.4 GHz, Oort 1988b, – et al. 1988), and have compact radio morphology (Ψ <1.0"; Oort et al. 1987). All this suggests that these blue radio galaxies have optically thick non-thermal cores on parsec scales. Their optical spectra are like starburst galaxies (Windhorst et al. 1987), and their radio emission is thus not directly related to the large scale stellar population. As a consequence, Ψ_{med} measurements becomes less meaningful for sub-mJy radio galaxies, because the compact nuclear radio source may be superposed on the 10–100 times fainter low SB continuum of the disk. Both quantities cannot be parametrized as simply as the "largest angular size" of a classical double source in a gE radio galaxy.

At local redshifts there is a change in radio morphology around the break in the RLF around log $P^{*}_{1.4}$ =25.0 W Hz^{-1} . Above this radio luminosity, most sources are either classical doubles or triples associated with gE's or quasars. Below P^{*} there is a mixture of complex and compact radio sources, mostly associated with these actively starforming galaxies. Oort et al. (1987) unambiguously separated the z and log P dependence of the radio source linear sizes in gE galaxies. These follow one single z-dependence over the entire radio luminosity range (P> P^{*}):

$$D(z,\;logP_{1.4})\;\propto\;P_{1.4}^{0.3}\;(1+z)^{-2.7}\quad for\;\;H_o=50,\;\;q_o=0 \tag{3}$$

This determination suggests a steeper z-dependence of linear sizes than was previously assumed, and has been confirmed by Kapahi (1988) and Singal (1988). Models for the maximum distance to which the radio source produced by an AGN can expand, were made recently by Gopal-Krishna and Wiita (1987). They suggest that the pressure balance between the galaxy halo and the intergalactic medium defines a transition surface, the position of which is redshift dependent. From this, they predict a redshift dependence of the linear size approximately $\propto(1+z)^{-3}$, in good agreement with the observations.

3. LOW FREQUENCY RADIO SOURCE COUNTS

During the last decade, several deep surveys have been made with various aperture synthesis radio telescopes. At mJy levels important contributions were made with the WSRT 3 km array and the Cambridge One Mile Telescope. At sub-mJy levels, deeper surveys have been done with the VLA and the WSRT, currently down to rms noise values of ~3–10 μJy , at 6 and 21 cm respectively.

Figure 3 shows the differential 1.41 GHz source counts normalized to those expected in a homogeneous, static, Euclidean Universe (see Fig. 3). The 1.4 GHz counts showed a significant change in their slope below 5 mJy (Windhorst 1984; – et al. 1985). This "upturn" was confirmed by independent surveys of the same (Oort and Windhorst 1985) and different areas (Condon and Mitchell 1984; Oort 1987a). Because the median angular sizes of weak radio sources turned out to be so small (Fig. 1), the deeper surveys showed that the resolution correction of the earlier surveys lead to a 15–25% overestimation of the original count. Section 2 showed that a plain omission of the resolution correction is not justified either. From eq. (2) we computed a more realistic correction for missing low SB sources at the mJy level, which amounts to ~35% of the original value (i.e., ~1.16 instead of 1.45 for mJy sources), and revised the original counts accordingly.

Including the 1200 sources summarized in the amalgamated Westerbork 1.4 GHz source count (Katgert et al. 1988), the resulting count is now well constrained from 10,575 radio sources in 108 *independent* bins from 24 surveys (Fig. 3). In general, these appear to be statistically consistent (Fig. 4). Error bars were omitted in Figure 3 to prevent it from becoming optically thick. But Figure 4a shows the residuals w.r.t a weighted polynomial fit (long dashed line) to all bins with more than about 15 objects. (On average, there are ~ 100 sources per bin). The χ^2 's are shown in Figure 4b, with their signs preserved. The observed r.m.s. dispersion of the counts with respect to this best fit is 20%. For the small-area surveys this increases to 31%, while that expected from their $N^{-1/2}$ errors is 19%. Remaining systematic errors can be of order 10–15% survey-to-survey ($\lesssim$5% for the flux scale, $\lesssim$5% for the flux determining algorithm, and ~10% uncertainty in the resolution correction; see Windhorst 1989b). Although the amplitude of the normalized differential counts is now formally known to within 2%, subtle systematic errors in the overall counts may persist at the ~5% level. But the net result is that each of the small-area counts may have some *intrinsic* variance (~15–20%) in excess of the expected $N^{-1/2}$ and systematic errors. This may be due to real field-to-field anisotropies (see section 4).

Figure 3 demonstrates that the revised resolution correction does not make the upturn in the sub-mJy source counts disappear. Hence, this gradual change in slope in the mJy counts is *not* due to instrumental effects, but is a property of the Universe. It starts below 10 mJy, and is visible in all survey fields that contribute significantly below a few mJy. The upturn is significant, because it cannot easily be attributed to the canonical gE radio galaxies or quasars, which dominate the counts only for $S_{1.4} \geq 5$ mJy. Neither can it be explained by normal spirals or Seyferts, unless these objects have undergone significant cosmological evolution in the recent past (Windhorst 1984). Various models have been developed to explain the (sub-)mJy counts, such as evolving normal spirals (Condon 1984b, 1989b), local non-evolving low luminosity radio galaxies (Wall et al. 1986; Subrahmanya and Kapahi 1983), or (possibly evolving) actively starforming galaxies (Kron et al. 1985; Windhorst et al. 1984, 1987; Oort 1987a; Franceschini et al. 1989). To explain the steep slope of the upturn, Windhorst (1984) and Oort (1987a) argued that the population of actively starforming galaxies probably underwent cosmological evolution similar to that of gE's and quasars. If these two evolving populations indeed peacefully coexist (without evolving into each other), extrapolation of their RLF's(z) beyond the

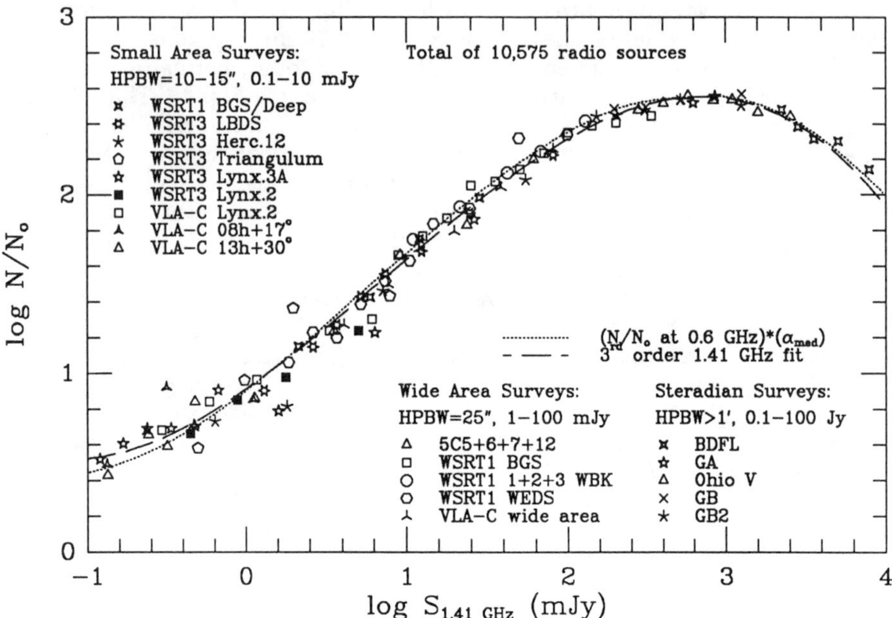

Fig. 3. The normalized differential 1.41 GHz counts ($N_o = S_{1.4}^{-2.5}$). Plotted are 108 independent bins with 10,575 sources from 24 surveys.

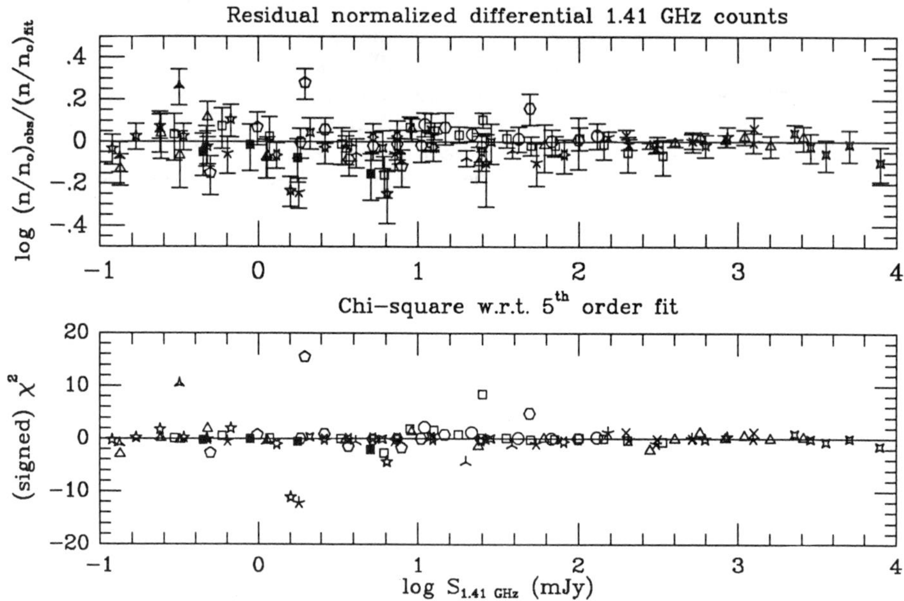

Fig. 4. a) Residual normalized differential 1.41 GHz counts w.r.t. a 5th order polynomial fit. b) χ^2 vs. $S_{1.4}$ with its sign preserved.

redshift to which it was measured indicates that 100% of the 1.4 GHz counts is reached for fairly low values of the redshift cut-off (z_{max} ~1.5 ±0.5; this is about a 95% value; the 5% tail in N(z) cannot be precisely constrained in these models). Since the overall 1.4 GHz counts are now known within 5%, the "only" remaining uncertainty in these models is the adopted RLF(z) and q_0 .

4. THE REDSHIFT DISTRIBUTION, w(θ), AND ANISOTROPIES

In this section we will summarize the evidence for (an)isotropy in the distribution of weak radio sources on the sky. Superclustering on scales of tens of Mpc could have an important effect on the radio source distribution on scales of degrees, even out to cosmological distances.

Each of the 6 small-area surveys (section 3) covers $\lesssim 1$ deg^2 at different, but high Galactic latitudes. In general, their 1.4 GHz counts appear to be statistically consistent (Fig. 4a). Close inspection of Figure 4b shows that in 4 out of the 6 small-area surveys the formal χ^2 -value exceeds that expected for a random distribution at the 99.0–99.99% level (for typically 3–8 degrees of freedom). However, most of the excess χ^2 only comes from 1–2 bins per survey, and usually not the faintest ones. It is therefore not likely due to systematic errors or true field-to-field variations, because these would show up primarily in the faintest bins, or across all bins of a survey (since the RLF is $\gtrsim 10$ dex wide). We believe these excesses must be largely statistical fluctuations.

This is not the whole story, though. Oort (1987a) recently discovered some evidence for clustering in the distribution of weak radio sources in several small-area surveys. In all fields, his angular correlation function, w(θ), was appropriately corrected for primary beam effects and normalized to Poissonian in exactly the same way. It showed indeed no power in excess of Poissonian for several of the brighter survey fields, or a sum of such fields. As an example, Figure 5b shows his angular covariance function for the Hercules field. However, his w(θ) for (somewhat) fainter sources in the deeper Lynx fields was far from uniform, both in Lynx.2, Lynx.3A (Fig. 5a), and the 327 MHz Lynx field.

A clue for this may be obtained from the redshift distribution for the complete, updated sub-samples of Kron et al. (1985) in both Lynx and Hercules (Fig. 6). The approximate limits for complete or representative spectroscopy are indicated. Both N(z)'s are plotted with bins of 0.01 in z. There appear to be some identical groups of redshifts – with empty spaces in between – on scales of $\lesssim 1$ degree out to z~0.6. The large field galaxy surveys have shown similar clumpiness with much better statistics (Ellis 1987; Koo and Kron 1988).

Apparent field-to-field anisotropies in the radio source densities could arise if: (1) the fields are small compared to the largest expected structures (1 deg corresponds to ~35 Mpc at the median redshift); (2) N(z) is dominated by the effects of superclustering; (3) the radio source redshift cut-off is not a very large number; and (4) the RLF within a supercluster is constant at a given z (Windhorst, 1986). The average spacing in z between significant peaks in the two field galaxy surveys is about 0.04. Small-area radio surveys sample radio galaxies at the few mJy level with a z_{med} ~0.6–0.75 (Fig. 9), and z_{max} is possibly as low as 1.5–2.0 (section 3). Assuming that the space density of superclusters has *at best* remained constant with increasing redshift, a small-area deep radio survey would sample *at most* 3–4 dozen sheets out to z~2. The statistical fluctuations

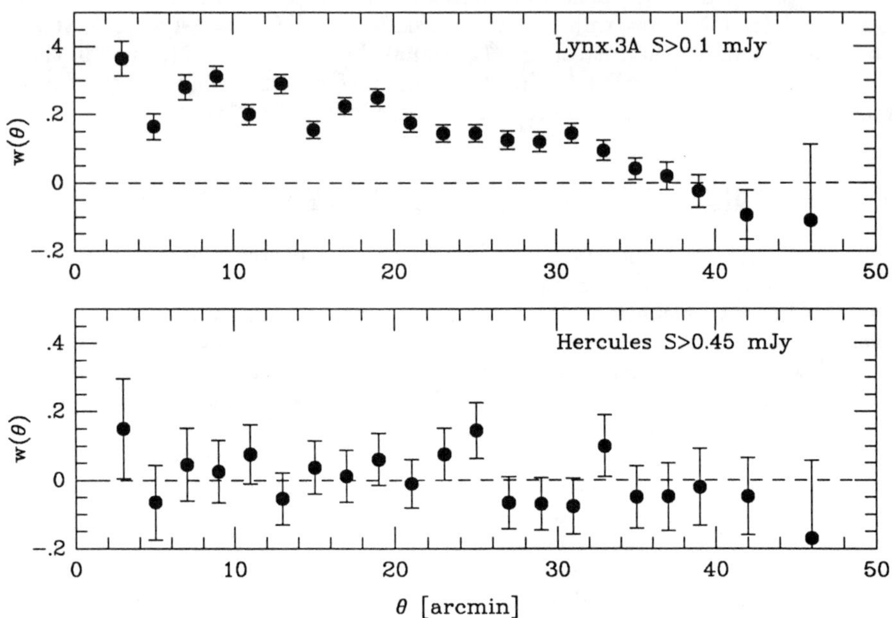

Fig. 5. Angular correlation function, w(θ), for LBDS radio sources in: a) the Lynx.3A deep survey; b) in the Hercules field (Data from Oort 1987a).

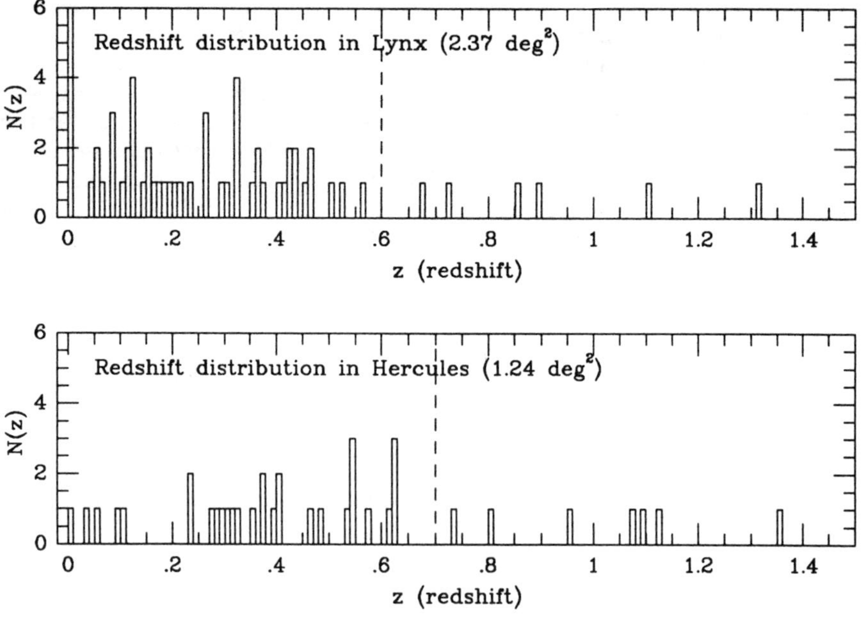

Fig. 6. Spectroscopic redshift distribution of LBDS in: a) the entire Lynx field; b) the Hercules field. Dashed: approximate completeness limits.

in the radio source surface density could thus be easily of order ~15% on scales of degrees. This is consistent with the ~15–20% intrinsic variance that may be present the small-area source counts (section 3). High z superclustering can also explain the difference between w(θ) in the Lynx and Hercules fields. Figure 6 suggests that a natural fluctuation may have produced an apparently larger number of "foreground" superclusters in Lynx, compared to Hercules, resulting in the apparent excess power in Oort's (1987a) w(θ) for the Lynx fields, and the apparent excess χ^2 in the deepest counts in Lynx.3A.

Although these results are tantalizing, our feeling is that they can never be measured much more significantly than at the 99% level, because the required larger surveys of greater depth will necessarily wash out any real anisotropies.

5. THE LOW FREQUENCY SPECTRAL INDEX – FLUX RELATION

Figure 7 shows the median spectral index measured in bright radio surveys selected at low frequency (0.4 GHz). After removal of systematic errors in several of the source samples, Kapahi and Kulkarni (1986) showed that the median 0.4–5 GHz spectral index for low frequency samples stabilizes to ~0.9 down to $S_{0.4}$ ~100 mJy. Gopal-Krishna's (1988 and references therein) suggested that the α_{med} - $S_{0.4}$ relationship flattens at lower fluxes. The three point spectral indices of weak LBDS radio sources are to first order straight between either 0.6, 1.4 or 4.9 GHz with $< \alpha >$=0.74±0.02 ($S_\nu \propto \nu^{-\alpha}$; Donnelly et al. 1987; Oort 1988a; see Windhorst 1989c for a review). Synchrotron self-absorption occurs in less than 10% of their complete sub-samples, when selected at ~0.5 GHz. Reliable surveys below a few mJy are not avaliable at 408 MHz. In Figure 7 we therefore compare the median spectral index of the weak sources selected at 0.6 GHz to the brighter surveys at 0.4 GHz. The high frequency flux for the spectral index determination is available from homogeneous 5.0 GHz measurements (except for the LBDS where part of the sources had only 1.4 GHz data, but less than 10% of the LBDS sources show significant high frequency components at 5 GHz). Figure 7 suggests that α_{med} - $S_{0.6}$ does indeed turn downward to flatter spectra at weaker radio fluxes.

Figure 8 reviews the 0.6 GHz source counts of the LBDS, and several large area 0.6 GHz surveys (see Windhorst 1989c). All counts are statistically consistent, although not determined as well as at 1.4 GHz. A best fit to the 0.61 GHz counts was transformed to 1.41 GHz by convolving with the α_{modal}-$S_{0.6}$ relation of Figure 7, and a dispersion that slowly increases from 0.2 in α for the bright surveys to 0.35 as measured in the deep surveys. The result is shown as the dotted line in Figure 3. The prediction is so similar to the observed 1.4 GHz counts that the true α_{med} - $S_{0.6}$ relationship ought to be similar to the dashed line in Figure 7, with likely systematic deviations of a few x 0.01 in α. The dashed line in Figure 7 is not a best fit to the data, but the relationship required to transform the 0.6 GHz counts to 1.4 GHz, and v.v. (with a slightly different dispersion on the flat side). The normalized differential source counts (Fig. 3 and 8) and the α_{med} - $S_{0.6}$ relation show the same general trend. As pointed out by Gopal-Krishna (1988), the most straightforward explanation is that the redshift distribution of radio sources causes the same global behavior.

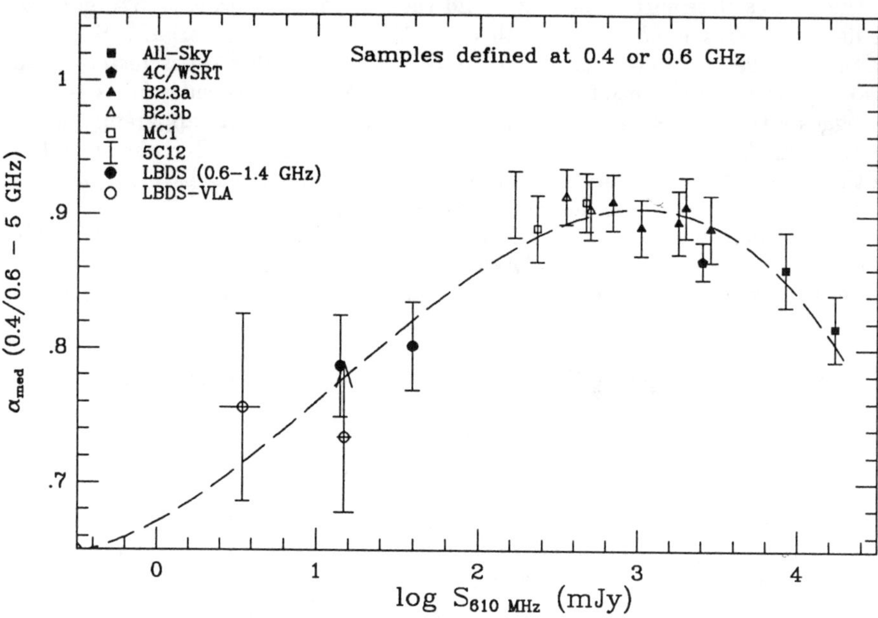

Fig. 7. Median spectral index – $S_{0.6}$ relation for sources selected at 0.6 or 0.4 GHz ($S_{0.4}$ transformed to $S_{0.6}$ with this relation; ν_2=5 GHz).

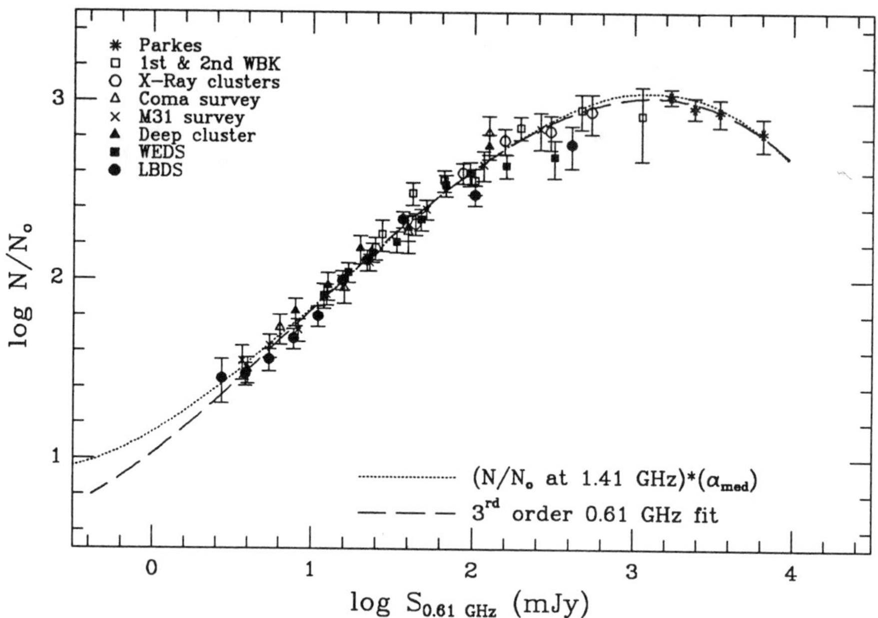

Fig. 8. Normalized differential 0.61 GHz source counts. The dotted line is the 1.4 GHz counts, convolved with α_{med} - $S_{1.4}$ (as in Fig. 7).

6. RADIO SPECTRAL INDEX VERSUS REDSHIFT AND RADIO POWER

When combined with redshift information (Fig. 6 and 9), the α_{med} - $S_{0.6}$ relation can be used to disentangle the effects of radio power and redshift (or radio K-correction) on the observed spectral index. The large median redshift in the bright surveys (~ 1 Jy) causes a substantial radio K-correction, especially because high frequency curvature will affect the value of α of powerful sources when measured between ~ 0.5 and 5.0 GHz (Gopal-Krishna 1988). Therefore, Table I lists the median spectral index for different samples selected at 0.4 or 0.6 GHz, but with spectral indices computed between these frequencies and homogeneous 1.4 GHz data. Table I also lists the measured median redshifts (see Fig. 9; or in two cases interpolated from Windhorst 1984), and the median 1.4 GHz radio powers (K-corrected with the observed α_{med}). The observed low frequency ν_1 of ~0.5 GHz corresponds to a median selection frequency of ~0.9±0.3 GHz in the restframe, while the observed ν_2=1.4 GHz corresponds to a restframe frequency of ~2.3±0.5 GHz (except for the Gopal-Krishna 1 Jy sample).

TABLE I Median spectral index, α_{med}(0.5-1.4 GHz), versus redshift and 1.4 GHz radio power

Survey (Reference)	ν_{1sel}, ν_2 (GHz)	$S^{1.4}_{med}$ (mJy)	z_{med}	log$P^{1.4}_{med}$ (W/Hz)	α^{obs}_{med} (m.e.)	α^{pred}_{med} (z,log P)	α^{pred}_{med} (z only)	α^{pred}_{med} (log P)
LBDS-VLA (Donnelly ea 1987)	0.6, 1.4	2.0	0.6	24.69	0.76 0.07	0.756	0.79	0.73
LBDS-VLA (Oort 1988a)	0.6, 1.4	7.5	0.40	24.85	0.72 0.05	0.731	0.75	0.74
LBDS (Windhorst 1989c)	0.6, 1.4	12.0	0.75	25.71	0.799 0.025	0.804	0.83	0.79
LBDS (F<23) (Windhorst 1989c)	0.6, 1.4	12.0	0.35	24.93	0.732 0.03	0.725	0.74	0.75
LBDS (F>23) (Windhorst 1989c)	0.6, 1.4	12.0	1.0	26.04	0.836 0.03	0.843	0.87	0.80
LBDS/WSRT (Windhorst 1989c)	0.6, 1.4	50.0	0.9	26.55	0.860 0.02	0.846	0.85	0.83
1 Jy-sample (Gopal-Krishna 88)	0.4, 5.0	450	1.2	27.84	0.910 0.015	0.914	0.90	0.90
1 Jy (z>0.95) (Lilly 1989)	0.4, 1.4	450	1.5	28.13	0.960 0.03	0.950	0.94	0.91
4C/WSRT (Katgert-M. 1980)	0.4, 1.4	1200	0.7	27.65	0.828 0.017	0.851	0.81	0.89
4C/WSRT (m<21) (Katgert-M. 1980)	0.4, 1.4	1200	0.3	26.79	0.782 0.025	0.767	0.72	0.84
4C/WSRT (m>21) (Katgert-M. 1980)	0.4, 1.4	1200	1.0	28.06	0.900 0.014	0.899	0.87	0.91

Column 5 assumes H_0 = 50 km s^{-1} Mpc^{-1} , q_0 =0, and radio K-correction from column 4, 6.

Laing and Peacock (1980) noted a positive correlation between low frequency spectral index and either radio power or redshift, but could not decide between both because of the limited dynamic range of the 3CR. We cannot compare the Laing and Peacock α_{med} -z and α_{med} -P relations directly to the data in Table I, because of the different selection frequency. When we redetermine similar relations from the actual data in Table I, we do not obtain very good fits (cols. 8 and 9, respectively; rms deviations are ~0.04 in α). But the following bivariate dependence:

$$a_{med}(z, P_{1.4}) = 0.659 + 0.516 \, log(1+z) + 0.0275 \, log \, (P_{1.4}/P^*) \qquad (4)$$

(with $P_{1.4}$ in W Hz^{-1} for H_0 =50, q_0 =0 and log P^*=25.0) provides a superior fit to all data, as shown by column 7 in Table I (rms deviations of 0.01 in α). The dependency on redshift is much stronger than that on radio power. E.g., for a typical radio source with log $P_{1.4}$ =26.5 at z=0.75, the spectral steepening is about 0.13 due to the z-dependence, and only 0.04 due to the radio power. Although the log P dependency does not completely vanish, it is about 3 x weaker than the Laing and Peacock relation, because most of the correlation appears to be with redshift.

To understand the physical meaning of eq. (4), we must take into account the spectral steepening $\Delta\alpha$ induced to α_{obs} by the K-correction alone, and the fact that the log P-term itself may have a hidden dependency on redshift. The latter would result if the cosmological evolution of radio sources is caused by pure luminosity evolution (see, e.g. Condon 1984b), so that P(z)=P(z=0).$(1+z)^p$, with p~3–4 (Windhorst 1984). In that case, eq. (4) reduces to:

$$a^{rest}_{med}(z, P_{1.4}) = 0.659 + (0.52 + 0.027p - K_\alpha) log(1 + z) + 0.027\ log(P_{1.4}(0)/P^*) \quad (5)$$

where log $P_{1.4}$ (0) is the restframe radio power reduced to z=0, and $K_\alpha = (\alpha_2 - \alpha_1)$ / $log_{10}(\nu_2/\nu_1)$ is the *spectral* K-correction coefficient between ν_2 and ν_1, respectively. Eq. (4) and (5) are strictly only valid between ~ 0.5 and 1.4 GHz, or ~ 0.9 and 2.3 GHz in the restframe. For a typical synchrotron source like Cygnus-A, the radio spectrum steepens by maximally 0.5 (from 0.75 to 1.25) between frequencies of 1 and several GHz, the latter value depending, amongst others, on the source age (van der Laan and Perola, 1969). Spectral curvature is expected to be smaller over this rather narrow frequency range ($\alpha_2 - \alpha_1 \lesssim 0.2$), but K_α may still be as large as 0.5. Hence, without luminosity evolution (p=0) the apparent z-dependence of the spectral index is largely due to the expected radio K-correction, and *no intrinsic* dependence on redshift is required. In the case of strong luminosity evolution (p>3), the K-correction does not explain all of the apparent spectral steepening with redshift. In that case, the apparent dependence of α_{med} on redshift is in a sense an artifact of the cosmological evolution of the radio source population, and the true physics lies in the relations between α_{med} and log $P_{1.4}$, plus log $P_{1.4}$ and redshift. The former may be explained by stronger synchrotron losses for more powerful sources (because they radiate their energy more quickly), the latter by an epoch dependent fueling that was somehow more efficient in the past. Inverse Compton scattering against the microwave background cannot be the cause of the spectral steepening, since the expected losses would be a stronger function of redshift. Eq. (5) also explains why recently very high redshift galaxies have been found so successfully amongst very steep spectrum strong radio sources.

7. VERY HIGH REDSHIFT GALAXIES IN FAINT RADIO SAMPLES

In our Palomar 200" Four-shooter survey we found nearly all optical counterparts of a sample of 453 weak radio sources in the extended LBDS down to V$\lesssim$26.5 mag. We have acquired low resolution spectroscopy for 175 radio sources with $J^+ \lesssim 24$ mag. For one object, Herc202, we recently measured a redshift of z=2.390. It is a classical example of a very steep spectrum ($\alpha^{0.6}_{1.4}$ =1.10) compact radio source (Ψ <1.0"), that is not variable on timescales of years. Its optical counterpart is a compact V~23.5 galaxy that is barely resolved on our CCD frames

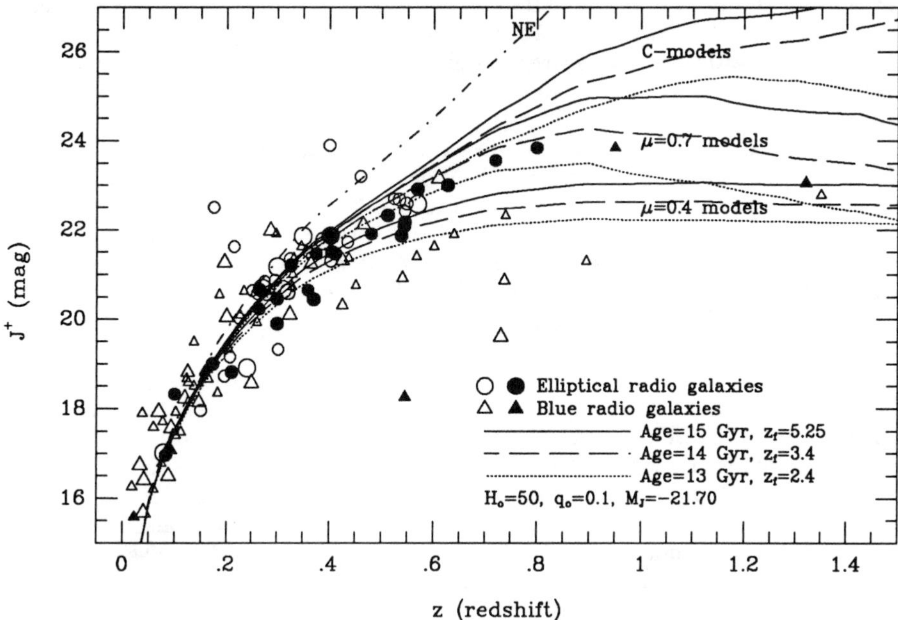

Fig. 9. Updated J^+-mag – redshift diagram for the entire LBDS. Open symbols: unresolved radio sources ($\Psi < 3"$), filled: $\Psi \geq 3"$.

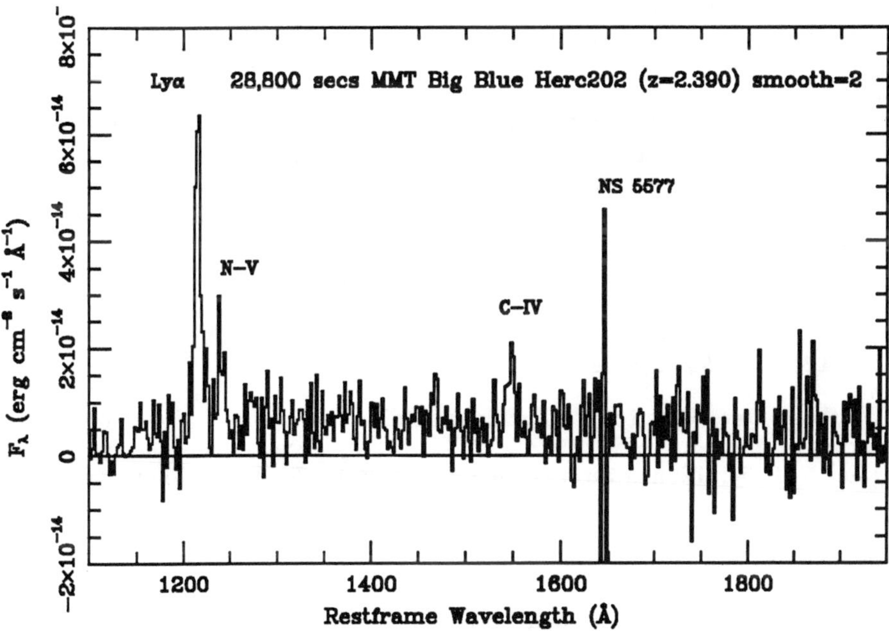

Fig. 10. 8-hour MMT spectrum (restframe) of Herc202, a compact galaxy at z=2.390 (V=23.5) associated with a steep spectrum LBDS source.

(FWHM~1.6" in 1.3" seeing). Its size is $\gtrsim$2.6" down to r=28.5 mag/arcsec^2, or $\gtrsim$35 kpc at z=2.390. Figure 10 shows an 8 hour spectrum of Herc202, shifted back to the restframe. It has weak emission lines, most noticeably Ly-α, N V, and C IV. The Ly-α line is resolved (FWHM~ 21Å), its deconvolved velocity width is about 1100 km/sec. This is too large for a completely virialized galaxy, and might be due to a weak QSO-like central engine, or a cooling flow. A 15 ksec narrow-band Ly-α exposure by Koo on the Lick 3 meter demonstrates that Herc202 is clearly elongated in Ly-α (size ~ 6.5", or ~ 85 kpc). Its optical-IR colors are fairly blue (Gunn g-r=0.03, r-i=0.31, r-H=2.20), all consistent with a young (~1–2 Gyr) stellar population at z=2.4. Hence, Herc202 is quite unlike the protogalaxy-like objects recently found in the 3CR and 1 Jy samples, perhaps more similar to Lilly's (1988) galaxy, although the emission lines of Herc202 are weaker.

A comparison with the 28 hour IUE spectrum of M87 (Buson et al. 1989) shows some rather surprising results (joint publication on these two galaxies is in preparation by Windhorst et al. and Buson et al.). Both restframe spectra are very similar: they have Ly-α and N V of nearly the same equivalent width ($W_\lambda \sim 50-80$Å), although Herc202 has stronger C IV. Both galaxies have similar compact radio structure (Biretta et al. 1989). Incidentally, they also have about same restframe frequency spectral index (α=0.74), and radio power (log $P_{1.4}$ (z=0)~25.68, both from eq. (5) with p=4). There are two marked differences in the UV continuum, though. Herc202 has m(1900-4900Å) =1.0 mag, while M87 is much redder in this color (~2.25 mag; Burstein et al. 1988). Herc202 thus has a much larger star formation rate at z=2.4 contributing to its UV continuum, much like NGC205. However, Herc202 does *not* have the hot UV upturn below 1400 Å seen in M87 and other metal rich ellipticals.

All this appears to be consistent with Herc202 being a genuinely young, but completely formed elliptical galaxy with a current age of at least 14 Gyr. At z$\simeq$2.4, it simply has not had enough time to develop a (P)AGB, consistent with Burstein et al.'s (1988) alternate explanation that the PAGB might be responsible for the hot UV upturn in nearby, old metal rich ellipticals.

In conclusion, Herc202 is the third LBDS radio *galaxy* redshift we measured to have z>1 (out of 175 spectroscopic redshifts for objects with J^+ <24 mag). It is an unusual example of an apparently "normal", but very high redshift galaxy telling us something about its likely nearby counterparts.

ACKNOWLEDGEMENTS

We thank Dave Burstein for illuminating discussions. This research was supported by grant AST–8821016 of the National Science Foundation, NASA-IUE grant NAG 5-1172, and through an Alfred P. Sloan Research Fellowship.

REFERENCES

Biretta, J. A., Owen, F. N., and Cornwell, T. J. 1989, *Ap. J.*, **342**, 128.
Burstein, D., et al. 1988, *Ap. J.*, **328**, 440.
Buson, L. M., Bertola, F., Burstein, D. et al. 1989, in preparation.
Condon, J. J. 1984b, *Ap. J.*, **287**, 461.

Condon, J. J., and Mitchell, K. J. 1984, *A. J.*, **89**, 610.
Condon, J. J. 1989b, *Ap. J.*, in press.
Danese, L., et al. 1987, *Ap. J. (Letters)*, **318**, L15.
Donnelly, R. H., Partridge, R. B., and Windhorst, R. A. 1987, *Ap. J.*, **321**, 94.
Ellis, R. S. 1987 in *IAU Symposium 124, Observational Cosmology*, ed. A. Hewitt, G. Burbidge and L. Z. Fang (Dordrecht: Reidel), p. 367.
Franceschini, A., et al. 1989, *Ap. J.*, **344**, 35.
Gopal-Krishna, and Wiita, P. J. 1987, *M.N.R.A.S.*, **226**, 531.
Gopal-Krishna 1988, *Astr. Ap.*, **192**, 37.
Katgert, P., Oort, M. J. A., and Windhorst, R. A. 1988, *Astr. Ap.*, **195**, 21.
Kapahi, V. K., and Kulkarni, V. K. 1986, *Astr. Ap.*, **165**, 39.
Kapahi, V. K. 1989, *A. J.*, in press.
Kellermann, K. I., and Wall, J. V. 1987, in *IAU Symposium 124*, op. cit., p. 545.
Koo, D. C. 1986, in *Spectral Evolution of Galaxies*, ed. C. Chiosi and A. Renzini (Dordrecht: Reidel), **ASSL 122**, p. 419.
Koo, D. C., and Kron, R. G. 1988a, in *Towards Understanding Galaxies at Large Redshifts*, ed. R. Kron & A. Renzini (Dordrecht: Reidel), **ASSL 141**, p. 209.
Kron, R. G., Koo, D. C., and Windhorst, R. A. 1985, *Astr. Ap.*, **146**, 38.
Laing, R. A., and Peacock, J. A. 1980, *M.N.R.A.S.*, **190**, 203.
van der Laan, H., and Perola, G. C. 1969, *Astr. Ap.*, **3**, 486.
Lilly, S. J. 1988, *Ap. J.*, **333**, 161.
McCarthy, P. J., et al. 1987, *Ap. J. (Letters)*, **319**, L39.
Oort, M. J. A., and Windhorst, R. A. 1985, *Astr. Ap.*, **145**, 405.
Oort, M. J. A. 1987a, *Ph.D. thesis*, University of Leiden.
Oort, M. J. A., Katgert, P., Steeman, F. W. M., and Windhorst, R. A. 1987, *Astr. Ap.*, **179**, 41.
Oort, M. J. A. 1988a, *Astr. Ap.*, **192**, 42.
Oort, M. J. A. 1988b, *Astr. Ap.*, **193**, 5.
Oort, M. J. A., Steemers, W. J. G., and Windhorst, R. A. 1988, *Astr. Ap. Suppl.*, **73**, 103.
Peacock, J. A., and Miller, L. 1988, in *Optical Surveys for Quasars*, Eds. P. S. Osmer et al., *Astronomical Society of the Pacific Conference Series, Vol. 2*, p. 194 .
Singal, A. K. 1988, *M.N.R.A.S.*, **233**, 87.
Subrahmanya, C. R., and Kapahi, V. K. 1983, in *IAU Symposium 104, The Early Evolution of the Universe and its Present Structure*, ed. G. O. Abell and G. Chincarini (Dordrecht: Reidel), p. 47.
Wall, J. V., et al. 1986, in *Highlights of Astronomy, Vol.* **VII**, ed. J. -P. Swings (Dordrecht: Reidel), p. 345.
Windhorst, R. A. 1984, Ph.D. Thesis, University of Leiden.
Windhorst, R. A., et al. 1985, *Ap. J.*, **289**, 494.
Windhorst, R. A. 1985, in *Reports on Astronomy, IAU Transactions, Vol.* **XIX A**, ed. R. M. West (Dordrecht: Reidel), p. 681.
Windhorst, R. A. 1986, in *Highlights of Astronomy, Vol.* **VII**, ed. J. -P. Swings (Dordrecht: Reidel), p. 355.
Windhorst, R. A., Dressler, A., and Koo, D. C. 1987, in *IAU Symposium 124*, op. cit., p. 573.
Windhorst, R. A. 1989b, *Astr. Ap.*, submitted.
Windhorst, R. A. 1989c, *Astr. Ap.*, submitted.